MODERN HYDRONIC HEATING

* Note: This product is available in two versions. The Hydronics Design Toolkit is software that is packaged with one version. There is also a version available without this software for users who do not wish to use a software package.

MODERN HYDRONIC HEATING

For Residential and Light Commercial Buildings

John Siegenthaler, P.E.

Mohawk Valley Community College
Utica, New York

Delmar Publishers

I(T)P An International Thomson Publishing Company

Albany • Bonn • Boston • Cincinnati • Detroit • London • Madrid • Melbourne
Mexico City • New York • Pacific Grove • Paris • San Francisco • Singapore • Tokyo
Toronto • Washington

NOTICE TO THE READER

Publisher does not warrant or guarantee any of the products described herein or perform an independent analysis in connection with any of the product information contained herein. Publisher does not assume, and expressly disclaims, any obligation to obtain and include information other than that provided to it by the manufacturer.

The reader is expressly warned to consider and adopt all safety precautions that might be indicated by the activities herein and to avoid all potential hazards. By following the instructions contained herein, the reader willingly assumes all risks in connections with such instructions.

The publisher makes no representation or warranties of any kind, including but not limited to the warranties of fitness for particular purpose or merchantability, nor are any such representations implied with respect to the material set forth herein, and the publisher takes no responsibility with respect to such material. The publisher shall not be liable for any special, consequential, or exemplary damages resulting, in whole or part, from the readers' use of, or reliance upon, this material.

The information presented in this text is believed to be accurate and up to date. However, neither the author of the text, the programmer of the Hydronics Design Toolkit software, nor Delmar Publishers, Inc. assume any responsibility for, nor express any warranty on the use of any information, diagrams, methods, data, or software presented or described in the text, software, or any associated ancillary materials.

Furthermore, no specific products that may be mentioned in the text or software are endorsed or recommended for a specific purpose by the author or Delmar Publishers, Inc.

All diagrams presented are conceptual in nature, and do not represent ready-to-install drawings. All such diagrams should be examined and further detailed for the specific requirements of a given system, and relevant building or mechanical codes.

Delmar Staff:
Senior Administrative Editor: Vernon Anthony
Project Editor: Eleanor Isenhart
Production Coordinator: Karen Smith
Art/Design Coordinator: Cheri Plasse
Cover Design: Charles Cummings Advertising/Art Inc.

COPYRIGHT © 1995
By Delmar Publishers
a division of International Thomson Publishing Inc.
The ITP logo is a trademark under license

Printed in the United States of America

For more information, contact:

Delmar Publishers
3 Columbia Circle, Box 15015
Albany, New York 12212-5105

International Thomson Editores
Campos Eliseos 385, Piso 7
Col Polanco
11560 Mexico D F Mexico

Nelson Canada
1120 Birchmount Road
Scarborough, Ontario
Canada M1K 5G4

International Thomson Publishing Europe
Berkshire House 168 - 173
High Holborn
London WC1V7AA
England

International Thomson Publishing GmbH
Königswinterer Strasse 418
53227 Bonn
Germany

International Thomson Publishing - Japan
Hirakawacho Kyowa Building, 3F
2-2-1 Hirakawacho
Chiyoda-ku, Tokyo 102
Japan

Thomas Nelson Australia
102 Dodds Street
South Melbourne, 3205
Victoria, Australia

International Thomson Publishing Asia
221 Henderson Road
#05 - 10 Henderson Building
Singapore 0315

1 2 3 4 5 6 7 8 9 10 XXX 01 00 99 98 97 96 95

Library of Congress Cataloging-in-Publication Data

Siegenthaler, John.
    Modern hydronic heating for residential and light commerical
buildings / by John Siegenthaler.
        p.   cm.
    Includes bibliographical references and index.
    ISBN 0-8273-6595-0
    1. Hot-water heating.   I. Title.
TH7511.S54 1995   94-37155
697'.4—dc20   CIP

CONTENTS

PREFACE

This book is about the state of the art in hydronic (water-based) heating systems for residential and light commercial buildings. It was written to provide comprehensive and unbiased information for heating technology students as well as heating professionals working in the trade. It is aimed at those willing to work with their head as well as their hands.

During my years working as an HVAC consulting engineer, it has become apparent that an information gap exists for those interested in upgrading their knowledge of designing and installing high quality hydronic heating systems for smaller buildings. I believe that residential and small commercial buildings deserve better heating systems than they often get. After all, even small heating systems affect the comfort and well-being of numerous people over many years.

Much of what has been written on hydronics is aimed at engineers, and intended for use in larger commercial or institutional buildings. It is often impractical to scale down such large system designs for use in smaller buildings. Manufacturer's information, while often well done, seldom takes the heating technician through the entire design process. Since most heating systems for smaller buildings are non-engineered, the installing contractor often has sole responsibility for ensuring that the components are properly matched, sized, and installed. In the absence of proper training this can lead to the same mistakes being repeated over and over. It also leads to design stagnation, where installers are only willing to deal with a certain type of equipment or system design, regardless of the situation with which they are faced. Many overlook the profitable opportunities offered through creative use of hydronic heating.

Organization

The first chapters of the book acquaint the reader with the fundamental physical processes involved in hydronic heating. Topics such as basic heat transfer, heating load calculations, and properties of fluids are discussed. A good understanding of these basics enhances the technician's design and troubleshooting skills.

Later chapters use the fundamental principles for overall system design. The reader is referred back to relevant sections of earlier chapters during the design process to reinforce the importance of the basics. Case studies at the end of some of the later chapters show complete system piping and control wiring diagrams.

Chapter 1 provides an overview of hydronic heating, and the basic concepts involved. It emphasizes *comfort* as the ultimate goal of the heating professional. It is intended to encourage an attitude of craftsmanship and professionalism as the reader continues through the text.

Chapter 2 discusses heating load calculations. Experience indicates that such calculations can be a stumbling block for students, as well as for those in the trade who want to jump right into design and layout without first determining what the system needs to provide. Without proper load information, *any* type of heating system can fail to deliver the required comfort. A complete method for determining design heating loads is presented. New work sheets and data are also provided.

Chapter 3 surveys a wide spectrum of hydronic heat sources. These include conventional gas and oil-fired boilers, as well as newer devices such as hydronic heat pumps, condensing boilers, thermal storage systems, and solar energy collectors. The pros and cons of each type of heat source are discussed. Emphasis is placed on matching the temperature and flow requirements of the heat source with those of the distribution system.

Chapter 4 describes, in simple terms, the physical properties of water including specific heat, density, viscosity, and solubility of air. Two very important equations that relate heat and heat flow to water temperature and flow rate are introduced.

Chapter 5 is a show and tell chapter that covers the proper use of piping, fittings, and valves in hydronic systems. The use of copper tubing as well as newer materials such as PEX and polybutylene is discussed. Common valves are illustrated and their proper use is emphasized. Several new or specialty fittings and valves are also discussed.

Chapter 6 is the key analytical chapter of the text. It introduces a new method for calculating the head loss of a fluid as it flows through a piping system. This new concept, called hydraulic resistance, builds an analogy between fluid flow in piping circuits and the principles of current, voltage, and resistance in electrical circuits. A method is developed that allows the reader to analyze both simple or complex piping circuits. The goal of the chapter is to determine the system resistance curve for a specific piping system. This curve is later used for circulator selection.

Chapter 7 presents both qualitative and quantitative information on circulators for smaller hydronic systems. The pump curve is introduced, and shown to be a crucial element in properly matching a circulator to a piping system. The intersection of the pump curve of a candidate circulator, and system resistance curve of the piping system, reveals the operating flow rate of the circulator/piping system combination. Quantitative and graphical methods are shown for finding this point. The chapter also places strong emphasis on pump cavitation and how to avoid it.

Chapter 8 surveys several types of hydronic heat emitters including finned-tube baseboard convectors, fan-coils, panel radiators, and radiant baseboards. The advantages and disadvantages of each type are discussed. Performance and sizing information is also given.

Chapter 9 is a major chapter dealing with control components and systems for hydronic heating. Several fundamental controls such as thermostats and boiler limit controls are covered. The chapter then goes on to describe custom control systems that may be required in certain applications. Ladder diagrams are used as a framework on which to design such control systems. The chapter concludes with a look at the latest electronic and microprocessor-based controls for hydronics.

Chapter 10 is devoted entirely to hydronic radiant floor heating. This topic is receiving ever-increasing attention in the trade and popular presses. Many installers of hydronic heating have, or soon will be asked to design and install such systems for homeowners and businesses. This chapter presents several methods of installation as well as new analytical design information. It equips those in the trade to work within this rapidly expanding market.

Chapter 11 covers several types of distribution piping configurations including series loop, one-pipe diverter tee, multi-zone, two-pipe reverse-return, and primary/secondary systems. The methods of analyzing piping systems and circulators given in Chapters 6 and 7 are now called upon to design each type of piping system. Both good and bad applications for each type of piping system are given.

Chapters 12 and 13 deal with the specialized topics of expansion tanks and air removal. Both survey the latest types of hardware, and show how to properly select and install it.

Chapter 14 deals with accommodating auxiliary heating loads such as domestic water heating, intermittent garage heating, or swimming pools. Emphasis is placed on using the space-heating boiler to supply these loads, and how to manage boiler capacity to handle multiple loads. Sizing of heat exchangers is covered.

Hydronics Design Toolkit

This text is available in two versions: one with software and one without. The software package is called the Hydronics Design Toolkit. This collection of 19 program modules was developed by the author and his associate Mario Restive *specifically for this book*. Routines for room heating load, pump selection, baseboard sizing, expansion tank selection, and many more analytical procedures covered in the text are included. These programs allow the reader to rapidly evaluate the calculations described in the text, while also minimizing the chance of errors. A disk icon in the margin of the text allows quick reference to material dealing with use of the Hydronics Design Toolkit.

By using "what-if" scenarios, the reader can quickly see how a proposed system will react to changes in design. *Potential problems can be spotted and avoided before they show up in the field*. The Hydronics Design Toolkit is intended to function not only as an academic aid, but also as a day-to-day design assistant for those working in the trade.

Other Features

Other features of this text include:

- Hundreds of *new* and *standardized* schematic drawings for both piping and control systems. These drawings are presented in a consistent format throughout the text rather than simply reprinted from several inconsistent sources. An appendix of the graphic symbols is provided for reference.
- Many photographs of the latest in hydronic heating hardware, as well as photos of work in progress as hydronic systems are installed.
- A full description of all variables, along with their proper units, is given whenever an equation is introduced. All

inputs are expressed in customary English units.
- Many example calculations organized in the form of a situation statement, solution procedure, and discussion of the results.

Much of the material in this text has been classroom tested over the past three years at Mohawk Valley Community College in Utica, New York. The text and software has formed the basis of a one semester course dedicated solely to hydronic heating. Student input has been helpful in determining the appropriate presentation level of the material, and in fine-tuning the software.

Acknowledgments

My deepest gratitude goes to longtime associate Mario Restive of Mohawk Valley Community College who diligently worked to create the computer code for the Hydronics Design Toolkit. Without his help, this crucial aspect of the project would not exist.

I also extend my sincere thanks to Roy Collver of tekmar Controls, Steve Lapp of Grundfos Pumps, Professor Darious Spence of Northern Virginia Community College, and Roger McDonald of Brookhaven National Labs for their many hours of work in fine-tuning the manuscript. Their many helpful suggestions have unquestionably improved the final product.

The support of the administration of Mohawk Valley Community College is also gratefully appreciated.

Finally I would like to thank all the members of the Delmar staff for their support and guidance with this text. All first-time authors would be fortunate to be part of such a team.

John Siegenthaler, P.E.

Dedication

This book is dedicated to my wife Joyce, and my children Dale, Chris, and Heidi, who have lovingly supported its development and patiently awaited its conclusion.

FUNDAMENTAL CONCEPTS 1

OBJECTIVES

After studying this chapter you should be able to:

- Describe the advantages of hydronic heating
- Define heat and describe how it is measured
- Describe three methods by which heat travels
- Explain thermal equilibrium within a hydronic heating system
- Define four basic hydronic subsystems
- Explain the differences between a radiator and a convector
- Explain the differences between an open-loop and a closed-loop hydronic system.
- Summarize the basic components of a hydronic heating system and explain how they operate.

1.1 WHAT IS A HYDRONIC HEATING SYSTEM?

Hydronic heating systems use water (or water-based solutions) to move **thermal energy** from where it is produced to where it is needed. The water within the system is neither the source of the heat nor its destination, only its transportation system. Thermal energy is absorbed by the water at a **heat source**, conveyed by the water through the distribution piping, and finally released into a heated space by a **heat emitter**. In most cases the same water remains in the system year after year.

Water has many characteristics that make it ideal for such an application. It is readily available in most locations, nontoxic, nonflammable, and has one of the highest heat storage abilities of any known material. All three states of water (solid, liquid, and vapor) are used for various building heating and cooling applications. Hydronic systems, however, make use of the *liquid state only*. The practical temperature range for water in residential and light commercial hydronic systems is from about 50° to 250 °F. At the upper end of this range the water is maintained in a liquid state by system pressurization. The lower end of the range can be extended well below 32 °F by the addition of antifreeze. Such a solution is

called a **brine**, and would be used in specialty applications such as snowmelting.

Hydronic heating began as an outgrowth of steam heating. Early hydronic systems relied on the **buoyancy** of hot water to create circulation from the boiler to the heat emitters. Due to its lower density, hot water would rise from a boiler through supply pipes into heat emitters. As it cooled, the water would flow downward toward the boiler through return pipes. Such systems required careful pipe sizing and installation since the buoyancy-driven flows were rather weak. System designs were significantly limited by today's standards. The emergence of electrically-powered **circulators** made it possible to move water at higher flow rates through more elaborate piping systems.

Modern hydronics technology enables heating systems to deliver the right amount of heat precisely when and where it is needed. Often several **loads** such as **space heating** and **domestic hot water** are supplied by a single hydronic system. Well designed and installed hydronic systems can provide unmatched comfort and fuel efficiency for the life of the building.

1.2 ADVANTAGES OF HYDRONIC HEATING

This section discusses several advantages of hydronic heating. Among these are:

- Comfort
- Energy savings
- Design flexibility
- Clean operation
- Quiet operation
- Unobtrusive installation

Comfort

Appropriately, the first advantage listed above is comfort. *Providing comfort should be the primary objective of any heating system designer or installer.* All too often this objective is compromised by other factors, the most

1

common of which is cost. Even small residential heating systems affect the health, productivity, and general contentment of several people for many years. It only makes sense to plan and install them accordingly.

Unfortunately many consumers don't think about heating systems in this way. Many see them as a necessary but uninteresting part of a building. Often when construction budgets are tightened, it's the heating system that is compromised to save money for other, more "impressive" building attributes. Heating professionals should take the time to discuss comfort as well as price with their clients before decisions on system type are made. Often people who have lived with uncomfortable heating systems simply don't realize what they have been missing. Many would welcome the opportunity to have truly comfortable heating, and would willingly spend more money (if necessary) to achieve it.

Maintaining comfort is not a matter of supplying heat to the body, but of controlling the manner and rate at which the body loses heat. A normal adult engaged in light activity produces heat at a rate of approximately 400 **British thermal units** per hour. Of this, about 48% is released by **thermal radiation** to colder surfaces, 30% by **convection** to surrounding air, and 22% by evaporation from the skin. When the interior environment allows heat to leave the body at a rate greater than or less than the rate at which it is produced, some degree of discomfort is felt.

The interior environment affects the way the body loses heat. For example, most people will feel uncomfortable in a room with many cool surfaces such as large windows, even if the room's air temperature is about 70 °F. For optimum comfort, the interior environment must provide the proper balance of air temperature, average surface temperature, and relative humidity, to accommodate the way the body releases heat.

Properly designed hydronic systems can control both the air temperature and surface temperatures of a building to maintain optimum comfort. Modern electronic controls can maintain room air temperature to within ± 1 °F of the **setpoint temperature**. To control surface temperatures, some hydronic heat emitters such as radiant floors and wall panel radiators transfer a large portion of their heat output as radiant energy rather than as warmed air. Since the human body is especially responsive to its radiant environment, these heat emitters enhance the feeling of comfort. Comfortable interior humidity levels are also easier to maintain in buildings that use hydronic heating, especially when low temperature heat emitters are used.

Several factors such as activity level, age, and sex determine what is a comfortable environment for a given individual. When several people are living or working in a common environment, any one of them might feel too hot, too cold, or just right. Heating systems that allow different areas of a building known as zones to be maintained at different temperatures can adapt to the comfort needs of several individuals. Both forced-air and hydronic heating systems can be zoned. Hydronic **zoning** is usually simpler and better controlled than its forced-air counterpart.

Energy Savings

Ideally a building's rate of heat loss would not be affected by how the heat is replaced. Experience, however, has shown that otherwise identical buildings can have significantly different heating loads based on the types of heating systems installed. Although the results are certain to vary with each case, hydronically heated buildings have, in general, shown lower heating energy usage than equivalent structures with forced-air heating systems.

A number of factors contribute to this finding. One of these is that hydronic systems do not affect room air pressure while operating. Small changes in room air pressure can occur when the blower of a central forced-air heating system is operating. Increased room air pressure often results from the lack of an adequate return air path from the rooms to the furnace. This condition drives heated air out through every small crack, hole, or other opening in the exterior surfaces of the room. A study that compared several hundred homes, some with central forced-air systems, others with baseboard convectors, found air leakage rates averaged 26% higher and energy usage averaged 40% greater in the homes with forced-air heating.

Another factor affecting building energy use is air temperature **stratification**. This refers to the tendency of warm air to rise toward the ceiling while cool air settles to the floor. In extreme situations the difference in air temperature from floor to ceiling can exceed 20 °F. Stratification tends to be increased by high ceilings, poor air circulation, and heating systems that supply air at high temperatures. Since the occupied area of a room is near the floor level, maintaining comfort there may lead to significantly higher air temperatures near the ceiling (see Figure 1–1). The warmer air increases heat loss through the ceiling. It also increases building air leakage due to **stack effect**.

Heating systems that transfer the majority of their heat by thermal radiation, which includes several types of hydronic systems, reduce air temperature stratification, and thus reduce heat loss through ceilings. Often comfort can be maintained at lower room air temperatures when a space is radiantly heated. This leads to further energy savings.

Zoned hydronic systems also save energy by allowing unoccupied or utility areas of a building to be kept at lower temperatures.

Design Flexibility

Hydronic heating offers almost unlimited design possibilities for both common and highly specialized heating requirements. A single system can often supply space

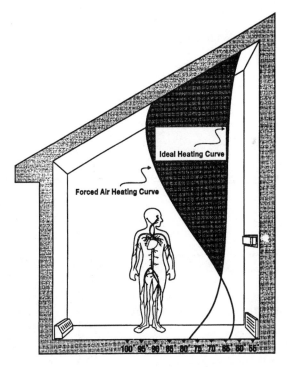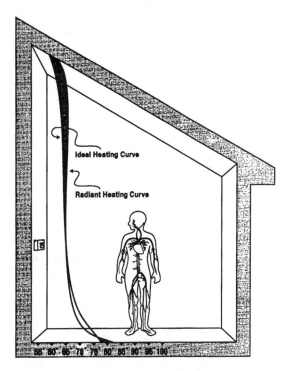

Figure 1–1 Comparison of air temperatures from floor to ceiling for forced air heating (left), and radiant floor heating (right). Courtesy of Wirsbo Company.

heating, domestic hot water, and specialty loads such as spa heating. Such systems can reduce installation costs because redundant components such as multiple heat sources, exhaust systems, electrical hookups, and fuel supply pipes are eliminated.

In some buildings the manner in which spaces are used can require different heat emitters in different areas. Imagine, for example, a small veterinary clinic where areas occupied by animals will be heated by a radiant floor, and office areas by individually controlled wall panel radiators. Such situations are relatively easy to accommodate with hydronics.

Clean Operation

A common complaint about forced-air heating is its ability to move dust and other airborne particles such as pollen and smoke throughout a building. In buildings where air filtering equipment is either low quality or poorly maintained, dust streaks around ceiling and wall diffusers are often evident. Eventually duct systems require internal cleaning to remove dust and dirt that has accumulated over several years of operation.

Few hydronic systems involve forced-air circulation. Those that do create room air circulation rather than building air circulation. This reduces the dispersal of airborne particles, which is a major benefit in situations where air cleanliness is imperative, such as for people with allergies, in health care facilities, or laboratories.

Quiet Operation

A properly designed and **deaerated** hydronic distribution system is almost totally silent. The loudest sound in the system is often the gas or oil burner. Modern systems that use constant circulation with variable water temperature avoid expansion noises that can occur when high temperature water is injected directly into a room temperature heat emitter.

Unobtrusive Installation

Hydronic heating systems are easily integrated into the structure of most small buildings without compromising their function or the aesthetic character of the space. Interestingly, the underlying reason for this is the high **heat capacity** of the water itself.

Consider the difficulty often encountered when a ducting system has to be "buried" out of sight within the framing of a typical house, especially if all surfaces are to be finished. In many cases the best that can be done is to encase the ducting in exposed soffits. It is reasonable to assume such situations might lead to compromises in duct sizing and/or placement.

Now consider that a 1-inch diameter pipe carrying water at a flow rate of about 8 gallons per minute around a hydronic system with a 20 °F temperature drop from boiler outlet to return, can transport the same heat as a 10 inch by 18 inch duct supplying 130 °F air sized for a

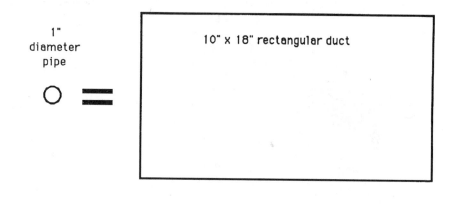

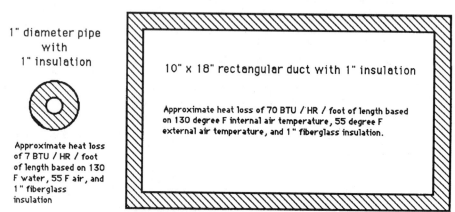

Figure 1–2 A 1-inch diameter pipe in a hydronic system is capable of transporting heat at the same rate as a 10 inch by 18 inch duct in a forced-air system. Notice the ducting also will require considerably more insulation material. If the water in the pipe and the air in the duct are the same temperature, there is far greater heat loss from the duct because of its larger surface area.

face velocity of 1,000 feet per minute. Certainly it will be easier to route such a pipe through the same house. Furthermore, if the distribution system is to be insulated, which is now a code requirement in some areas, considerably less material and labor will be used insulating piping rather than ducting. Even when insulated with the same material, the heat loss of a duct will still be several times greater than that of the pipe (see Figure 1–2).

Hydronic systems using flexible **polymer** piping are often easier to retrofit into existing buildings than equivalent forced-air systems. The flexible tubing can be routed through closed framing spaces much like electrical cable. A typical retrofit might involve running a nominal ½-inch flexible supply and return pipe from a central **manifold** to a heat emitter in each room. This approach also allows the possibility of controlling each room as a separate zone (see Figure 1–3).

For situations where utility space is at a premium, there are several types of wall-hung boilers that can easily be mounted in a closet. In most of these installations there is also room for a domestic hot water tank heated by the same boiler. The entire heating and hot water system could occupy less than 10 square feet of floor area.

1.3 HEAT AND HEAT TRANSFER

Before attempting to design any type of heating system, it is crucial to understand the entity being manipulated: heat.

What we commonly call heat could also be described as energy in thermal form. Other forms of energy such as electrical, chemical, mechanical, and nuclear can be converted into heat through various processes and devices. In fact all heating systems consist of devices that convert one form of energy into another.

From a microscopic point of view, heat is our perception of atomic vibrations within a material. Our means of expressing the intensity of these vibrations is called temperature. The more intense the vibrations, the greater the temperature of the material, and the greater its heat content. Any material above absolute zero temperature (−460 °F) contains some amount of heat.

There are several units for expressing a quantity of heat. In this country the most common is the British thermal unit (Btu). A Btu is defined as the amount of heat required to raise 1 pound of water by 1 °F. This unit is very convenient for use with hydronic heating systems.

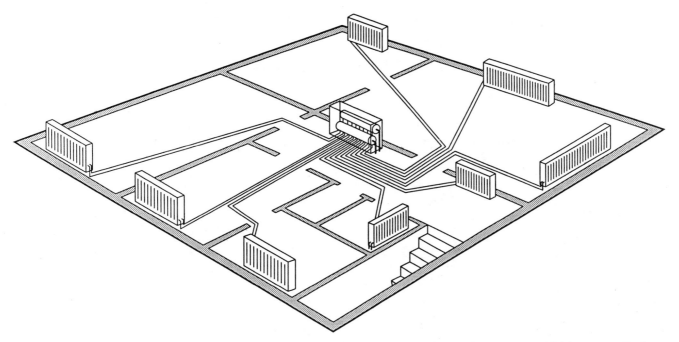

Figure 1–3   Separate supply and return piping from each of several radiators to a central manifold system. Such a system is especially well-suited for retrofit jobs using flexible polymer tubing.

Heat always moves from an area of higher temperature to an area of lower temperature. In a hydronic system this phenomenon takes place at several locations. First, heat moves from the hot flames and exhaust gases within the combustion chamber of a boiler into the cooler metal walls of water-filled compartments surrounding the chamber. It then moves through the metal walls into the cooler water within. After being transported to a heat emitter by the moving water, the heat passes through the cooler metal walls of the radiator into the still cooler air and objects of a room. Finally it moves through the exposed surfaces of a room into the outside air (see Figure 1–4). In every instance the heat moves from an area of higher temperature to an area of lower temperature. *Without this temperature difference there would be no heat movement.*

The rate of heat transfer is governed by a number of factors, some of which we can control. It can be expressed in several different units. In this book the rate of heat transfer will usually be expressed in British thermal units per hour, often abbreviated as Btu/hr, or Btuh. It is very important to distinguish between the *quantity* of heat present in an object (measured in Btu), and the *rate* at which heat moves in or out of the object (measured in Btu/hr). These terms are often misquoted by people, including those in the heating trade.

One factor that affects the rate of heat transfer is the temperature difference that exists between where the heat is, and where it is moving. *Temperature difference is the driving force that causes heat to move.* The greater the temperature difference, the faster the heat will flow. In most instances the rate of heat transfer through a material is directly proportional to the temperature difference across the material. Thus if we could double the temperature difference across a material, we would also double the rate of heat transfer through the material.

Another factor is the type of material through which the heat moves. Some materials such as copper allow heat to move through very quickly. Others such as polyurethane foam greatly reduce the rate of heat transfer. This factor is more fully explained later in the chapter.

Thermal Equilibrium

Whenever a material is not gaining or losing heat and remains in a single physical state (solid, liquid, or gas) its temperature remains the same. This is also true if the material happens to be gaining heat from one object while simultaneously releasing heat to another object at the same rate. These principles have many practical applications in hydronic heating. For example, if you observed the operation of a hydronic heating system for an hour, and found no change in the temperature of the water leaving the boiler even though it was firing continuously, what could you conclude? Answer: Since there is no change in the water's temperature, it did not undergo any net gain or loss of heat. Therefore the rate at which the boiler was injecting heat into the water was the same as the rate at which the radiators were extracting heat from the water. Under these conditions the system was in **thermal equilibrium**. *All heating systems will inherently find their point of thermal equilibrium and remain in operation*

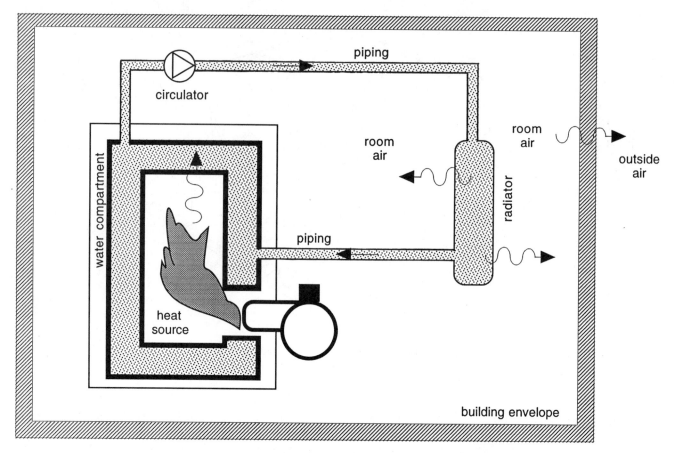

Figure 1–4 Heat movement with a hydronic heating system

at that point. The goal of the system designer is to ensure that thermal equilibrium is established at conditions that properly heat the building, and do not adversely affect the operation or longevity of the system's components.

Three Modes of Heat Transfer

Thus far two important principles of heat transfer have been discussed. First, heat moves from an area of higher temperature to an area of lower temperature. Second, the rate of heat transfer depends on temperature difference and the type of material. To gain a more detailed understanding of how the thermal components of a hydronic system work, we need to classify heat transfer into three modes: **conduction**, **convection**, and **thermal radiation**.

Conduction is the type of heat transfer that occurs through solid materials. Recall that heat has already been described in terms of atomic vibrations. Heat transfer by conduction can be thought of as a dispersal of these vibrations from a point where there is a source of heat, out across trillions of atoms that are bonded together to form a solid material. The index that denotes how well a material transfers heat is called its **thermal conductivity**. The higher a material's thermal conductivity, the

faster heat can move through it. Heat moving from the inner surface of a pipe to its outer surface is an example of conduction heat transfer. Heat moving through the metal housing of a radiator is another.

The rate of heat transfer by conduction is directly proportional to both the temperature difference across the material and its thermal conductivity. It is inversely proportional to the thickness of a material. Thus if one were to double the thickness of a material while maintaining the same temperature difference between its sides, the rate of heat transfer through the material would be cut in half. In certain locations within hydronic systems designers try to enhance conduction, while in other locations they try to minimize it. For example, the greater the conduction of heat from a tube embedded in concrete for purposes of heating the floor, the better it delivers heat to the building. By contrast, the slower the rate of heat conduction through the insulation of a hot water storage tank, the better it retains heat during standby periods. Equations for calculating heat flow by conduction are presented in Chapter 2.

Convection heat transfer occurs when a fluid at some temperature moves along a surface at a different temperature. In this context the term fluid can refer to either a gas (such as air) or a liquid (such as water).

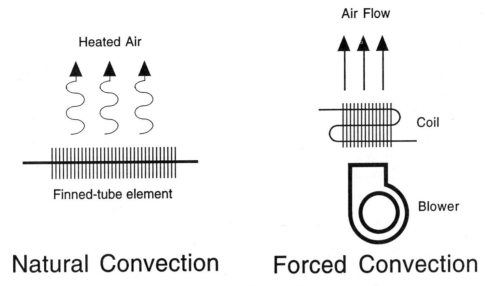

Figure 1-5   Two types of convective heat transfer

Consider the example of water at 100 °F flowing along a surface at 150 °F. As the cooler water molecules contact the hotter surface, they absorb heat from it. These molecules are constantly being churned about as the water moves along. The heated molecules are carried away from the surface into the bulk of the water stream, and are replaced by cooler molecules that repeat the process. It is convenient to think of convection heat transfer as a process in which the fluid is "scrubbing" heat off the surface of a solid, and carrying it off with the wash water, so to speak.

Intuition and experience indicate the speed at which the fluid moves over the surface greatly affects the rate of heat transfer. We have all experienced the greater cooling that results when air is blown across our skin rather than lying stagnant against it. This is frequently called wind chill. Even though the air temperature may not be extremely cold, the speed at which it moves over our skin results in much faster convective heat transfer from our skin to the air. To achieve this same rate of heat transfer with calm air would require it to be at a much colder temperature (the wind-chill temperature).

When a fluid's motion is created by either a circulator or a blower, the term **forced convection** is used to more accurately describe the source of the fluid's motion. When buoyancy differences within the fluid itself cause it to move along a surface, the term **natural convection** is used. As a rule, heat transfer by natural convection is considerably slower than by forced convection. Figure 1-5 illustrates these concepts.

Within a hydronic system, convection moves heat from water to the inner surfaces of a heat emitter. Some heat emitters such as finned-tube baseboard are especially designed to release heat from their outer surfaces into the surrounding air by convection. Such devices are correctly referred to as convectors.

Thermal radiation is probably the least understood of the three modes of heat transfer. Thermal radiation is a form of **electromagnetic energy** that is closely related to visible light. Like visible light, thermal radiation travels outward in all directions from its source in straight lines, and cannot bend around corners. It does not need any material to travel through. Consider, for example, people sitting by a campfire. Their faces often feel warm even though the air temperature around them may be quite cool. Thermal radiation emitted by the fire travels through the air and is absorbed by their exposed skin. Likewise thermal radiation emitted from the surface of a warm radiator can pass through room air with very minimal disruption, and then be absorbed in large part by the objects it strikes. In both these examples *the radiation becomes heat only after being absorbed by a material.*

The difference between thermal radiation and visible light is the wavelength of the radiation. Anyone who has watched molten metal cool has noticed how the bright yellow color fades to duller shades of red, until finally the metal's surface no longer glows. As the surface of the metal cools below about 1,000 °F, our eyes no longer detect visible light from the surface. Our skin, however, still senses that the surface is very hot. Though unseen, thermal radiation in the infrared portion of the electromagnetic spectrum is still being strongly emitted from the metal's surface.

Any surface will emit thermal radiation to any other cooler surface within sight of it. A radiator that is warmer than our skin or clothing surfaces will transfer heat to us by thermal radiation. Likewise, our skin or clothing surfaces will give off thermal radiation to surrounding surfaces at lower temperatures.

The term **mean radiant temperature** (MRT) is used to describe the area-weighted average temperature of all surfaces in a room. As the mean radiant temperature of a room increases, the air temperature required to maintain comfort decreases. In this situation heat released from the body by radiation is reduced, so the amount released by convection must increase. The converse is also true.

A generally accepted rule of thumb is that for each degree the MRT of a room is lowered, the air temperature must be increased by 1.4 °F to maintain the same level of comfort. This explains why a person in a room with an air temperature of 70 °F can still feel cool if surrounded by cold surfaces such as large windows on a cold winter night.

As thermal radiation strikes the surface of an object, part of it is absorbed by the material and part is reflected off the surface. Most solid opaque materials do not transmit any significant amount of thermal radiation. The actual percentage of incoming radiation that is absorbed or reflected is determined by the optical characteristics of the surface, and the wavelength of the radiation. From a practical point of view, however, very little if any thermal radiation emitted by warm surfaces in a typical room escapes from the room as radiation. This results from the fact that even reflected radiation will very likely strike another surface in the room and be further absorbed, and so on and so on, until it is fully absorbed.

The rate at which radiant heat is transferred between two surfaces depends upon their temperature difference, an optical property of each surface called its emissivity, and the angle between the surfaces. It is a more complicated relationship than is the case with conduction or convection. For our purposes, however, it is not necessary to delve into a detailed mathematical treatment of radiative heat transfer.

Most hydronic heat emitters deliver their heat to the room through a combination of convection and thermal radiation. The percentages delivered by each mode will depend on many factors such as surface orientation, shape, type of finish, and temperature. When a particular heat emitter has been designed to transfer the larger part of its heat output as thermal radiation, it is appropriately called a radiator.

1.4 FOUR BASIC HYDRONIC SUBSYSTEMS

In a hydronic heating system, water is heated by a heat source and conveyed by way of a **distribution system** to heat emitters, where the thermal energy is released to the building. A control system regulates these elements in response to the heating load of the building. The overall hydronic system thus consists of four interrelated subsystems:

1. Heat source   (boiler, heat pump, solar collector)
2. Distribution system   (piping, valves, circulators)
3. Heat emitters   (baseboard convectors, wall panel radiators, radiant floors)
4. Control system   (thermostat, aquastats, switches)

This section shows how these subsystems are connected to form a simple hydronic heating system. The discussion begins with the concept of a simple piping loop. Several essential components are described and situated into the loop, one at a time. Finally, a composite schematic drawing shows all components in place within the loop. By the end of this chapter you should have a basic understanding of what the major components of a hydronic system are, and what they do. The details of proper component selection and sizing are left to later chapters.

The Basic Loop

Any hydronic system can be fundamentally described as a loop, or piping circuit, as it is sometimes called. If the loop is sealed off from the atmosphere at all locations (as is true with most hydronic systems) it is referred to as a **closed-loop system**. If the loop is open to the atmosphere *at any point,* it is called an **open-loop system**.

The loop contains components from each of the subsystems previously listed. Figure 1–6 shows the simplest form of such a loop. Three of the four basic hydronic subsystems are depicted: the heat source, the distribution system, and a single heat emitter.

Upon demand for heat, water is circulated around the loop by the circulator. It carries heat from the heat source to the heat emitter where it is released into the heated space.

In an ideal application, the rate at which the water absorbs heat from the heat source would exactly match the rate at which the heat is released by the heat emitters. This rate would also, ideally, match the rate of heat loss from the building. Unfortunately, in real applications such ideal conditions seldom exist.

Consider, for example, the common practice of selecting a boiler with a heating output greater than the building's heating requirement on the coldest day of the year. During a mild day, a boiler selected by this method can deliver heat to the building much faster than the building loses heat. Continuous operation of the boiler under these circumstances would quickly overheat the building. Obviously some method of controlling the system's heat output is needed.

Temperature Controls

Figure 1–7 adds two simple control devices, the **room thermostat** and the heat source **aquastat**, to the basic system. The room thermostat determines when the building

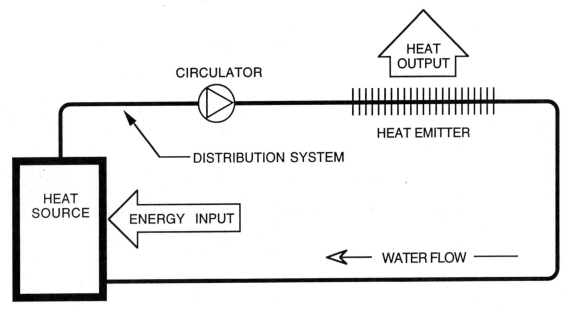

Figure 1-6 The basic hydronic heating circuit

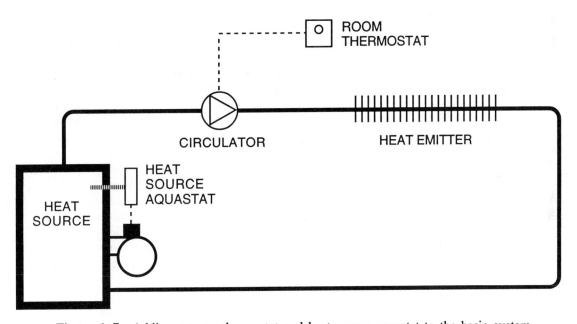

Figure 1-7 Adding a room thermostat and heat source aquastat to the basic system

requires heat based on indoor air temperature. When the indoor air temperature drops slightly below what the thermostat is set for, it turns on the system's circulating pump to convey heat from the heat source to the heat emitters. In some systems the room thermostat simultaneously turns on both the circulator and the heat source. The room thermostat is simply a temperature operated switch. As such it is either on or off.

The heat source aquastat ensures that the water temperature within the heat source remains within preset upper and lower limits. It functions by turning the heat producing element of the heat source, (i.e., burner, heat-

ing elements, compressor) on and off as needed. Like the room thermostat, the aquastat is a temperature operated switch. Consider, for example, that the aquastat on a boiler is set for 160 °F with a 10 °F **differential**. When the water temperature inside the boiler drops to 150 °F (160 °F setpoint – 10 °F differential) the burner is turned on by the aquastat. Assuming the room thermostat keeps the system operating, the burner remains on until the water in the boiler reaches its upper setpoint temperature of 160 °F.

Together these components provide the necessary control action to reasonably match the output of the heat source to the heating requirement of the building. Control

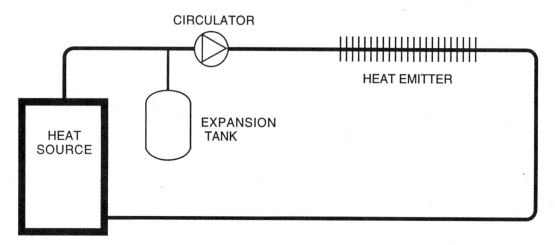

Figure 1–8 Adding an expansion tank to the basic circuit

components of this type have been used on nearly all hydronic systems for many years. More sophisticated controls, which allow the output of the heat source to follow the building's heating requirement with greater accuracy, have been introduced in recent years. These will be discussed in Chapter 9.

Expansion Tank

As water is heated it also expands. This increase in volume is an extremely powerful but predictable characteristic that must be accommodated in any type of closed-loop hydronic system. Figure 1–8 adds an **expansion tank** to the basic loop. This tank contains a captive volume of air. As the water expands, this air is compressed resulting in a slight increase in system pressure. As the water cools, its volume decreases allowing the compressed air to re-expand, and the system pressure returns to its original value.

Most modern hydronic systems use diaphragm-type expansion tanks. Such tanks contain their captive air in a special balloon-like chamber. Older hydronic systems often used expansion tanks without diaphragms. Such tanks had to be considerably larger than diaphragm-type tanks.

Both the size and the placement of the expansion tank are important aspects of system design. The specifics of expansion tanks are discussed in Chapter 12.

Pressure Relief Valve

Consider the fate of a hydronic system installed without safety devices in which a defective control device fails to stop the heat source from producing heat after its upper temperature limit is reached. The water in the system would get hotter and hotter. In a closed-loop system, the pressure would steadily increase due to the water's expansion. This pressure could eventually exceed the

pressure rating of the weakest component in the system. Most residential system components have pressure ratings of at least 60 psi, and may withstand two or more times this pressure before bursting. The consequences of a system component bursting at these high pressures and temperatures could be devastating. For this reason, *all closed-loop hydronic systems must be protected by a pressure relief valve.* Such a valve is universally required by plumbing and boiler codes across the U.S. Its function is to provide a relatively controlled and safe release of hot water in the event of excessive pressure in the system.

Pressure relief valves are designed and labeled to open at a specific pressure. Residential and light commercial hydronic systems are typically equipped with pressure relief valves rated at 30 pounds per square inch (psi) pressure. In systems with boilers, the relief valve is almost always threaded directly into the boiler. A pressure relief valve has been added to the heat source in Figure 1–9.

Make-up Water System

Most hydronic systems experience very minor water losses due to evaporation at valve packings, pump seals, air vents, and other components. These losses are normal and must be made up to maintain adequate system pressure. The common method for replacing the water is through a **make-up water system** consisting of a **pressure reducing valve, backflow preventer**, and shut-off valve.

Because the pressure in a municipal water main or private water system is usually higher than code mandated relief valve settings for hydronic systems, such water sources cannot be directly connected to the loop. A pressure reducing valve (also called a boiler feed water valve) is used to reduce the water main pressure, and to maintain a constant *minimum* pressure within the system. This

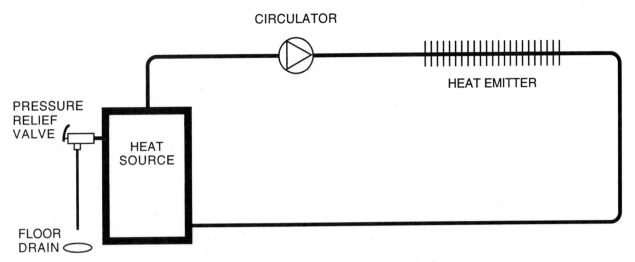

Figure 1–9 Adding a pressure relief valve to the heat source

valve allows water to pass into the system whenever system pressure falls below the setting on the valve. This often occurs when air is vented from the system during startup or following service. Water also comes into the loop to make up for the minor losses previously mentioned. The pressure setting is adjustable on most pressure reducing valves. Determining the proper setting is covered in Chapter 12.

The backflow preventer stops any water that has entered the hydronic system from returning and possibly contaminating the water supply system. Most municipal codes require such a device on any heating system connected to a public water supply. Even in the absence of such a code requirement, using a backflow preventer is a wise decision, especially if any antifreeze or corrosion inhibitors are to be used in the system.

The shut-off valve is installed to isolate the hydronic system from the pressurized water supply main whenever the system needs to be drained for service.

The pressure reducing valve, backflow preventer, and shut-off valve have been added to the basic loop in Figure 1–10.

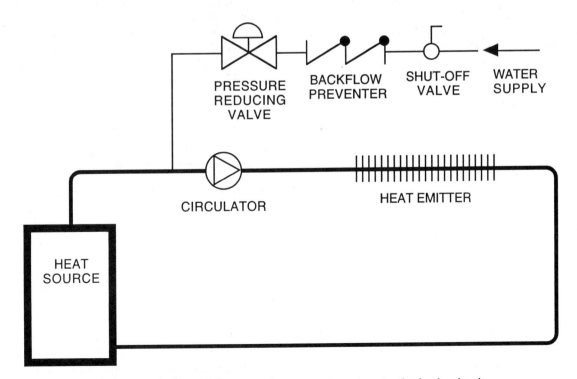

Figure 1–10 Adding a make-up water system to the basic circuit

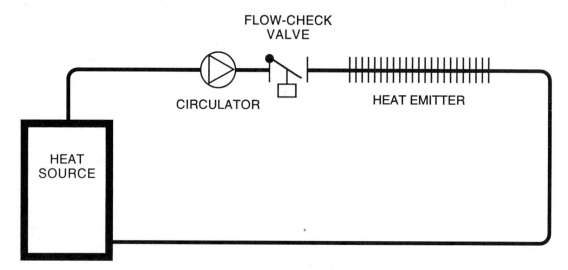

Figure 1-11   Adding a flow-check valve to the basic circuit

Flow-check Valve

Another component often used in hydronic systems is a **flow-check valve**. This valve can serve one or two basic purposes, depending on the system type.

In single zone systems it prevents buoyancy driven **thermosiphon** flow from occurring. This refers to the undesirable migration of hot water upward from the heat source into the distribution system when the circulator is off. It is especially undesirable during warm weather when the boiler is being fired to produce domestic hot water.

In multi-zone systems that use separate circulators for each zone circuit, a separate flow-check valve is used in each circuit to eliminate thermosiphoning, as well as to prevent backward flow of water from an active zone circuit through an inactive circuit. Figure 1-11 shows the placement of a flow-check valve within a single zone system. Placement of flow-check valves in multi-zone systems is covered in Chapter 11.

Air Purger

An **air purger** is designed to separate air from the system's water and eject it from the system. The device functions by creating a low pressure region through which all system water must flow. In the reduced pressure region, dissolved air comes out of solution much like the bubbles that appear when a can of soda is opened. These bubbles rise into the collection chamber of an automatic **air vent**. A float mechanism in the air vent allows the air to be ejected by system pressure. The float mechanism automatically reseals after the air leaves.

The process of **deaerating** the system water is enhanced as the water is heated. For best results, the air purger should be located where water temperatures are the highest—in the supply line from the heat source. The air

purger is shown in Figure 1-12. Air elimination is coved in detail in Chapter 13.

1.5   THE IMPORTANCE OF SYSTEM DESIGN

Figure 1-13 is a composite drawing showing all the components previously discussed. By assembling these components we have formed the basis of a simple hydronic heating system. It must be emphasized, however, that just because all the components are present does not guarantee the system will function properly. Combining these components is not simply a matter of choosing a favorite product for each and connecting them (the "catalog engineering" approach). Major subsystems such as the heat emitters and the heat source have certain temperature and flow rate ranges that must be observed if they are to function properly.

For example, suppose the heat source in Figure 1-13 was a hydronic heat pump, and the heat emitters were finned-tube baseboard convectors. A typical hydronic heat pump is not able to produce water temperatures above approximately 130 °F. Baseboard convectors are often sized at water temperature greater than 150 °F. If one were to assemble a system with these components and start it up, they would initially find the heat pump injecting heat into the loop faster than the baseboard could release it into the building. Thermal equilibrium would not exist. The water temperature in the loop would continue to rise until the heat pump would reach its upper temperature limit and shut itself off based on high refrigerant head pressure. At the same time insufficient heat would be transferred to the building due to the relatively low water temperature in the baseboard convectors. In

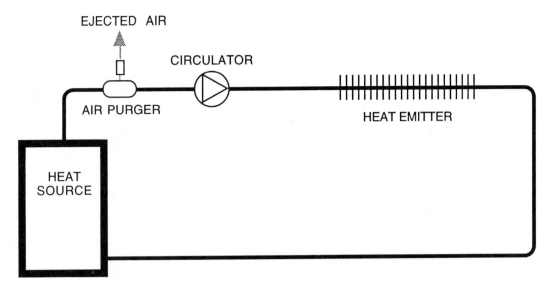

Figure 1–12   Adding an air purger to the basic circuit

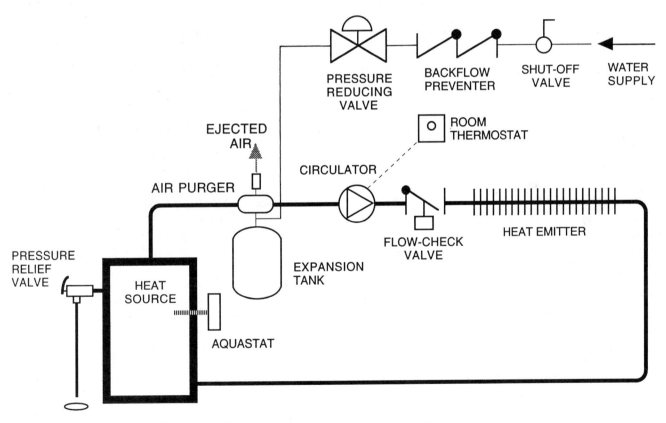

Figure 1–13   Composite drawing showing all the components

this example the baseboard convectors represent a significant "bottleneck" to the heat flow.

This hypothetical example illustrates why a designer must consider the temperature compatibility of the heat source and the heat emitters. An equally valid example could be made for flow rate compatibility. The objective is to achieve a stable, dependable, affordable, and efficient overall system.

Failure to respect the operating characteristics of these components will result in installations that either underheat, overheat, waste energy, or otherwise disappoint the building occupants. The chapters that follow present detailed information on heat sources and heat emitters with emphasis on temperature and flow rate requirements. By properly matching components, calamities such as those just described can be avoided.

SUMMARY

The primary objective of any heating system is to provide the highest possible comfort for the building occupants. Other important objectives include making efficient use of fuel, providing quiet and dependable operation, and maximizing the service life of all system components. Hydronic heating systems offer tremendous *potential* for attaining all these objectives. The key to turning potential into reality is a solid understanding of the processes and components introduced in this chapter. Indeed a lack of understanding of these basics is often the cause of improper system design, or difficulty in diagnosing the cause of poor system performance. Before proceeding into more detailed aspects of design and installation, take time to test your understanding of these all-important fundamentals.

KEY TERMS

Air purger
Air vent
Aquastat
Backflow preventer
Brine
British thermal unit
Buoyancy
Circulators
Closed-loop system
Conduction
Convection
Deaerated
Differential
Distribution system
Domestic hot water
Electromagnetic energy
Expansion tank
Flow-check valve
Forced convection
Heat emitter
Heating capacity
Heating load
Heat source
Loads
Make-up water system
Manifold
Mean radiant temperature
Natural convection
Open-loop system
Polymer
Pressure reducing valve
Pressure relief valve
Room thermostat
Setpoint temperature
Space heating
Stack effect
Stratification
Thermal conductivity
Thermal energy
Thermal equilibrium
Thermal radiation
Thermosiphon
Zoning

CHAPTER 1 QUESTIONS AND EXERCISES

1. Why do the dimensions of ducting in forced-air heating systems have to be so much larger than the diameter of piping in a hydronic heating system of equal heating capacity?
2. Why is it preferable to surround a person with warm surfaces as opposed to just warm air?
3. What type of heat transfer creates the wind-chill effect we experience during winter?
4. A certain block of material conducts heat at a rate of 100 Btu/hr. One side is maintained at 80 °F, and the other at 70 °F. Describe what happens to the rate of heat transfer when:
   a. the 100 °F side is raised to 130 °F.
   b. the thickness of the block is doubled.
   c. the 100 °F side is raised to 130 °F, and the thickness is cut in half.
5. At approximately what temperature does a surface start emitting visible radiation?
6. How does thermal radiation differ from visible light? How are the two similar?
7. Describe thermal equilibrium within a hydronic system.
8. What is the function of an aquastat on a hydronic heat source?
9. What is the difference between an open-loop system and a closed-loop system?
10. Are there any closed-loop hydronic heating systems on which a pressure relief valve is not required? Why?
11. List two types of hydronic heat emitters other than finned-tube baseboard convectors.
12. What is the function of a flow-check valve in a single-zone hydronic system?
13. What is a brine? In what type of hydronic heating application would it be used?
14. Why is it necessary to have a backflow preventer in the make-up water line?
15. What are the customary units for heat and heat flow rate in the U.S.?
16. Define a zoned hydronic system.
17. What made water flow within early hydronic systems before circulators were used?
18. What causes dissolved air to come out of solution within an air purger?
19. How are the expansion tanks used in modern hydronic systems different from those used in older systems?
20. What are some common ways small amounts of water can leak out of a hydronic system?

HEATING LOAD ESTIMATES 2

OBJECTIVES

After studying this chapter you should be able to:

- Describe what a design heating load is and why it is important to hydronic system design
- Explain the difference between room heating loads and building heating loads
- Determine the thermal envelope of a building
- Calculate the effective total R-value of a building surface
- Estimate infiltration heat loss using the air change method
- Determine the heat loss of foundations and slab floors
- Explain what degree days are and how they are used
- Estimate the annual space heating energy usage of a building

2.1 INTRODUCTION

This chapter shows how to estimate the **design heating load** of individual rooms and entire buildings. Determining these loads is the starting point in designing all hydronic systems. *The care given to this step will directly affect system cost, efficiency, and most importantly, customer satisfaction.*

Even for readers already familiar with heating load estimates, this chapter may provide a more streamlined approach to this routine chore. It presents the fundamental tools for estimating the heating load using a minimum of mathematics. It also discusses computer-aided load estimating. The ROOMLOAD and BSMTLOAD programs that are part of the Hydronics Design Toolkit are described. An example of a complete heating load estimate for a residential building is given. The chapter concludes with a method for estimating the annual space heating energy usage of a building.

2.2 DEFINITION OF DESIGN HEATING LOAD

It is important to have a clear understanding of what a design heating load is before attempting to calculate it. *The design heating load of a building is an estimate of the rate at which a building loses heat during the near minimum outdoor temperature.* This definition contains a number of key words that need further explanation.

First, it must be emphasized that a heating load is a calculated *estimate* of the rate of heat loss of a room or building. It is not possible to fully account for the hundreds of construction details and thermal imperfections in an object as complex as a building, even a simple house. Among the factors that add uncertainty to the calculations are:

- Imperfect installation of insulation materials
- Variability in R-value of insulation materials
- Shrinkage of building materials leading to greater air leakage
- Complex heat transfer paths at wall corners and other intersecting surfaces
- Effect of wind direction on building air leakage
- Overall construction quality
- People traffic into and out of the building

Second, the heating load is a *rate* of heat flow from the building to the outside air. It is often misstated, even among heating professionals, as a number of Btus rather than a rate of flow in Btus per hour (Btu/hr). This is like stating the speed of a car in miles rather than miles per hour. Recall from the discussion of thermal equilibrium in Chapter 1 that when the rate of heat flow into a system matches the rate of heat flow out of a system, the temperature of that system remains constant. In the case of a building, *when the rate at which the heating system injects heat into the building matches the rate at which the building loses heat, the indoor temperature will remain constant.*

Finally, the design heating load is estimated assuming the outside temperature is *near* its minimum value. This temperature, specifically called the **97.5% design dry bulb temperature**, is not the absolute minimum temperature for the location. Rather, it is the temperature the outside air is at or above during 97.5% of the year. Although outside temperatures do occasionally go below this design temperature, the duration of these low temperature periods is so short that buildings tend to "coast" through them using heat stored in their thermal mass. When outside temperatures are above the design temperature, the rate of heat flow from the building will be less.

Use of the 97.5% design temperature for heating system design helps prevent excessive oversizing of the heat source.

Building Heating Load versus Room by Room Heating Load

One method of calculating the heating load of a building yields a single number that represents the design heating load of the building as a whole. This is called the **building heating load**, and is useful for selecting the heating capacity of the heat source, and performing estimates for annual heating energy usage. However, this single number is not necessarily useful for designing the heating distribution system. This is because *individual rooms often have heating loads that are not proportional to their floor area.* For example, a small room with a large window area can have a greater heating load than a large room with only a moderate amount of windows. Even when the building heating load is properly determined, failure to properly distribute the heat to individual rooms in proportion to their heating loads can lead to overheating in one room and underheating in another. For this reason it is necessary to perform heat loss estimates on a room by room basis. Once these individual **room heating loads** are determined, they can simply be added up to obtain the total building heating load.

Why Bother with Heat Load Calculations?

Would a mason order concrete without first knowing the measurements of the slab to be poured? Would a carpenter begin cutting rafters before knowing the measurements of the roof? Obviously the answer to these questions is no. Yet many so-called "heating professionals" routinely select equipment for a heating system with only a guess as to the building's heating load.

Statements that attempt to justify this approach range from: "I haven't got time for doing all those calculations," to "I'd rather be well oversized than get called back on the coldest day of the winter," to "I did a house like this one a couple of years ago, it can't be much different." But rather than argue with experience, let us look at some of the consequences of not properly estimating the building heating load.

- If the estimated heat loss is too low, the building will be uncomfortable during cold weather (a condition few homeowners are willing to tolerate). An expensive call back will be the eventual result.
- Greatly overestimated heating loads lead to systems that not only cost more to install, but also cost more to operate due to lower efficiency. The reduced efficiency can waste hundreds, if not thousands, of dollars of fuel over the service life of the system.

Most building owners never realize when their heating system is significantly oversized and simply accept the fuel usage of the system as normal. Most trust the heating professional they hire to properly size and select equipment in their best interest. A heating professional who wants to accomplish this needs to follow one simple rule: *Before attempting to design a heating system of any sort, always calculate the design heating load of each room in the building using creditable methods and data.*

2.3   CONDUCTION HEAT LOSSES

In Chapter 1, conduction heat transfer was described as a method by which heat moves through a solid material whenever a temperature difference exists across that material. The rate at which heat is conducted depends on its thermal conductivity. It also depends on the temperature difference across the material. The fundamental relationship between these quantities is given in Equation 2.1.

(Equation 2.1)

$$Q = A\left(\frac{k}{\Delta x}\right)(\Delta T)$$

where:

Q = rate of heat transfer through the material (Btu/hr)
k = thermal conductivity of the material (Btu/°F • hr • ft)
Δx = thickness of the material in the direction of heat flow (ft)
ΔT = temperature difference across the material (°F)
A = area heat flows across (ft²)

Example 2.1: Determine the rate of heat flow by conduction through the panel shown in Figure 2–1.

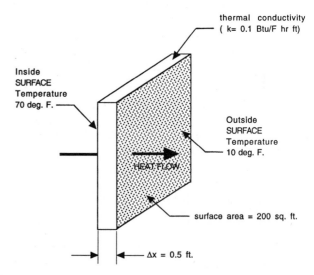

Figure 2–1   Heat flow through a material by conduction

Solution: Substituting the data in Equation 2.1:

$$Q = A\left(\frac{k}{\Delta x}\right)(\Delta T)$$

$$Q = 200\,\text{ft}^2\left(\frac{0.1\dfrac{\text{Btu}}{{}^\circ\text{F}\bullet\text{hr}\bullet\text{ft}}}{0.5\,\text{ft}}\right)(70^\circ\text{F} - 10^\circ\text{F})$$

$$Q = 2,400\ \text{Btu/hr}$$

Thermal Resistance of a Material

Since building designers are usually interested in *reducing* the rate of heat flow from a building, it is more convenient to think of a material's resistance to heat flow rather than its thermal conductivity. A material's **thermal resistance** or **R-value** can be defined as its thickness in the direction of heat flow divided by its thermal conductivity.

(Equation 2.2)

$$\text{R-value} = \frac{\text{thickness}}{\text{thermal conductivity}} = \frac{\Delta x}{k}$$

The greater the R-value of a material, the slower heat will pass through it when a given temperature differential is maintained across it. Equations 2.1 and 2.2 can be combined to yield an equation that is convenient for use in estimating heating loads:

(Equation 2.3)

$$Q = \left(\frac{A}{R}\right)(\Delta T)$$

where:

Q = rate of heat transfer through the material (Btu/hr)
ΔT = temperature difference across the material (°F)
R = R-value of the material (°F • ft² • hr/Btu)
A = area across which heat flows (ft²)

Some interesting facts can be shown from this equation. First, *the rate of heat transfer through a given material is directly proportional to the temperature difference (ΔT) maintained across its thickness.* If this temperature difference were doubled, the rate of heat transfer through the material would also double.

Second, *the rate of heat transfer through a material is inversely proportional to its R-value.* For example, if its R-value were doubled, the rate of heat transfer through the material would be cut in half (assuming the temperature difference across the material remained constant).

Finally, the larger the object's surface area, the greater the rate of heat transfer through it. For example a 4 ft by 4 ft window with an area of 16 ft² would transfer heat twice as fast as a 2 ft by 4 ft window with an area of 8 ft².

Example 2.2: Determine the rate of heat transfer through the 6-in thick wood panel shown in Figure 2–2.

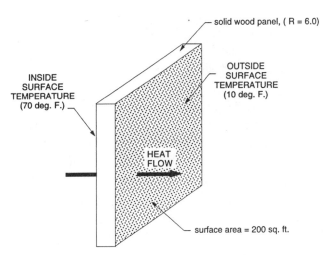

Figure 2–2   Heat flow through a panel by conduction (using R-value)

Solution: Substituting the data in Equation 2.3:

$$Q = \left(\frac{A}{R}\right)(\Delta T) = \left(\frac{200}{6}\right)(70 - 10) = 2,000\ \text{Btu/hr}$$

The R-value of a material is also directly proportional to its thickness. For example, if a 1-in thick panel of polystyrene insulation has an R-value of 5.4, a 2-in thick panel would have an R-value of 10.8, and a 1/2-in thick piece would have an R-value of 2.7. This fact is very useful when the R-value of one thickness is known, and the R-value of a different thickness needs to be determined.

The R-value of a material is slightly dependent on the material's temperature. In general, as the temperature of the material is lowered, its R-value increases. This results from a number of complex physical effects such as the slowed motion of air within the tiny cells of the material. For most building materials, the change in R-value is fairly small over the temperature ranges the material typically experiences, and thus may be assumed constant.

Total R-value of an Assembly

Since the walls, floors, and ceilings of buildings are rarely constructed of a single material, it is often necessary to determine the **total R-value of an assembly** made up of several materials in contact with each other. This is accomplished by simply adding up the R-values of the individual materials as illustrated in Figure 2–3.

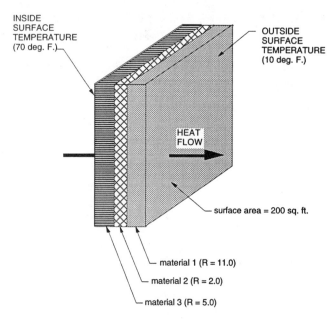

Figure 2–3 Wall assembly consisting of several materials with different R-values

The total R-value of this assembly is the sum of the individual R-values:

$$R_{total} = 11.0 + 2.0 + 5.0 = 18.0$$

The heat flow across the panel can now be calculated using the total R-value substituted into Equation 2.3:

$$Q = \left(\frac{A}{R}\right)(\Delta T) = \left(\frac{200}{18}\right)(70 - 10) = 667 \text{ Btu/hr}$$

The R-values of some common building materials can be found in Appendix B. To find the R-values of other materials, consult manufacturer's literature or the ASHRAE Handbook of Fundamentals.

Air Film Resistances

In addition to the thermal resistances of the solid materials that make up an assembly, there are two invisible, but not insignificant, thermal resistances. These are called the inside and outside **air film resistances**. They represent the insulating effects of thin layers of air that cling to all building surfaces. The R-values of these air films are dependent on surface orientation, air movement along the surface, and reflective qualities of the surface. Values for these resistances can also be found in Appendix B. These R-values include the effect of conduction, convection, and radiation heat transfer, but are stated as a conduction-type thermal resistance for simplicity.

The following example illustrates how the total R-value of the wall assembly shown in Figure 2–4 is affected by the inclusion of the air film resistances.

Notice that the temperatures used to determine the ΔT in this example are now the indoor and outdoor air temperatures rather than the surface temperatures as used in the previous examples.

Example 2.3: Determine the rate of heat transfer through the assembly shown in Figure 2–4. Assume still air on the inside of the wall, and 15 mph wind on the outside. See Appendix B for values of the air film resistances.

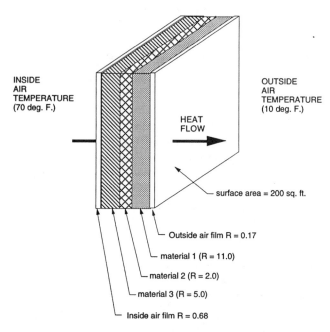

Figure 2–4 Air film resistances add R-value to a wall assembly

Solution: The total R-value of the assembly is again found by adding the R-values of all materials, plus the R-values of the inside and outside air films:

$$R_{total} = 0.17 + 11.0 + 2.0 + 5.0 + 0.68 = 18.85$$

The rate of heat transfer through the assembly can again be found using Equation 2.3:

$$Q = \left(\frac{A}{R}\right)(\Delta T) = \left(\frac{200}{18.85}\right)(70 - 10) = 637 \text{ Btu/hr}$$

Effect of Framing Members

The rate of heat flow through assemblies made of several materials stacked together like a sandwich can now be calculated. In such situations, and when edge effects are not considered, the heat flow through the wall is uniform over each square foot of surface area. Unfortunately, few buildings are constructed with walls, ceilings, and other surfaces made of simple stacked layers of

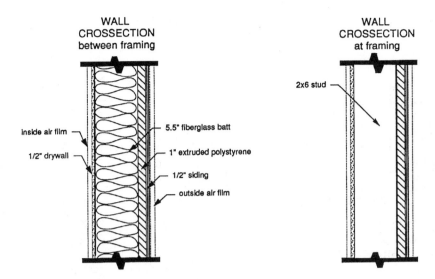

Material	R-value	Material	R-value
Inside air film	0.68	Inside air film	0.68
1/2" drywall	0.45	1/2" drywall	0.45
6" fiberglass batt	19.0	2 × 6 stud	5.5
1" foam sheathing	5.4	1" foam sheathing	5.4
3/4" wood siding	0.8	3/4" wood siding	0.8
outside air film	0.17	outside air film	0.17
TOTAL	26.5	TOTAL	13.0

Figure 2–5   Calculating the effective total R-value of a wall assembly, including the effect of framing members

materials. This is particularly true of wood-framed buildings with wall studs, ceiling joists, window headers, and other framing members. Wooden framing members that span across the insulation cavity of a wall tend to reduce its **effective total R-value**. This is because the thermal resistance of wood (about 1.0 per inch of thickness) is usually less than the thermal resistance of insulation materials used to fill the wall cavity.

It is possible to adjust the R-value of an assembly to include the thermal effects of framing. This requires that the assembly's total R-value be calculated both between the framing and at the framing. The resulting total R-values are then weighted according to the percentage of solid framing in the assembly. Although this percentage could be calculated for each assembly, such calculations can be tedious, and often result in insignificant changes in the results. Typical wood-framed residential and light commercial construction have from 10% and 20% of the wall area as solid framing across the insulation cavity. The lower end of this range would be appropriate for 24 inches on-center framing, with insulated headers over windows and doors; the upper end typical of 16 inches on-center framing with solid headers.

To illustrate these calculations, the effective total R-value of the wall assembly shown in Figure 2–5 will be calculated assuming 15% of the wall is solid framing.

$$R_{effective} = 0.85(26.5) + 0.15(13.0) = 24.5$$

To obtain the total heat loss of a room, the effective total R-value of each different **exposed surface** within the room must first be determined using a procedure similar to that shown in Figure 2–5. Fortunately, many buildings have the same type of exterior wall construction, ceiling construction, etc. used for each room. In such cases the effective R-value of the assembly need only be calculated once.

It is good practice to make a sketch of the cross section of each exposed assembly (i.e., wall, ceiling, floor), similar to that shown in Figure 2–5. The R-value of each material used in the assembly can then be looked up in Appendix B, and the effective total R-value calculated. The sheets showing the cross sections of the assemblies and the R-values of their materials can be saved in a file, and used as a reference on other buildings with similar construction.

2.4   FOUNDATION HEAT LOSSES

Heat flow from a basement or exposed slab edge is determined by complex interactions between the building,

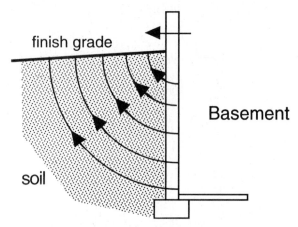

Figure 2–6   Theoretical heat flow paths through a uniformly insulated basement wall and adjacent soil

the surrounding soil, and the air above grade. It is impractical to attempt to model the complexities of this heat transfer for routine estimates of foundation heat loss. Several approximating methods, however, have been developed from these complex models that are suitable for routine heating load calculations. One such method can be used for estimating the heat loss of a partially buried basement, the other for the heat loss from the exposed edge of a floor slab.

Heat Loss through Basement Walls

An important factor affecting the heat loss of a basement wall is the height of soil against the outside of the wall. The deeper the wall is buried, the lower the rate of heat loss through the wall. Figure 2–6 illustrates the theoretical heat flow paths through a uniformly insulated basement wall and adjacent soil. Notice the heat flow paths from the inside basement air to the outside air become longer for the lower portions of the wall. As the path length increases, so does the effective thermal resistance of the soil.

The method presented for approximating the heat loss through the foundation wall is based on data published in the Air Conditioning Contractors of America Manual J. The method divides the wall into horizontal strips based on the height of finish grade. Figure 2–7 illustrates the concept.

The equations for determining the effective R-value for the wall strips shown in this figure are:

(Equation 2.4a)
$$R = R_{\text{foundation wall}} + R_{\text{added}}$$

(Equation 2.4b)
$$R = 8 + 1.126(R_{\text{added}})$$

(Equation 2.4c)
$$R = 11.5 + 1.09(R_{\text{added}})$$

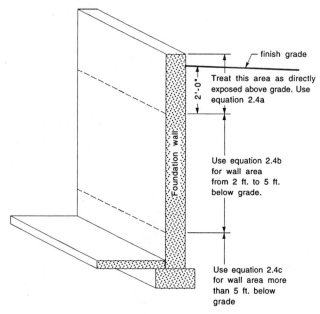

Figure 2–7   Dividing the basement wall into horizontal strips to estimate heat loss

In Equations 2.4a through 2.4c, R_{added} represents the R-value of any insulation *added* to the foundation wall over the particular strip for which the effective R-value is being determined. Once the R-values of each wall strip have been determined, Equation 2.3 can be used to calculate the heat loss of each area. *It should be emphasized that ΔT in Equation 2.3 will still be the difference between the basement air temperature and the outside air temperature.*

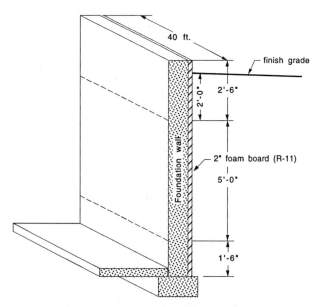

Figure 2–8   Basement wall used in Example 2.4

Example 2.4: Estimate the heat loss through a 10-in thick concrete basement wall 9 ft deep with 6 ins exposed above grade. The wall is 40 ft long, and has 2 ins of extruded polystyrene insulation (R-11) added on the outside. The basement temperature is 70 °F. The outside air temperature is 10 °F. Assume the R-value of the 10-in concrete wall is 2.0. A drawing of the wall is shown in Figure 2–8.

Solution: For the upper portion of the wall, use Equation 2.4a to determine the total R-value. Then use Equation 2.3 to calculate the heat loss.

$$R = 2.0 + 11 = 13$$

$$A = 40\text{ft}(2.5\text{ft}) = 100\text{ft}^2$$

$$\Delta T = (70°\text{ F} - 10°\text{ F}) = 60°\text{ F}$$

$$Q = \left(\frac{A}{R}\right)(\Delta T) = \left(\frac{100}{13}\right)(60) = 462 \text{ Btu/hr}$$

For the middle section of the wall, use Equation 2.4b to determine the R-value. Then use Equation 2.3 to calculate the heat loss.

$$R = 8 + 1.126(R_{added}) = 8 + 1.126(11) = 20.4$$

$$A = 40\text{ft}(5\text{ft}) = 200\text{ft}^2$$

$$\Delta T = (70°\text{ F} - 10°\text{ F}) = 60°\text{ F}$$

$$Q = \left(\frac{A}{R}\right)(\Delta T) = \left(\frac{200}{20.4}\right)(60) = 588 \text{ Btu/hr}$$

For the bottom portion of the wall, use Equation 2.4c to calculate the R-value. Again use Equation 2.3 to calculate the heat loss.

$$R = 11.5 + 1.09(R_{added}) = 11.5 + 1.09(11) = 23.5$$

$$A = 40\text{ft}(1.5\text{ft}) = 60\text{ft}^2$$

$$\Delta T = (70°\text{ F} - 10°\text{ F}) = 60°\text{ F}$$

$$Q = \left(\frac{A}{R}\right)(\Delta T) = \left(\frac{60}{23.5}\right)(60) = 153 \text{ Btu/hr}$$

The total heat loss for this wall, for the stated temperatures, is the sum of the heat losses from each wall strip:

$$Q_{total} = 462 + 588 + 153 = 1,203 \text{ Btu/hr}$$

Discussion: Note that the same ΔT (the difference between the basement air temperature and the outside air temperature) was used in the calculations for all three wall strips.

Heat Loss through Basement Floors

The following equation can be used to estimate downward heat loss for basement floor slabs more than 2 feet below finish grade:

(Equation 2.5)

$$Q_{\text{basement floor}} = 0.024(A)(\Delta T)$$

where:

$Q_{\text{basement floor}}$ = rate of heat loss through the basement floor slab (Btu/hr)
A = floor area (ft²)
ΔT = a. basement air temperature – outside air temperature (°F) *for unheated slabs*
b. slab temperature – outside air temperature (°F) *for heated slabs*

Example 2.5: Determine the rate of heat loss from a 30 ft by 50 ft basement floor slab when the outside temperature is 10 °F, and the basement temperature is 65 °F. Assume:

a. the floor has a temperature approximately equal to the inside air temperature.
b. the floor is hydronically heated, and has an average temperature of 95 °F.

Solution: Substituting the data into Equation 2.5 for each case:

a. $Q_{\text{basement floor}}$ = 0.024(30 • 50)(65 – 10) = 1,980 Btu/hr

b. $Q_{\text{basement floor}}$ = 0.024(30 • 50)(95 – 10) = 3,060 Btu/hr

Heat Loss through Slab-on-Grade Floors

Many residential and light commercial buildings have concrete floor slabs rather than crawl spaces or full basements. Heat flows from the floor slab downward to the surrounding earth, and outward through any exposed edges of the slab. Of these two paths, the outward flow of heat through the slab edge tends to dominate. This is because the outer edge of the slab is exposed to either the outside air or to soil only slightly warmer than outside air. Soil under the slab tends to become thermally stable once the building is maintained at normal comfort temperatures, and hence downward heat losses are relatively small. Areas of the floor close to the exposed edges will have higher rates of heat loss compared to areas several feet in from the edge. It is therefore important to have higher R-value insulation under the slab near its edges.

As with basement walls, the actual paths of heat flow from a slab to the soil and outside air are complex and not easily modeled. For this reason approximating methods have been developed to estimate the rate of heat loss from a floor slab. Equation 2.6 can be used to estimate the *combined downward and edgewise heat loss* from slab-on-grade floors.

(Equation 2.6)

$$Q_{slab} = \frac{(L)(\Delta T)}{1.21 + 0.214(R_{edge}) + 0.0103(R_{edge})^2}$$

where:

Q_{slab} = rate of heat loss through the slab edge and floor area near slab edge (Btu/hr)

L = *exposed edge length* of the floor slab (ft)

R_{edge} = R-value of edge insulation added to the slab edge (from R-0 to R-11)

ΔT = a. difference between inside and outside air temperature (°F) *for unheated slabs*

b. difference between the slab temperature and outside air temperature (°F) *for heated slabs*

Example 2.6: Determine the rate of heat loss from the room slab shown hatched in Figure 2–9. The slab has 1.5 ins of extruded polystyrene edge insulation. The inside temperature is 70 °F, the outside air temperature is 10 °F. Assume the slab is not heated.

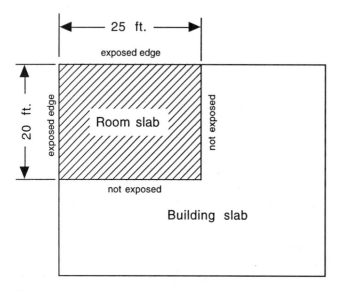

Figure 2–9 Room with exposed slab edge used in Example 2.6

Solution: Note all four sides of the room's slab are not exposed to the outside. The total linear footage of *exposed* slab edge is 20 + 25 = 45 feet. The R-value of extruded polystyrene insulation is 5.4 per inch of thickness, therefore the R-value of 1.5 inches is 1.5(5.4) = 8.1. Equation 2.6 can now be evaluated:

$$Q_{slab} = \frac{(45)(70 - 10)}{1.21 + 0.214(8.1) + 0.0103(8.1)^2} = \frac{2,700}{3.619}$$

$$= 746 \text{ Btu/hr}$$

The Thermal Envelope of a Building

Heat flows through all building surfaces that separate heated space from unheated space. Together these surfaces (walls, windows, ceilings, doors, foundation, etc.) are called the **thermal envelope** of the building. It is only necessary to calculate heat flow though the surfaces that together constitute this thermal envelope. Other walls, ceilings, etc. that have heated space on both sides and thus are not exposed surfaces should not be included in these calculations. Figure 2–10 illustrates a part of the thermal envelope for a house.

2.5 INFILTRATION HEAT LOSSES

In addition to conduction losses, heat is also carried out of buildings by air leakage, often called **air infiltration**. The faster air leaks through a building, the greater the amount of heat it carries away. The average thermal envelope of a building contains hundreds of imperfections through which air can pass. Some leakage points, such as fireplace flues, exhaust hoods, and visible cracks around windows and doors, are obvious. Others are small and out of sight, but when taken together represent significant amounts of leakage area. These include cracks where walls meet the floor deck, air leakage through electrical outlets, small gaps where pipes pass through floors or ceilings, and more.

It would be impossible to assess the location and magnitude of all the air leakage paths in a typical house. Instead, an estimate of air leakage can be made based on

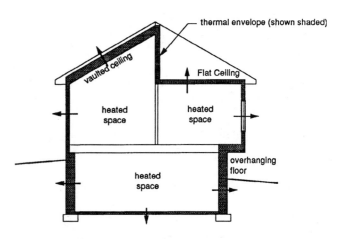

Figure 2–10 A portion of the thermal envelope of a house (shown shaded)

Floor area (sq. ft.)	900 or less	900-1500	1500-2100	over 2100
"Best" quality	0.4	0.4	0.3	0.3
"Average" quality	1.2	1.0	0.8	0.7
"poor" quality	2.2	1.6	1.2	1.0

Add for EACH fireplace:

"Best" quality	0.1
"Average" quality	0.2
"poor" quality	0.6

Figure 2-11   Air change rates suggested in ACCA Manual J

the air sealing quality of the building. This approach is known as the **air change method** of estimating heat loss due to air infiltration. Based on a descriptive evaluation of air sealing quality, and experience, the designer chooses a *rate* at which the interior air volume of the building (or an individual room) is exchanged with outside air. For example, a leakage rate of 1.0 air change per hour indicates the entire volume of heated air in a space is assumed to be replaced each hour with outside air.

The infiltration rates shown in Figure 2-11 are suggested in ACCA Manual J based on a somewhat subjective evaluation of the air sealing quality of the house.

- Best quality air sealing refers to buildings with continuous infiltration barriers such as Tyvek® housewrap, continuous interior vapor barriers, high quality windows, well-caulked joints in exterior walls, no combustion air used by appliances, and dampers on all appliance vents.
- Average quality air sealing refers to buildings with plastic vapor barriers, average quality windows and doors, major penetrations caulked, non-caulked or taped electrical fixtures, combustion air used by appliances, and dampered appliance vents.
- Poor quality air sealing refers to buildings with no air infiltration or vapor barriers, lower quality windows and doors, no caulking at penetrations of thermal envelope, combustion air requirements for heat source, and no dampers on appliances.

Remember that these values are at best estimates, and they represent a wide range of variation. For example, in a 2,000 square foot house, the air leakage rate associated with the poor quality air sealing will result in four times more infiltration heat loss than the rate associated with best quality air sealing.

Experience is also helpful in estimating infiltration rates. A visit to the building during construction gives the heating system designer a feel for the air sealing quality being achieved by the builder. An inspection visit to an existing building, though not as revealing as when the build-

ing is under construction, can still help the designer assess air sealing quality. Things to look for or consider include:

- Visible cracks around windows and doors
- Poorly installed or maintained weatherstripping on doors and windows
- Slight air motion near electrical fixtures
- Smoke that disappears upward into ceiling lighting fixtures
- Deterioration of paint on the downwind side of building
- Poor quality duct system design (especially a lack of proper return air ducting)
- Interior humidity levels (drier buildings generally indicate more air leakage)
- Presence or lack of flue damper on gas- or oil-fired heating system
- Windy versus somewhat sheltered building site
- Storm doors and entry vestibules help reduce air infiltration
- Casement and awning windows are generally tighter than double hung or sliding windows
- Cold air leaking in low in the building generally means warm air leaking out higher in the building

Estimating Infiltration Heat Loss

Air leakage rates can be converted into rates of heat loss using Equation 2.7:

(Equation 2.7)
$$Q_i = (0.018)(N)(V)(\Delta T)$$

where:

Q_i = estimated rate of heat loss due to air infiltration (Btu/hr)

N = number of air changes per hour, estimated based on air sealing quality (1/hr)

V = interior volume of the heated space (room or entire building) (ft³)

0.018 = set value representing the heat capacity of air (Btu/ft³/°F)

ΔT = inside air temperature minus the outside air temperature (°F)

Example 2.7: Determine the rate of heat loss by infiltration for a 30 ft by 50 ft building with 8 ft ceiling height and average quality air sealing. The inside temperature is 70 °F. The outside temperature is 10 °F.

Solution: Referring to Figure 2-11 for a 1,500 square foot house with average air sealing quality, one finds a suggested air leakage rate of 0.8 air changes per hour. Substituting this value along with the other data into Equation 2.7:

$$Q_{\text{infiltration}} = (0.018)(0.8)(30 \cdot 50 \cdot 8)(70 - 10)$$

$$= 10{,}368 \approx 10{,}400 \text{ Btu/hr}$$

As the conduction heat losses of buildings are reduced through use of higher R-value insulation, air infiltration tends to becomes a larger percentage of the total heat loss. When higher insulation levels are specified for a building, greater efforts at reducing air infiltration should also be undertaken. These would include the use of high quality windows and doors, close attention to caulking, sealed infiltration barriers, etc.

2.6   PUTTING IT ALL TOGETHER

Finding the total heat loss of a room or building is simply a matter of adding the conduction heat losses to the infiltration heat losses. It is convenient to do this on a room by room basis. The resulting numbers can then be used to size heat emitters during the design of a hydronic distribution system. The sum of the room heat losses can be used to select the capacity of the heat source.

A simplified table for performing a room heat load calculation has been developed, and is shown along with instructions in Figure 2–12. Figure 2–13 summarizes basement heat loss calculations. To obtain the total heating load of a building with a basement, add the total room loads to the basement load.

2.7   EXAMPLE OF A COMPLETE HEATING LOAD ESTIMATE

This section will demonstrate a complete heating load estimate for an example house having the floor plan shown in Figure 2–14. The room heat load form shown in Figure 2–12 will be used to keep the calculations organized. Other assumed values for the purposes of these calculations are as follows:

- Unit R-value for all windows: R-3.0
- Unit R-value for exterior door: R-5
- R-value of foundation edge insulation: R-11
- Rate of air infiltration in vestibule and mechanical room: 1 air change/hr

Room-by-Room Heat Load Worksheet

Indoor desired air temperature   (Ti)   = _____ deg. F.

Outdoor design air temperature   (To)   = _____ deg. F.

Design temperature difference = ΔT = (Ti- To)   = _____ deg. F.

Room Name		1 Exposed Walls	2 Windows	3 Exposed Doors	4 Exposed Ceilings	5 Exposed Floor	6 Exposed Slab Edge	7 Room Volume	8 #AC / hr.	9 Infil. Loss	10 TOTAL ROOM LOAD
	1	A=	A=	A=	A=	A=	L=	V=			total of row 3 for columns 1,2,3,4,5,6,9
	2	R=	R=	R=	R=	R=	R=				
	3	A/R (ΔT)	A/R (ΔT)	A/R (ΔT)	A/R (ΔT)	A/R (ΔT)	Q slab			Q infil.	

Instructions to complete table:

Row #	Column #	Description of required entry
1	1	Enter NET area of exposed WALLS (subtract for windows and doors) sq. ft.
1	2	Enter area of WINDOWS in sq. ft.
1	3	Enter area of exposed DOORS in sq. ft.
1	4	Enter area of exposed CEILINGS in sq. ft.
1	5	Enter area of FLOORS over unheated space in sq. ft.
1	6	Enter linear feet of exposed slab edge in feet
1	7	Enter volume of room (width x length x depth) in cubic ft.
1	8	Enter rate of air change for this room in air changes per hour
2	1 - 5	Enter the total R-value of the various exposed surfaces
2	6	Enter the R-value of any slab edge insulation used
3	1 - 5	Enter the results of the calculation: $Q = \left(\dfrac{A}{R}\right)(\Delta T)$ (equation 2.3) ΔT is obtain from the top of the sheet.
3	6	Enter the results obtained from equation 2.6
3	9	Enter the results obtained from equation 2.7
3	10	Add the values in row 3 for columns 1,2,3,4,5,6,9 to get total room load.

Figure 2–12 Table and instructions for determining room heating load

Basement Heating Load Worksheet

Indoor desired air temperature (T_i) = _____ deg. F.
Outdoor design air temperature (T_o) = _____ deg. F.
Design temperature difference = $\Delta T = (T_i - T_o)$ = _____ deg. F.

		1 Depth (ft.)	2 Length (ft.)	3 Area (sq. ft.)	4 R-value	5 Q (Btu / hr.)
Top strip	1					
Middle strip	2					
Bottom strip	3					
Floor	4					
	5				Q_{total}	

Row	Column	Instructions
1,2,3	1	Enter the depth of each strip of wall (in ft.), see diagram
1,2,3	2	Enter the length of foundation wall (in ft.)
1,2,3	3	Calculate the area of each strip by multiplying the depth x length (in ft. sq.)
1	4	R-value of the foundation wall + the R-value of any added insulation (use equation 2.4a)
2	4	Enter calculated R-value of middle wall strip (use equation 2.4b)
3	4	Enter calculated R-value of bottom wall strip (use equation 2.4c)
1,2,3	5	Calculate the heat loss of each strip using equation 2.3
4	3	Enter the floor are of the basement (in sq. ft.)
4	5	Enter the calculated heat loss through the floor using equation 2.5 (in Btu / hr)
5	5	Add all values of Q in rows 1 through 4 for total basement heating load (in Btu / hr)

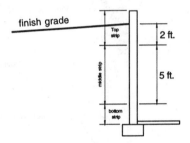

Figure 2–13   Table and instructions for determining basement heating load

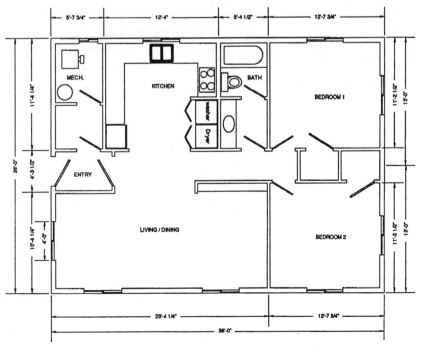

Figure 2–14   Floor plan of example house for which heating load will be estimated

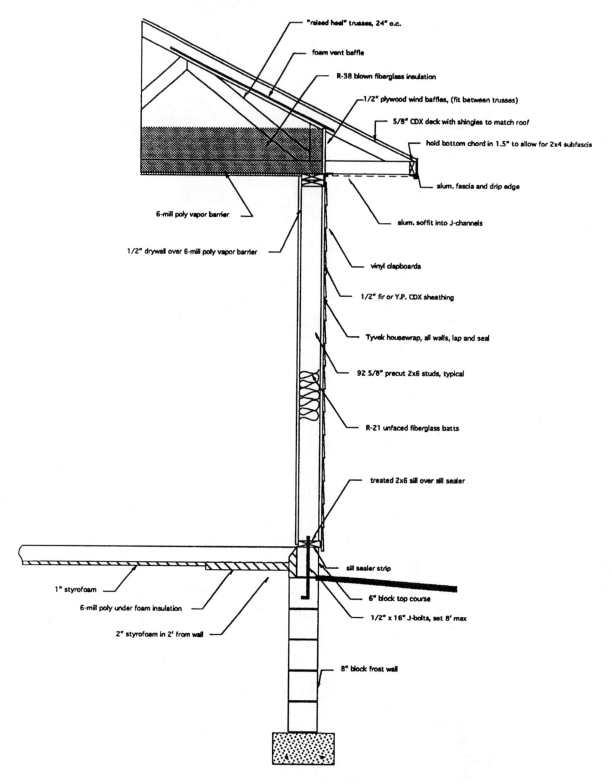

Figure 2-15　Wall cross section for example house

- Rate of air infiltration in all other rooms: 0.5 air change/hr
- Outdoor design temperature: –5 °F
- Desired indoor temperature: 70 °F
- Window height: 4 ft
- Exterior door height 6'–8"

Determining the Total R-value of the Thermal Envelope Surfaces

Walls: A wall cross section is shown in Figure 2–15. It is made of 2 in by 6 in wood framing 24 ins on center with R-21 fiberglass batt insulation. The inside finish is $1/2$-in drywall. The outside of the wall is sheathed with $1/2$-in plywood covered with a Tyvek® infiltration barrier, and vinyl siding.

The following is a tabulation of the R-values of the materials that make up the wall both between the framing and at the framing:

Material:	R-value between Framing	R-value at Framing
inside air film	0.68	0.68
$1/2$-inch drywall	0.45	0.45
stud cavity	21.0	5.5
$1/2$-inch plywood sheathing	0.62	0.62
Tyvek® infiltration barrier	~0	~0
vinyl siding	0.61	0.61
outside air film	0.17	0.17
Total	23.53	8.03

Assuming 15% of wall is solid framing, the effective total R-value of the wall is:

$$R_{\text{effective}} = 0.85(23.53) + 0.15(8.03) = 21.2$$

Ceilings: The ceiling consists of $1/2$-in drywall covered with approximately 12 ins of blown fiberglass insulation. The roof trusses displace a minor amount of this insulation. We will again calculate the effective total R-value of the ceiling assembly shown in Figure 2–16.

The following is a tabulation of the R-values of the materials that make up the ceiling both between the framing and at the framing:

Material Description	R-value between Framing	R-value at Framing
bottom air film	0.61	0.61
$1/2$-inch drywall	0.45	0.45
insulation	40	28.3
framing	0	3.5
top air film	0.61	0.61
Total	41.67	33.47

Assuming 10% of ceiling is solid framing, the effective total R-value of the ceiling is:

$$R_{\text{effective}} = 0.90(41.67) + 0.10(33.47) = 40.9$$

Notice this value is very close to the R-value between framing, and thus the wooden truss chords have very minimal effect.

Room by Room Calculations

Using information from the floor plan and the total R-values of the various surfaces, the design heating load of each room will now be determined. This will be done

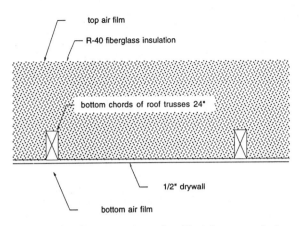

Figure 2–16   Cross section of ceiling in example house

by breaking out each room from the overall building and determining all necessary information, such as areas of walls, windows, etc. This information will then be entered into the table of Figure 2–12 along with the results of the various equations that have been presented.

To expedite the process, many of the relevant dimensions are shown on the floor plan of the individual rooms. If such were not the case, the dimensions would have to be estimated using an architectural scale along with the floor plan. As a matter of convenience, the room dimensions are taken to the centerline of the common walls. When working with architectural dimensions it is convenient to convert inches and fractions of inches into decimal portions of feet since these dimension will be added and multiplied together. Example 2.8 illustrates this.

Example 2.8: Convert the dimension 10'–4 $1/4$" to decimal feet.

Solution:
$$10' - 4\frac{1}{4}" = 10\,\text{ft} + \left(\frac{4.25}{12}\right)\text{ft}$$
$$= 10\,\text{ft} + 0.35\,\text{ft} = 10.35\,\text{ft}$$

The conversion from feet and inches into decimal feet has been performed prior to the calculation of areas and lengths in Figures 2–17 through 2–21.

Obtaining the Total Building Heating Load

The total building heating load is obtained by summing the heating loads of each room. In this case the overall building heating load is:

Living/dining room	4,221 Btu/hr
Bedroom 1	3,004 Btu/hr
Bedroom 2	3,004 Btu/hr
Kitchen	1,865 Btu/hr
Bathroom	664 Btu/hr
Mechanical room/entry	2,452 Btu/hr
Total	15,210 Btu/hr

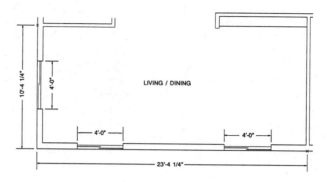

Gross exterior wall area = (10.35 ft. + 23.35 ft.)(8 ft.) = (33.708 ft.)(8 ft.) = 269.7 sq. ft.
Window area = 3 (4 ft × 4 ft) = 48 sq. ft.
exterior door area = 0 sq. ft.
Net exterior wall area = Gross exterior wall area - Window and door area = 221.7 sq. ft.
Ceiling area = (10.35 ft.)(23.35 ft.) = 241.7 sq. ft.
Exposed slab edge length = (10.35 ft. + 23.35 ft.) = 33.7 linear ft.
Room volume = (10.35 ft.)(23.35 ft.)(8.0 ft.) = 1933.4 cubic ft.

		1	2	3	4	5	6	7	8	9	10	
Room Name		Exposed Walls	Windows	Exposed Doors	Exposed Ceilings	Exposed	Exposed Slab Edge	Room Volume	#AC / hr.	Infil. Loss	TOTAL ROOM LOAD	
LIVING & DINING	1	A= 221.7	A= 48	A= 0	A= 241.7	A= 0	L= 33.7	V= 1933	0.5		total of row 3 for columns 1,2,3,4,5,6,9	
	2	R= 21.2	R= 3	R= N/A	R= 40.9	R= N/A	Re= 11					
	3	A/R (ΔT) 784	A/R (ΔT) 1200	A/R (ΔT) 0	A/R (ΔT) 443	A/R (ΔT) 0	Q slab 525			Q infil. 1305	4257 $\frac{BTU}{HR}$	

Figure 2–17   Floor plan of living/dining room from example house of Figure 2–14

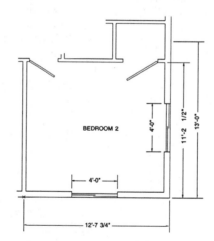

Gross exterior wall area = (12.65 ft. +13.0 ft.)(8 ft.) = (25.65 ft.)(8 ft.) =205.2 sq. ft.
Window area = 2 (4 ft × 4 ft) = 32 sq. ft.
Exterior door area = 0 sq. ft.
Net exterior wall area = Gross exterior wall area – Window and door area =173.2 sq. ft.
Ceiling area = (12.65 ft.)(13.0 ft.) = 164.5 sq. ft.
Exposed slab edge length = (12.65 ft. + 13.0 ft.) = 25.65 linear ft.
Room volume = (12.65 ft.)(13.0 ft.)(8.0 ft.) = 1315.6 cubic ft.

		1	2	3	4	5	6	7	8	9	10	
Room Name		Exposed Walls	Windows	Exposed Doors	Exposed Ceilings	Exposed	Exposed Slab Edge	Room Volume	#AC / hr.	Infil. Loss	TOTAL ROOM LOAD	
BEDROOM 2	1	A= 173.2	A= 32	A= 0	A= 164.5	A= 0	L= 25.7	V= 1316	0.5		total of row 3 for columns 1,2,3,4,5,6,9	
	2	R= 21.2	R= 3	R= N/A	R= 40.9	R= N/A	Re= 11					
	3	A/R (ΔT) 613	A/R (ΔT) 800	A/R (ΔT) 0	A/R (ΔT) 302	A/R (ΔT) 0	Q slab 401			Q infil. 888	3004 $\frac{BTU}{HR}$	

Figure 2–18   Floor plan of bedrooms from example house of Figure 2–14

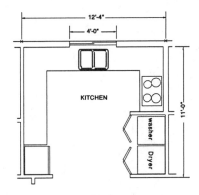

Gross exterior wall area = (12.33 ft.) (8 ft.) = 98.6 sq. ft.
Window area = 1 (4 ft x 4 ft) = 16 sq. ft.
Exterior door area = 0 sq. ft.
Net exterior wall area = Gross exterior wall area - Window and door area =82.6 sq. ft.
Ceiling area = (12.33 ft.) (11.0 ft.) = 135.6 sq. ft.
Exposed slab edge length = 12.33 linear ft.
Room volume = (12.33 ft.) (11.0 ft.) (8.0 ft.) = 1085 cubic ft.

Room Name		Exposed Walls (1)	Windows (2)	Exposed Doors (3)	Exposed Ceilings (4)	Exposed (5)	Exposed Slab Edge (6)	Room Volume (7)	#AC / hr. (8)	Infil. Loss (9)	TOTAL ROOM LOAD (10)
KITCHEN	1	A= 82.6	A= 16	A= 0	A= 135.6	A= 0	L= 12.3	V= 1085	0.5		total of row 3 for columns 1,2,3,4,5,6,9
	2	R= 21.2	R= 3	R= N/A	R= 40.9	R= N/A	Re= 11				
	3	A/R (ΔT) 292	A/R (ΔT) 400	A/R (ΔT) 0	A/R (ΔT) 249	A/R (ΔT) 0	Q slab 192			Q infil. 732	1865 $\frac{BTU}{HR}$

Figure 2-19 Floor plan of kitchen from example house of Figure 2–14

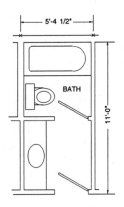

Gross exterior wall area = (5.38 ft.) (8 ft.) = 43.0 sq. ft.
Window area = 0 sq. ft.
Exterior door area = 0 sq. ft.
Net exterior wall area = Gross exterior wall area - Window and door area =43.0 sq. ft.
Ceiling area = (5.38 ft.) (11.0 ft.) = 59.2 sq. ft.
Exposed slab edge length = 5.38 linear ft.
Room volume = (5.38 ft.) (11.0 ft.) (8.0 ft.) = 473 cubic ft.

Room Name		Exposed Walls (1)	Windows (2)	Exposed Doors (3)	Exposed Ceilings (4)	Exposed (5)	Exposed Slab Edge (6)	Room Volume (7)	#AC / hr. (8)	Infil. Loss (9)	TOTAL ROOM LOAD (10)
BATHROOM	1	A= 43	A= 0	A=0	A= 59.2	A= 0	L= 5.38	V= 473	0.5		total of row 3 for columns 1,2,3,4,5,6,9
	2	R= 21.2	R= 3	R= N/A	R= 40.9	R=N/A	Re= 11				
	3	A/R (ΔT) 152	A/R (ΔT) 0	A/R (ΔT) 0	A/R (ΔT) 109	A/R (ΔT) 0	Q slab 84			Q infil. 319	664 $\frac{BTU}{HR}$

Figure 2-20 Floor plan of bathroom from example house of Figure 2–14

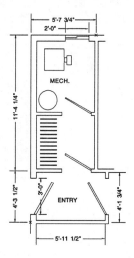

Gross exterior wall area = (4.29 ft. + 11.35 ft. + 5.65 ft.) (8 ft.) = (21.28 ft.) (8 ft.) = 170.2 sq. ft.
Window area = (2 ft. x 4 ft.) = 8 sq. ft.
Exterior door area = (6.66 ft x 3 ft) = 20 sq. ft.
Net exterior wall area = Gross exterior wall area - Window and door area =142.2 sq. ft.
Ceiling area = (4.29 ft. + 11.35 ft.) (5.65 ft.) = (15.64 ft. x 5.65 ft.) = 88.4 sq. ft.
Exposed slab edge length = 21.28 linear ft.
Room volume = (15.64 ft.) (5.65 ft.) (8.0 ft.) = 707 cubic ft.

Room Name		Exposed Walls	Windows	Exposed Doors	Exposed Ceilings	Exposed	Exposed Slab Edge	Room Volume	#AC / hr.	Infil. Loss	TOTAL ROOM LOAD
		1	2	3	4	5	6	7	8	9	10
	1	A= 142.2	A= 8	A= 20	A= 88.4	A= 0	L= 21.3	V= 707	1.0		total of row 3 for columns 1,2,3,4,5,6,9
	2	R= 21.2	R= 3	R= 5	R= 40.9	R= N/A	Re= 11				
	3	A/R (ΔT) 503	A/R (ΔT) 200	A/R (ΔT) 300	A/R (ΔT) 162	A/R (ΔT) 0	Q slab 332			Q infl. 955	2452 $\frac{BTU}{HR}$

Figure 2–21   Floor plan of mechanical room and entry from example house of Figure 2–14

A number of simplifying assumptions have been used in this example that are worth pointing out.

- The rooms were divided at the centerline of the common partitions. This simply ensures all the exterior wall area is assigned to individual rooms, and not ignored as being between the rooms.
- Since the two bedrooms are essentially identical in size and construction, it was only necessary to find the heat load of one, then double it when determining the total building load.
- The exterior dimensions of the rooms were used to determine areas, volumes, etc. This makes the calculations somewhat conservative in that the exterior wall area of a room is slightly greater than the interior wall area.
- A value of 0.5 air changes per hour was selected based on the relatively energy efficient construction shown on the drawing of the wall cross section. Such a value is consistent with observed rates of air leakage in small, relatively well-sealed houses. A greater air leakage rate was assigned to the entry vestibule since it is a traffic path into or out of the house. This higher rate was also applied to the mechanical room because a combustion-type heat source is used.
- The areas and volumes of the closets between the bedrooms were equally divided up among the two bedrooms. Since a separate heat emitter would not be supplied for each closet, this part of the load is simply included with the bedroom load where the heat emitter would be located.

2.8 COMPUTER-AIDED HEATING LOAD CALCULATIONS

After gathering the data and performing all the calculations needed for a building heating load estimate, it becomes obvious that a great deal of information goes into finding the result(s). If the designer then decides to go back and change one or more of the numbers, more time is required to unravel the calculations back to the point of the change, and then recompute the new result. Even for people used to handling this amount of information, these calculations can become tedious, and are

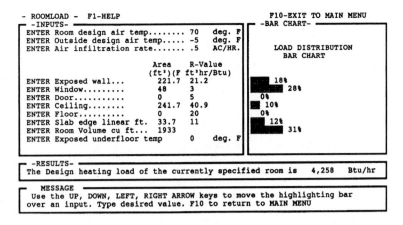

```
- ROOMLOAD -  F1-HELP                            F10-EXIT TO MAIN MENU
  -INPUTS-                                        -BAR CHART-
    ENTER Room design air temp........ 70   deg. F
    ENTER Outside design air temp..... -5   deg. F
    ENTER Air infiltration rate....... .5   AC/HR.    LOAD DISTRIBUTION
                                                        BAR CHART
                                  Area    R-Value
                                (ft²)(F ft²hr/Btu)
    ENTER Exposed wall...       221.7   21.2          ▮▮ 18%
    ENTER Window.........        48      3              ▮▮▮ 28%
    ENTER Door..........         0       5           0%
    ENTER Ceiling........       241.7   40.9          ▮ 10%
    ENTER Floor..........        0      20           0%
    ENTER Slab edge linear ft.  33.7    11            ▮ 12%
    ENTER Room Volume cu ft...  1933                  ▮▮▮ 31%
    ENTER Exposed underfloor temp  0    deg. F

  -RESULTS-
    The Design heating load of the currently specified room is   4,258   Btu/hr

  MESSAGE
    Use the UP, DOWN, LEFT, RIGHT ARROW keys to move the highlighting bar
    over an input. Type desired value. F10 to return to MAIN MENU
```

Figure 2-22   Screen shot of the ROOMLOAD program from the Hydronics Design Toolkit

always subject to calculation errors. The time required to perform accurate heating load calculations is probably the single greatest reason why some "heating professionals" decide not to bother with them.

Like many routine design procedures, heating load calculations have made the transition from pencil and paper to desktop computer. The advantages of using a load estimating program are many:

- Rapid (almost instant) calculations to quickly study the effect of changes
- Automated referencing of R-values, and weather data, in some programs
- Significantly less chance of error due to number handling
- Ability to print professional reports for customer presentations
- Ability to store project files for possible use in similar future projects
- Ability to import area and R-value information directly from CAD-based architectural drawings

There are presently several software packages available for estimating the heating loads for residential and light commercial buildings. These packages vary in cost from less than $10 (shareware) to over $400. The higher cost programs usually offer more features such as material reference files and customized output reports. The lower cost programs offer no frills output, but can nonetheless yield accurate results.

The ROOMLOAD Program

The Hydronics Design Toolkit contains a program called ROOMLOAD that can quickly calculate the design heating load of a room. Figure 2-22 shows the main screen of the ROOMLOAD program as it would appear for the Living/Dining room load determination in the example house of Section 2.7. As the user types in information such as area and R-value for the various

components, the program instantly calculates the design heating load for the specified room. It also displays a bar graph that indicates the percentage of total heat loss through the various paths discussed in this chapter.

The BSMTLOAD Program

The Hydronics Design Toolkit also contains a program called BSMTLOAD that uses the methods presented in this chapter to determine the design heating load of a user-specified basement. It offers the options of having both masonry- and wood-framed walls in the basement. The effect of any doors and windows in the basement can also be included. A screen shot of the BSMTLOAD program is shown in Figure 2-23.

2.9   ESTIMATING ANNUAL HEATING ENERGY USAGE

Building owners often ask heating professionals for an estimate of the annual heating cost of their building with the proposed heating system. Such an estimate can be made after the design heating load of the building and the seasonal efficiency of the heat source is determined.

The Degree Day Method

The seasonal energy used for space heating depends on its geographical location. The most common method of correlating the effect of geographical location on heating energy usage is the concept of **degree days**. The number of degree days that accumulate over a 24-hour day is the difference between 65 °F and the average outdoor air temperature for that 24-hour period. See Equation 2.8. The average temperature is calculated by averaging the high and low air temperature for the 24-hour period.

```
- BSMTLOAD -  F1-HELP                                    F10-EXIT TO MAIN MENU
┌ -INPUTS- ─────────────────────────────────────────────────────────────────┐
│   ENTER Basement design air temp,(40 to 90)..............   65    deg. F    │
│   ENTER Outside design air temp,(-30 to 50)..............   0     deg. F    │
│   ENTER Air infiltration rate, (0.1 to 1.0)..............   .2    AC/HR.    │
│   ENTER Basement wall height,(5 to 10)...................   8     ft        │
│   ENTER Height of wall exposed above grade,(.5 to 6)......   1     ft        │
│   ENTER Perimeter of exposed basement wall, (10 to 200)....  140   ft.       │
│   ENTER R-value of additional wall insulation,(0 to 25)....  10    F ft²hr/Btu│
│   ENTER Floor area of basement, (100 to 3000)..............  1200  ft²       │
│                                                                             │
│                                         Area       R-Value                  │
│                                         (ft²)      (F ft²hr/Btu)            │
│   ENTER Framed wall.....................   0          20                    │
│   ENTER Window..........................   10         3                     │
│   ENTER Exposed door....................   20         5                     │
│   ENTER Exposed slab edge... (linear ft.)  0          5                     │
├ -RESULTS- ──────────────────────────────────────────────────────────────── ┤
│   The Design heating load of the currently specified basement  9,099 Btu/hr │
├ MESSAGE ──────────────────────────────────────────────────────────────────┤
│   Use the UP, DOWN, LEFT, RIGHT ARROW keys to move the highlighting bar      │
│   over an input. Type desired value. F10 to return to MAIN MENU             │
└─────────────────────────────────────────────────────────────────────────────┘
```

Figure 2–23 Screen shot of the BSMTLOAD program from the Hydronics Design Toolkit

(Equation 2.8)

$$DD_{daily} = (65 - T_{ave})$$

where:

DD_{daily} = number of degree days accumulated in a 24-hour period

T_{ave} = average of the high and low outdoor temperatures for the 24-hour period (°F)

The total degree days for a given month or an entire year can be found by totalling the daily degree days over the desired time period. Many heating reference books contain tables of monthly and annual degree days for many major cities in the U.S. and Canada. A sample listing is given in Figure 2–24.

Degree day data has been used to estimate fuel usage for several decades. Such data is often recorded by fuel suppliers, utility companies, and local weather stations. Degree day statistics are frequently listed in the weather section of newspapers.

When the concept of degree days was first conceived, fuel was relatively inexpensive, and buildings were poorly insulated. Their heat loss was substantially greater than it is for most current buildings. *Estimating seasonal energy consumption using degree days relies on the assumption that buildings need supplemental heat input whenever the outside temperature drops below 65 °F.* While this may be true for many older, poorly insulated buildings, it is often incorrect for newer, better-insulated buildings. These new buildings often maintain comfortable interior temperatures, without heat input from their space heating systems, even when outdoor temperatures drop into the 50s or 40s. This is a result of greater internal heat gains from appliances, lights, people, and sunlight, combined with much lower rates of heat loss. Interior heat gains eliminate the need for a corresponding amount of heat from the building's heating system. Passive solar buildings are a good example. Many require

absolutely no heat input on sunny days, even when the outside temperature is below zero!

If annual heating energy usage is based solely on degree days, it will usually be significantly overestimated for well-insulated buildings with significant internal gains. To compensate for the effect of internal heat gains, ASHRAE devised a factor known as C_d. This factor is intended to correct for the effects of internal heat gains in typical residential buildings. ASHRAE presents suggested C_d factors in a graph that plots C_d versus total heating degree days in a given climate. This is shown in Figure 2–25.

It is important to realize the C_d value given by this graph is representative of average conditions over a large number of buildings. Any given building could have a value higher or lower than this average. For example, a passive solar building in a sunny climate such as Denver,

CITY	deg F • days (base 65)
Atlanta, GA	2990
Baltimore, MD	4680
Boston, MA	5630
Burlington, VT	8030
Chicago, IL	6640
Denver, CO	6150
Detroit, MI	6290
Madison, WI	7720
Newark, NJ	4900
New York, NY	5219
Philadelphia, PA	5180
Pittsburgh, PA	5950
Portland, ME	7570
Providence, RI	5950
Seattle, WA	4424
Syracuse, NY	6720
Toronto, ONT	6827

Figure 2–24 Annual heating degree days for some major cities

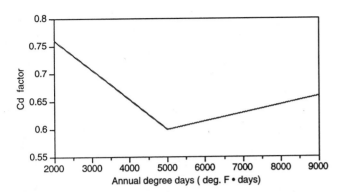

Figure 2–25 Mean values of C_d versus annual heating degree days suggested by ASHRAE. Data source: 1989 ASHRAE Fundamentals Handbook.

C_d = correction factor from Figure 2–25 (unitless)
ΔT_{design} = design temperature difference for which design heating load was calculated (°F)

Example 2.10: A house has a calculated design heating load of 55,000 Btu/hr when its interior temperature is 68 °F and the outdoor temperature is –10 °F. The house is located in a climate having 7,000 annual heating degree days. Estimate the annual heating energy that will be required by the building.

Solution: This data is substituted into Equation 2.9 as follows:

$$E_{annual} = \frac{(55,000)(7,000)(24)(0.63)}{1,000,000[68-(-10)]}$$

$$= 74.6 \text{ MMBtu/yr}$$

Colorado, is likely to have a much lower C_d value because it may get a major part of its heating energy from the sun. On the other hand, a building with small window areas, in a shaded location, or one with very little internal heat generation may have a large value of C_d. This variation will directly affect the estimate of annual fuel usage.

Equation 2.9 can be used to *estimate* annual heating energy required by the building:

(Equation 2.9)

$$E_{annual} = \frac{(Q_{design})(DD)(24)(C_d)}{1,000,000(\Delta T_{design})}$$

where:

E_{annual} = estimated annual heating energy *required by the building* (**MMBtu**)
Q_{design} = design heating load of the building (Btu/hr)
DD = annual total heating degree days at the building site (degree days)

The delivered cost of this energy can be determined using the FUELCOST program in the Hydronics Design Toolkit. This program determines the unit cost of several fuels in units of $/MMBtu based on their purchase cost and the seasonal efficiency of the heating system. A screen shot of the FUELCOST program is shown in Figure 2–26.

SUMMARY

This chapter has presented straightforward methods for estimating the design space heating load of residential and light commercial buildings. This step is an essential part of overall system design. It should never be skipped in favor of immediately laying out piping systems, etc. Many presently installed heating systems that yield only marginal comfort testify to the importance of knowing the load *before* designing the system.

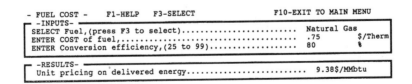

Figure 2–26 Screen shot of the FUELCOST program from the Hydronics Design Toolkit

KEY TERMS

97.5% design dry bulb temperature
Air change method
Air film resistance
Air infiltration
Building heating load
Degree days
Design heating load
Effective total R-value
Exposed surface
MMBtu
R-value
Room heating load
Thermal resistance
Thermal envelope
Total R-value of an assembly

CHAPTER 2 QUESTIONS AND EXERCISES

Note: Questions and exercises requiring the Hydronics Design Toolkit are indicated with marginal symbol.

1. One wall of a room measures 14 ft long and 8 ft high. It contains a window 5 ft wide and 3.5 ft high. The wall has an effective total R-value of 15.5. Find the rate of heat flow through the wall when the inside air temperature is 68 °F and the outside temperature is 5 °F.

2. Using the data in Appendix B, determine the total effective R-value of the wall shown in Figure 2–27. Assume that 15% of the wall area is wood framing.

3. The rate of heat flow through a wall is 500 Btu/hr when the inside temperature is 70 °F and the outside temperature is 10 °F. What is the rate of heat flow through the same wall when the outside temperature drops to –10 °F and the inside temperature remains the same?

4. The perimeter of the basement of a house measures 120 ft. The wall is 8 ft tall with 1 ft exposed above grade. The wall is insulated with 2 ins of extruded polystyrene insulation (R = 11.0). The basement air temperature is maintained at 65 °F. Determine the rate of heat loss through the basement walls when the outside air temperature is 10 °F. Assume the R-value of the uninsulated wall is 2.0.

5. Determine the rate of heat loss through the floor slab shown in Figure 2–28. The exposed edges are insulated with 1.5 ins of extruded polystyrene. The outside air temperature is 0 °F and the slab is maintained at 90 °F.

6. Using the form shown in Figure 2–12, determine the design heat loss of the room shown in Figure 2–29. Assume the exterior walls are constructed as shown in the figure for Exercise 2. Assume the outdoor design temperature is –10 °F and the desired indoor temperature is 70 °F.

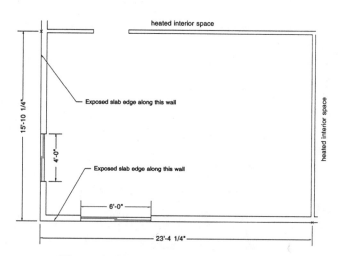

Figure 2–28 Floor plan for Question 5

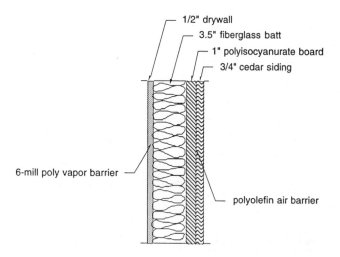

Figure 2–27 Wall section for Question 2

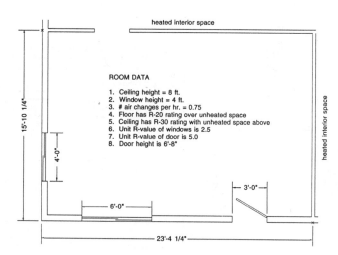

ROOM DATA

1. Ceiling height = 8 ft.
2. Window height = 4 ft.
3. # air changes per hr. = 0.75
4. Floor has R-20 rating over unheated space
5. Ceiling has R-30 rating with unheated space above
6. Unit R-value of windows is 2.5
7. Unit R-value of door is 5.0
8. Door height is 6'-8"

Figure 2–29 Floor plan for Question 6

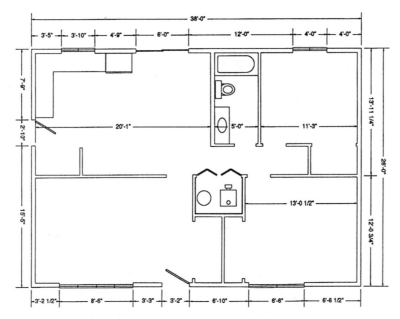

Figure 2–30   Floor plan for Question 7

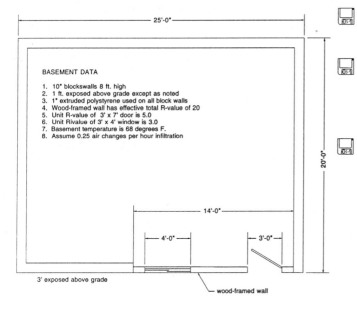

BASEMENT DATA

1. 10" blockswalls 8 ft. high
2. 1 ft. exposed above grade except as noted
3. 1" extruded polystyrene used on all block walls
4. Wood-framed wall has effective total R-value of 20
5. Unit R-value of 3' x 7' door is 5.0
6. Unit Rivalue of 3' x 4' window is 3.0
7. Basement temperature is 68 degrees F.
8. Assume 0.25 air changes per hour infiltration

3' exposed above grade

wood-framed wall

Figure 2–31   Foundation plan for Question 9

7. Using the table shown in Figure 2–12 and the information presented in this chapter, determine the design heat loss of the entire house shown in Figure 2–30. Estimate any dimension you need by proportioning it to the dimensions shown.

8. Using the ROOMLOAD program in the Hydronics Design Toolkit, determine the design heat loss of the room described in Question 6. Print the results.

9. Using the BSMTLOAD program in the Hydronics Design Toolkit, determine the design heat loss of the basement shown in Figure 2–31. Assume an inside temperature of 65 °F and an outside temperature of 0 °F.

10. Using the ROOMLOAD program in the Hydronics Design Toolkit, recalculate all room loads for the example house described in Section 2.7 with the following changes:

Window R-value = 1.6
Walls are 2 in by 4 in studs with 3.5 inches of cellulose insulation
Ceiling is insulated with 6 inches of blown fiberglass
Slab edge insulation is 1-inch extruded polystyrene

Note: To properly include these changes, it will be necessary to recalculate the effective total R-value of the walls and ceiling before using the ROOMLOAD program. Use Appendix B to determine R-values for the stated materials.

3 HYDRONIC HEAT SOURCES

OBJECTIVES

After studying this chapter you should be able to:

- Evaluate several types of hydronic heat sources for a given application
- Describe several types of boilers and how they are constructed
- Define and work with several boiler performance indices
- Discuss the consequences of excessively oversizing a boiler
- Describe favorable applications for condensing boilers
- Identify the advantages of modular boiler systems in certain applications
- Discuss the operation of electric thermal storage systems as hydronic heat sources
- Explain the operating principles of a hydronic heat pump
- Predict how operating conditions will affect the operation of a hydronic heat pump
- Estimate the seasonal fuel usage of a boiler using the appropriate efficiency rating
- Evaluate the suitability of solar energy systems as hydronic heat sources
- Contrast the pros and cons of wood-fired boilers

3.1 INTRODUCTION

This chapter provides an overview of several types of heat sources suitable for use in hydronic heating systems. It covers some of the standard design considerations for each type of heat source, notes the applications where each type of heat source has certain technical or economic advantages, and illustrates some basic schematic configurations for certain types of equipment. Related topics such as exhaust handling, combustion air requirements, and methods of piping are also discussed. Several performance ratings for heating output and efficiency are presented with emphasis on using the proper rating for sizing the heat source, as well as predicting its annual energy usage.

This chapter is not meant to be an exhaustive reference on each type of heat source. The details necessary for proper installation of a specific device should be obtained directly from the manufacturer's installation instructions. Sometimes the building code or mechanical equipment codes of a given municipality will influence both equipment selection and installation. For this reason, always check with local codes officials for any specific limitations affecting a proposed hydronic heat source.

Finally, although the heat source is certainly a major piece of equipment, it is still only part of the hydronic system. Even the best heat source can be misapplied resulting in disappointing performance or shortened service life. As you read about the different heat sources, pay particular attention to temperature and flow rate limitations that ultimately affect how the heat source interfaces with the rest of the system.

3.2 CLASSIFICATION OF HYDRONIC HEAT SOURCES

Within this chapter, hydronic heat sources are classified as follows:

- Gas- and oil-fired boilers
- Electric resistance boilers
- Electric thermal storage equipment
- Hydronic heat pumps
- Renewable energy heat sources

Of these, gas- and oil-fired boilers are currently used in the vast majority of residential and commercial hydronic heating systems in the U.S. This is largely due to lower energy costs associated with natural gas and fuel oil in the major hydronic market areas.

Electrically powered hydronic heat sources, such as electric boilers and hydronic heat pumps, have not gained a market share comparable to fossil-fuel boilers to date. However, buildings serviced by electric utilities with low kilowatt-hour rates, or time-of-use rate structures, can be good candidates for electric heat source equipment. Depending on the specifics of the local electric rate structure, such systems can have very competitive operating costs compared to traditional fossil fuel systems.

Hydronic heat sources that rely on renewable energy currently represent a very small portion of the market. Nevertheless, when properly applied, they can be both technically and economically viable and will likely gain an increased market share in the future.

3.3 GAS AND OIL-FIRED BOILER DESIGNS

Most boilers that operate on either natural gas or fuel oil can be classified according to their physical construction and heat exchanger material. This section gives a brief description of all major types of gas and oil-fired boilers. It also discusses some lesser known but nonetheless state of the art boiler designs.

Cast-iron Sectional Boilers

The cast-iron **sectional boiler** is the most common design used in residential and light commercial buildings in the U.S. In this type of boiler, water is heated in cast-iron chambers called sections. The sections are bolted together to form a boiler block. Hot combustion gases that are formed in the combustion chamber near the bottom of the boiler rise through cavities between the sections, transferring heat to the water within. The greater the number of sections, the greater the rate of heat output from the boiler. Figure 3–1 shows the construction of a typical **wet base** sectional boiler.

Most cast-iron sectional boilers for residential and small commercial use are sold as **packaged boilers.** This implies that components such as the circulator, burner, and controls are selected and assembled onto the boiler

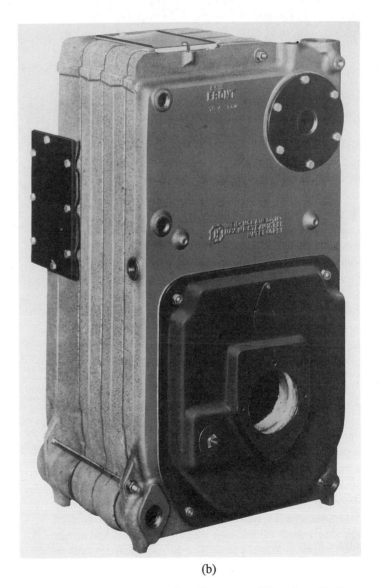

(a)   (b)

Figure 3–1 Example of a wet base boiler section (a), and boiler block assembly (b). Courtesy of Burnham Boiler Corporation.

Figure 3–2 Example of a packaged cast-iron sectional boiler with factory-installed oil burner, circulator, and controls. Courtesy of Dunkirk Radiator Corporation.

by the manufacturer. The components selected by the manufacturer assume the boiler will be used in a typical residential system. In some cases one or more of these components may not be the best choice for a specific application. For example, the small circulator supplied as part of a packaged boiler may not be able to produce the flow rate and pressure differential required by an extensive radiant floor heating system. For this reason some manufacturers also sell a "bare-bones" version of the boiler without such components. Figure 3–2 depicts a typical packaged oil-fired boiler for residential applications.

Cast-iron sectional boilers are limited to use in closed-loop hydronic systems. Properly designed closed-loop systems rid themselves of excess air in the system water after a few days of operation. At this point the potential for corrosion damage to the cast-iron sections from the water is very small. *Improperly designed or installed systems that allow a continued presence of dissolved air in the system's water will rapidly corrode cast iron.* This can result from persistent water leaks, chronic opening of pressure relief valves, misplaced air vents, and other abnormalities. When properly applied, cast-iron boilers can last for decades. Lifetimes in excess of 30 years are not uncommon. Given today's rapid changes in technology, it is likely that such boilers will be technologically obsolete before they fail due to corrosion.

The amount of metal in cast-iron sectional boilers makes them relatively heavy. A residential size boiler can weigh from 350 to 500 pounds. These boilers also typically contain 10 to 15 gallons of water. The combination of metal and water weight gives such boilers the ability to absorb a significant amount of heat into their own materials. This high **thermal mass** is undesirable in situations where the boiler is significantly oversized and thus experiences extended periods of off-time between firing cycles. During this time, residual heat remaining in the boiler's metal and water following burner shut-down can be carried up the flue by air currents passing through the combustion chamber. Part of this residual heat is also transferred through the boiler's housing, or **jacket** as it is commonly called, to the air surrounding the boiler.

Some manufacturers design the cast-iron sections for vertical passage of flue gases. The sections shown in Figure 3–1 are an example of this approach. To clean the flue passages in these boilers, the sheet metal top panel, flue connection, and flue gas collector assembly must be removed before a cleaning brush can be maneuvered between the sections.

Another type of design uses horizontal flue gas passages between the sections with a hinged door on the combustion chamber. The exhaust gases usually make two or more passes across the sections before exiting the boiler. This allows more heat to be transferred to the water. Such designs are also relatively easy to clean through the front door panel. An example of a cast-iron sectional boiler with horizontal exhaust gas flow is shown in Figure 3–3.

Figure 3–3 Cast-iron sectional boiler with horizontal flue passages and a hinged combustion chamber door for simplified cleaning. Courtesy of Buderus Hydronic Systems.

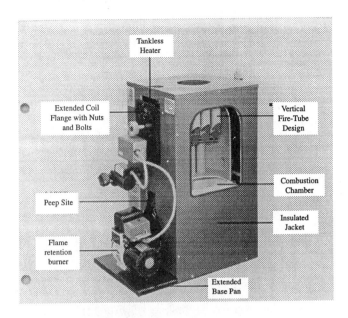

Figure 3–4 Cut-away view of a vertical fire-tube steel boiler. Courtesy of Columbia Boiler Company.

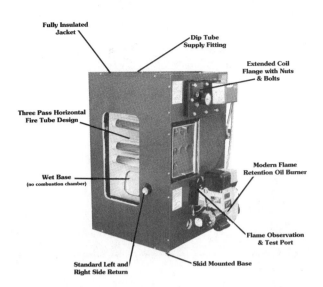

Figure 3–5 Example of a three-pass horizontal fire-tube boiler. Courtesy of Columbia Boiler Company.

Steel Fire-tube Boilers

Steel has been used in boiler construction for decades. In a steel **fire-tube boiler,** water surrounds a group of steel tubes through which the hot combustion gases pass. Spiral baffles known as turbulators are inserted into the fire-tubes to increase heat transfer by inducing turbulence and slowing the passage of the exhaust gases. The fire-tubes are welded to steel bulkheads at each end to form the overall heat exchanger assembly.

In some boilers the tubes are oriented vertically. Figure 3–4 shows an example of a vertical fire-tube boiler. Notice the cut away turbulators within the fire tubes. In this design the fire-tube assembly sits on top of the combustion chamber.

Other boiler designs are built around horizontal fire-tubes. Some boilers route the flue gases through multiple tubes before they reach the flue pipe connection. This allows additional heat to be extracted from the flue gases. In general, multiple pass horizontal fire-tube boilers will have slightly higher efficiencies compared to single pass vertical fire-tube boilers. This is due to the greater contact area between the boiler water and fire tubes. Horizontal fire-tube designs also have the advantage of easier tube cleaning through a removable access panel on the front to the boiler. Figure 3–5 shows a cutaway of a three pass horizontal fire-tube boiler.

Like cast-iron, steel is also subject to corrosion from system water containing dissolved air. Because of this, steel fire-tube boilers are also limited to closed-loop system applications.

Although steel fire-tube boilers generally have less metal weight than comparative models of sectional cast-iron, their water content is often greater. Residential size boilers often contain 15 to 30 gallons of water. This again results in a boiler with a relatively high thermal mass, and potential for higher off-cycle heat losses.

Copper Water-tube Boilers

Some boilers contain water in copper tubes that are directly exposed to the combustion gases. Design variations include both vertical and horizontal tube arrangements. The tubes are usually manufactured with fins that greatly increase the heat transfer area exposed to the flue gases. A cross section of a finned tube is shown in Figure 3–6a. The high thermal conductivity of copper relative to steel or cast-iron allows significantly less surface area to yield the same rate of heat transfer. This makes for smaller and significantly lighter boilers.

Copper water-tube boilers require constant water circulation while being fired to prevent damage from thermal stress. This requirement must be incorporated into system piping and control design.

Due to their corrosion resistance, some copper tube boilers are suitable for direct heating of domestic water and swimming pools. They are also suitable for space heating applications. The light construction and low water content result in a low thermal mass boiler with fast warm-up. An example of a copper tube boiler with a forced combustion system is shown in Figure 3–6b.

Other Boiler Designs

Figure 3–7 shows the heat exchanger of a relatively new boiler design that does not fit any of the previous classifications. This boiler uses a steel water jacket coiled into a spiral around a central combustion chamber. Exhaust

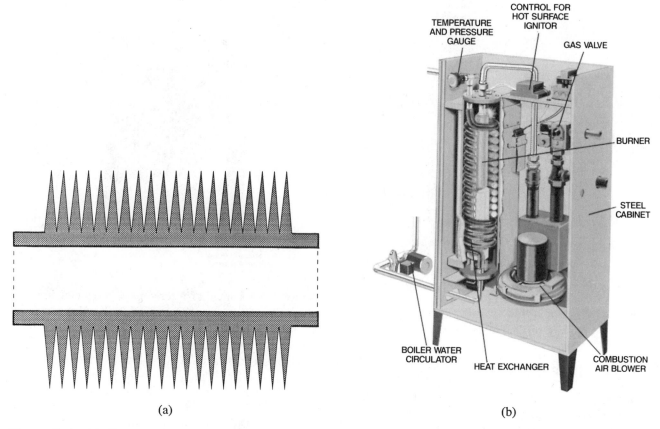

(a)

(b)

Figure 3-6 (a) Cross section of a finned copper tube. (b) Cut-away view of a copper water tube boiler. Courtesy of Glow Core Corporation.

gases travel outward around the spiral until finally reaching the flue connection. This boiler contains only about 2.5 gallons of water, making for very low thermal mass and fast warm up. The low thermal mass also reduces off-cycle heat losses, improving overall efficiency. High combustion efficiency is maintained using a low thermal mass combustion chamber that heats up very quickly to provide a high temperature combustion zone for the fuel. The assembly illustrated in Figure 3–7 is mounted into an insulated metal jacket to complete the unit.

Many conventional cast-iron and steel boilers can be damaged by condensation of vapors contained in exhaust gases. Condensation can begin to form when return water temperatures drop into the range of 130 °F on gas-fired boilers. This is well above the temperature at which some hydronic distribution systems operate.

One boiler design intended to prevent condensation formation while still operating at relatively low temperatures is shown in Figure 3–8. This design uses a stainless steel combustion chamber surrounded with radial cast-iron fins. A separate stainless steel shell isolates the boiler water from the fins. Upon firing, the low thermal mass combustion chamber quickly heats up to provide a high temperature radiant environment for efficient combustion. The combustion gases reverse direction at the

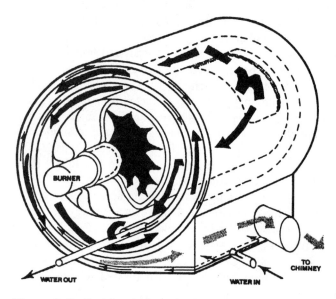

Figure 3-7 Interior design of a low thermal mass boiler using a spiral-shaped steel heat exchanger. Courtesy of Energy Kinetics, Inc.

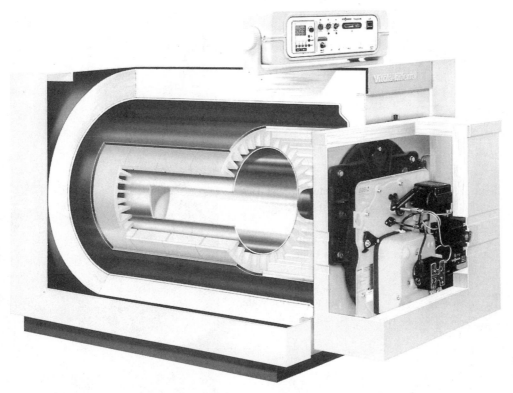

Figure 3–8   Cut-away view of a combination cast-iron/steel boiler that can operate at low temperatures without condensation of exhaust gases. Courtesy of Viessmann Manufacturing.

back of this chamber, travel forward toward the front of the boiler, then make another reversal to flow through the radial cast-iron fins. The thermal conductivity of the fins, in combination with the contact area between the fins and outer stainless steel shell allows them to attain a temperature high enough to prevent the formation of condensation. This design allows the boiler to be connected directly to low temperature distribution systems, such as those used for radiant floor heating, without the need for mixing valves and their associated controls.

Combined Boiler/Domestic Hot Water Tank Assemblies

A large portion of residential hydronic systems supply both space heating and domestic hot water. Although several methods have been used to provide domestic hot water from a space heating boiler, it is generally accepted that the most efficient approach uses a storage tank that is indirectly heated by the boiler. Several manufacturers now offer products that combine a boiler and hot water storage tank into a single unit. This approach saves room compared to the use of a separate tank and boiler. It also reduces installation costs because the piping between the boiler and hot water tank is factory installed. Figure 3–9 shows two examples of such a product.

Additional Boiler Terminology

Further terminology is used to describe the specifics of how a boiler is designed and constructed. One of the common design approaches results in a **wet-base boiler**. In this type of boiler the combustion chamber is completely surrounded by the water-filled heat exchanger. This is a very common configuration for oil-fired boilers. It has the advantage of exposing more of the section's surface area to the combustion chamber. An example of wet-base boiler design is shown in Figure 3–10.

Another common configuration is called a **dry-base boiler**. It is commonly used for atmospheric gas-fired boilers. Combustion occurs on the surface of burner tubes suspended below the water-filled sections. Hot exhaust gases are drawn upward between boiler sections, or through fire tubes, and enter a flue gas collector at the top of the boiler. Room air is always directly available at the burner tubes. *The use of an automatic flue damper on this type of boiler is especially important for attaining high efficiency.* An example of a dry-base boiler is shown in Figure 3–11.

A relatively new approach in the gas-fired boiler industry, wall hung boilers were developed to conserve space in buildings such as apartments or condominiums, and to eliminate the need for a traditional chimney. Most boilers of this type are designed to mount directly on *exterior*

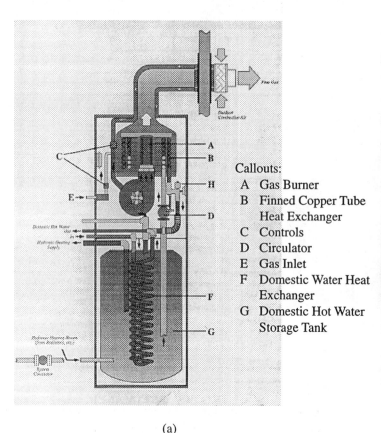

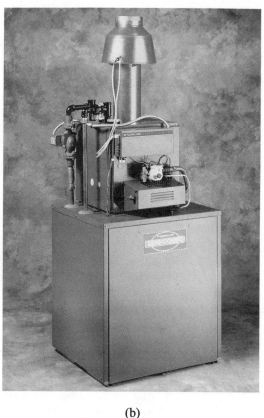

Callouts:
A   Gas Burner
B   Finned Copper Tube
    Heat Exchanger
C   Controls
D   Circulator
E   Gas Inlet
F   Domestic Water Heat
    Exchanger
G   Domestic Hot Water
    Storage Tank

(a)   (b)

Figure 3–9   Examples of combined space heating and domestic water heating units. (a) Courtesy of Heatmaker Corporation. (b) Courtesy of Teledyne-Laars Corporation.

walls. Exhaust gases and air for combustion travel through a specially-designed coaxial sleeve that penetrates the wall. An example of a wall hung boiler is shown in Figure 3–12.

3.4   CONDENSING VERSUS NON-CONDENSING BOILERS

Most boilers are operated such that condensation of water vapor produced by combustion of fossil fuels does not occur within the boiler or its flue piping. This prevents corrosion of cast iron or steel components exposed to the flue gases. The water vapor is exhausted up the chimney along with other combustion by-products. The residual heat contained in the water vapor is lost.

Following the energy crisis of the early 1970s, higher fuel costs motivated boiler manufacturers to research new methods for improving boiler efficiency. The performance objective was simple: Obtain the maximum heat output for each unit of fuel consumed. One of the outcomes of this research was the development of the condensing gas-fired boiler.

Analysis of the combustion process for natural gas reveals that significant amounts of heat can be recaptured from the exhaust stream if the water vapor it contains can be condensed *within the boiler.* This latent heat can then be transferred to the system's water and used to supply the heating load. To condense the water vapor, the exhaust gases must be cooled below the vapor's **dewpoint temperature.** This temperature corresponds to the condition where the exhaust gases contain the maximum amount of water vapor they are capable of holding. Any cooling of the gases below this temperature forces some of the water vapor to condense into liquid droplets.

A good analogy to the condensing process within a boiler is seen in the exhaust from a car on a cold winter day. When the car is first started, the temperature of the exhaust pipe, muffler, etc., is low enough to cause water vapor produced during combustion to condense within the exhaust system. Under these conditions, water can often be seen dripping from the tailpipe. As the exhaust system warms above the dewpoint of the exhaust gases, the water vapor no longer condenses, and the dripping from the tailpipe stops.

Under typical conditions, each therm of natural gas (100,000 Btu energy content) produces approximately

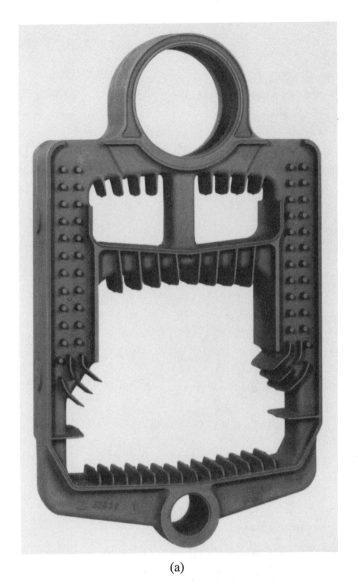

(a)

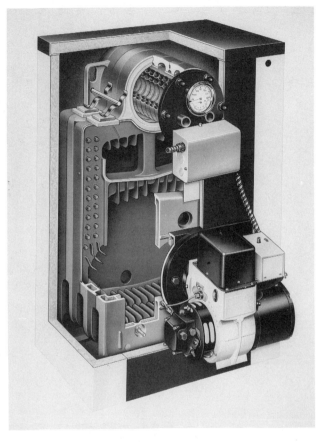

(b)

Figure 3–10   Example of a wet-base boiler (a) boiler section, (b) completed boiler assembly. Courtesy of Vaillant Corporation.

(a)

(b)

Figure 3–11   Example of dry-base boiler construction used for a gas-fired boiler (a) dry base boiler section, (b) assembled boiler block resting on burner assembly with flue gas collector at top. Courtesy of Burnham Corporation.

(a)

(b)

Figure 3–12 Example of a wall hung gas-fired boiler (a) internal construction of boiler, (b) boiler mounted on exterior wall. Courtesy of Burnham Corporation.

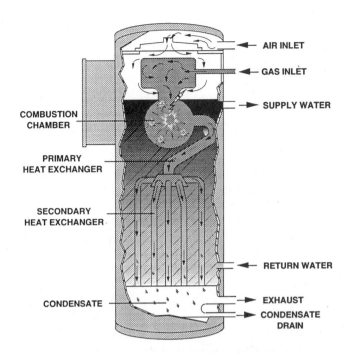

Figure 3-13 Cut-away view of a condensing boiler. Condensate forms near bottom of lower (secondary) heat exchanger. Courtesy of HydroTherm division of Mestek Corporation.

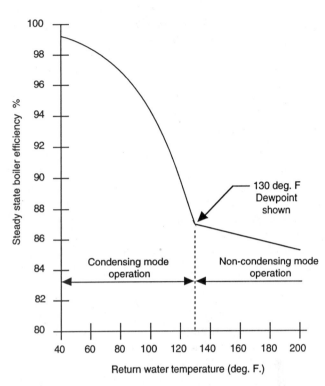

Figure 3-14 Efficiency of a boiler in both condensing and noncondensing modes

1.15 gallons of water vapor during combustion. If *all* this vapor could be condensed within the boiler, approximately 10,100 Btus of heat would be recovered from the exhaust gases for each therm of gas consumed. This represents about 10% of the original energy content of the natural gas, which, in a non-condensing boiler, is wasted as part of the exhaust stream.

In condensing boilers, the water vapor is condensed in a secondary heat exchanger through which the exhaust gases must pass after leaving the primary heat exchanger. The secondary heat exchanger provides additional contact area between the exhaust gases and the boiler water. The additional area allows more heat to be extracted from the exhaust gases, cooling them down to or below dewpoint temperature. A cut-away view of a modern condensing boiler is shown in Figure 3-13.

The amount of condensate formed within the boiler is highly dependent on the return water temperature of the hydronic distribution system the boiler is connected to. The lower the return water temperature, the greater the amount of condensate, and thus the higher the boiler's efficiency. Figure 3-14 shows the variation in efficiency for a typical boiler operating in either a **condensing mode** (e.g., return water temperature below the dewpoint temperature), or **non-condensing mode** (e.g., return water temperature above the dewpoint temperature).

Notice the rapid increase in efficiency to the left of the dewpoint temperature. The dewpoint temperature varies depending on the hydrogen content of the fuel and

the carbon dioxide content of the exhaust gases. Typical dewpoint temperatures for gas-fired boilers range between 120 °F and 140 °F.

Early model condensing boilers were often plagued by corrosion problems due to the acidic nature of the condensate. This corrosion was not necessarily limited to just the secondary heat exchanger where the condensate formed. In some cases, migration of condensate from the secondary heat exchanger into the primary heat exchanger during the off-cycle resulted in further corrosion damage. Manufacturers learned from these early experiences and have incorporated new designs and better materials to greatly extend the life of condensing boilers. Most second generation condensing boilers now use high grades of stainless steel or proprietary electroplated coatings to effectively eliminate corrosion problems.

Oil-fired condensing boilers are not available at present for a number of reasons. First, the hydrogen content of fuel oil is lower than that of natural gas. This results in less water vapor being formed during combustion, which is hence not available for condensing. Second, the higher sulfur content of fuel oil leads to higher concentrations of sulfuric acid in the exhaust gas condensate. This increases the risk of corrosion. Third, soot formation under low exhaust temperature conditions is much more of a problem with oil-fired equipment than with gas-fired equipment due to the oil's higher carbon content. Finally, the dewpoint temperature of an oil-fired boiler is lower than that of a gas-fired boiler due to the lower hydrogen content of fuel oil. This would require lower return water temperatures relative to gas-fired boilers to ensure condensing mode operation.

Condensing Versus Non-condensing: Which Should You Use?

Unfortunately there is no universal answer to this question. In each potential application the answer depends on the type of hydronic distribution system, installation cost, and energy savings. Since condensing boilers can add $1,000 to $1,500 to the installation cost of a residential system, their economic viability depends on recovering this extra cost in fuel savings within a reasonable time period. A few general guidelines should be considered in making the decision.

First, Figure 3–14 shows the operating efficiency of a condensing boiler is strongly dependent on the return water temperature from the distribution system. *Low temperature systems such as radiant floor heating, dedicated domestic water heating, swimming pool heating, or snow melting, are good candidates for condensing boilers.* Hydronic systems that lower supply water temperature based on increasing outside air temperature may also be candidates for condensing boilers. The key consideration is how long the system's return water temperature will be at or below the dewpoint temperature.

Traditional hydronic baseboard systems are often sized around supply water temperatures of 160 °F to 200 °F. Return water temperatures in such systems are typically 140 °F to 190 °F. A condensing boiler in such an application would only operate in the condensing mode during brief warm-up periods following cold starts. During the rest of the operating time it would operate in a non-condensing mode, with an efficiency comparable to that of a conventional boiler. The performance gain expected of the condensing boiler would not be realized to the extent necessary to provide a reasonable payback.

If a condensing boiler is being considered, be sure the installation location has provisions for non-attended drainage of the condensate. *In a typical residential application, several gallons of condensate can be formed each day during cold weather.* This condensate must be disposed of properly. Most plumbing codes in the U.S. allow condensate to be routed to the sewer system. A condensate pump may be required depending on the elevation of the boiler relative to the drainage piping. Never assume the condensate will simply evaporate or dissipate down through cracks or small holes in a concrete floor slab. The acid in the condensate will react with concrete and eventually cause it to deteriorate.

Be certain the exhaust piping used with a condensing boiler is compatible, both from a temperature and moisture standpoint, with its intended application. CPVC and other synthetic materials are currently specified by manufacturers for such applications. *Galvanized steel flue pipe should never be used with condensing boilers.* The moist, acidic nature of the exhaust would rapidly corrode it. This in turn could allow toxic exhaust gases to leak directly into the building.

Exhaust piping from condensing boilers should not be tied into existing chimney flues. The cool temperature of the exhaust may not be sufficient to create proper draft in the flue, even with the assist of power venting. Leakage of toxic gases such a carbon monoxide from cracked or unlined flues, other appliance connections, or cleanouts obviously must be avoided. However, the existing flue of a dormant chimney (e.g., a chimney not used to vent any other device) can be used as a chase through which the exhaust and supply air piping of the condensing boiler can be routed. Specifications for this type of installation are covered in the **National Fuel Gas Code** ANSI Z223.1.

Be sure not to create low points in the exhaust piping that could act like plumbing traps and eventually fill with water that condenses within the exhaust system. Eventually such an accumulation of water could prevent boilers with **draft proving switches** from operating. If a low point were located in an area exposed to outdoor temperature, the condensate in the exhaust pipe could freeze and burst the pipe, leading to water damage and/or leakage of exhaust gases. A common recommendation is to pitch the exhaust pipe at least $1/4$ inch per foot, with the downward end toward the boiler. The outside portion of the exhaust system should not terminate directly behind shrubbery, under porches or decks, or near windows. *As always check for specific installation requirements of the manufacturer.*

3.5 DOMESTIC HOT WATER TANKS AS HYDRONIC HEAT SOURCES

In certain applications it is possible to provide space heating and domestic hot water from a single gas-fired hot water tank. Such systems are particularly suitable for apartments with relatively small heating loads and limited space for mechanical equipment. They also provide a good option for home additions where tying into existing heating or hot water systems is impractical. Figure 3–15 shows a typical arrangement for this type of system.

When first introduced, these systems created controversy between manufacturers and plumbing code officials. Safety concerns regarding possible contamination of the potable water from components used for space heating, or stagnant water in inactive zone circuits, were the main issue. Because such open loop systems have a constant supply of fresh water containing dissolved oxygen, ferrous metal components such as cast iron circulators, valves, etc., would quickly corrode, producing discoloration of domestic water and potential health hazards.

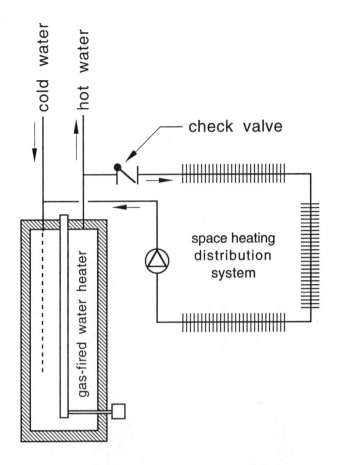

Figure 3–15 Schematic of a gas-fired domestic water heater used for space heating and domestic hot water loads

Lead based solders such as 50/50 tin lead, commonly used for the non-potable piping in hydronic heating systems, cannot be used in such systems. Substitutes include 95/5 tin/antimony or silver solder. All components must be suitable for prolonged contact with fresh (oxygen containing) water. This is not a problem for the copper tube and brass valves used in most conventional hydronic systems. Cast-iron components such as flow-check valves, circulators, and air scoops, however, must now all be replaced with components made of either bronze, stainless steel, or in some cases high temperature synthetic materials. These components are usually more expensive than their cast-iron counterparts.

Water quality is also an important issue in such systems. Since fresh water travels throughout the system, scaling, sediment, or other problems associated with fresh water will now affect all portions of the system. Certain components such as circulators and heat exchangers can be fouled by such deposits resulting in added maintenance and expense. The use of anti-corrosion water treatments is ruled out since they would contaminate water used for domestic purposes.

Another consideration is the efficiency of a gas-fired domestic water heating tank relative to a modern gas-fired boiler. Domestic water heating tanks usually have less

heat exchanger area relative to a boiler of equivalent capacity. This reduces efficiency, especially in high temperature operation. Since the hot water within the tank represents a considerable thermal mass, there is always considerable energy stored in the tank. During the off cycle this energy can be steadily purged away by room air passing up through the combustion chamber of the storage tank, and then out the flue. An automatic flue damper will reduce this off-cycle heat loss.

The relatively small **heating capacity** of electric hot water tanks limits their space heating ability to small buildings or additions. The limited heating capacity can also be a problem when simultaneous demand for both hot water and space heating occurs. For example, during the wake-up period in a typical household, the heating system may have to bring the building temperature up from nighttime setback, while several occupants take showers within a short period of time. In such situations it is common to have controls that temporarily suspend heat output to the heating system until the tank has recovered to an acceptable temperature for supplying domestic hot water. One way of improving this situation is to bring the building out of setback before the peak demand for domestic hot water begins.

If use of a hot water tank for heating and domestic hot water is being considered, be certain local codes accept the approach. Some local codes may also require special equipment such as timers that periodically circulate system water to prevent bacterial growth. Since it is fairly new, not all codes specifically address it, nor are all code officials familiar with it.

Several other approaches for providing space heating and domestic hot water from a single device are available. The other approaches use conventional boilers with greater capacity than residential domestic water heaters. These are discussed in detail in Chapter 14.

3.6 POWER VENTING EXHAUST SYSTEMS

A new method for exhausting combustion gases without using a chimney is rapidly gaining popularity. This method, known as **power venting,** can be used with both gas- and oil-fired boilers, and is an absolute necessity on condensing boilers due to the low temperature exhaust gases.

Power venting systems use a small blower to force exhaust gases out through an exhaust pipe. On some boilers the exhaust blower is installed within the boiler by the manufacturer, as shown in Figure 3–16. In other cases the power venting unit is purchased as a separate assembly, and mounted external to the boiler, as shown in Figure 3–17. The exhaust pipe leading from the blower often exits the building through a sidewall rather than the roof. This reduces costs and simplifies installation.

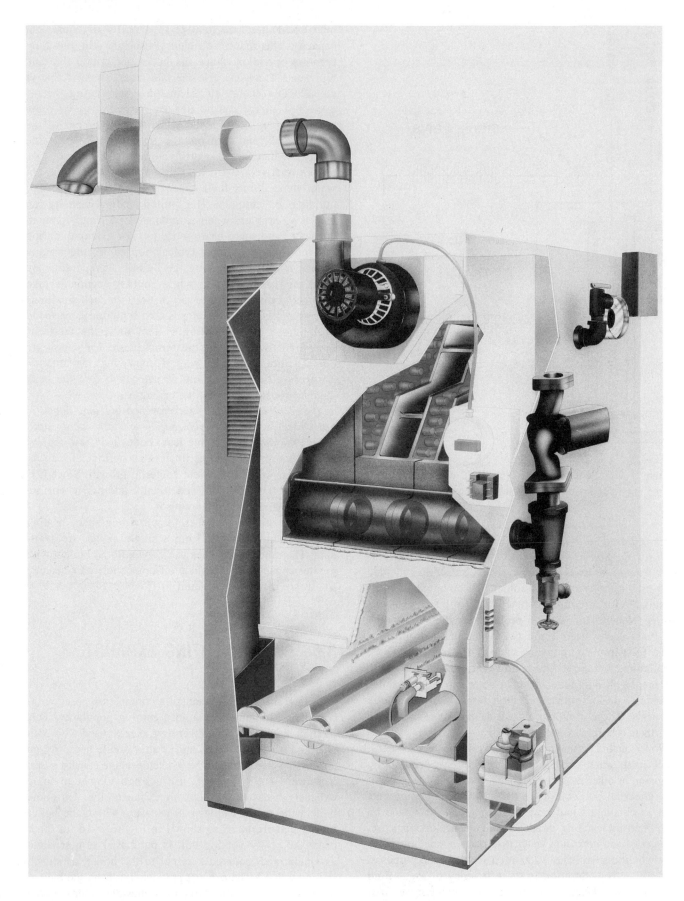

Figure 3–16   Example of a boiler with a self-contained power venting system. Courtesy of Utica Boilers.

(a)

(b)

Figure 3-17 Example of an external power venting system installed on an oil-fired boiler (a) interior flue and blower assembly, (b) outdoor vent terminal. Courtesy of Tjernlund Corporation.

All power venting systems use a device called a draft proving switch to verify a proper negative pressure in the exhaust system before allowing the burner to fire. In the rare but possible event that the exhaust system becomes restricted or the exhaust blower fails to operate, the draft proving switch will not allow the burner to fire. This makes power venting systems safer than conventional natural draft chimneys that cannot automatically stop the burner in the event of draft loss. Power venting systems are also less likely to be affected by winds or other negative pressure conditions induced within the building.

The flue pipe used in power venting applications generally falls into three categories:

- Galvanized steel flue pipe is used for oil-fired systems due to higher exhaust temperatures.
- Specialized high temperature synthetic pipe and fittings can be used where flue temperatures do not exceed 480 °F. This is typical of gas fired equipment in the mid- to high efficiency range where condensation in the exhaust system might occur.
- CPVC piping is often used for full-condensing boilers.

As always, be sure to use the manufacturer's recommended materials.

Direct Vent/Sealed Combustion Systems

A variation of power venting used on compact boilers is known as a direct vent/**sealed combustion.** This approach uses a specially designed coaxial air intake/exhaust assembly through which outside air is drawn for combustion and exhaust gases are routed back outside. A small exhaust blower provides the required pressure differential in the exhaust system. No inside air is required for combustion. Figure 3–18 illustrates the concept applied to a small wall hung boiler. Notice the annular passage surrounding the center flue pipe through which outside air is drawn. The

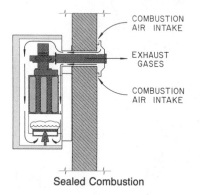

Sealed Combustion

Figure 3–18 A coaxial air intake/exhaust system used with a wall hung boiler. Courtesy of Burnham Corporation.

incoming air provides a naturally cool metal surface in the vicinity of the wall materials.

Sealed combustion is especially desirable in buildings that are well sealed. In such cases, air infiltration alone may not provide sufficient air for proper combustion. Sealed combustion systems also save energy since outside air rather than heated inside air is used for combustion. In coaxial air intake/exhaust systems such as that shown in Figure 3–18, the incoming air is preheated by heat recovered from the exhaust gases prior to reaching the combustion chamber. This improves combustion efficiency. Compact boilers using direct vent/sealed combustion systems are often used in condominiums and apartments where space for mechanical equipment is minimal.

Direct vent/sealed combustion has also been used on floor-mounted boilers. This is illustrated in Figure 3–19. In such systems outside air is routed to the combustion chamber through PVC piping. The exhaust gases are routed back outside through piping made of CPVC or other temperature-resistant plastic material.

3.7 COMBUSTION AIR REQUIREMENTS

Any type of fuel-burning boiler requires a proper supply of air for safe and efficient combustion, exhaust gas dilution, and boiler room ventilation. Provisions for ensuring such requirements will be met are specified in the National Fuel Gas Code, (NFPA No. 54-1984/ANSI Z223.1) from the National Fire Protection Association. The provisions of the NFPA code are widely accepted in most state and local building codes.

Under the NFPA code, the requirements for combustion and ventilation air depend on whether the boiler is located in a **confined space** or **unconfined space.** As defined in the code, unconfined space must have a *minimum* of 50 cubic feet of volume per 1,000 Btu/hr of gas input rating of all gas-fired appliances in that space. Any space not meeting the above requirements is considered confined space under the NFPA code.

Boilers located within confined space may draw air from adjacent rooms or from outside. When air will be drawn from adjacent rooms, the total volume of the boiler room and the adjacent rooms must meet the above stated criteria for unconfined space. The boiler room must have two openings connecting it with the adjacent space. The *minimum* **free area** of *each* opening shall be 1 square inch per 1,000 Btu/hr gas input rating of all gas-fired equipment within the boiler room, but not less than 100 square inches. The top of the upper opening must be within 12 inches of the boiler room ceiling. The bottom of the other opening must be within 12 inches of the floor.

When air is drawn from outside, the size of the required openings will vary depending on the type of fuel

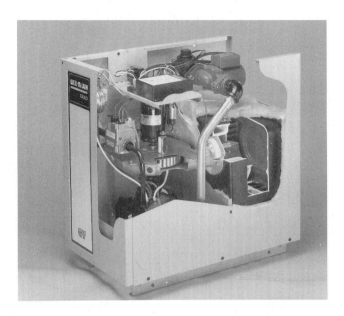

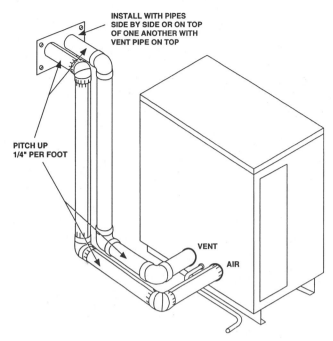

INSTALL WITH PIPES SIDE BY SIDE OR ON TOP OF ONE ANOTHER WITH VENT PIPE ON TOP

PITCH UP 1/4" PER FOOT

VENT

AIR

Figure 3–19   Example of sealed combustion/power vented floor-mounted boiler. Courtesy of Weil McLain Company.

used and the orientation of supply air ducts. The NFPA code requirements are as follows:

1. For gas-fired equipment, using vertical supply air ducting, or direct through-the-wall openings: Provide two openings between the boiler room and the outside, *each* having a minimum *free area* of 1 square inch per 4,000 Btu/hr of gas input rating of all gas-fired equipment within the boiler room.
2. For gas-fired equipment, using horizontal supply air ducting: Provide two openings between the boiler room and the outside, *each* having a minimum *free area* of 1 square inch per 2,000 Btu/hr of gas input rating of all gas-fired equipment within the boiler room.
3. For oil-fired equipment with either horizontal or vertical supply air ducting: Provide two openings between the boiler room and the outside, *each* having a minimum *free area* of 1 square inch per 5,000 Btu/hr of all oil-fired equipment in the boiler room.

In all of the above cases, one opening, or end of duct, must be within 12 inches of the boiler room ceiling, and the other within 12 inches of the floor. These openings are intended to prevent the boiler room from overheating by allowing natural convection air flow. They also allow any gases or fumes that might be present in the boiler room to escape.

When ducts are used to bring outside air into the boiler room, their cross-sectional area must equal the free area of the openings they connect to. The minimum dimension of any rectangular supply air duct allowed by the NFPA code is 3 inches.

It should be emphasized that all of the above vent area requirements refer to free area. This is the true *unobstructed* area of the opening after accounting for the effects of any screens or louvers covering the opening. Louver manufacturers usually list the free area for each type of louver they offer. The table below lists typical free area to gross area ratios for different types of opening covers. This data should be used for comparison purposes only:

Type of opening cover:	F/G ratio:
• $\frac{1}{4}$-inch screen	0.8
• metal louvers	0.6
• wooden louvers	0.2

Equation 3.1 can be used to determine the required area of the opening or duct.

(Equation 3.1)

$$A_{\text{opening}} = \frac{A_{\text{free}}}{F/G \ \text{ratio}}$$

where:

A_{opening} = required cross-sectional area of each opening or duct (ft^2)

A_{free} = free area required by the NFPA code for a given installation (ft^2)

F/G ratio = ratio of free area to gross area of the louver, or screen as specified by its manufacturer

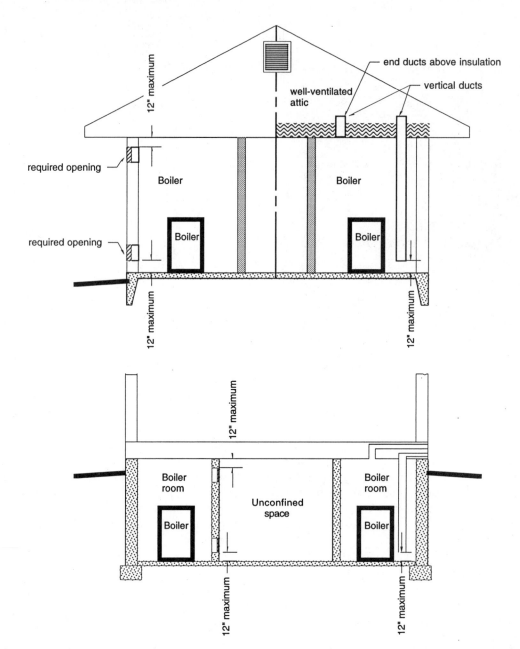

Figure 3–20 Several possible arrangements for combustion/ventilation air supply to a boiler

Example 3.1: In an application requiring 50 square inches of free area for each of two supply air ducts, and where metal louvers will cover the intake of each duct, the required area of each opening would have to be a minimum of:

$$A_{opening} = \frac{50}{0.6} = 83.3 \text{ in}^2$$

Figure 3–20 shows some possible arrangements for providing combustion/ventilation air. Consult your local code for any requirements in addition to, or more restrictive than, the above.

3.8 BOILER HEATING CAPACITY

The heating capacity of a boiler refers to its rate of useful heat output in Btu/hr. The word *output* is also used as a synonym for heating capacity. Exactly what portion of the total heat produced by a boiler is useful has led to several different, but standardized, definitions of heating capacity. Each definition contains underlying assumptions about the boiler and how it is applied in a system.

The performance information listed in boiler manufacturer's literature often makes reference to one or more of the IBR ratings. IBR stands for the Institute of Boiler and Radiator Manufacturers. This organization was the forerunner of today's Hydronics Institute. The two IBR ratings for boilers capacity are **IBR Gross Output,** and **IBR Net Output.**

The IBR Gross Output rating of a boiler is the rate at which heat is transferred to the system's water under steady state operation. Implicit to the definition of IBR gross output is the assumption that all heat released from the boiler jacket is wasted, (e.g., does not offset a portion of the building's heating load).

An IBR Net Output rating is obtained from the IBR Gross Output rating by deducting 15%. This 15% of the gross output is assumed to be lost from the distribution piping, or needed to warm-up the thermal mass of the system following an off-cycle. The latter of these losses is called the pickup allowance. Boilers with low thermal mass, sized close to the building design heating load, will have minimal pickup loads. In effect, the IBR Net Output rating deducts for an assumed heat loss from the distribution piping.

Still another indicator of boiler output is called the **DOE heating capacity.** According to the definition of DOE heating capacity, all heat losses from the boiler jacket are assumed to help heat the building. This assumption is valid if the boiler is contained within heated space. Because it includes jacket heat losses, the DOE heating capacity of any boiler will always be greater than the IBR Gross Output rating. One or more of these capacity ratings are usually listed on the boiler's data plate, and product literature. Figure 3–21 shows a typical table of performance information furnished by a boiler manufacturer. This table includes, among other data, the IBR net rating, DOE heating capacity, and oil firing rate.

The preferred capacity rating for use in boiler selection depends on where the boiler and distribution piping are located. If the boiler and distribution piping are both located in heated space, the DOE heating capacity is appropriate since losses from both the boiler jacket and distribution piping will offset a portion of the building's heating load. When the boiler is located in unheated space, but the distribution piping is mostly run within heated space, the IBR Gross Output rating can be used. If the boiler and distribution piping are both located in unheated space, such as a cold basement, crawl space, or garage, the IBR Net Output rating should be used.

Keep in mind the deduction for heat loss from the distribution system factored into the IBR Net Output rating is a fixed percentage of the IBR Gross Output rating. As such, it may not accurately represent these losses in all situations. For example, in cases involving new home construction, many state energy codes now mandate that hot water piping passing through unheated space must be insulated to specified standards. In such cases the actual heat loss from the piping would be considerably reduced, and the IBR Net Output rating would underestimate available heating capacity of the boiler. The prudent designer will make a separate calculation of piping heat loss to assess its effect on required boiler capacity. Such calculations are detailed in Chapter 8.

3.9 EFFICIENCY OF GAS AND OIL-FIRED BOILERS

The word efficiency is often used in describing boiler performance. In its simplest mathematical form, *efficiency is the ratio of a desired output quantity, divided by the necessary input quantity to produce that output.*

(Equation 3.2)

$$\text{Efficiency} = \frac{\text{Desired output quantity}}{\text{Neccessary input quantity}}$$

When using Equation 3.2, *both input and output quantities must be measured in the same units.* The resulting efficiency will then be a decimal between 0 and 1,

Boiler Model Number (1)	I=B=R Oil Burner Input (2)		D.O.E. Heating Capacity MBH	I=B=R Net Ratings (3)		Natural Draft Chimney Size	Nozzle Furnished 140 PSIG	A.F.U.E. Rating
	G.P.H.	MBH		Water MBH	Sq. Ft.			
SFH-365	.65	91	79	68.7	458	8×8×15	.60 80°B	86.0
SFH-3100	1.00	140	117	101.7	678	8×8×15	.85 80°B	81.0
SFH-4100	1.00	140	120	104.4	696	8×8×15	.85 80°B	86.0
SFH-3125	1.25	175	144	125.0	834	8×8×15	1.10 60°B	80.0
SFH-4125	1.25	175	149	129.6	864	8×8×15	1.10 80°B	82.5
SFH-5125	1.25	175	151	131.3	875	8×8×15	1.10 80°B	86.0
SFH-4150	1.50	210	175	152.2	1015	8×8×15	1.25 80°B	81.0
SFH-6150	1.50	210	181	157.0	1049	8×8×15	1.25 80°B	86.0
SFH-5175	1.75	245	206	179.1	1194	8×8×15	1.50 80°B	81.5
SFH-5200	2.00	280	231	200.9	1339	8×8×15	1.75 80°B	81.0

Figure 3–21 Typical listing of boiler performance information provided by manufacturer. Courtesy of Utica Boilers.

corresponding to a percentage of 0 to 100. The higher the efficiency of a boiler, the more heat is extracted from a given amount of fuel. All other factors being equal, the higher the efficiency, the lower the operating cost of the boiler.

Over the years, the concept of efficiency has been the basis for many new, and sometimes confusing, definitions of boiler performance. The terms steady state efficiency, cycle efficiency, and annual fuel utilization efficiency all refer to the ratio of an output quantity divided by an input quantity. However, all are based on different reference conditions. *Using the wrong definition of efficiency can lead to erroneous performance predictions.* Each variation on the concept of efficiency will be defined so the appropriate values can be used for their intended purposes.

Steady State Efficiency

A boiler operates at **steady state efficiency** only when it is being fired continuously under non-varying conditions. All inputs such as fuel composition, air-to-fuel ratio, air temperature, entering water temperature, and so forth, must remain constant. Under these conditions, the steady state boiler efficiency can be calculated using Equation 3.3:

(Equation 3.3)

$$\text{Steady State Efficiency} = \frac{\text{Heat output rate of boiler (Btu/hr)}}{\text{Energy input rate (Btu/hr)}}$$

Example 3.2: A boiler runs continually under steady state conditions for one hour. During this time it consumes 0.55 gallons of #2 fuel oil, and delivers 65,000 Btus of heat. Determine the boiler's steady state efficiency. *Note:* One gallon of #2 oil = 140,000 Btu **chemical energy content.**

Solution: In this example it is necessary to convert the fuel consumption rate of 0.55 gallons per hour into an equivalent energy input rate in Btu/hr, before taking the ratio of the heat output rate to the energy input rate.

$$\text{Steady State Efficiency} = \frac{\left(65{,}000\,\dfrac{\text{Btu}}{\text{hr}}\right)}{\left(0.55\,\dfrac{\text{gallon}}{\text{hour}}\right)\left(140{,}000\,\dfrac{\text{Btu}}{\text{gallon}}\right)}$$

$$= 0.844 \text{ or } 84.4\%$$

In reality, steady state conditions seldom exist for long in either residential or commercial systems. Theoretically these conditions would exist only while the heating load of the building was equal to, or exceeded, the heating capacity of the boiler. This usually only occurs for a very small percentage of the heating season, if at all, particu-

larly with boilers that are grossly oversized. Just because a boiler operates at high steady state efficiency under nearly ideal laboratory conditions does not imply this efficiency can be maintained after it is installed. *Erroneous predictions will arise if steady state efficiency values are used in estimating seasonal performance and subsequent operating costs.* This would be like estimating the average gas mileage of a car using only the highway miles per gallon rating.

The steady state efficiency of the boiler can be estimated by dividing its DOE heating capacity by its fuel input rate.

$$\text{Steady State Efficiency} \approx \frac{\text{DOE heating capacity (Btu/hr)}}{\text{fuel input rate (Btu/hr)}}$$

Boiler manufacturers often list the fuel input rate, and DOE heating capacity in their technical literature. For gas-fired boilers, the fuel input rate is listed in Btu/hr or 1000s of Btu/hr (MBtu/hr). For oil-fired boilers the fuel input rate is listed in gallons per hour (gph). For #2 fuel oil, the fuel input rate in Btu/hr can be obtained by multiplying the gallons per hour firing rate by 140,000 Btu/gallon.

Combustion Efficiency

By measuring the temperature and carbon dioxide (CO_2) content of the exhaust gases leaving a boiler under steady state conditions, it is possible to determine its **combustion efficiency.** The measured CO_2 content of the exhaust gas is compared to the theoretical maximum CO_2 content that would result if the complete combustion of the fuel occurred. The resulting combustion efficiency is then read from a standard combustion chart for the fuel being burned. The higher the CO_2 content, and the lower the exhaust temperature, the greater the combustion efficiency. Although attained in a different way, combustion efficiency is essentially a field-measured value for steady state efficiency. *Because it is measured under continuous firing conditions, combustion efficiency should not be used to predict seasonal fuel usage.*

Cycle Efficiency

Since boilers seldom operate continuously, they experience heat losses during each off-cycle that ultimately waste part of the heat produced by the fuel. To account for such losses another efficiency measurement called **cycle efficiency** is used. Cycle efficiency is defined as the ratio of the total heat output from a boiler divided by the chemical energy value of the fuel consumed, measured over a period of time while the boiler cycles on and off to meet the heating load. Mathematically this is expressed as Equation 3.5:

(Equation 3.5)

$$\text{Cycle efficiency} = \frac{\text{Total useful heat output of boiler over a time period (Btu)}}{\text{Energy content of fuel consumed over same time period (Btu)}}$$

Cycle efficiency is always less than steady state efficiency. It is highly dependent on the percent of the time the burner is actually firing. This percentage is called the **run fraction** of the boiler, and can be calculated using Equation 3.6:

(Equation 3.6)

$$\text{Run fraction} = \frac{\text{Total time burner is on}}{\text{Total elapsed time}}$$

Example 3.3: The burner of a boiler fires for five minutes, then remains off for 20 minutes before restarting. Determine its run fraction over this time period.

Solution: $$\text{Run fraction} = \frac{5 \text{ minutes}}{(5 \text{ minutes} + 20 \text{ minutes})}$$

$$= 0.2 \text{ or } 20\%$$

Equation 3.6 is useful when the actual firing time is known. In other cases it is necessary to predict what the run fraction would be under certain heating load conditions. This can be done using Equation 3.7.

(Equation 3.7)

$$\text{Run fraction} = \frac{\text{Heating requirement of building (Btu/hr)}}{\text{Steady state heating output of boiler (Btu/hr)}}$$

Example 3.4: Find the run fraction of a boiler with an output of 85,000 Btu/hr while supplying a building heating load of 40,000 Btu/hr.

Solution: $$\text{Run fraction} = \frac{40,000 \text{ Btu/hr}}{85,000 \text{ Btu/hr}}$$

$$= 0.47 \text{ or } 47\%$$

Figure 3–22 shows how the cycle efficiency of a boiler decreases as its run fraction decreases. Notice that the decrease in efficiency is relatively small for run fractions from 100% down to about 30%. However, below 30% run fraction the cycle efficiency drops very rapidly. This implies inefficient use of fuel for the useful heat produced under low run fraction conditions.

Low cycle efficiency conditions can easily occur on days with moderate outside temperature, or even during cold weather if the building has significant internal heat gain from people, equipment and sunlight. *Boilers with heating capacities significantly greater than their building's design heating load will operate at needlessly low cycle efficiency even during the coldest weather.* The owner of such a system pays the penalty for this oversizing both

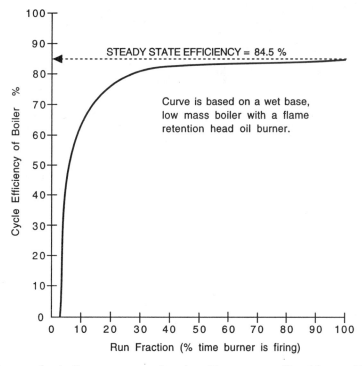

Figure 3–22   Cycle efficiency of a boiler versus run fraction. Data source: Brookhaven National Laboratory, BNL 60816 & 80-38-HI.

in a higher initial cost, and more importantly, in greater fuel usage over the life of the system.

Example 3.5: A building has a design heating load of 100,000 Btu/hr when the outside temperature is 0 °F, and the inside temperature is maintained at 70 °F. The selected boiler has a heating capacity of 150,000 Btu/hr. Using the graph shown in Figure 3–22, estimate the cycle efficiency when the outside temperature is 40 °F and internal heat gains total 20,000 Btu/hr.

Solution: The heat required of the boiler is the building's current heating load minus the current internal heat gains. This can be calculated as:

$$\text{Heating load} = 100,000 \left(\frac{70 - 40}{70 - 0} \right) - 20,000$$

$$= 22,860 \text{ Btu/hr}$$

The run fraction required of the boiler can be calculated as:

$$\text{Run fraction} = \frac{22,860 \text{ Btu/hr}}{150,000 \text{ Btu/hr}} = 0.152 \text{ or } 15.2\%$$

From the graph in Figure 3–22 the corresponding cycle efficiency is about 72%. Thus a boiler with a steady state efficiency of 84.5% would be operating at only 72% efficiency for the stated conditions. Keep in mind that spring and fall weather combined with internal heat gains can result in several thousand hours of partial load operation during a single heating season.

There are several factors that should be considered in an attempt to keep a boiler's cycle efficiency as high as possible. These include:

- *Do not needlessly oversize the boiler.* The author suggests an absolute maximum oversizing factor of 20% of the design heating load. Minimum oversizing allows the boiler to run for longer periods of time, and thus at higher efficiency. *An accurate heating load estimate is an absolute necessity before selecting a boiler.*
- In general, boilers with low thermal mass tend to have higher cycle efficiencies due to lesser amounts of heat remaining in the boiler following burner shut down.
- If a high thermal mass boiler is used, it should be very well insulated by its manufacturer, and be equipped with an automatic device that prevents air flow through the boiler during the off-cycle. A tight sealing automatic flue damper is especially important on conventional dry base gas-fired boilers.
- A control strategy known as heat purging may be incorporated into the system. This entails running the circulator for several minutes after the burner has shut

down to carry residual heat out of the boiler and dissipate it either into the building or a domestic water heating tank. Some boiler manufacturers provide built-in controls for boiler purging. In other cases a purging control system can be assembled from standard control components.

- The use of reset control systems that lower boiler water temperature as outside temperature increases, reduces the residual heat content of a boiler following burner shut-down.

The Hydronics Design Toolkit contains a program called BINEFF for approximating the seasonal efficiency of a boiler based on bin temperature data. This program determines the overall seasonal effect of boiler oversizing for a particular climate. The program requires data for the cycle efficiency of a boiler as a function of its run fraction such as graphed in Figure 3–22. This data should be obtained from the boiler manufacturer. A screen shot of the BINEFF program is shown in Figure 3–23.

The BINEFF program can be used to make a comparison of seasonal boiler efficiency versus the boiler's oversizing factor for a particular climate and load. Figure 3–24 shows such a comparison for a boiler whose cycle efficiency is described by Figure 3–22, and located in a building in Syracuse, New York.

Annual Fuel Utilization Efficiency (AFUE)

In 1978, the U.S. Department of Energy established a standard for predicting the seasonal energy usage of boilers and furnaces. This DOE standard currently applies to all gas and oil-fired boilers sold in the U.S. having fuel input rates up to 300,000 Btu/hr fuel input rating. Its intent is to provide a uniform basis of comparison for all residential and light commercial size boilers sold in the U.S.

This standard defines a performance indicator called the **Annual Fuel Utilization Efficiency,** or AFUE for short. The AFUE rating is now the commonly used basis for expressing seasonal boiler efficiency within the U.S. Minimum values for AFUE are often cited as part of the heating equipment performance standards in state energy codes.

AFUE values are expressed as a percent similar to steady state efficiency. Unlike steady state efficiency, or cycle efficiency, however, AFUE values are intended to account for the many effects associated with part-load operation over an entire heating season. As such, *the AFUE value is intended for use in approximating seasonal fuel usage.* The following equation can be used:

(Equation 3.8)

$$\text{Approximate seasonal fuel usage} = \frac{\text{seasonal heating requirement}}{\text{decimal value of AFUE}}$$

```
- BINEFF -      F1-HELP     F3-SELECT               F10-EXIT TO MAIN MENU
 = -INPUTS- ==============================================================
|  ENTER Outside design air temperature,(-30 to 40).......  0           deg. F  |
|  ENTER Desired inside comfort temperature(50 to 85).....  70          deg. F  |
|  ENTER Design heating load of building,(10K to 300K)....  50000       btu/hr  |
|  ENTER Heating balance point temp of building,(25 to 65)  65          deg, F  |
|  ENTER D.O.E. heating capacity of boiler,(25K to 300K)..  70000       btu/hr  |
|  ENTER Steaty state boiler efficiency,(65 to 90)........  84          %       |
|  ENTER Cycle efficiency of boiler at 75% run fraction,..  81          %       |
|  ENTER Cycle efficiency of boiler at 50% run fraction,..  77          %       |
|  ENTER Cycle efficiency of boiler at 25% run fraction,..  60          %       |
|  ENTER Cycle efficiency of boiler at 10% run fraction,..  35          %       |
|  SELECT Fuel type,(press F3 to select)..................  #2 Fuel Oil         |
|  SELECT Weather data location,(press F3 key to select)..  Syracuse   NY       |
 ========================================================================

 = -RESULTS- =============================================================
|  Total heat delivered by boiler........   135 MMbtu/season              |
|  Total energy content of fuel consumed..  218 MMbtu/season              |
|  Seasonal efficiency of boiler..........  62.0%                         |
|  Total units of fuel consumed...........  1556 gal                      |
 ========================================================================

 =  MESSAGE ==============================================================
|  Use the UP, DOWN, LEFT, RIGHT ARROW keys to move the highlighting bar   |
|  over an input. Type desired value. F10 to return to MAIN MENU. F3 to SELECT |
 ========================================================================
```

Figure 3–23   Screen shot of BINEFF program from the Hydronics Design Toolkit

Boiler capacity ───────────── Design heating load	Seasonal boiler efficiency
1.0	68%
1.5	58%
2.0	50%
2.5	45%
3.0	39%

Figure 3–24   Decreasing seasonal efficiency as boiler capacity is increased relative to the design heating load. Data is based on a boiler with cycle efficiency described by Figure 3–22, located in a building in Syracuse, New York.

Example 3.6: Assume the calculated seasonal heating requirement of a given home is 60,000,000 Btus (also stated as 60 MMBtus). The home is equipped with a gas-fired boiler having an AFUE rating of 82%. Approximate the seasonal heating energy usage of the home.

$$\text{Approximate seasonal fuel usage} = \frac{60 \text{ MMBtu}}{0.82} = 73.2 \text{ MMBtu}$$

The test procedure for establishing the AFUE of a boiler uses measured on-cycle and off-cycle performance data obtained from certified laboratory testing, as input to a computer program. This program calculates the AFUE rating based on complex engineering principles and assumptions about exhaust gas composition, and boiler sizing relative to its load. Those interested in the industry standard test procedure for establishing the AFUE value should consult ASHRAE Standard 103-82.

It should be noted that AFUE values are established under the assumption that the boiler is installed in heated space. As such, all heat losses from the boiler's jacket are considered as useful heat output to that space. If the boiler is installed in unheated space, its actual seasonal efficiency will likely be less than its AFUE value.

The rating standard also assumes the boiler's capacity is approximately 70% greater than the design heating load of the building. Boilers that have been sized closer to design load should yield true seasonal efficiencies somewhat higher than their AFUE values.

It is important to remember that *true seasonal boiler efficiency is significantly affected by the system the boiler is a part of, and how it is sized relative to the design heating load of the building.* The AFUE rating is based on a fixed set of assumptions for these system effects, and therefore cannot be expected to be accurate in all cases. Because of this, *AFUE values should only be used for relative boiler comparisons, or to approximate seasonal fuel usage.*

3.10 MODULAR BOILER SYSTEMS

When designing a boiler system for a moderate-sized commercial building, or perhaps a large sprawling house, the designer has two basic options: 1) Select a single large boiler with sufficient capacity to meet the design heating load of the building, or 2) Select two or more smaller boilers that together can meet this design load. The second option, known as a **modular boiler system**, has a number of advantages compared to the first. These include:

- ability to adjust heat output to closely match load requirements
- higher efficiency operation due to longer boiler on-time
- partial heating capacity even if one of the boilers is shut down for service

Nearly all major boiler manufacturers currently offer pre-engineered modular boiler systems ranging from two to nine or more individual boilers. Figure 3–25 depicts the usual arrangement of the boilers.

The most significant feature of a modular boiler system is its ability to closely match heat output to the current heating load of the building, while also maintaining high efficiency on all boilers. The underlying reasons are based on the fact that a boiler operated at high run frac-

tions will have a greater efficiency than if run for short periods of time.

Consider the situation illustrated in Figure 3–26. Two identical buildings have design heating loads of 300,000 Btu/hr. One building has a single 300,000 Btu/hr boiler. The other has a modular boiler system consisting of three individual 100,000 Btu hr boilers. At design load conditions the single large boiler operates continuously, and thus attains high steady state efficiency. Likewise all three modular boilers operate continuously at high steady state efficiency. Next consider what happens as the weather warms up so that the building's load is reduced to 200,000 Btu/hr. The single large boiler operates approximately 40 minutes per hour, or 66% run fraction. The efficiency of the boiler drops slightly due to a longer off-cycle. Meanwhile, in the modular system only two of the boilers are operating, but they operate continuously, and thus at high steady state efficiency. The differences become even more pronounced as the weather warms, and the heating load on the building further decreases. When the building load has dropped to 50,000 Btu/hr, the single large boiler is operating only about 10 minutes per hour, or a run fraction of 17%. Under these conditions the efficiency of a typical boiler is significantly lowered. In the modular system, one of the three boilers operates a total of 30 minutes per hour, or 50% run fraction. Its

Figure 3–25 Typical arrangement of a four-boiler modular system. Courtesy of HydroTherm division of Mestek.

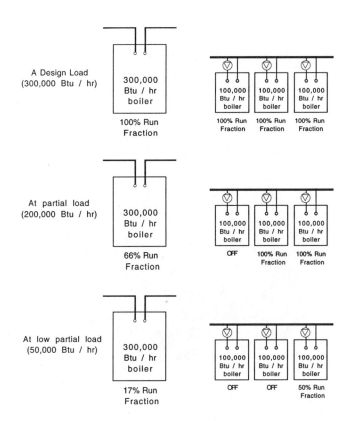

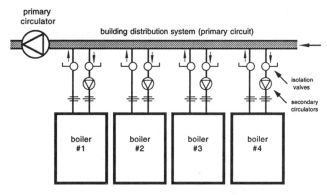

Figure 3-27 A modular boiler arrangment using primary/secondary piping

Figure 3-26 Comparison of a single large boiler with a three-boiler modular system at three different load conditions.

operating efficiency is only slightly reduced. The other boilers in the modular system remain off. *The more the load is reduced, the greater the efficiency advantage of the modular boiler system compared to the single larger boiler.*

The number of boilers making up the modular boiler system also affects seasonal efficiency. In general, the greater the number of boilers, and the smaller their individual heating capacities, the greater the seasonal efficiency of the modular boiler system. The greater the number of boilers, the better the system can "track" the building's heating load. Smaller heating capacity means the boilers that are required to fire will have longer on-cycles with corresponding higher cycle efficiencies.

The "brains" operating a modular boiler system usually consists of an electronic control that performs a number of operations. These include determining the proper water temperature for the system at any given time, turning boilers on and off as needed to maintain that temperature, and varying the firing sequence of the boilers to accumulate approximately the same run time hours on each. Such control systems play a very important role in maintaining the high efficiency of a modular boiler system. They are discussed in Chapter 9.

Another critical requirement of maintaining high efficiency in a modular boiler system is preventing heated system water from flowing through nonoperational boilers. If this occurs, heat will be continuously purged out

of the inactive boilers by air currents passing through their combustion chambers and then up the chimney.

There are a number of ways of preventing hot water circulation through nonoperational boilers in a modular boiler system. The most common method, known as **primary/secondary piping**, is depicted in Figure 3–27.

Notice the individual circulators on each boiler, and how they are connected to the building distribution piping. The short section of large diameter pipe between the supply and return connections of an individual boiler has a negligible pressure drop. Because of this, there is virtually no tendency for system water to circulate through the boiler when its own circulator is off. When a boiler is fired, its own circulator turns on to provide the needed pressure difference to induce water circulation. The small circulators on each boiler are not designed to pump water through the entire building distribution system, just the boiler itself. Each boiler and associated circulator constitute a "secondary" piping loop connected to the main (or "primary") loop. Water distribution through the building is handled by the larger system pump located in the primary loop. Other design approaches using primary/secondary pumping are discussed in Chapter 11.

Modular boiler systems should not be considered as being solely for commercial buildings. Larger houses, particularly older ones with limited opportunity for energy conservation can have heating loads large enough to consider a two- or three-boiler modular system. The savings in fuel due to the higher seasonal efficiencies should be compared to any additional cost associated with the modular boiler system.

3.11 ELECTRIC BOILERS

Any device that uses only electrical energy to produce hot water for space heating could be considered an electric boiler. All such devices contain one or more elements that heat up when an electrical current passes through them. These elements are mounted in an enclosure through which system water flows.

When an electrical current passes through a heating element, the rate of heat output can be calculated using the following equation:

(Equation 3.9)

$$P = iv$$

where:

i = current (amps)
v = voltage (volts)
P = power output (watts)

The conversion factors that allow one to change the output rating of electrical heating devices (usually expressed in watts or kilowatts) to the more common Btu/hr ratings are:

$$1 \text{ watt} = 1 \text{ w} = 3.413 \text{ Btu/hr}$$

$$1 \text{ kilowatt} = 1{,}000 \text{ watts} = 1 \text{ kw} = 3{,}413 \text{ Btu/hr}$$

The second conversion factor is simply the first conversion factor multiplied by 1,000. These factors are used so commonly in heating system design that they should be memorized.

Several manufacturers offer packaged electric boilers in sizes suitable for use in residential or commercial hydronic systems. All heating elements, controls, etc., are typically assembled by the manufacturer so that only piping connections and electrical power supply wiring needs to be connected at the site. Figure 3–28 is representative of a typical wall mounted electric boiler.

Electric boilers usually contain two or more heating elements that together produce the rated heat output. With the proper controls, each individual element can be operated as an independent stage of heat. Only those stages that are needed to satisfy the load at any given time need to be on. This approach is similar in concept to modular boilers except the heating capacity of each stage can be much smaller with electrical heating elements.

Electric boilers do have some advantages compared to fossil fuel boilers. Since combustion is not involved, no air supply is required for combustion or draft, and no exhaust system is necessary. There is no concern about allowing low temperature water to pass directly into the boiler since there are no exhaust gases to condense. On-site fuel storage is not required. Periodic maintenance is minimal since there is no soot to remove, fuel filters to replace, etc. If used in conjunction with a weather-responsive control system, the water temperature in an electric boiler could be directly varied by controlling the on-time of the element(s), or using a solid-state power control device. These techniques eliminate the need for a mixing valve in variable water temperature systems. Since nearly all buildings have an electric service regard-

Figure 3–28 Example of an electric boiler. Courtesy of Burnham Corporation.

less of how they are heated, the choice of an electric boiler could eliminate the need for a natural gas service, and its associated monthly service charge.

The economic consideration of whether to use an electric boiler versus a fossil fuel boiler must consider the net cost of electricity versus competing fossil fuels. As would be expected, the economic viability of an electric boiler is highest in areas where utility rates are relatively low. In commercial or industrial buildings where utilities often impose a demand charge in addition to an energy charge, the economics of an electric boiler are questionable because the boiler would significantly increase the building's electrical demand. The Hydronics Design Toolkit contains two programs called FUELCOST and ECONOMICS that allow an electric boiler to be compared to other options from the standpoint of both owning and operating cost.

An electric hot water tank might also serve as a hydronic heat source. The limiting factor is usually the maximum wattage of the heating elements. In tanks intended for residential water heating, element wattage is usually 3.8 to 4.5 kw. The maximum wattage of elements used in these tanks is 6 kw. Using the previous conversion factors, 6 kw is equivalent to about 20,500 Btu/hr. This may be suitable for a moderate addition to an existing home, but would be significantly undersized for most houses.

Some state energy codes impose more rigid energy conservation standards on electrically heated buildings.

Any extra cost associated with meeting these standards should be factored into any economic comparison.

Finally, if an electric boiler is being considered, be sure the building's electric service entrance is adequate to handle the load. This is particularly true in older buildings that may only have 60 amp service entrances. A 200 amp/240 volt residential service is generally considered minimum for such applications.

3.12 ELECTRIC THERMAL STORAGE (ETS) SYSTEMS

Before discussing the details of **electric thermal storage (ETS) systems** as hydronic heat sources, it is appropriate to look at the factors that lead to their development.

Many electric utilities currently face the challenge of meeting the growing demands for electricity without spending hundreds of millions of dollars building new generating facilities. As the national demand for electrical energy increases, all utilities must plan ahead to ensure they can supply the extra energy, especially during critical peak demand periods. A fundamental approach to this planning is to reduce peak demand, while simultaneously increasing demand during traditional low demand periods.

One way to do this is to offer customers optional **time-of-use rates.** In there simplest form, time-of-use rates offer significantly lower cost electricity during specified off-peak periods. Such periods are usually comprised of a set number of hours with definite beginning and ending times. In some cases, the rates may vary several times during the day, as well as seasonally, during weekends, and even on holidays.

Since space heating is usually the dominant electrical load in all-electric buildings, it represents a major opportunity for savings if the energy could be purchased during off-peak periods. This concept requires that the heat produced during these periods be stored for subsequent use during on-peak periods. Since water already has excellent heat storage properties, hydronic systems using electric heating elements combined with thermal storage tanks are ideal for such applications.

Time-of-use rates also create an opportunity for low cost hydronic-based cooling through the use of electrically driven chillers operated during the off-peak periods, combined with chilled water storage coupled to a suitable hydronic distribution system.

Classification of ETS Systems

ETS systems are classified according to their ability, by design, to meet heating loads using energy purchased and stored during the previous off-peak period. The two types of designs are called **full storage systems** and **partial storage systems.**

In full storage systems, all energy used during the next on-peak period is purchased and stored during the off-peak cycle. The storage tank must hold sufficient thermal energy to supply the building's load through the entire on-peak period, which may be as long as 17 hours. Storage tank volumes of several hundred to more than 1,000 gallons are typical for such applications. Because they must generate sufficient heat for a 24-hour period during the off-peak period only, the heating capacity of the elements must be significantly greater than the design heating load of the building. The duration of the off-peak period is specified by each utility offering a time-of-use rate. Typical durations are 7 to 12 hours.

In partial storage systems only a portion of the on-peak heating energy will be purchased and stored during the off-peak period. Once this energy is depleted, the heating elements must operate as necessary to maintain comfort. All operation of the elements during the on-peak period uses electricity at the more expensive on-peak rate. However, because it stores only a portion of the on-peak heat, a partial storage system often has fewer heating elements and a smaller storage tank than a full storage system. This approach offers lower initial cost in exchange for a somewhat higher operating cost.

Figure 3–29 shows a comparison of hourly electrical heating demand for a house using electrical resistance heating, and a comparable house using a full storage ETS system. Notice the high electrical demand of the ETS system during the off-peak period when electrical rates are lowest. This is the charging period in which the water of a hydronic ETS system is being heated. Ideally the heat for the entire day is obtained during this time. At

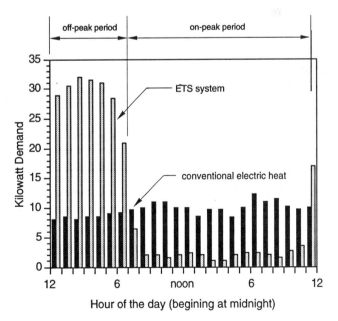

Figure 3–29 Electric demand of full storage ETS system versus convential electric heating. Data based on a design day in upstate New York.

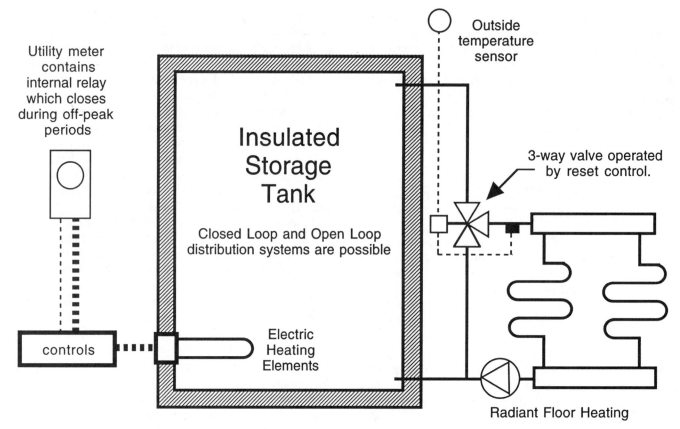

Figure 3–30   Schematic of a hydronic ETS system coupled to a radiant floor distribution system

approximately 7 AM the electric heating elements are shut off. During the middle of the day the small amount of energy consumed by the ETS system is for distribution system components such as pumps or, in some cases, blowers. This demand is very small compared to that of the conventional electric heating system.

The economics of partial storage versus full storage systems are very dependent on the specific project. An economic analysis that considers the trade-off between initial cost of the equipment, and the on-peak/off-peak rate differential is essential in making an informed decision. The system with the lowest energy cost may not necessarily be the best choice. Rather one should look for the lowest total owning and operating cost over an estimated system lifetime.

ETS System Layout

ETS systems do not represent exotic technology. They are composed of mostly standard hydronic components and designed using fundamental concepts covered in other chapters of this book. A basic schematic for an ETS system using electric heating elements mounted in a storage tank is shown in Figure 3–30. In this example the stored hot water is reduced in temperature by a three-way mixing valve before being routed to a radiant floor heating system.

Storage Tanks for ETS Systems

In both types of ETS systems, the key component is a properly sized, well-insulated storage tank. A wide variety of tanks are potentially usable. They range from pressure-rated steel tanks to non-pressure-rated high density polyethylene and polypropylene tanks. The choice must address factors such as temperature limit, corrosion, life expectancy, ease of installation, ease and type of insulation.

Pressure-rated steel tanks are a good choice for closed-loop, pressurized systems. Although their cost in dollars per gallon is usually higher than non-pressure-rated thermoplastic tanks, they allow more flexibility in design. The cost of such tanks can be reduced by selecting standard size tanks that are manufactured in quantity. For example, tanks designed for bulk storage of liquid propane are mass produced in nominal sizes of 500 and 1,000 gallons. They are available through most local propane distributors and are currently priced in the range of $2 to $3 per gallon. These tanks are ASME certified for pressures much higher than most hydronic systems will ever operate at, and come with a variety of threaded openings.

Non-pressure-rated tanks of high density polyethylene and polypropylene are widely used for handling chemicals in various industrial processes. Some polypropylene tanks have continuous service temperature ratings as high as 220 °F. Such tanks would only be suitable for open loop hydronic systems since they cannot be pressurized. *Proper support of all surfaces of thermoplastic tanks is critical.* If this is not done, the polyethylene will "creep," eventually deforming the tank to the point of possible failure. Although less expensive than steel tanks, non-pressure-rated thermoplastic tanks introduce many constraints on the remainder of the system design.

3.13   HYDRONIC HEAT PUMPS

A simple definition of a **heat pump** is *a device for converting low temperature heat into higher temperature heat.* The low temperature heat is obtained from some material at some location, and must be concentrated and moved to another material at another location.

Heat pumps are familiar devices to most heating professionals. Many mechanical contractors who install both heating and cooling systems have either installed, or at least have some knowledge of, the type of system known as an air-to-air heat pump. Such systems have been on the market since the early 1970s, and are commonly used for heating and cooling in moderate climates. The ability to supply both heating and cooling from a single piece of equipment makes the heat pump a year-round energy system.

Categories of Heat Pumps

Heat pumps are classified as either air-source or water-source units. The word source refers to the material from which heat will be absorbed. A hydronic heat pump absorbs heat from some lower temperature body of water, and thus is called a water-source heat pump. The material into which heat is released is known as the sink. For hydronic heat pumps, heat is always released into water or a water-based solution. Because water is both the source of the heat, and the material into which the higher temperature heat is released, a hydronic heat pump is also sometimes called a water-to-water heat pump. Figure 3–31 shows the installation of a water-to-water heat pump that is part of a hydronic heating system.

The Basic Refrigeration Cycle

The basis of operation of all heat pumps, hydronic or otherwise, is the **refrigeration cycle**. During this cycle, a chemical compound called the refrigerant circulates around a closed loop passing through all major components of the heat pump. To describe how this cycle works, a quantity of refrigerant will be followed through the major components of a water-to-water heat pump. The four stations in the refrigerant cycle are shown in Figure 3–32.

Figure 3–31   A water-to-water heat pump installed as part of a hydronic heating system

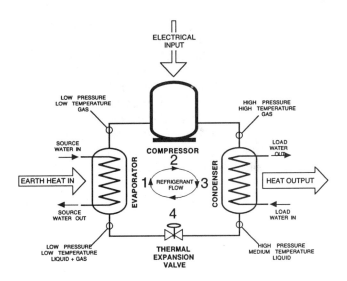

Figure 3–32   The basic refrigerant cycle within a water-to-water heat pump

The cycle begins at station 1 as cold liquid refrigerant evaporates within a water-to-refrigerant heat exchanger known as the evaporator. The water flowing through the evaporator is being supplied from some low temperature heat source such as a well. Because the liquid refrigerant in the evaporator is colder than the water, it absorbs heat, and changes from a liquid to a gas.

At station 2 the refrigerant gas flows through the heat pump's compressor where its volume is greatly reduced and its pressure and temperature are greatly increased.

At station 3 the hot refrigerant gas travels through another refrigerant-to-water heat exchanger known as the condenser. Here it releases heat into a flowing steam of water that carries the heat off to some type of heating load. As it releases heat, the refrigerant changes from a gas back to a liquid. It is still at a high pressure and relatively warm.

In order to begin the cycle over again, the refrigerant must be returned to the same cold liquid state from which this description began. At station 4 the refrigerant passes through an expansion device, (usually called a thermal expansion valve, or TXV), that greatly lowers its pressure. The lowered pressure induces an immediate drop in temperature, and the refrigerant is ready to begin the cycle again.

The refrigerant cycle is not unique to heat pumps. It is used in refrigerators, freezers, room air conditioners, dehumidifiers, water coolers, soda vending machines, and other heat-moving machines. The average person is certainly familiar with these devices, but usually takes for granted how they operate.

Water Sources for Hydronic Heat Pumps

Low temperature source water for a hydronic heat pump can come from a variety of locations. For typical residential and light commercial heating applications, heat pump water sources include wells, lakes, or piping loops buried in the earth. In more limited applications, the low temperature heat might even be extracted from a suitable waste water stream, or even a solar storage tank.

Wells, ponds, or lakes are referred to as **open loop sources** since at some point the water is directly exposed to the atmosphere. After the water passes through the heat pump it is discharged back to another well or the body of water from which it came. A piping loop buried in the earth would be considered a closed loop source since the same water recirculates between the piping loop and the evaporator in the heat pump.

Open Loop Sources

Most hydronic heat pumps require 2 to 3 gallons per minute (gpm) of water flow per ton (12,000 Btu/hr) of heating capacity. For a typical four-ton residential system, this would require several thousand gallons of water per day to flow through the heat pump during cold weather. While this volume of water is no problem for a lake or pond, it can place a tremendous load on a residential well. If for any reason the well cannot keep up with the water demand, the heat pump will eventually shut itself off, and the building will be without heat, and probably without water as well. It is essential the water source being considered can consistently provide the necessary quantity of water to the heat pump. *In the case of a well, always have a certified well-driller verify the well's sustained recovery rate before committing it to supply water to a heat pump.*

Since the water must be returned to the environment without contamination, it cannot be chemically treated prior to entering the heat pump. It is imperative that water quality be compatible with the materials it will contact within the heat pump. Water supplies with high concentrations of calcium, iron, hydrogen sulfide, silts, or other materials may prove unsuitable for such an application. *Always have the prospective water source analyzed by a competent water testing agency before committing to this type of open-loop system.*

The minimum source water temperature usable by the heat pump is also an important consideration. This minimum water temperature should always be checked with the heat pump manufacturer. Typical values range between 36 °F and 40 °F. Minimum temperatures a few degrees above freezing are necessary to prevent ice formation on the cold surfaces of the evaporator. The water temperature in most drilled wells usually remains fairly close to the annual average air temperature in a given location. Such data is often available and should be verified with local well drillers or heat pump installers.

To summarize, always verify the water quantity, chemical composition, and the likely minimum temperature before committing to the use of an open loop water source for a hydronic heat pump.

Closed Loop Sources

When open loop water sources are impractical, a closed-loop **earth heat exchanger** can be used to supply low temperature water to a hydronic heat pump. This heat exchanger usually consists of several hundred feet of buried plastic pipe through which water, or a mixture of water and non-toxic antifreeze, is circulated. As this fluid passes through the evaporator of the heat pump, it is chilled to a temperature lower than that of the soil surrounding the earth heat exchanger. It then flows out through the earth heat exchanger where it reabsorbs heat from the surrounding soil. After returning from the earth heat exchanger, the fluid is again chilled by the evaporator and ready to begin another pass through the earth heat exchanger. This process is illustrated in Figure 3–33.

Once it is filled and purged of all air, no fluid enters or leaves the earth heat exchanger. The use of a closed loop earth heat exchanger eliminates the need for a water well or other large body of water at the site. It also eliminates the potential scaling or corrosion problems of pumping poor quality water directly through the heat pump.

The type of pipe used for an earth heat exchanger is of critical importance to its success and longevity. Only two types of thermoplastic pipe are currently recommended for earth heat exchangers: high density polyethylene and polybutylene. Both types are available in a range of sizes, and coil lengths up to several hundred feet. These types

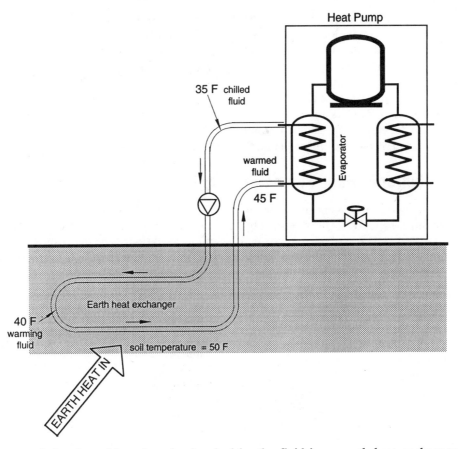

Figure 3–33   Conceptual drawing of how heat is absorbed by the fluid in an earth heat exchanger and extracted by the evaporator of the heat pump

of pipe can flex to accommodate the stresses imposed by variations in soil temperature and moisture content. PVC pipe does not have these qualities, and should not be used for buried portions of any earth heat exchanger.

Any joints in the polyethylene or polybutylene earth heat exchanger are made by heat fusion. In this process, the two pipe ends are heated to over 500 °F then joined to form a joint stronger than the pipe itself. Polyethylene is joined by butt fusion in which the pipe ends are directly fused to each other without a coupling. Butt fusion is often used by utilities when installing polyethylene piping for natural gas service. Polybutylene pipe is joined by socket fusion in which the pipe end and polybutylene fittings are fused together.

There can be no compromise in the materials or methods used to fabricate earth heat exchangers. When properly installed, earth heat exchangers should last for several decades.

Figure 3–34 shows a cross section of a typical horizontal earth heat exchanger used by the author on several hydronic heat pump systems in upstate New York.

This configuration requires approximately 180 feet of trench (or 720 feet of pipe) per ton of heat pump capacity at 30 °F minimum fluid temperature when used in heavy/wet soils. For installations in damp/sandy soils, the loop length must be increased to 260 feet of trench (or 1,040 feet of pipe) per ton. The longer loop requirement is due to the poorer heat transfer properties of sandy soils having a lower moisture content. Bear in mind that these

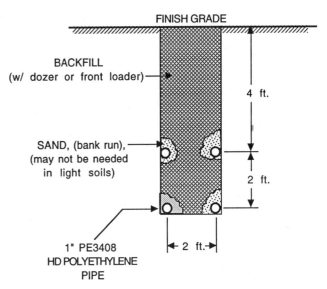

Figure 3–34   Cross section of a "four-pipe square" earth heat exchanger

loop lengths were calculated for a specific location (upstate New York), specific minimum loop temperature, (30 °F), and piping geometry/depth, (four-pipe square configuration at 4 feet and 6 feet depths). The results will vary for other locations, soil types, piping layouts, etc.

Complete details on designing closed earth loops are beyond the scope of this text. Readers interested in a complete sizing procedure for earth loops should consult one of the publications listed at the end of the chapter.

Performance of Hydronic Heat Pumps

The efficiency of a hydronic heat pump is called its **coefficient of performance, or COP.** COP is defined by Equation 3.10.

(Equation 3.10)

$$COP = \frac{\text{Heat output of heat pump (Btu/hr)}}{\text{Electrical input to run heat pump (Btu/hr)}}$$

The higher the COP of a heat pump, the more heat it produces for a given amount of electrical energy consumed. A heat pump with a COP of 3.0 will yield three times more heat output than the electrical energy required to run it. This additional heat is not created by the heat pump, but rather absorbed from the source water using the refrigeration process previously described. Ground source hydronic heat pumps can have COPs ranging from 2.0 to upwards of 4.0 depending on the temperatures of both the source water and the heating system distribution water. Since a heat pump can output significantly more heat than an electric boiler on the same amount of electricity, it follows that its operating cost will be significantly lower by comparison.

Figure 3–35 depicts the energy flows in a hydronic heat pump connected to a low temperature water source such as an earth heat exchanger. Notice the relative widths of the arrows representing the two energy flows into the heat pump, and the single arrow representing

energy flow out. The wider the arrow, the greater the rate of energy flow.

Both the coefficient of performance and the heating capacity of a heat pump vary as the temperature of the source water and distribution water change. Heat pump manufacturers rate the performance of their products at specific temperatures and water flow rates. Figure 3–36 shows how the heating capacity of a typical hydronic heat pump varies with the temperatures of the source water and distribution water. This graph has been normalized to show an output of 1.00 at a source water temperature of 40 °F, and entering load water temperature of 100 °F. For a given heat pump, the output of 1.00 might, for example, represent a 36,000 Btu/hr heating capacity at these conditions. The vertical axis of the graph indicates the multiplier for this reference heating capacity as the temperature of either the source water or distribution water change.

Example 3.7: Assume a given heat pump is rated at 32,000 Btu/hr output at a 40 °F entering source water temperature and 100 °F entering load water temperature. Find its heating capacity at 35 °F entering source water temperature and 120 °F entering load water temperature.

Solution: Referring to Figure 3–36, the intersection of the vertical line from 35 °F source temperature with the line for 120 °F entering load water temperature indicates a multiplier of 0.88. The heating capacity of the heat pump under these conditions would therefore be:

Heating Capacity = (0.88) 32,000 = 28,200 Btu/hr

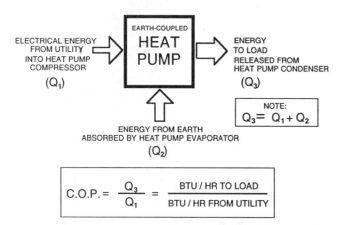

Figure 3–35   Energy flows through a hydronic heat pump

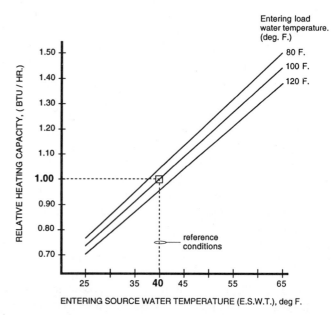

Figure 3–36   Variation in heating capacity of a hydronic heat pump with varying source water and load water temperatures

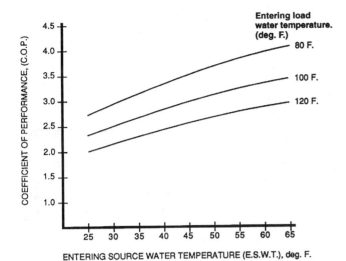

Figure 3–37 Variation in COP of a hydronic heat pump with changes in source water and load water temperature

Figure 3–36 illustrates the fact that the higher the source water temperature, and the lower the load water temperature, the greater the heating capacity of the heat pump. Conversely, the colder the source water is and the hotter the load water must be, the lower the heating capacity of the heat pump.

The COP of a hydronic heat pump is also affected by both the source and load water temperatures. The higher the source water temperature and the lower the load water temperature, the higher the COP of the heat pump. This effect is shown in Figure 3–37.

3.14 SYSTEM DESIGN CONSIDERATIONS FOR HYDRONIC HEAT PUMPS

Hydronic heat pumps have unique operating requirements that must be addressed during system design. Two primary requirements are sustained water flow and limitations on water temperature within the evaporator and condenser.

Sustained Water Flow

Whenever a heat pump is operating, refrigerant is removing heat from the water within the evaporator, and releasing heat into the water in the condenser. If the flow of water through either the evaporator or condenser is stopped or significantly slowed, the heat pump will automatically shut down to prevent physical damage.

A flow restriction or stoppage in the evaporator will allow ice crystals to form on the cold heat exchanger surfaces separating the water and refrigerant. Since the evaporator contains only a small amount of water, the ice

forms rapidly, and can result in a hard freeze in as little as one minute of operation without water flow. This could rupture the copper tubing in the evaporator, resulting in a costly repair. Because of this possibility, manufacturers install a temperature-sensing safety switch near the water outlet of the evaporator. This switch monitors the water temperature leaving the evaporator and will stop the compressor if the temperature drops below a preset limit, thus preventing a hard freeze.

Maintaining water flow through the condenser whenever the heat pump is operating is equally important. A flow restriction in this stream will cause the refrigerant head pressure to increase rapidly. Eventually this high pressure will trip a high pressure safety switch that shuts down the compressor.

Heat pump manufacturers usually recommend specific minimum water flow rates for both the evaporator and condenser sides of their products. *In the absence of manufacturers' recommendations, the system designer should provide for approximately three gpm of water flow per ton of capacity, through both the evaporator and condenser.*

The system designer must be careful not to install devices such as zone valves, or three-way mixing valves in series with the heat pump condenser. If such valves were present, they could restrict or totally stop water flow through the condenser causing an automatic shut down of the heat pump's compressor. One method for combining a hydronic heat pump with traditional zone controls is the use of a **buffer tank** between the heat pump and the distribution system. Figure 3–38 illustrates this concept.

The buffer tank acts as a thermal reservoir between the heat pump and the distribution system. Heat may be added to the tank by the heat pump, and removed by the

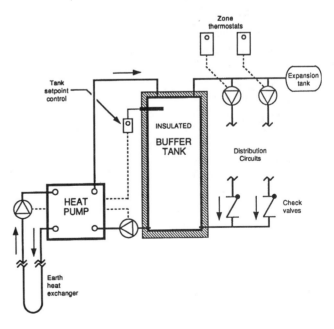

Figure 3–38 Use of a buffer tank to allow for zoned distribution with a hydronic heat pump

distribution system at different rates. The temperature of the buffer tank is maintained within a preset range using a temperature control on the tank to operate the heat pump as necessary.

As the building requires heat, water from the buffer tank can be circulated to the distribution system regardless of whether the heat pump is on or off. The buffer tank must be insulated since it contains heated water. Its size depends on the capacity of the heat pump, the temperature differential through which the tank must drop prior to starting the heat pump, and the minimum on-cycle time for the heat pump. These factors can be determined using fundamental concepts presented in Chapter 4.

The use of a buffer tank is equally applicable to chilled water cooling systems. As with heating, the buffer tank allows the heat pump to operate independently of the distribution system. If a buffer tank is used to store chilled water, it must be insulated, and have a continuous vapor barrier to prevent moisture migration toward the cold tank surfaces. Without a vapor barrier, condensation of water vapor on the tank surfaces could present a significant problem.

Water Temperature Limitations

All hydronic heat pumps have a range of acceptable water temperatures for both their evaporator and condenser. These temperatures usually depend upon the refrigerant used in the heat pump, as well as the type of expansion device used to control the flow of liquid refrigerant into the evaporator.

The temperature of source water entering the evaporator usually depends on the source. A heat pump supplied from a drilled water well experiences relatively consistent water temperature year-round. Well water temperature data is available for specific locations from most water source heat pump manufacturers. Typical values are between 45 °F and 60 °F for the northern half of the U.S. Earth heat exchangers yield fluid temperatures from 25 °F to 70 °F depending on the location, depth, length, and time of year. The lower end of this range can only be used when a suitable antifreeze is added to the water in the earth heat exchanger. Such low temperature operation also requires a heat pump equipped with a thermal expansion valve.

The temperature range of the condenser water also depends on the type of refrigerant used in the heat pump. Heat pumps using R-22 refrigerant have a practical upper temperature limit of 130 °F. R-500 based heat pumps can reach 150 °F to 155 °F, but with some loss of low temperature performance. As previously shown in Figures 3–36 and 3–37, the lower the condenser water temperature can be kept, the higher the heating capacity and COP of the heat pump will be.

The type of distribution system selected by the designer will have the greatest effect on condenser water temperature. Slab-type hydronic radiant floor systems are well suited for use with water-to-water heat pumps because they can utilize water in the relatively low temperature range of 80 °F to 110 °F. By comparison, finned tube baseboard systems usually require relatively high water temperatures for proper operation. *Such systems are not suitable for use with hydronic heat pumps.*

Hydronic heat pumps represent an important alternative to conventional boilers. They have been successfully used in many home heating applications. In general, however, they are less tolerant of design errors than conventional boilers. Good performance and dependable service will only be achieved through careful design and installation.

3.15 RENEWABLE ENERGY HEAT SOURCES

The late 1970s through the early 1980s was a time when America was infatuated with renewable energy. Various approaches such as solar heating, woodburning, and wind energy conversion were popular topics both technically and socially. By 1985 popular interest in renewable energy systems began to decline in the face of lower oil prices and dwindling government support. Many lessons, both good and bad, were learned about integrating renewable energy into the American economy.

On the plus side, engineers learned valuable lessons about what worked and what did not. Products such as solar collectors progressed through various stages of development. Well designed products evolved. Good design methods were developed.

On the negative side, many concepts, products, and system designs, though well intentioned, did not survive. They succumbed to the inevitable consequences of over-complexity, use of unsuitable materials, and the realities of having to compete with traditional energy sources. The public's zeal for renewable energy was in many cases stunted by market opportunists, unrealistically high pricing, and poorly designed systems that presented major service problems. Still, in light of the rising prices of conventional fuels, renewable energy will likely play an increasingly important role in America's future.

Two types of renewable energy systems are worthy of discussion as hydronic heat sources: active solar energy systems and wood-fired boilers.

3.16 ACTIVE SOLAR ENERGY SYSTEMS

Although many different designs for active solar heating systems have been developed, this discussion will be limited to systems using **flat plate solar collectors**. This

Figure 3–39   Construction of a typical flat plate solar collector. Courtesy of American Energy Technologies, Inc.

is the most available type of collector, and has proven to be the most cost effective in areas of the U.S. that represent the major hydronic heating markets. Figure 3–39 illustrates the construction of a typical flat plate solar collector.

Individual flat plate solar collectors are grouped together to form an array. Such an array if often located on a south facing roof. To maximize wintertime heat output, the array should have a slope angle equal to the local latitude +10 to 15 degrees. A roof-mounted array consisting of six flat plate collectors is shown in Figure 3–40. Ground mounted collector arrays are also possible, but are

usually more expensive due to the need for more extensive rack structures. They are also more subject to accidental damage or vandalism.

Because a major portion of a building's heating load occurs when solar energy is not available (i.e., at night, or during overcast weather), water filled thermal storage tanks are incorporated into the system to absorb excess solar heat during the day for use during these periods. Heat from the collectors is transferred to the thermal storage tank by circulating tank water through the collectors, or by circulating an antifreeze solution between the collectors and a heat exchanger coupled to the tank.

Closed Loop/Antifreeze Systems

Figure 3–41 shows a simplified schematic of a closed loop/antifreeze-type solar energy system.

An external heat exchanger is the link between the collector array and the storage tank. Since a portion of the collector circuit is exposed to outdoor temperatures, it must be protected against freezing. This is most often done by filling this circuit with a glycol-based antifreeze solution. This solution transports heat from the collector array to the heat exchanger. A second circulator moves water between the storage tank and the heat exchanger. Both circulators must operate whenever solar energy is being collected.

An advantage of the antifreeze-based system is that the piping between the collectors and storage tank can be installed in virtually any orientation, inside or outside.

Figure 3–40   An array of six flat plate solar collectors that supply heat to a hydronic distribution system

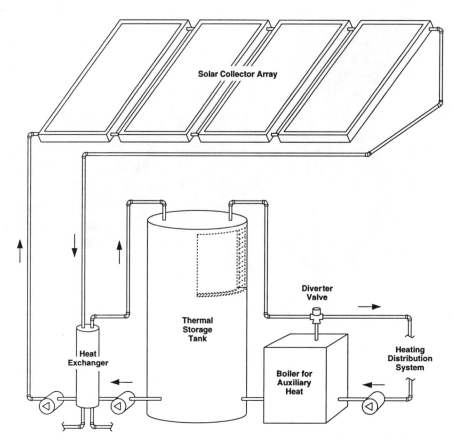

Figure 3–41   Closed loop/antifreeze-type solar energy system for space heating

Direct Circulation Systems

There are several advantages to circulating the system's water directly through the solar collectors. First, the expense of the heat exchanger between the collector array and storage tank is eliminated, as is the antifreeze solution. Components required by the closed collector loop design such as an expansion tank, pressure relief valve, and air purging device are also eliminated. The cost of the heat exchanger and associated components can amount to several hundred dollars in residential scale systems.

A second advantage of direct circulation systems is the elimination of a performance penalty associated with use of a heat exchanger between the collector array and thermal storage tank. In closed-loop/antifreeze-type systems, a temperature difference of 10 °F to 15 °F often exists between the antifreeze solution in the collector loop and the water in the thermal storage tank, whenever the collectors are operating. This temperature difference is necessary to transfer heat across the heat exchanger at a suitable rate. Unfortunately this temperature difference forces the collectors to operate at higher temperatures

than would be necessary with a direct circulation system. Heat losses from the collectors are increased at the higher temperatures, and collector efficiency is lowered.

Because they do not contain antifreeze, *all direct circulation systems must have automatic freeze protection of system components not contained within heated space.* This is accomplished by draining all water from exposed piping as freezing conditions approach. Several methods have been devised to facilitate this draining.

Draindown Systems

One design approach, known as a **draindown system,** uses an electronic control to sense near freezing conditions at the collectors and signals motorized valves to open, thus draining water from the collectors and exposed piping. *All collectors as well as exposed piping must be pitched a minimum of 1/4 inch per foot to allow complete drainage.* Any failure of the outdoor temperature sensor, electronic control, or motorized valves could prevent this type of system from draining and cause a very costly "hard freeze," possibly rupturing the absorber plates in the collectors.

Draindown systems also introduce fresh (oxygen containing), water into the system when refilled following a

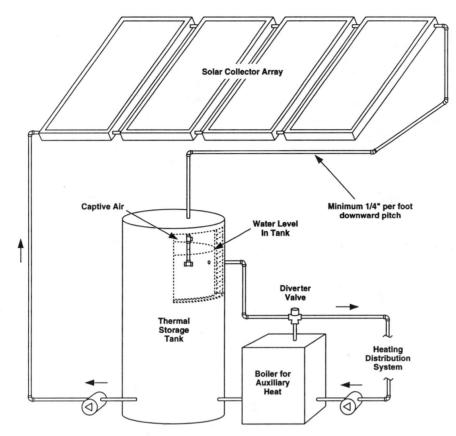

Figure 3–42   Example of a gravity drainback solar energy system for space heating

draindown. Because of this they must be designed as open-loop systems, and cannot contain any cast-iron or steel components.

Gravity Drainback Systems

The second method for draining a direct circulation system is known as a **gravity drainback system**. A schematic of this type of system is shown in Figure 3–42.

In this approach, gravity alone drains all water out of the collectors and exposed piping whenever the circulator stops. No motorized valves or electronic controls are needed. *Again it is absolutely necessary that the collectors and all exposed piping be pitched a minimum of ¹/₄ inch per foot toward the storage tank to allow complete drainage.* The collectors and exposed piping are refilled with tank water, not fresh water, each time the collector circulator starts. This circulator must be sized to be able to lift the water to the top of the collector array.

Notice the captive volume of air at the top of the solar storage tank in Figure 3–42. This air will quickly replace the water in the collector array and piping above the water level in the tank whenever the collector circulator stops. When the circulator stops air enters the return pipe through an open tee located above the water level in the tank. If properly sized,

this air volume can also function as the expansion tank for both the collector loop and the distribution system.

The return line from the collectors to the tank should be sized for a minimum flow velocity of 2 feet/second. This allows the water returning from the collectors to entrain air bubbles and return them back to the storage tank. As the return pipe becomes filled with water, a siphon is established over the top of the collector loop. This approach minimizes gurgling noises in the piping and increases the water flow rate through the collectors, slightly improving their efficiency.

The gravity drainback approach holds several advantages. First, since it relies on gravity alone for system drainage, it is not prone to freezing due to failure of a control device. Second, unlike a draindown system, it recirculates the same system water over and over. Since no water actually leaves the system during drainback, no fresh water is required to replace it. This allows drainback systems to be designed as closed loop systems using less expensive cast-iron circulators and related components. Finally, several components associated with the other design approaches are no longer necessary, and thus costs are lower. Of all methods for designing active solar heating systems, the direct circulation/gravity drainback system is most preferred by the author. Several such system have been designed and installed in severe winter climates.

Design Considerations for Active Solar Systems

An important design objective for any active solar energy system is to maintain collector temperature as low as possible. The lower the collector temperature, the higher its efficiency, and hence the greater the amount of available solar energy it will collect.

Slab-type radiant floor heating is one of the best hydronic distribution systems for use with active solar collectors. Its low water temperature requirements allow maximum utilization of the collected energy. Hydronic fan-coil units will also work provided they are sized to meet design heating loads using relatively low temperature water (110 °F to 120 °F). Conventional hydronic baseboard systems require relatively high temperature water, and thus are a poor choice for use with solar collectors.

Active solar heating systems usually provide only part of a building's seasonal heating energy. This percentage depends upon geographical location, collector area in relation to building load, and many other factors. Practical designs always use some type of auxiliary heat source to supply heat not obtainable from the solar collectors. Any of the hydronic heat sources discussed earlier in this chapter could potentially be used. Since design heating load may occur during inclimate weather when no solar energy is being collected, the auxiliary heat source must be sized to handle the full design heating load of the building.

The auxiliary heat source should not maintain the thermal storage tank at temperatures suitable for use by the distribution system during non-solar periods. The tank should be allowed to cool down to room temperature during such periods, thus giving up all its transferable heat. Provided the tank is located within heated space, the heat released as the tank cools off still offsets a portion of the building's heating load. This energy will eventually be replaced by the solar collectors, rather than the auxiliary heat source. The use of auxiliary energy to maintain tank temperature during non-solar periods can waste solar energy due to delayed collector start-up when solar radiation again becomes available. This delayed start-up is caused by the differential temperature control system commonly used with active solar heating systems. With this type of control, the collector circulator starts only when collector temperature is 8 to 10 degrees above storage tank temperature. Another reason for not maintaining the tank temperature with auxiliary energy is that storing energy in thermal form (heated water) is less efficient than maintaining this energy in its original chemical or electrical form until just before it is needed by the load. Proper system design allows the distribution water to bypass the solar storage tank when its temperature is too low for use by the heat emitters. This is readily accomplished using a three-way diverting valve operated by a control that senses storage tank temperature.

Sizing Active Solar Energy Systems

Detailed sizing and design methods for active solar energy systems have been developed and refined over the past decade and are widely available. Readers interested in detailed sizing methods should consult one of the references at the end of the chapter.

As with any other type of hydronic heat source, cost considerations are an important part of system design. The cost of an active solar collector system is influenced by many factors such as installed collector costs, seasonal availability of solar radiation, and distribution temperature requirements. The total owning and operating cost of a solar collector system over its expected life must be compared to those of competing heat sources.

3.17   WOOD-FIRED BOILERS

Following the energy crisis of the mid 1970s there was a dramatic rise in the number of wood-burning devices being installed for home heating. During this time, wood-burning took on an almost "patriotic" connotation. Many wood-burning boilers for use with hydronic distribution systems emerged on the market. They ranged from simple firebox heat exchangers to sophisticated forced-draft boilers capable of producing combustion temperatures in excess of 2,000 °F. As was the case with solar energy systems, some inferior products appeared on the market and have since disappeared. At present, several U.S. and foreign manufacturers have wood-fired boilers available.

The decision to use a wood-fired boiler involves a time commitment much more demanding than for conventional heat sources. It could be said that burning wood for home heating is a way of life, rather than just a choice of fuel source. Many one-time wood-burning homeowners have lost their enthusiasm for wood burning when these requirements were realized after the system was installed. Many households simply have no one at home during an average workday to tend a wood-fired boiler. Long-term satisfaction with wood-burning under such circumstances is questionable. This observation is not meant to discourage those considering wood-burning boilers, but rather to encourage thorough and realistic comparisons with competing fuel sources, before making a decision.

Types of Wood-fired Boilers

Wood-fired boilers come in a number of designs. Some rely on the natural draft created by a standard chimney, others use a small blower to force air into the combustion chamber. Many boilers that can burn wood can also burn coal if properly set up. These are often called **solid fuel boilers.** Still others can operate as either a wood-fired boiler, or an oil-fired boiler depending on the

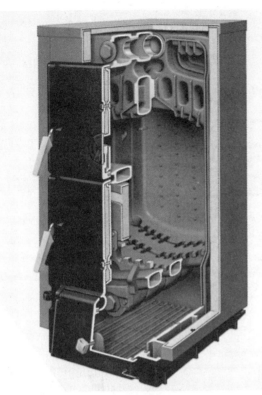

Figure 3–43 Modern solid fuel boiler. Courtesy of Buderus Hydronic Systems.

Figure 3–44 Example of a combination wood/oil boiler. Courtesy of Burnham Corporation.

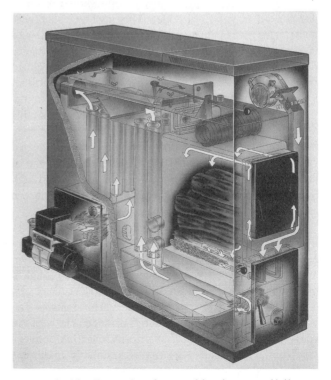

Figure 3–45 Example of a combination wood/oil boiler with separate combustion chambers for each fuel. Courtesy of HS Tarm Stoveworks.

intensity of heat output from the wood fire. These are called combination wood/oil boilers.

Figure 3–43 illustrates a modern solid fuel boiler capable of burning either wood or coal. This boiler can partially control the heat output of the fire through the use of a thermostatically controlled air damper. The combustion chamber is built of cast-iron sections using a wet base design similar to a typical oil-fired boiler. However, the combustion chamber is much larger to accommodate a load of wood that will burn for several hours. An ash cleanout drawer can be seen at the bottom of the boiler.

Because a wood-fired boiler can not load its own fuel as needed, several manufacturers offer a boiler equipped with an oil burner to either supplement the heat output of the wood fire or provide the total heat output if no wood fire is being maintained. An example of this type of combination boiler is shown in Figure 3–44.

When a single combustion chamber is used for both fuels it is of necessity a compromise between the ideal chamber shape for oil firing, and the space requirements of wood burning. The efficiency of such a boiler in oil-fired operation generally is not as high as with a modern oil-only boiler. Some manufacturers use separate combustion chambers for each fuel to correct for these limitations. An example of such a boiler is shown in Figure 3–45. The dual combustion chamber design allows each chamber to be optimized for its particular fuel and space requirements.

Another concern with boilers using an oil burner permanently mounted into a solid fuel combustion chamber is the accumulation of soot and fly ash from wood-fired operation on the flame head of the oil burner. The flame head is exposed to an environment much dirtier than in a conventional oil-fired boiler. This creates the need for more frequent cleaning in order to maintain high burner efficiency. One remedy for this problem is mounting the oil burner so that its flame head is not permanently exposed to the combustion chamber while a solid fuel is being burned. Some manufacturers use a double door system to allow the boiler to be quickly converted from wood to oil-fired operation by simply selecting which door is closed over the combustion chamber. One door is a standard firebox door, the other is equipped with an oil burner supplied by a flexible oil line. Of course this design means the boiler is either wood-fired or oil-fired, but not both at the same time.

The Dual Boiler Approach

When both wood and oil will be used, the best performance, with the least problems, is often achieved using separate boilers for each fuel. This allows each boiler to be designed for optimum efficiency and maintenance with its own specific fuel. This approach is also common for retrofit situations. There are several ways of piping the two boilers into the system.

One approach is to pipe the wood-fired boiler in series with the oil or gas-fired boiler. This is shown in Figure 3–46a. The design intent is for the wood-fired boiler to handle the heating load until its output decreases due to fuel depletion. At this point, if heat is still being called for, a temperature control brings on the conventional boiler to supplement and eventually take over for the wood-fired boiler. The series piping arrangement is simple in terms of piping and control. However, the efficiency of the combined system suffers from the fact that heated water always passes through both boilers, even if one of them is not being fired. This allows air moving through the combustion chamber of the inactive boiler to absorb heat and carry it up the chimney.

Another arrangement is to pipe the wood-fired boiler in parallel with the conventional boiler. In this arrangement, illustrated in Figure 3–46b, each boiler has its own circulator and check valve. Water only flows through a boiler while it is being fired. This greatly reduces off-cycle heat loss in the inactive boiler. The controls can function the same as with the series arrangement with the wood-fired boiler having first priority over the conventional boiler provided it maintains a suitably high supply temperature. In this arrangement, the circulator on each boiler must be large enough to circulate the water through the entire distribution system. If the controls allow both boilers to operate simultaneously, the system's

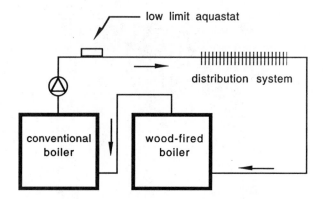

Figure 3–46a Series piping of wood-fired boiler and conventional gas or oil-fired boiler

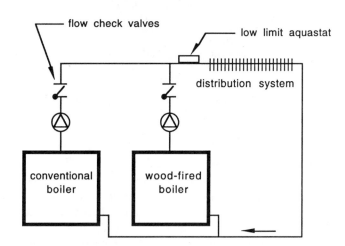

Figure 3–46b Parallel piping of wood-fired boiler and conventional gas or oil-fired boiler

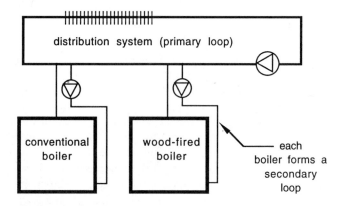

Figure 3–46c Primary/secondary piping of wood-fired boiler and conventional gas or oil-fired boiler

flow rate would increase significantly. This could lead to excessive flow velocities and associated noise.

A final configuration makes use of the same primary/secondary piping arrangement previously discussed for modular boiler systems. As shown in Figure 3–46c, both boilers are set up as secondary loops connected into the building's main distribution loop. Each boiler has its own small circulator. The building loop also has its own circulator. The flow rate in the distribution system is virtually unaffected by whatever boiler(s) are operating. This design also minimizes off-cycle heat loss within the inactive boiler(s). The wood-fired boiler should be installed upstream of the conventional boiler so that it recieves the coolest inlet temperature.

Efficiency of Wood-fired Boilers

The seasonal efficiency of wood-burning boilers is not as well documented as those of more traditional hydronic heat sources. At present there are no U.S. government standards or minimum efficiency values for wood-burning boilers. As a guideline one might expect a quality air-tight wood-fired boiler to achieve a 40% to 65% combustion efficiency under steady operation. Modern forced-draft wood-fired boilers can attain steady state efficiencies in the range of 80%. This value, however, is highly dependent on the moisture content of the firewood. Use of unseasoned wood, with a high moisture content, could easily cut this figure in half.

The efficiency of a wood-burning boiler is very dependent upon the combustion temperatures existing in the firebox at any given time. To achieve high efficiency, the pyrolytic gases released from the burning wood must themselves ignite and release their chemical energy content. If the temperature of these gases reaches approximately 1,100 °F, and sufficient oxygen is available, com-

bustion will occur. If the pyrolytic gases do not ignite due to low flame temperature or insufficient oxygen, they are exhausted out of the wood-burner and will partially condense within the chimney forming creosote. Forced draft systems, with combustion chamber temperatures approaching 2,000 °F, can burn nearly all the pyrolytic gases, and hence creosote formation is essentially eliminated. *High combustion temperature is a key to high efficiency and clean operation of wood-fired boilers.*

Another consideration is the ability to control the heat output from a wood-fired boiler. Unlike oil, gas, or electrically powered boilers which can be turned on and off as the heating load requires, wood-fired boilers must essentially burn off a loaded quantity of wood before heat output ceases. Many wood-fired boilers have some type of thermostatic air damper for marginal control over the air fed into the firebox. Remember, however, that an oxygen-starved flame smolders away at very low efficiency and leads to increased creosote formation. It is very inefficient for wood-burning boilers to stand by with a smoldering flame while waiting for the next call for heat. If a wood-fired boiler cannot be well controlled at a low firing rate, its operation may have to be limited to only the coldest weather.

Forced Combustion Combined with Thermal Storage

One approach aimed at eliminating low efficiency standby periods is to completely burn off the loaded quantity of firewood at a high firing rate, and transfer the heat to a large thermal storage tank. The heat would then be delivered by a hydronic distribution system as needed. This concept is similar to the buffer tank design discussed with hydronic heat pumps. It allows combustion to occur at a high rate and hence high efficiency. It also reduces

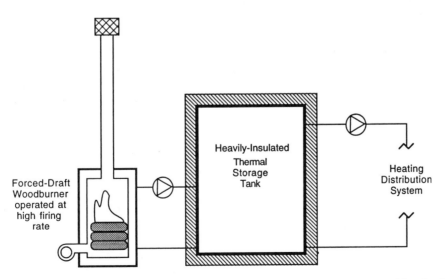

Figure 3–47   Schematic concept of a forced-combustion, high capacity woodburner combined with a thermal storage tank

the actual firing time to only a few hours per day, which is more easily worked into the homeowner's schedule. This concept is illustrated in Figure 3–47.

The boiler used in such a system must have a heat output significantly greater than the design load of the building. This is necessary because the boiler must produce sufficient heat in a period of a few hours to supply the building load for 24 hours. A forced-draft type boiler is often used to achieve the high firing rate. Such a boiler may be loaded with a large volume of wood, and then not reloaded during the firing cycle. The size of the thermal storage tank will depend on the amount of heat to be stored, as well as the length of the firing cycle. It can be calculated based on methods presented in Chapter 4. The cost of this type of system can be substantial because of the high capacity boiler and insulated thermal storage tank.

New technologies will undoubtedly emerge for the improved handling and burning of solid fuels such as wood, and other forms of "biomass." Promising methods such as wood pelletization, in which wood pellets are automatically fed by an auger into a combustion chamber, and wood gasification, which yields a gas suitable for use in a gas-fired boiler, are already in use in large scale industrial systems. Such systems are economically justified at facilities such as paper mills and lumber mills where wood by-products are readily available on site, and can be utilized as fuel. These concepts have also been demonstrated in smaller, residential scale systems, but at present remain too complex and expensive to make a major penetration into the market.

SUMMARY

This chapter has examined several types of hydronic heat sources, pointing out some of the advantages and disadvantages of each. In some cases the type of heat source used will be specified by the customer. In other cases it will be dictated by the availability of certain fuel types and the cost of these fuels. In all cases the system designer should consider the effects of the proposed heat source on the distribution system used. Incompatibilities do exist, and are easily conceived by careless "catalog engineering." *Always remember that system performance and reliability, and not that of an individual component, will determine the success of the design.*

KEY TERMS

Annual Fuel Utilization Efficiency (AFUE)
Boiler jacket
Buffer tank

Chemical energy content (of fuel)
Coefficient of performance (COP)
Combustion efficiency
Condensing mode
Confined space
Cycle efficiency
Dewpoint temperature
Direct vent
DOE heating capacity
Draft proving switch
Draindown system
Dry base boiler
Earth heat exchanger
Electric thermal storage (ETS) system
Fire tube boiler
Flat plate solar collector
Free area (of an opening)
Full storage system
Gravity drainback system
Heat pump
Heating capacity
IBR gross output
IBR net output
Modular boiler system
National Fuel Gas Code
Non-condensing mode
Off-cycle losses
Open-loop source
Packaged boiler
Partial storage system
Power venting
Primary/secondary piping
Refrigeration cycle
Run fraction
Sealed combustion
Sectional boiler
Solid fuel boiler
Steady state efficiency
Thermal mass (of a boiler)
Time-of-use rate
Unconfined space
Water tube boiler
Wet base boiler

CHAPTER 3 QUESTIONS AND EXERCISES

Note: Questions and exercises requiring the Hydronics Design Toolkit are indicated with marginal symbol.

1. Define the thermal mass of a boiler. Describe some desirable and undesirable traits of low thermal mass boilers. Do the same for high thermal mass boilers.

2. What happens when a condensing boiler operates at a return water temperature above the dewpoint temperature?

3. Name some limitations that must be considered if a domestic water heater will also be used for space heating. Where is an application such as this justifiable?

4. What is the function of a draft proving switch in a power exhaust system? Describe how it operates.

5. A boiler with a gas input rating of 100,000 Btu/hr will be installed in a basement. Using the NFPA 54 requirements described in Section 3.7, determine the required floor area of the basement if it is to be considered unconfined space. Assume the height of the basement is 8 feet.

6. A boiler with a gas input rating of 100,000 Btu/hr will be installed in a boiler room within a basement. The combustion air will be brought in through horizontal ducts. The ends of the ducts are covered with metal louvers having a free/gross area ratio of 0.6. Determine the size of each duct opening in square inches.

7. A boiler produces 60,000 Btus in 30 minutes. In the process it consumes 0.5 gallons of #2 fuel oil (140,000 Btu per gallon). Determine its average efficiency during this period.

8. In one hour, a boiler fires for a total of 12 minutes. Use Figure 3–22 to determine the boiler's cycle efficiency for this period.

9. A house requires 75 MMBtu of heat for an entire heating season. Assume this is provided by an oil-fired boiler with an AFUE of 80%. Estimate how many gallons of #2 fuel oil would be required over the season.

10. What is the advantage of primary/secondary piping when connecting boilers in a modular boiler system?

11. An electric boiler outputs 40,000 Btu/hr. It is supplied by a 240 volt circuit. Estimate the electrical current flow to the boiler.

12. Over a 24-hour period a house has an average heating load of 30,000 Btu/hr. If the house is supplied by a full storage ETS system operating for an off-peak period of 8 hours per night, what is the required total heating capacity of the elements? *Hint:* The entire heat for the 24-hour period must be obtained in the 8-hour off-peak period.

13. A hydronic heat pump has an output of 48,000 Btu/hr. The electrical input rate is 4.5 kw. Determine the COP of the heat pump under these conditions.

14. A hydronic heat pump has an output of 36,000 Btu/hr while its source water enters at 40 °F and its distribution water enters at 100 °F. Using Figure 3–36 estimate its heat output when its source water temperature drops to 35 °F and its load water enters at 110 °F.

15. Why is a buffer tank required if a hydronic heat pump is combined with several independent heating zones?

16. Describe some key design requirements when matching a hydronic heat pump to a hydronic distribution system. Describe what can go wrong if these requirements are not met.

17. What are some advantages of a drainback solar system compared to a draindown system?

18. Why is it necessary to burn wood at high combustion temperatures to achieve high efficiency? What can happen if high temperature combustion does not occur?

19. Using the BINEFF program in the Hydronics Design Toolkit, estimate the seasonal average efficiency of a 120,000 Btu/hr boiler installed in a house with a 73,000 Btu/hr load (at 70 °F inside and –10 °F outside temperatures). The boiler performs according to the curve shown in Figure 3–22. It is installed in Albany, New York. What happens to the seasonal average efficiency if the boiler capacity is reduced to 90,000 Btu/hr output?

FOR FURTHER READING

Ground Source Heat Pump Design:

Closed-loop/Ground Source Heat Pump Systems, Installation Guide. 1988. International Ground Source Heat Pump Association, P.O. Box 1688, Stillwater, OK 74076-1688.

Geothermal Heat Pump Options Manual. 1987. Edison Electric Institute, 1111 19th St. N.W., Washington, DC 20036.

Solar Energy System Sizing:

Beckman, et al. 1977. *Solar Heating Design by the F-chart Method.* New York: J. Wiley and Sons.

Duffie, et al. 1980. *Solar Engineering of Thermal Processes.* New York: Wiley Interscience.

4 PROPERTIES OF WATER

OBJECTIVES

After studying this chapter you should be able to:

- Describe several fluid properties relevant to hydronic system design and operation
- Explain why water is an excellent material for transporting heat
- Demonstrate how heat and temperature are related
- Calculate the amount of heat stored in a substance
- Explain the difference between a Btu and a Btu/hr
- Estimate the output of a heat emitter from measurements of temperature and flow rate
- Predict when water will boil based on pressure and temperature
- Explain how a fluid's dynamic viscosity affects its flow rate in a hydronic system
- Describe how air is dissolved into and released from water
- Look up or calculate the values of several fluid properties for use in later chapters

4.1 INTRODUCTION

Water is the key material in any hydronic heating system. Its properties make possible the high efficiency and comfort associated with such systems. This chapter will define and explain several properties of water that are relevant to hydronic heating in the simplest possible terms. A working understanding of these properties is invaluable in making design decisions, as well as for troubleshooting hydronic systems.

4.2 SPECIFIC HEAT AND HEAT CAPACITY

One of the most important properties of water in regards to its use in hydronic systems is **specific heat**. Simply put, *the specific heat of any material is the amount of heat required to raise the temperature of one pound of that material by one degree F.* The specific heat of water is approximately 1.0 Btu per pound per degree F. To raise the temperature of one pound of water by one degree F will require the addition of one Btu of heat. This relationship between temperature, material weight, and energy content also holds true when a substance is cooled. For example, to lower the temperature of one pound of water by one degree F will require the removal of one Btu of energy.

The specific heat of any material varies slightly with its temperature. For water this variation is quite small over the temperature range in which most residential and small commercial hydronic systems operate. The graph in Figure 4–1 shows this variation. For the purpose of

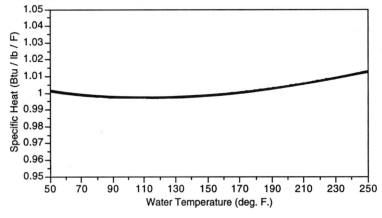

Figure 4–1 The specific heat of water as a function of temperature

Material	Specific Heat (Btu / lb. / deg. F.)	Density (lb / cubic ft.)	Heat Capacity (Btu / cubic ft. / deg. F.)
water	1.0	62.4	62.4
concrete	0.21	140	29.4
steel	0.12	489	58.7
wood (fir)	0.65	27	17.6
ice	0.49	57.5	28.2
air	0.24	0.074	0.018
gypsum	0.26	78	20.3
sand	0.1	94.6	9.5
alcohol	0.68	49.3	33.5

Figure 4–2 The specific heats of some common materials

designing residential or small commercial hydronic systems this slight variation can be safely ignored, and the specific heat of water can be considered to remain constant at 1.0 Btu/lb/°F.

The value of a material's specific heat also changes when the material changes phase (i.e., between liquid and vapor, or liquid and solid). For example, the specific heats of ice and water vapor are about 0.48 Btu/lb/°F, and 0.489 Btu/lb/°F respectively. Both values are considerably lower than that of liquid water.

The specific heat of water is high in comparison to other common materials. In fact water has one of the highest specific heats of any known material. The table in Figure 4–2 compares the specific heats of several common materials. Also listed is the **density** of each material, and the product of specific heat times density, which is referred to as the **heat capacity** of the material.

At first glance this data may be deceiving. For example, why is the specific heat of air greater than the specific heat of steel? The answer is that even though steel has a lower specific heat, it is much denser than air (489 lb/ft³ for steel versus about 0.074 lb/ft³ for air). Therefore, the heat capacity of steel, which reflects both its specific heat and density, is many times larger than that of air. By comparing the heat capacity of water and air, it can be shown that *a given volume of water can hold almost 3,500 times as much heat as the same volume of air, for the same temperature rise in each material.* This makes water much more efficient than air for transporting heat.

The Hydronic Design Toolkit contains a routine called FLUIDS that can be used to find the specific heat of water and several water-based fluids used in hydronic heating systems over a wide range of temperatures.

4.3 SENSIBLE HEAT VERSUS LATENT HEAT

The heat absorbed or released by a material while it remains in a single phase is called **sensible heat**. The word *sensible* refers to the fact that the presence of the heat can be *sensed* by a temperature change in the material. As heat is absorbed by the material its temperature increases. As heat is released, its temperature drops.

This process of sensible heat transfer is not the only way a material can absorb or release heat. If the material changes phase while absorbing heat, no change in temperature occurs. The heat absorbed in such a process is called **latent heat**. For water to change from a solid (ice) to a liquid requires the absorption of 144 Btu/lb while its temperature remains constant at 32 °F. When water changes from a liquid to a vapor it must absorb approximately 970 Btu/lb, again while its temperature remains constant. The amount of energy required to vaporize water is very large, which explains why water evaporating from the surface of an object (such as your skin) is a very effective means of cooling it.

4.4 SENSIBLE HEAT QUANTITY EQUATION

The definition of specific heat forms the basis of an equation that can be used to determine the *quantity of heat* stored in a given amount of material that undergoes a specific temperature change. This relationship, called the sensible heat quantity equation, is given as Equation 4.1:

(Equation 4.1)

$$h = wc(\Delta T)$$

where:

h = *quantity of heat* absorbed or released from the material (Btus)
w = weight of the material (lbs)
c = specific heat of the material (Btu/lb/°F)
ΔT = temperature change of the material (°F)

Two conversion factors that are often required when using this equation for water are:

• 1 gallon of water weighs approximately 8.33 lbs
• 1 cubic foot of water weighs approximately 62.4 lbs

When the first of these values is factored into Equation 4.1, along with the specific heat of water = 1.0 Btu/lb/°F, the resulting equation becomes very convenient for use with water:

(Equation 4.2)

$$h = 8.33v(\Delta T)$$

where:

h = *quantity of heat* absorbed or released from the water (Btus)
v = volume of water involved (gallons)
ΔT = temperature change of the material (°F)

Example 4.1: A storage tank contains 500 gallons of water initially at 70 °F. How much energy must be added or removed from the water to:

a. raise the water temperature to 180 °F?
b. cool the water temperature to 40 °F?

Solution:

a. Using the sensible heat quantity equation for water (Equation 4.2),

$$h = 8.33(500)(180 - 70) = 458,000 \text{ Btu}$$

b. Using the same sensible heat quantity equation for water (Equation 4.2),

$$h = 8.33(500)(70 - 40) = 125,000 \text{ Btu}$$

4.5 SENSIBLE HEAT RATE EQUATION

Many times the system designer needs to know the *rate of heat transfer* to or from a stream of water flowing through a device such as a heat source or heat emitter. This can be done by introducing time into the sensible heat quantity equation. The resulting equation is called the sensible heat rate equation:

(Equation 4.3)

$$Q = Wc(\Delta T)$$

where:

Q = *rate of heat transfer* into or out of the fluid stream (Btu/hr)
W = rate of flow of the liquid (lb/hr)
c = specific heat of the liquid (Btu/lb/°F)
ΔT = temperature change of the liquid within the device it exchanges heat with (°F)

For water only this equation can be rewritten as:

$$Q = \left(\frac{\text{gallons}}{\text{minute}}\right)\left(\frac{8.33 \text{ lb}}{\text{gallon}}\right)\left(\frac{60 \text{ minutes}}{\text{hour}}\right)\left(\frac{1 \text{ Btu}}{\text{lb} \bullet °F}\right)(\Delta T °F)$$

which simplifies to:

(Equation 4.4)

$$Q = 500f(\Delta T)$$

where:

Q = *rate of heat transfer* into or out of the water stream (Btu/hr)
f = **flow rate** of water through the device (gpm)
500 = constant rounded off from 8.33 • 60
ΔT = temperature change of the water through the device (°F)

Example 4.2: Water flows into a convector at 140 °F. It leaves at 125 °F. The water's flow rate is 4.5 gpm. What is the rate of heat transfer from the water to the convector?

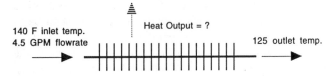

140 F inlet temp.   Heat Output = ?
4.5 GPM flowrate   125 outlet temp.

Figure 4–3 Convector for Example 4.2

Solution: Use the sensible heat rate equation for water (Equation 4.4)

$$Q = 500(4.5)(140 - 125) = 33,750 \text{ Btu/hr}$$

Example 4.3: A heating system is proposed that will deliver 50,000 Btu/hr to a house while circulating only

2.0 gpm of water. What temperature drop will the water have to go through to accomplish this?

Solution: This situation again involves a rate of heat transfer from a water stream. Equation 4.4 will be used. It will be rearranged to solve for the temperature drop term (ΔT).

$$\Delta T = \frac{Q}{500\,f} = \frac{50,000}{500(2.0)} = 50 \ \degree F$$

Although a hydronic distribution system could be designed to yield a 50 °F temperature drop, this value is much larger than normal. Such a system would require very careful sizing of the heat emitters, and a fairly high boiler outlet temperature. This example shows the usefulness of the sensible heat rate equation to quickly evaluate the feasibility of an initial design concept.

4.6 VAPOR PRESSURE AND BOILING POINT

*The **vapor pressure** of a liquid is the minimum pressure that must be applied at the liquid's surface to prevent it from boiling.* If the pressure on the liquid's surface drops below its vapor pressure, the liquid will *instantly* begin to boil. If the pressure is maintained above the vapor pressure, the liquid will remain a liquid and no boiling will take place.

The vapor pressure of a liquid is strongly dependent on its temperature. The higher the liquid's temperature, the greater its vapor pressure becomes. Stated in other terms: The greater the temperature of a liquid, the more pressure must be exerted on its surface to prevent it from boiling.

Vapor pressure is stated as an **absolute pressure**, and has the units of psia. On the absolute pressure scale, zero pressure represents a complete vacuum. On earth, the weight of the atmosphere exerts a pressure on any liquid surface open to the air. At sea level, this atmospheric pressure is about 14.7 psi. Stated another way, the absolute pressure exerted by the atmosphere at the earth's surface is approximately 14.7 psia. This implies that vapor pressures lower than 14.7 psia would represent partial vacuum conditions from our perspective in the earth's atmosphere.

Did you ever wonder why water boils at 212 °F? What is special about this number? Actually there is nothing special about it. It just happens that at 212 °F. the vapor pressure of water is 14.7 psia (equal to atmospheric pressure at sea level). Therefore water boils at 212 °F at elevations near sea level. If a pot of water is carried to an elevation of 5,000 ft above sea level and then heated, its vapor pressure will eventually rise until it equals the lower atmospheric pressure, and boiling will begin at about 202 °F. This is why foods cooked in boiling water take longer to cook at high altitudes.

Because water has been widely used in many heating and power production systems, its vapor pressure versus temperature characteristics have been well established for many years. This data is available in the form of steam tables. The vapor pressure of water between 50 °F and 250 °F can also be read from the graph in Figure 4–4.

The relationship shown in Figure 4–4 can also be generated using Equation 4.5 for water temperatures between 50 °F and 250 °F.

(Equation 4.5)

$$P_v = 0.771 - 0.0326(T) + 5.75 \cdot 10^{-4}(T)^2$$

$$- 3.9 \cdot 10^{-6}(T)^3 + 1.59 \cdot 10^{-8}(T)^4$$

for: ($50 \le T \le 250$)

where:

P_v = vapor pressure of water (psia)
T = temperature of the water (°F)

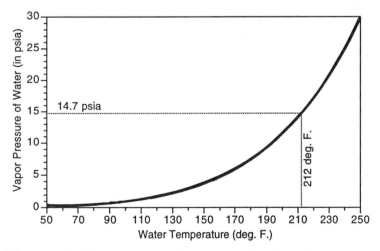

Figure 4–4 Vapor pressure of water as a function of temperature

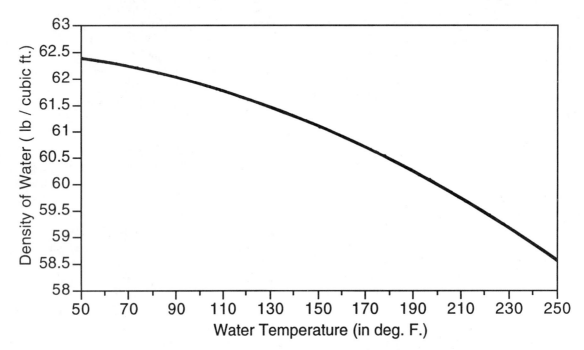

Figure 4–5 The density of water as a function of temperature

A good understanding of vapor pressure is important for any hydronic system designer. This knowledge will later be used to avoid pump cavitation and other problems created when water flashes from a liquid to a vapor. *Hydronic systems should be designed to ensure that the absolute pressure of the water at all points in the system stays safely above the water's vapor pressure at all times.*

4.7 DENSITY

The **density** of a substance is the number of pounds of the substance needed to fill a volume of one cubic foot. For example, it would take about 62.4 pounds of water at 50 °F to fill a one cubic foot container, so its density is said to be 62.4 lb/ft³.

The density of water is dependent on temperature. Like most substances, as the temperature of water increases its density decreases. As temperature increases, each molecule of water requires more space. This expansion effect is extremely powerful, easily bursting even steel pipes if not properly accommodated. A graph showing the relationship between temperature and density of water is shown in Figure 4–5.

This relationship can also be represented by Equation 4.6 for temperatures between 50 °F and 250 °F.

(Equation 4.6)

$$D = 62.56 + 3.413 \cdot 10^{-4}(T) + 6.255 \cdot 10^{-5}(T)^2$$

for: $(50 \leq T \leq 250)$

where:

D = density of water (lb/ft³)
T = water temperature (°F)

The change in density of water or water-based antifreeze solutions directly effects the size of the expansion tank required on all closed-loop hydronic systems. Proper tank sizing requires data on the density of the system fluid both when the system is filled and when it reaches its maximum operating temperature. The Hydronics Design Toolkit contains a program called FLUIDS that can be used to find this data for several water-based fluids over a wide range of temperatures. Detailed methods for applying this data for sizing expansion tanks is presented in Chapter 12.

4.8 VISCOSITY

The **viscosity** of a fluid indicates its resistance to flow. The higher the viscosity of a fluid, the more drag it creates as it flows through pipes, fittings, valves, or any other component in a piping circuit. Higher viscosity fluids also require more pumping power to maintain a given system flow rate as compared to fluids with lower viscosities.

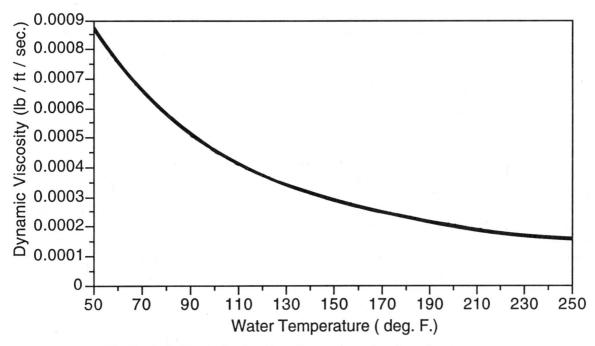

Figure 4-6   Dynamic viscosity of water as a function of temperature

Like all other properties discussed in this chapter, the viscosity of water is dependent on temperature. The viscosity of water and water-based antifreeze solutions decreases as their temperature increases. This implies that a given pump will be able to circulate warm water through a piping system at a slightly higher flow rate compared to cold water. This fact will be demonstrated in Chapter 6. The relationship between the viscosity of water and its temperature is shown in Figure 4-6.

This relationship can also be described by Equation 4.7 for temperatures between 50 °F and 250 °F.

(Equation 4.7)

$$\mu = 0.001834 - 2.73 \cdot 10^{-5}(T) + 1.92 \cdot 10^{-7}(T)^2$$

$$- 6.53 \cdot 10^{-10}(T)^3 + 8.58 \cdot 10^{-13}(T)^4$$

for: $(50 \leq T \leq 250)$

where:

μ = dynamic viscosity of water (lb_{mass}/ft/sec)
T = temperature of the water (°F)

Both Figure 4-6 and Equation 4.7 indicate the *dynamic* viscosity of water as a function of temperature. Another type of viscosity, called *kinematic* viscosity is sometimes discussed in other texts on fluid mechanics. However, it is not used in this book.

The physical units of dynamic viscosity can sometimes be confusing. They are a mathematical consequence of how viscosity is defined, and do not lend themselves to a practical interpretation of what the viscosity number

actually is. The units used for dynamic viscosity within this book are (lb_{mass}/ft/sec). These units are compatible with all equations in this book that require the use of viscosity data. The reader is cautioned, however, that several other units for viscosity may be used in other sources of viscosity data (such as manufacturer's literature). *These other units must first be converted to lb_{mass}/ft/sec) using the appropriate conversion factors, before being used in any equations in this book.*

The addition of glycol-based antifreeze to water greatly increases its viscosity. The greater the concentration of antifreeze, the higher the viscosity of the solution. This increase in viscosity can result in significantly lower flow rates within a hydronic system if it is not accounted for during design. The FLUIDS program in the Hydronics Design Toolkit can also be used to find the dynamic viscosity of water as well as several water-based antifreeze solutions over a wide temperature range.

4.9   DISSOLVED AIR IN WATER

Water has the ability to contain air much like a sponge can contain a liquid. Molecules of the various gases that make up air, including oxygen and nitrogen, can exist *in solution* with water molecules. Even when water appears perfectly clear it can contain air in solution. In other words, *just because there does not appear to be bubbles in the water does not mean air is not present.*

When hydronic systems are first filled with water, **dissolved air** is present in the water throughout the system. If allowed to remain in the system, this air can create numerous problems including noise, cavitation, corrosion, and improper flows in radiators. A well-designed hydronic system must be able to automatically capture dissolved air and expel it from the system. A proper understanding of what affects the air content of water is crucial in designing a method to get rid of it.

The amount of air that can exist in solution with water is strongly dependent on the temperature of the water. As water is heated, its ability to hold air in solution rapidly decreases. However, the opposite is also true. As heated water cools off, it attempts to reabsorb air from any source it can. Heating water to release dissolved air is like squeezing a sponge to expel a liquid. Allowing the water to cool is like letting the sponge expand to soak up more liquid.

The pressure maintained on the water also has a marked effect on its ability to hold dissolved air. When the pressure of the water is lowered, its ability to contain dissolved air decreases, and vice versa.

Figure 4–7 shows the maximum amount of air that can be contained (dissolved) in water as a percent of the water's volume. The effects of both temperature and pressure on dissolved air content can be determined from this graph. To understand this graph, first consider a single curve such as the one designated 45 psi. This curve represents a *constant pressure* condition for the water. Notice that as the water temperature increases, the curve descends rapidly. This descent indicates a decrease in the water's ability to hold air in solution. The exact change in air content can be read off the vertical axis. This same trend holds true for all other constant pressure curves. Now consider what happens if the water temperature remains constant, say at 150 °F, and the pressure is lowered from 90 psi to 15 psi. The lower the pressure the lower the amount of air in solution.

It is important to keep in mind that Figure 4–7 indicates the *maximum possible* dissolved air content of water at a given temperature and pressure. Properly deaerated hydronic systems will quickly reduce the dissolved air content of their water to a small fraction of one percent. Methods of accomplishing this are discussed in Chapter 13.

4.10 INCOMPRESSIBILITY

When a quantity of water is put under pressure there is very little change in its volume. A pressure increase of 1

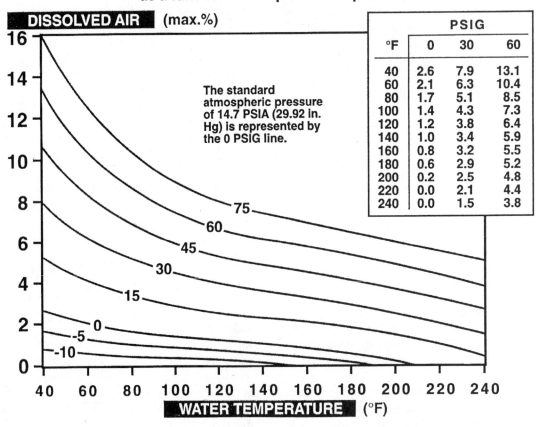

Solubility of Air in Water
as a function of temperature & pressure

DISSOLVED AIR (max.%)

°F	PSIG		
	0	30	60
40	2.6	7.9	13.1
60	2.1	6.3	10.4
80	1.7	5.1	8.5
100	1.4	4.3	7.3
120	1.2	3.8	6.4
140	1.0	3.4	5.9
160	0.8	3.2	5.5
180	0.6	2.9	5.2
200	0.2	2.5	4.8
220	0.0	2.1	4.4
240	0.0	1.5	3.8

The standard atmospheric pressure of 14.7 PSIA (29.92 in. Hg) is represented by the 0 PSIG line.

WATER TEMPERATURE (°F)

Figure 4–7 Solubility of air in water at various temperatures and pressures. Courtesy of Spirotherm Corporation.

psi will cause a volume reduction of only 3.4 *millionths* of the original water volume. This change is so small that it can be ignored in the context of designing hydronic systems. In effect we can treat water, and most other liquids, as being **incompressible**. In practical terms incompressibility means that liquids cannot be "squeezed together" without exerting a tremendously high pressure on them. Such pressure would have to be many times greater than the pressures exerted within hydronic heating systems.

Incompressibility implies that if a liquid's flow rate is known at any one point in a closed piping circuit, it must be the same at all other points. If the piping circuit contains parallel brach circuits, the flow can divide up through them, but the sum of all parallel brach flow rates must equal the total system flow rate. This simple concept, based on the concept of incompressibility, forms the basis for analyzing fluid flow in complex piping systems. It will be applied repeatedly in later chapters of this book.

Incompressibility also implies that liquids can exert tremendous pressure in any type of closed container when they expand due to heating. For example, a closed hydronic piping system *completely* filled with water, and not equipped with an expansion tank or pressure relief valve, would quickly burst apart at its weakest component as the water temperature increased.

SUMMARY

The properties of water presented in this chapter describe its heat absorption, boiling point, expansion, flow resistance, and air holding characteristics. Many of the questions that arise during system design can be answered based on a proper understanding of these fluid properties. Ignoring them can lead to serious problems such as low heat output, improper expansion tank size, and inadequate circulator selection.

The sensible heat quantity and rate equations are two of the most often used equations in hydronic system design. You will see them used frequently in later chapters.

KEY TERMS

Absolute pressure
Density
Dissolved air
Flow rate
Heat capacity
Incompressible
Latent heat
Sensible heat
Specific heat
Temperature drop

Vapor pressure
Viscosity

CHAPTER 4 QUESTIONS AND EXERCISES

Note: Questions and exercises requiring the Hydronics Design Toolkit are indicated with marginal symbol.

1. How many gallons of water would be required in a storage tank to absorb 250,000 Btu while undergoing a temperature change from 120 °F to 180 °F?
2. Water enters a convector at 180 °F and at a flow rate of 2.0 gpm. Assuming the convector releases heat into the room at a rate of 20,000 Btu/hr, what is the temperature of the water leaving the convector?
3. How much pressure must be maintained on water to maintain it as a liquid at 250 °F? State your answer in psia.
4. If you wanted to make water boil at 100 °F, how many psi of vacuum (below atmospheric pressure) would be required? *Hint:* psi vacuum = 14.7 − absolute pressure in psia.
5. Determine the air flow rate (in cubic feet per minute) necessary to transport 40,000 Btu/hr based on a temperature increase from 65 °F to 135 °F. Base the calculation on a specific heat for air of 0.245 Btu/lb/°F and an assumed density of 0.07 lb/ft³.
6. Assume you want to design a hydronic system that will deliver 80,000 Btu/hr. What flow rate of water is required if the temperature drop of the distribution system is to be 10 °F? What flow rate is required if the temperature drop is to be 20 °F?
7. Water flows through a series-connected baseboard system as shown in Figure 4–8. The inlet temperature to the first baseboard is 165 °F. Determine the outlet temperature of the first baseboard, and both the inlet and outlet temperatures for the second and third baseboards. Assume the heat output from the interconnecting piping is insignificant. *Hint:* Use the approximation that the inlet temperature for a baseboard equals the outlet temperature from the previous baseboard.

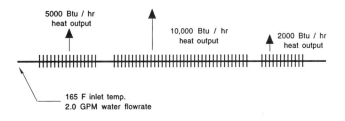

Figure 4–8 Series-connected baseboard convectors for Exercise 7

8. At 50 °F, 1,000 lbs of water occupies a volume of 16.026 ft³. If this same 1,000 lbs of water is heated to 200 °F, what volume will it occupy?

9. How much heat is required to raise the temperature of 10 lbs of ice from 20 °F to 32 °F? How much heat is then required to change the 10 lbs of ice into 10 lbs of liquid water at 32 °F? Finally, how much heat is needed to raise the temperature of this water from 32 °F to 150 °F? State all answers in Btus.

10. You are designing a thermal storage system that must supply an average load of 30,000 Btu/hr for 18 consecutive hours. The minimum water temperature usable by the system is 110 °F. Determine the temperature the water must be heated to by the end of the charging period if:
    a. the tank volume is 500 gallons.
    b. the tank volume is 1,000 gallons.

11. A swimming pool contains 18,000 gallons of water. A boiler with an output of 50,000 Btu/hr is operated eight hours per day to heat the pool. Assuming all the heat is absorbed by the pool water, with no losses to the ground or air, how much higher is the temperature of the pool after the eight hours of heating?

12. An 18,000 gallon swimming pool is to be heated from 70 °F to 85 °F. Assume an oil-fired boiler with an average efficiency of 80% will be used and that all heat is retained by the pool, with no losses to the ground or air. Determine:
    a. the number of Btus required to heat the pool.
    b. the number of gallons of #2 oil required by the boiler.
    c. at a cost of $1 per gallon of oil, the cost of bringing the pool temperature up.

13. A family of four uses 60 gallons of domestic hot water per day. The water enters the water heater at 50 °F and is heated to 140 °F by the time it leaves. Calculate the number of Btus per day used for this process.

14. Recalculate the Btus used for domestic water heating in Question 13 assuming the water is heated to only 120 °F. Estimate the annual energy savings of the reduced water temperature setting assuming the same load exists 365 days per year. Express the answer in MMBtus. (1 MMBtu = 1,000,000 Btu)

15. Assume a hydronic system is filled with 50 gallons of water at 50 °F and 30 psi pressure. The water contains the maximum amount of dissolved air it can hold at these conditions. The water is then heated and maintained at 200 °F and 30 psi pressure. Using Figure 4–7, estimate how much air will be expelled from the water.

16. Use the FLUIDS routine in the Hydronics Design Toolkit to determine the specific heat of a 50% ethylene glycol solution at 100 °F. Also find the specific heat of water at 100 °F. Assume that a flowing stream of each liquid had to deliver 40,000 Btu/hr of heat to a load with a 15 degree drop in temperature. What flow rate would be required for each fluid? Express your answers in pounds per hour.

17. Use the FLUIDS routine in the Hydronics Design Toolkit to determine the specific heat and density of a 30% ethylene glycol solution at 132.5 °F. Assume this fluid entered a radiator at 140 °F and exited the radiator at 125 °F, at a flow rate of 4.5 gpm. What is the rate of heat transfer from the fluid stream to the radiator? Express your answer in Btu/hr. Compare the results to those of Example 4.2. Explain the difference. *Note:* 1 ft³ = 7.49 gallons.

PIPING, FITTINGS, AND VALVES 5

OBJECTIVES

After studying this chapter you should be able to:

- Describe several types of piping materials used in hydronic systems
- Compare the temperature and pressure ratings of several piping materials
- Describe several types of piping support devices
- Describe the steps involved in soldering copper tube
- Discuss the issue of oxygen diffusion through plastic pipe
- Identify several common pipe fittings
- Specify the exact type of pipe fitting needed for a given situation
- Explain the use of several specialized pipe fittings
- Calculate the expansion movement of copper pipe
- Select the correct type of valve for a given application
- Identify several specialized valves and explain their use in hydronic systems
- Plan neat and efficient piping installations
- Recognize schematic symbols for piping system components

5.1 INTRODUCTION

This chapter presents basic information on piping, **fittings**, and valves used in residential and light commercial hydronic systems. It emphasizes the proper selection and subsequent installation of these components. The intent is to acquaint the reader with the wide range of available piping hardware, in preparation for overall system design in later chapters.

Much of the information deals with the benchmark hydronic piping material, copper tubing. Information on common and specialized copper fittings is presented. A short primer on soft soldering is also given.

Newer piping materials such as **cross-linked polyethylene** and **polybutylene tubing** are discussed. These materials have been successfully used for many years in Europe, and are gaining market share in the U.S. They open up a number of possibilities often unmatched by traditional materials.

This chapter also discusses the construction and intended application of several types of common valves. Several specialized valves designed specifically for hydronic systems are also covered.

Tips for professional installation of piping are also given. Small details such as level and plumb piping runs, wiped solder joints, and use of specialty fittings to reduce parts count, distinguish systems that merely work from those that convey professionalism and craftsmanship.

5.2 PIPING MATERIALS

This section compares several piping materials that are suitable for use in modern hydronic heating systems. The discussion is limited to those materials that are most practical in today's market. These include copper tube and several **polymer** materials. All the piping materials discussed have strengths and weaknesses. All have conditions attached to their use that must be followed if the system is to perform as expected and last as long as possible. No single material is ideal for all applications.

Piping material(s) must be selected based on considerations such as temperature and pressure ratings, availability, ease of installation, corrosion resistance, life expectancy, local code acceptance, and cost. The designer should not feel constrained to use a single type of pipe for the entire system, although this is certainly possible and quite often done. Rather, a given piping material should be applied in situations that best take advantage of its ability to overcome obstacles posed by other materials. For example, flexible polyethylene tubing may be ideal in retrofit situations where full access to framing cavities or other concealed building spaces may be impossible. Rigid copper tubing, however, has obvious advantages in areas where piping must be straight, or where components will be supported by the pipe. A combination of these materials may provide the ideal solution for a given installation.

Discussion will be limited to pipe sizes from 1/2-inch to 2-inch diameter. Nearly all the hydronic systems within the scope of this book can be constructed of piping within this size range. All analytic methods for piping design to be presented in later chapters, as well as in the Hydronics Design Toolkit software, also cover this range of pipe sizes.

Copper Tubing

Copper water tube was developed in the 1920s to provide an alternative to iron piping in a variety of uses. Its desirable features include:

- Good pressure and temperature rating for typical hydronic applications
- Good resistance to corrosion from water-based system fluids
- Smooth inner walls that offer low flow resistance
- Lighter than steel or iron piping of equivalent size
- Able to be joined by the well-known technique of soft soldering

In situations where good heat transfer through the walls of the pipe is desirable, copper is unsurpassed. Examples of such applications include heat exchangers or finned-tube convectors. The high thermal conductivity of copper can, however, be a disadvantage in situations where heat loss from piping has to be minimized. For example, an uninsulated copper tube carrying hot water through an unheated space will lose heat considerably faster than a plastic pipe in the same situation.

Copper tubing is available in a wide range of sizes and wall thicknesses. The type of tubing used in hydronic heating is called copper water tube. In the U.S., copper water tube is manufactured according to the ASTM B88 standard. In this category, **pipe size** refers to the **nominal inside diameter** of the tube. Nominal inside diameter means the actual inside diameter is approximately equal to the stated pipe size. The outside diameter of copper water tubing is always exactly ¹/₈ inch larger than the nominal inside diameter. Another category of copper tubing, designated as ACR tubing, is used in refrigeration and air conditioning systems. In this category, pipe size refers to the *outside* diameter of the tubing. ACR tubing is not used in hydronic heating or water distribution systems.

Copper water tube is available in three wall thicknesses designated as types K, L, and M in order of decreasing wall thickness. *The outside diameters of any given size of K, L, and M tubing are identical.* The inside diameter decreases as wall thickness increases. This allows all three types of tubing to match with the same fittings and valves.

Because the operating pressures of residential and light commercial hydronic heating systems are relatively low, the thinnest tube, type M, is most often used. This wall thickness provides several times the pressure rating of other common hydronic system components, as can be seen in Figure 5–1. In the absence of any local codes that require otherwise, type M copper water tube is the standard choice for hydronic heating systems.

Two common hardness grades of copper water tube are available. **Hard drawn tubing** is supplied in straight lengths of 10 and 20 feet. Because of its straightness and strength, hard drawn tubing is the most commonly used type of copper tubing for hydronic systems. So-called **soft temper tubing** is annealed during manufacturing to allow it to be formed with simple bending tools. It is useful in situations where awkward angles do not allow proper tubing alignment with standard fittings. Soft temper tubing comes in flat coils having standard lengths of 60 and 100 feet. The minimum wall thickness available in soft temper copper tubing is type L.

Supporting Copper Tubing

Copper tubing must be properly supported to prevent sagging or buckling. On horizontal runs of hard temper tubing, the following *maximum* support spacing is suggested:

- ¹/₂ in and ³/₄ in tube: 5 ft maximum
- 1 in and 1¹/₄ in tube: 6 ft maximum
- 1¹/₂ in and 2 in tube: 8 ft maximum

These distances do not include any allowance for the extra weight of piping components such as circulators,

Nominal Size	Inside Diameter	Outside Diameter	Rated Working Pressure*
1/2"	0.569"	0.625"	760 / 410 psi
3/4"	0.811"	0.875"	610 / 325 psi
1"	1.055"	1.125"	515 / 275 psi
1-1/4"	1.291"	1.375"	515 / 275 psi
1-1/2"	1.527"	1.625"	510 / 275 psi
2"	2.009"	2.125"	450 / 240 psi

* The rated working pressure is for fluid temperatures up to 200 degrees F. Above 200 deg. F. the working pressure decreases slightly. The first number is for drawn (straight length) tubing. The second number is for annealed type L tubing.

Figure 5–1 Physical data for selected sizes of type M copper water tube. Data provided by the Copper Development Association.

Figure 5–2   Examples of plastic-coated metal piping supports

expansion tanks, etc., that may be supported by the tubing between supports. In such cases the piping should be supported immediately adjacent to the component. On vertical runs, copper tubing should be supported at each floor level, or a maximum of every 10 feet.

A number of different supports are available for small tubing. In residential systems, horizontal piping runs are often supported by plastic coated hangers as shown in Figure 5–2. These simple devices allow the tubing to move slightly as it expands upon heating. The plastic coating reduces squeaking sounds as the tube moves through the hanger. Some pipe hangers have pointed ends that are driven in floor joists. Others are screwed into pilot holes. Still others have a perforated straps that can be nailed to framing.

Stand-off type piping supports as shown in Figure 5–3 are designed to be fastened to a solid surface. When tightened, they rigidly clamp the tubing in place. This is acceptable for relatively short runs, but can result in expansion noises on long runs. Stand-off supports can be used to support tubing running at any angle. They also allow sufficient space between the tube and mounting surface to accommodate most pipe insulations.

Copper tubing can also be supported at intermediate heights using clevis hangers held by threaded steel rods attached to the ceiling structure. Clevis hangers can be ideal when piping has to be supported away from wall or ceiling surfaces. An example of a clevis hanger is shown in Figure 5–4.

Figure 5–3   Example of a stand-off pipe support

Figure 5–4   Example of a clevis hanger

In areas where piping is closely spaced, or where several runs of piping and/or electrical conduit are in close proximity, a mounting rail system provides good adaptability. It consists of a steel mounting rail to which clamps for various sizes of pipe can be secured. Figure 5–5 shows the versatility of a mounting rail system in accommodating several pipe sizes.

Still another piping support system uses a specialized spring-operated gun to place high density polyethylene support clips over smaller diameter ($\frac{1}{2}$-inch and $\frac{3}{4}$-inch) tubing as shown in Figure 5–6. The clips hold the piping away from the framing to allow for insulation. They also absorb expansion movement. However, the polyethylene clips are rated for a maximum operating temperature of 180 °F, which may preclude their use in high temperature hydronic systems.

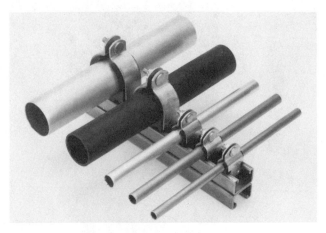

Figure 5–5   Example of a rail-type pipe mounting system. Courtesy of Hydra-Zorb Corporation.

Figure 5-6 Piping being installed with gun-driven polyethylene clips. Courtesy of Peter Mangone, Inc.

Soldering Copper Tubing

The common method of joining copper tubing in hydronic heating systems is soft **soldering** using a 50/50, tin/lead solder. The **working pressure** rating of the resulting joints is dependent on operating temperature and tube size, and can be found in Figure 5–7. These pressure ratings are well above the relief valve settings of nearly all residential and light commercial hydronic systems.

There are cases where 50/50 tin/lead solder is not suitable. The most notable of these is any piping carrying **domestic water**. *Plumbing codes no longer allow solders containing lead to be used for domestic water service.* The usual alternative is 95/5 tin/antimony solder. It contains

no lead and has higher pressure ratings than 50/50. The melting point temperature range of 95/5 solder is 452 °F to 464 °F. This is significantly higher and narrower than the 361 °F to 421 °F range of 50/50 tin/lead solder. Joints made with 95/5 require a longer heating time before the solder will flow. The narrow melting range of 95/5 also means the joint will solidify very quickly when heat is removed. In general, 95/5 solder is more difficult to work with than 50/50 tin/lead, and is often used only for domestic water piping.

Soldering Procedure

Proper soldering results in clean, neat, and water-tight joints. The attention given to such joints is often reflective of the overall professional skills of the installer. *Sloppy joints, even when water-tight, indicate poor craftsmanship.* With a little practice, making good soldered joints becomes second nature.

The following procedure describes the steps in making a good soldered joint:

Step 1. Be sure the tube is cut square. Use a wheel cutter or chop saw with a suitable blade. Avoid the use of a hacksaw whenever possible. See Figure 5–8.

Step 2. To remove any burrs, ream the end of the tube with the pointed blade of the wheel cutter, rounded file, or other type of deburring tool. See Figure 5–9.

Step 3. Assemble the joint and check it for proper fit and alignment. In most cases there will be no problem with the fit. However, sometimes a fitting may be damaged or defective. If the fitting wobbles noticeably on the pipe it should be replaced. Attempting to solder such a joint can result in a leak or poorly aligned joint. See Figure 5–10.

Step 4. Clean the socket of the fitting with a properly sized fitting brush. *Be sure to remove any pieces of the*

Solder	Service Temperature	Max. Allowable Water Pressure for (1/4" to 1" tube sizes)	Max. Allowable Water Pressure for (1-1/4" to 2" tube sizes)
50 / 50 Tin / Lead	100 F 150 F 200 F 250 F	200 psi 150 psi 100 psi 85 psi	175 psi 125 psi 90 psi 75 psi
95 / 5 Tin / Antimony	100 F 150 F 200 F 250 F	500 psi 400 psi 300 psi 200 psi	400 psi 350 psi 250 psi 175 psi

Figure 5-7 Pressure ratings versus operating temperature for soldered joints in copper tubing. Data provided by the Copper Development Association.

Figure 5–8 Getting a square cut of copper tubing. Courtesy of the Copper Development Association.

steel brush bristles remaining in the fitting after cleaning it. These small pieces of steel can create **galvanic corrosion** leading to failure of the joint. See Figure 5–11.

Step 5. Clean the outside of the tube with emery cloth or a power cleaning brush. All oxidation, scale, paint, or dirt should be removed from the tubing surface at least ½ inch farther back than the edge of the fitting socket. The tube should appear bright following cleaning. Be sure to clean all the way around the perimeter of the tube. See Figure 5–12.

Step 6. Apply paste **flux** to both the fitting socket and the portion of the tube that projects inside the socket. The flux chemically cleans the surface of the copper and helps prevent oxidation of the surfaces when heat is applied. *Always use a flux brush, not your finger, to apply the flux.* Do not apply excessive amounts of flux. Only a thin film is needed. After fluxing, slide the tube into the fitting and,

Figure 5–9 Deburring the end of the tube. Courtesy of the Copper Development Association.

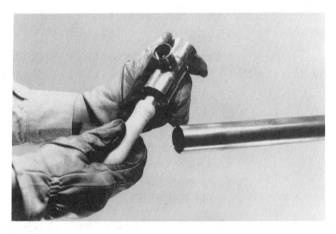

Figure 5–11 Cleaning the socket of the fitting with a steel fitting brush. Courtesy of the Copper Development Association.

Figure 5–10 Checking the fit of the fitting and tubing. Courtesy of the Copper Development Association.

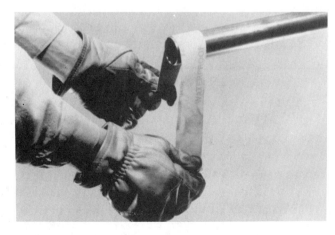

Figure 5–12 Cleaning the outside of the tubing with emery cloth. Courtesy of the Copper Development Association.

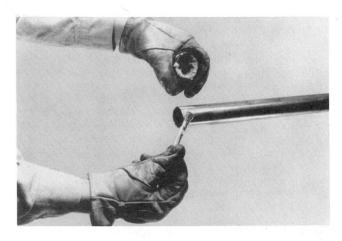

Figure 5-13 Applying paste flux to the outside of the tube. Courtesy of the Copper Development Association.

Figure 5-14 Applying paste flux to the inside of the fitting socket. Courtesy of the Copper Development Association.

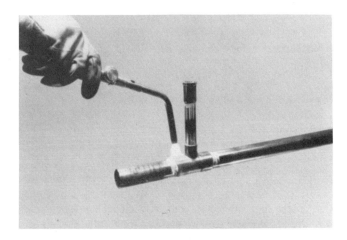

Figure 5-15 Heating the fitting. Courtesy of the Copper Development Association.

Figure 5-16 Testing the joint for solder melting temperature. Courtesy of the Copper Development Association.

whenever possible, rotate the fitting once or twice to further spread the flux. Finally, use a rag to remove any access flux from the surfaces adjacent to the joint. See Figures 5-13 and 5-14.

Step 7. Apply heat to the outside of the fitting socket using the torch. Keep the blue tip of the flame just above the surface of the fitting as shown in Figure 5-15. Move the flame around the outside of the fitting socket to promote even heating. Heating times differ considerably with the type of torch and gas being used, as well as pipe size, ambient temperature, and type of solder. When the flux begins to sizzle, test the joint by applying the end of the solder wire to the edge of the joint as shown in Figure 5-16. If it sticks but doesn't melt, additional heating is required. If the solder immediately melts, the joint is ready to draw in the molten solder by capillary action.

Once the solder begins to flow, it can be continually fed into the joint. Only a small amount of solder is needed. An experienced plumber knows by sight how much solder to feed into the joint. An inexperienced person often feeds excessive solder into the joint. The excess solder ends up as small loose balls inside the pipe. These can be carried around the system by fast moving fluid, becoming lodged in valves or circulators, causing malfunctions. The table in Figure 5-17 lists the approximate length of 1/8-inch solder wire required for joints in tube sizes from 1/2-inch to 2 inches.

A properly heated joint will quickly spread the molten solder around its perimeter. When the solder forms a narrow silver ring around the visible edge of the joint, adequate solder has been applied. If the solder begins to drip off the joint, excess solder is being used.

Step 8. After the solder is applied, remove the torch and carefully wipe the perimeter of the joint with a clean cloth rag to remove any excess solder or flux as shown in Figure 5-18. Be careful! All surfaces are still very hot and the solder being wiped is obviously still molten.

Tube size	Approx. length of 1/8" wire solder required per joint
1/2"	0.4"
3/4"	0.9"
1"	1.3"
1-1/4"	1.7"
1-1/2"	2.3"
2"	3.7"

Figure 5–17 Approximate length of ⅛-inch solder wire required for an average joint

Figure 5–19 Use of a soldering safety pad to prevent charred surfaces. Courtesy of the Mill Rose Company.

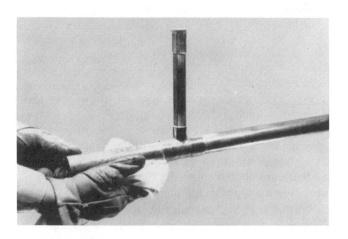

Figure 5–18 Wiping the hot joint. Courtesy of the Copper Development Association.

Always wear gloves, long-sleeved shirts, and safety goggles when soldering.

The joint should be allowed to cool naturally before being handled or stressed. Dipping the assembly in water for rapid cooling can cause high stresses in the metal and should be avoided. Cast fittings have been known to crack under such conditions. After the joint has cooled, a final wiping with a clean cloth will remove any remaining flux, preventing surface oxidation of the joint in the future.

More Tips on Soldering

- Experience shows it is best to make up piping subassemblies on a horizontal working surface whenever possible. The flow of solder is more easily controlled, and the work proceeds faster because it is done from a more comfortable position. After cooling, these piping assemblies can be joined into the overall system.
- When soldering has to take place next to combustible materials, slide a piece of thin sheet metal or a specially made soldering safety pad, as shown in Figure

5–19, between the joint and the combustible materials. This prevents charring wood next to the joint. *Charred wood next to piping joints doesn't make a big impression on homeowners.*

- Measure all tubing carefully before cutting. Allow for the take-up of the pipe into the fitting sockets at both ends. Figure 5–20 indicates the take-up allowances for copper and brass fittings in sizes from ½-inch through 2 inches.
- Whenever possible attempt to keep all piping plumb (vertical) or level (horizontal). Even a person with no plumbing background will notice the difference between plumb or level piping and pipes that simply connect point A to point B without regard to appearance. Install temporary piping supports if necessary during piping assembly to hold the pipe in proper alignment for soldering.
- *When soldering valves containing synthetic seals or washers, open the valve to hold the synthetic components away from metal surfaces. Even better, partially disassemble the valve so that only the metal body has to be heated.* This is especially true for zone valves that may not be designed to handle the heat of soldering. *Check the installation instructions before attempting to solder any valve containing nonmetal parts.*
- On fittings having both soldered and threaded connections, make up the soldered joint first. This prevents discoloration or burning of teflon tape or joint sealing compound during soldering.
- *Always keep one end of the piping assembly open to atmosphere during soldering.* If this is not done, the air trapped inside the pipe will undergo a pressure increase due to heating, and blow pinholes through the molten solder before it can solidify. When a piping loop is finally closed in, open a valve or other component near the final joint to allow the heated air to escape.

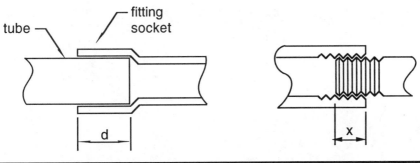

Tube size	Actual and approximate take-up distance d of solder-type fitting sockets		Typical thread engagement distance (x) for NPT pipe threads
	actual (d)	approximate (d)	
1/2"	0.5"	1/2"	1/2"
3/4"	0.75"	3/4"	1/2"
1"	0.91"	15/16"	9/16"
1-1/4"	0.97"	1"	5/8"
1-1/2"	1.09"	1-1/16"	5/8"
2"	1.34"	1-5/16"	11/16"

Figure 5–20   Typical take-up allowances for copper and brass fittings

- Before making up the joint, put a sharp bend in the wire solder at a distance in from the end corresponding to the lengths given in Figure 5–17. This provides a visual indication of how much solder has been fed to the joint.
- Finally, remember that copper is a premium architectural material due to its luster and durability. Take a few extra minutes to clean up any excess flux or shine up oxidized tubing or components. *Your customers will notice it.*

Polymer Piping Materials

Over the last three decades, several "plastic" piping materials have been developed as alternatives to metal piping in a variety of applications. Most of these materials consist of very long molecular chains of organic compounds derived from crude oil. They are technically referred to as polymers or polyolefins. Two of the most common polymer piping materials are polyethylene and polybutylene.

PEX Tubing

Cross-linked polyethylene tubing, also referred to as **PEX tubing**, is a product developed, refined, and extensively used in Europe over the last 25 years. It has proven itself as a reliable alternative to metal piping in a many hydronic heating and domestic water distribution systems. The cross-linking process arranges the polyethylene molecules into a three-dimensional network, increasing the pressure/temperature rating of the polyethylene. Several manufacturers currently offer PEX tubing rated for continuous operation at the following temperature/pressure combinations: 180 °F at 100 psi and 200 °F at 80 psi. These ratings allow PEX tubing to be used in many areas of hydronic heating. Any PEX tubing for use in hydronic heating applications should carry the ASTM F876 designation, proving it conforms to these temperature/pressure ratings.

PEX tubing is sold in continuous coils ranging from about 200 to more than 2,000 feet. The tubing is available in several diameters suitable for small- and medium-sized hydronic heating systems. See Figure 5–21. The most common nominal tube sizes are from 3/8 inch to 1 inch. Since most PEX tubing was originally manufactured, and in some cases still is manufactured in Europe, its diameter is often expressed in millimeters (mm). The following table correlates these metric sizes with their *approximate* counterpart in nominal U.S. pipe sizes:

8 mm ID ≈ 3/8 in ID
12 mm ID ≈ 1/2 in ID
16 mm ID ≈ 5/8 in ID
20 mm ID ≈ 3/4 in ID

The availability of long continuous coils, combined with the ability to bend around moderate curves without kinking, make PEX tubing ideal for use in radiant floor heating. It also allows the tubing to run through confined or concealed spaces in buildings where working with copper tubing would be difficult or impossible.

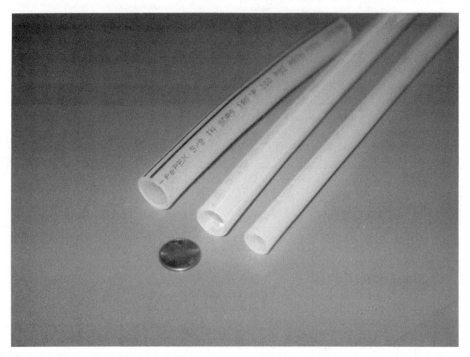

Figure 5-21   Examples of PEX tubing

Specialized fittings are available from the tubing manufacturers to transition from PEX tubing to a standard NPT pipe threads. An example of such a fitting is shown in Figure 5–22. These fittings allow PEX tubing to be directly connected to heat emitters, or other system components.

With reasonable care, PEX tubing is quite resistant to kinking during installation. If a kink does occur, it can be easily repaired in the field. The process requires the kinked area to be heated to approximately 275 °F using a hot air gun. At this temperature the crystalline structure of the polyethylene changes to an amorphous state, relieving the stresses created by the kink. The material resumes its normal shape. After cooling there is virtually no indication a kink ever existed. Because of this behavior, the material is said to have a "shape memory." Figure 5–23 shows a kink being removed from PEX tubing using this procedure.

Polybutylene Tubing

Another polymer material used for making tubing is polybutylene, or PB. Polybutylene tubing shares many of the same features as polyethylene tubing. It comes in long continuous coils and is quite flexible. It is also available in straight lengths up to 20 feet long that are useful when tubing runs must remain plumb or level for appearance. It has been successfully used for hot and cold water distribution, earth heat exchangers, and hydronic radiant floor heating applications.

Grades marked with the designation ASTM D-3309 are rated for continuous service at 180 °F at 100 psi pressure. Other grades are available that do not have this designation and should not be used for hydronic heating applications.

PB tubing can be joined with polybutylene fittings such as elbows and tees using either mechanical compression joints or a process called **socket fusion**. In the latter, the tube and fitting are heated to approximately 500 °F and then pushed together in their semi-molten state. The resulting joint is literally a single piece of polybutylene,

Figure 5-22   Brass fitting for transitioning from PEX tubing to standard NPT threads

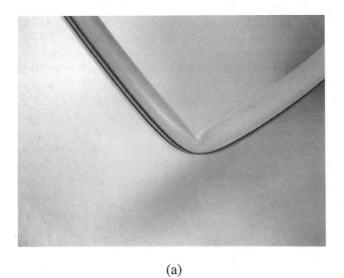

(a)

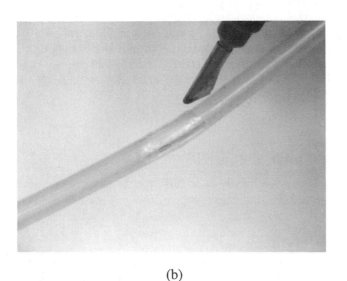

(b)

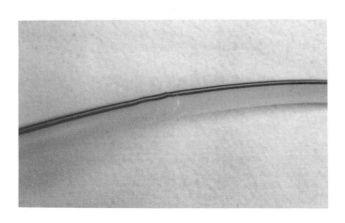

(c)

Figure 5–23 Procedure for removing a kink from PEX tubing (a) the kinked area, (b) kinked area heated using a hot air gun, (c) cooled tubing with kink removed

Figure 5–24 Closeup of a cross-section of a socket fusion joint between polybutylene tubing and a polybutylene fitting

and is stronger than the pipe itself. Once made, these fusion joints cannot be taken apart. A cross-section of a joint made using socket-fused polybutylene is shown in Figure 5–24.

Because polybutylene is not cross-linked, it cannot match the higher temperature and pressure ratings of PEX tubing with the ASTM F876 designation. This is not necessarily a problem, however, especially with low to medium temperature systems. PB tubing can be accidentally kinked by improper handling, and cannot be repaired by the heating process described for PEX tubing. Severe kinks could eventually create a leak, and have to be repaired either with mechanical couplings or by socket fusion.

Rubber-based Tubing

Another type of tubing currently available for hydronic heating applications is composed of various layers of rubber-based compounds and reinforcing mesh. Some varieties have temperature/pressure ratings of 200 °F at 100 psi. This type of tubing has been primarily used in hydronic radiant floor heating applications.

A major difference between rubber-based tubing and PEX or PB tubing is its flexibility. Most rubber-based tubing can literally be tied into a knot and still not collapse. However, this means the tubing must be properly supported if spanning a distance of more than a few inches. Most rubber-based tubing is joined using a combination of a barbed insert fitting combined with a mechanical compression clamp.

Oxygen Diffusion

Non-metallic pipes such as those made of PEX, polybutylene, or rubber compounds allow oxygen molecules from the surrounding air to slowly **diffuse** through their tube walls and be absorbed by the fluid within. This can occur even if the tubing is cast into a concrete slab. *Any time oxygen enters a hydronic system containing iron or steel components there is potential for corrosion.* It is therefore important to minimize or eliminate its entry whenever possible. Most manufacturers of polymer or rubber-based tubing intended for hydronic heating applications incorporate an **oxygen diffusion barrier** into their pipe. This barrier consists of one or more layers of a special compound that greatly reduces the ability of oxygen to pass through the tubing walls.

The necessity of an oxygen diffusion barrier on non-metallic piping has proven to be a controversial issue among several competing tubing manufacturers. Some insist it is necessary, others downplay its significance. In such situations, marketing issues make it difficult to separate speculation from fact. However, two important observations can be made. First, nearly every manufacturer of PEX, polybutylene, or rubber-based tubing currently offers tubing with an oxygen diffusion barrier, either standard or as an option. Second, there are a number of ways other than diffusion through which oxygen can enter a hydronic system. These include fresh water flowing into the system to compensate for minor leaks, misplaced air vents, and improper sizing or placement of expansion tanks. These other forms of oxygen entry can potentially be more of a corrosion threat than oxygen diffusion through non-metallic tubing. In other words, *the use of tubing with an oxygen diffusion barrier does not guarantee that corrosion won't occur.* It is reasonable to assume that, with all other factors being equal, tubing with an oxygen barrier does provide additional protection against corrosion. The slight (if any) additional cost of this protection is minor in comparison to the costs associated with repairing corrosion damage.

5.3 COMMON PIPE FITTINGS

Most of the fittings used in hydronic heating systems are the same as used in conventional domestic water supply systems. These would include:

- Couplings (standard and reducing)
- Elbows (90° and 45°)
- Street elbows (90° and 45°)
- Tees (standard and reducing)
- Threaded adapters (male and female)
- Unions

A cross section of most of these fittings is shown in Figure 5–25.

Fittings for copper tubing are available in either wrought copper or cast brass. Both are suitable for use in hydronic systems. During soldering, wrought copper fittings will heat up in less time than cast brass fittings. They also usually have smoother internal surfaces. Of the two, wrought fittings are more typical in the smaller pipe sizes used in residential and light commercial systems.

Fittings are manufactured with solder-type socket ends, threaded ends, or a combination of the two. Standard sockets for solder-type joints create a gap of between 0.002 inch and 0.005 inch between the outside of the tube and the inside of the fitting. This gap allows solder to flow between the tube and fitting by capillary action. The standard designation for solder-type sockets is the letter C. For example, a fitting designated as 3/4" C × C indicates both ends of the fitting have solder-type sockets to match 3/4-inch copper tubing. All fitting manufacturers in the U.S. use standardized dimensions and tolerances in constructing fittings to ensure compatibility with tubing.

In the U.S., threaded pipe fittings use standardized **national pipe threads** (NPT). External threads are designated as **male threads** (MPT or M). Internal threads are designated as **female threads** (FPT or F).

Specifying Fittings

There are hundreds of different fittings available for use with small diameter copper tubing. A standard method of describing fittings is needed to ensure that the designer, installer, and supplier all know exactly what type of fitting is being specified. Unfortunately, all people involved do not always adhere to a single standard. The following minimum information should always be given in describing a fitting:

- Name of fitting (i.e., elbow, tee, union)
- The nominal pipe size(s) of the fitting
- A designation for the type of connections (i.e., solder-type = C , male pipe threads = M, female pipe threads = F)

In the usual case where all connections on the fitting are the same pipe size, it is only necessary to specify this size once. When a two port fitting such as a coupling or bushing needs to have two different pipe sizes, *always specify the larger size first*, then an "×", followed by the small size. For example, specify a 1" × 3/4" reducer coupling, not a 3/4" × 1" coupling. *In the case of a tee, specify the larger of the two end ports, then the size of the other end port, and finally the size of the side port.* The designations C, M, and F are included with the nominal size

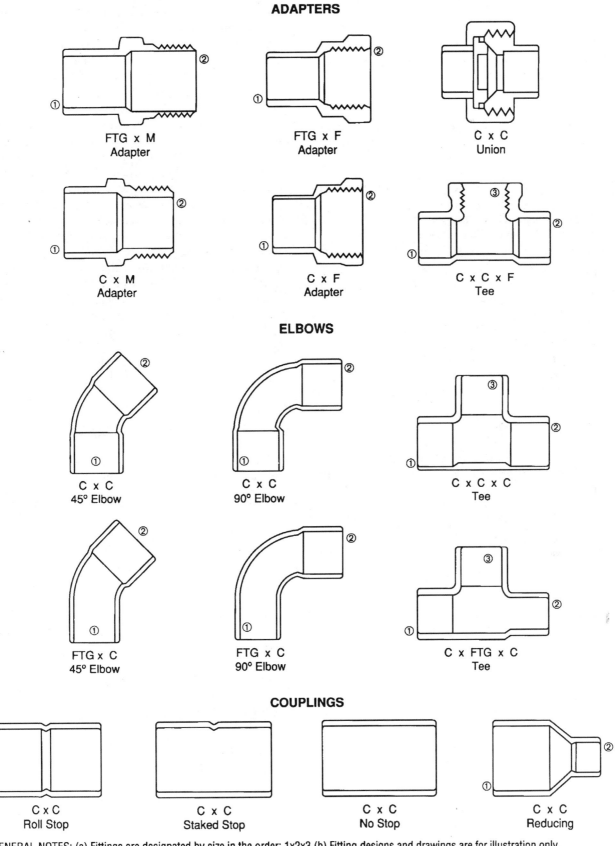

Figure 5-25 Cross-section of several common fittings used with copper tubing. Courtesy of the Copper Development Association.

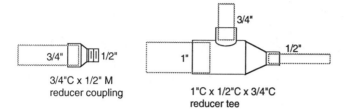

Figure 5-26 Example of proper specification of two reducer fittings

information. An example of the proper specification of two fittings is shown in Figure 5–26.

Although hundreds of combinations of fitting sizes and end connections are manufactured, not every imaginable combination can be obtained. As a general rule a single reducing fitting will not vary by more than two pipe sizes. Thus a 1" × ½" reducer coupling is available, but a 2" × ½" reducer coupling would likely not be, due to very limited applications. It is also very unlikely that all the fitting combinations manufactured will be stocked by a material supplier, due to inventory constraints. With experience, installers learn which specialty fittings are often used. They often place orders for box quantities of these specialty fittings to prevent job delays.

Tips on Using Fittings

A well-planned piping assembly will use the least number of fittings possible to minimize both labor and

materials. Such assemblies also give the job a more professional appearance.

The designer should have a knowledge of some specialized fittings in order to efficiently plan a piping layout. For example, the are several ways to attach a pressure gauge with a ¼-inch MPT threaded stem to a ¾-inch copper pipe. One approach uses a standard tee, a reducer coupling off the side of the tee, and a female adapter at the end of the reducer tee. Two short lengths of copper tubing are also required to hold these fittings together. In total, this assembly requires a total of six soldered joints and one threaded joint.

A better approach is to use a special tee designated as ¾"C × ¾"C × ¼"F. This tee has ¼-inch female pipe threads on its side port. The pressure gauge can be directly threaded into the side port. A total of two soldered joints and one threaded joint would be required. The slightly higher cost of this fitting is more than compensated for by the savings in labor and additional fittings. A comparison of these assemblies is shown in Figure 5–27. Similar situations occur in mounting devices such as expansion tanks, relief valves, and drain valves. Again, a small amount of planning yields a better looking job, often at a lower cost.

5.4 SPECIALIZED FITTINGS

Several specialized fittings have been developed for use in hydronic heating systems. These include the

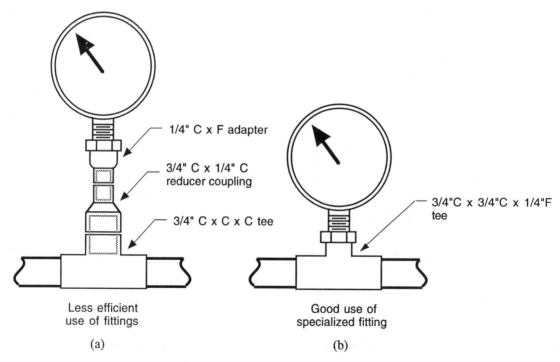

Figure 5-27 Comparison of fitting options for mounting a pressure gauge to a pipe (a) less efficient use of fittings, (b) good use of a specialty fitting

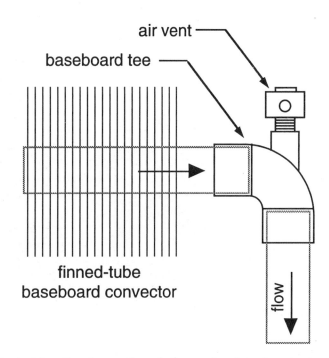

Figure 5–28   A baseboard tee (a) shown with air vent, (b) typical location for venting air from a baseboard convector

baseboard tee, diverter tee, and dielectric union. Each are worthy of individual discussion.

Baseboard Tees

A **baseboard tee** as shown in Figure 5-28 is commonly used to mount an air vent in locations where air could accumulate within hydronic systems. The fitting closely resembles a standard 90 degree elbow, but has an additional $1/8$-inch FPT. port intended for the air vent. Baseboard tees are available in sizes from $1/2$ inch to $1\,1/4$ inches in both wrought copper and cast brass, the latter being more common. The fitting's name is derived from its frequent use on the outlet of finned-tube baseboard convectors. These fittings can also be used to mount a threaded temperature sensor.

Diverter Tees

A **diverter tee** is designed to create flow through a branch piping path that passes through one or more heat emitters before rejoining the main piping circuit. It does so by either forcing a portion of the fluid entering its end port, out through the branch piping path (when mounted at the supply of the branch path), or by pulling fluid through the branch path by the suction effect it creates (when located at the return of the branch path). This is illustrated in Figure 5–29.

The outside of a diverter tee appears very similar to a reducing tee, as shown in Figure 5–30a. However, inside there is either a cone-shaped orifice (see Figure 5–30b) or a curved scoop. These partial obstructions generate the

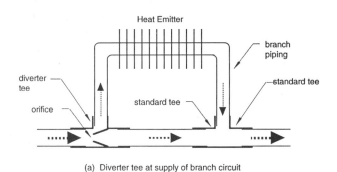

(a) Diverter tee at supply of branch circuit

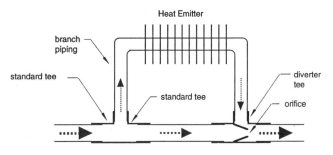

(b) Diverter tee at return of branch circuit

Figure 5–29   A diverter tee used to create flow in a branch piping circuit (a) diverter tee mounted at the supply of the branch piping path, (b) diverter tee mounted at the return of the branch piping path

(a)

(b)

Figure 5-30 Appearance of a diverter tee (a) side view showing reduced size of side port, (b) end view showing cone-shaped internal orifice

pressure differentials needed to create flow within the branch piping path.

A single diverter tee is often used with heat emitters that create relatively little flow resistance. The single diverter tee can be placed either at the beginning or end of the branch piping path as shown in Figure 5-31a. If the branch piping path is long, or contains components that create higher flow resistance, two diverter tees can be used to create a higher pressure differential as shown in Figure 5-31b. In either case *it is crucial that diverter tees be installed in the proper direction within the piping system*. This direction is indicated by an arrow on the side of the fitting, and/or a ring around one end port of

the diverter tee. In the latter case, *the fitting should always be installed with the ring facing toward the branch piping path*. The proper orientation of the fitting can also be determined by noting the direction of the cone-shaped orifices in Figure 5-31.

Within the heating trade, diverter tees are often called **Monoflo® tees** after the trademark brand from the Bell and Gossett Company. Another name sometimes used is venturi fitting, based on the suction effect of the tee when installed at the return of the branch piping path.

Diverter tees are extremely useful in designing hydronic systems that allow individual flow control through each heat emitter on a distribution circuit. This is accomplished

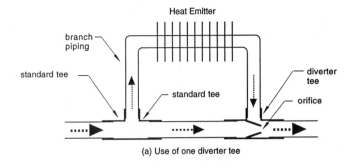

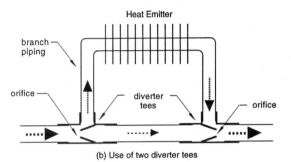

Heat Emitter

(a) Use of one diverter tee

Heat Emitter

(b) Use of two diverter tees

Figure 5–31 Cross-section of a branch piping path using (a) single diverter tee, (b) two diverter tees. Notice the increased flow rate, indicated by arrow width, when two diverter tees are used.

by regulating the flow rate through the brach piping path using a manual or thermostatic radiator valve. The design of these so-called **1-pipe systems** is discussed in later chapters. The Hydronic Design Toolkit software contains a program called MONOFLO for predicting the performance of such systems.

Dielectric Unions

In many hydronic systems, iron or steel components such as boilers and circulators are connected using copper tubing. If the copper and steel components are in direct contact with each other and if the system's fluid is slightly conductive, which is often the case, a process called galvanic corrosion will cause rusting of the steel or iron components. To prevent this from occurring, it is necessary to place an electrically insulating material, called a dielectric, between the two dissimilar metals. This can be accomplished with a **dielectric union**.

At a glance, dielectric unions appear similar to a standard union. Closer inspection, however, shows one side of the union has steel pipe threads, while the other side has a brass socket intended for soldering to copper tube. The steel and brass parts of the union are prevented from touching each other through the use of a synthetic O-ring and sleeve as shown in Figure 5–33.

Besides preventing galvanic corrosion, dielectric unions also provide for simple disassembly of a piping system, the same as with standard unions. Two available types of dielectric unions are shown in Figure 5–32.

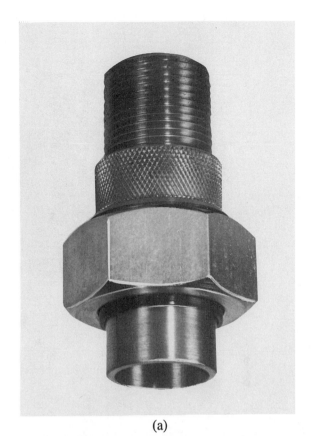

(a)

(b)

Figure 5–32 Examples of dielectric unions (a) with MPT threads on steel side, (b) with FPT threads on steel side. Courtesy of Watts Regulator Co.

5.5   THERMAL EXPANSION OF PIPING

All materials expand when heated and contract when cooled. In some hydronic systems, piping often undergoes wide temperature swings during cyclic operation using

Figure 5–33 Disassembled dielectric union showing insulating O-ring and sleeve

high temperature fluids. This causes significant changes in the length of the piping due to **thermal expansion**. The greater the *straight* length of the pipe and the greater the temperature change, the more movement takes place. If the piping is rigidly mounted, this expansion can cause annoying popping or squeaking sounds each time the pipe heats up or cools down. In extreme cases, the piping can even buckle due to the high thermal stresses that develop.

The expansion or contraction movement of *copper tubing* can be calculated using the following equation:

(Equation 5.1)
$$\Delta L = 0.0000094(L)(\Delta T)$$

where:

ΔL = change in length due to heating or cooling (inches)

L = original length of the pipe before the temperature change (inches)

ΔT = change in temperature of the pipe (°F)

0.0000094 = linear expansion coefficient of copper

Example 5.1: Determine the change in length of a ³/₄-in copper tube, 50 ft long, when heated from room temperature of 65 °F to 220 °F.

Solution:

$$\Delta L = 0.0000094(50 \cdot 12)(220 - 65) = 0.874 \text{ inches}$$

In many smaller hydronic systems, the piping is not rigidly supported or contains a sufficient number of elbows and tees to absorb the expansion movement without creating excessive stress in the pipe. Such situations only require plastic coated piping supports to prevent expansion noises. However, in situations requiring long straight runs of piping, rigidly supported at its ends, the movement must be absorbed with either an **expansion compensator** or **piping offset**.

An expansion compensator is designed to absorb length changes without creating large stresses in the pipe. This component is designed to be soldered or threaded inline with the pipe. One style allows the tubing to slide through a sealed joint as it changes length. Another design uses a metal bellows assembly to absorb the length change. Figure 5–34 shows two examples of expansion compensators.

An alternative method for accommodating thermal expansion is to construct a piping offset as shown in Figure 5–35. This offset flexes inward or outward to absorb piping movement without imposing excessive stress on the pipe. Suggested lengths for the legs of the piping offsets *for copper tubing* are given in the table.

5.6 COMMON VALVES

There are many types of valves used in hydronic systems. Some are common designs used in all categories

Figure 5-34 Two types of expansion compensators for copper tube. Courtesy of Amtrol, Inc.

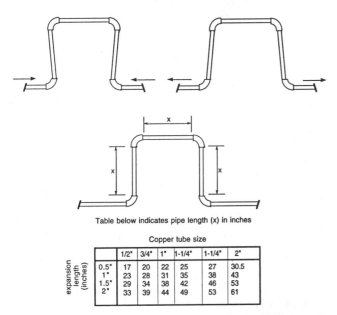

Table below indicates pipe length (x) in inches

Copper tube size

	1/2"	3/4"	1"	1-1/4"	1-1/4"	2"
0.5"	17	20	22	25	27	30.5
1"	23	28	31	35	38	43
1.5"	29	34	38	42	46	53
2"	33	39	44	49	53	61

(expansion length (inches))

Figure 5-35 Piping offset used for absorbing thermal expansion and contraction of piping

of piping including hydronic, steam, plumbing, pneumatics, and process piping. Others are designed for a very specific function in a hydronic heating system. The proper use of valves can make the difference between an efficient, quiet, and easily serviced hydronic system and one that wastes energy, creates objectionable noise, or even creates a major safety threat. *It is vital that designers and installers understand the proper selection and placement of several types of valves.*

Most of the common valve types are designed for one of the following duties:

- Component isolation
- Flow regulation

Component isolation refers to the use of valves to close off the piping leading to a device that may have to be removed or opened for servicing. Examples of such devices include circulators, boilers, heat exchangers, and

strainers. By placing valves on either side of such components, only minimal amounts of system fluid need to be drained or spilled during servicing. In many cases other parts of the piping system can remain in normal operation during this servicing.

Flow regulation requires a valve to set and maintain a given flow rate within a piping system, or portion thereof. Examples include adjusting the flow rates in **parallel piping** branches, or blending hot and cold fluids in a certain proportion. Valves used for flow regulation are specifically designed to remove mechanical energy from the fluid. This causes the fluid's pressure to drop as it passes through the valve. The greater the pressure drop, the slower the flow rate through the valve.

Gate Valves

Gate valves *are designed specifically for component isolation.* As such, they should be fully open or fully closed at all times.

The external appearance and internal design of a typical bronze gate valve are shown in Figure 5-36. When the handwheel is turned clockwise, the wedge moves down out of the upper body chamber until it totally closes off the fluid passage. Near the end of its travel, the wedge sits tightly against a seat forming a pressure tight seal.

In its fully open position, the valve's wedge is completely retracted into the upper body. It creates almost no interference with the fluid stream moving through the valve. This results in minimal loss of pumping energy during normal system operation, a desirable quality considering that a typical isolation valve remains open for most of its service life.

Gate valves should never be used to regulate flow. If the wedge is set to a partially open position, vibration and chattering are likely to occur. Eventually these conditions will erode the machined metal surfaces, possibly preventing a drip-tight seal when the valve is closed.

Globe Valves

Globe valves *are specifically designed for flow regulation.* Although their external appearance closely resembles that of a gate valve, there are major internal differences. The fluid's path through a globe valve contains several abrupt changes in direction, as can be seen in Figure 5-37. The fluid enters the lower valve chamber, flows upward through the gap between the orifice and disc, then exits sideways from the upper chamber. The gap between the disc and the orifice controls the amount of pressure drop created by the valve. *Be sure to install globe valves so the entering fluid flows into the lower body chamber, and thus upward toward the disc.* Reverse flow through a globe valve can cause unstable flow regulation and noise.

Globe valves should never be used for component isolation. The reason is that a fully open globe valve still

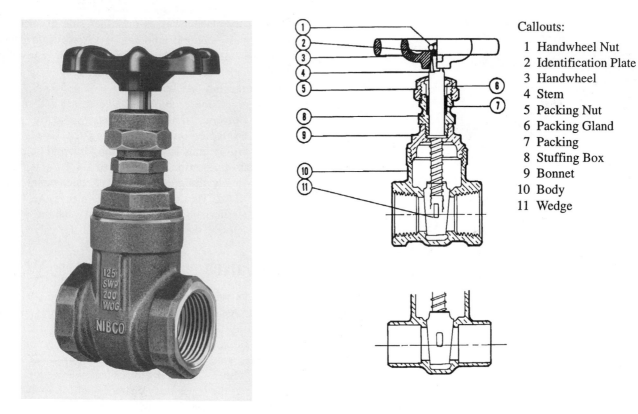

Callouts:

1. Handwheel Nut
2. Identification Plate
3. Handwheel
4. Stem
5. Packing Nut
6. Packing Gland
7. Packing
8. Stuffing Box
9. Bonnet
10. Body
11. Wedge

Figure 5-36 A typical bronze gate valve (a) external appearance, (b) internal construction. Photo and illustration courtesy of NIBCO, Inc., Elkhart, IN.

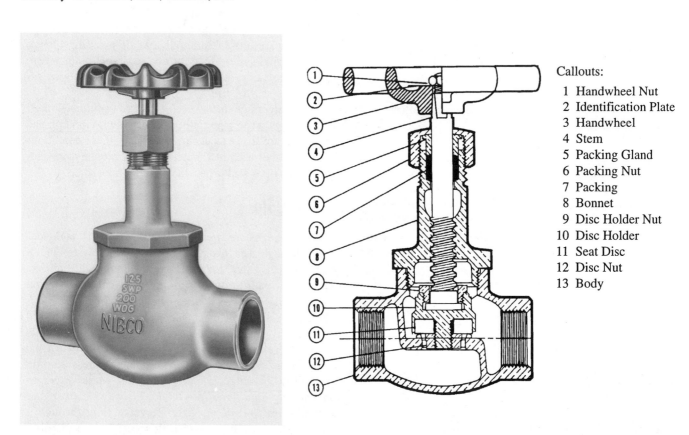

Callouts:

1. Handwheel Nut
2. Identification Plate
3. Handwheel
4. Stem
5. Packing Gland
6. Packing Nut
7. Packing
8. Bonnet
9. Disc Holder Nut
10. Disc Holder
11. Seat Disc
12. Disc Nut
13. Body

Figure 5-37 A typical bronze globe valve (a) external appearance, (b) internal construction. Photo and illustration courtesy of NIBCO, Inc., Elkhart, IN.

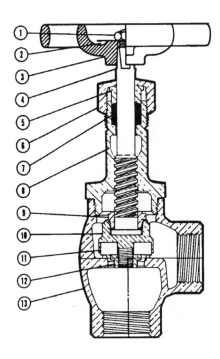

Callouts:
1 Handwheel Nut
2 Identification Plate
3 Handwheel
4 Stem
5 Packing Gland
6 Packing Nut
7 Packing
8 Bonnet
9 Disc Holder Nut
10 Disc Holder
11 Seat Disc
13 Seat Disc Nut
13 Body

Figure 5–38 A typical angle valve (a) external appearance, (b) internal construction. Photo and illustration courtesy of NIBCO, Inc., Elkhart, IN.

removes considerably more mechanical energy from the fluid compared to a fully open gate valve. Therefore, during the thousands of operating hours when component isolation is not needed, the globe valve unnecessarily wastes pumping energy. Using a globe valve for component isolation is like driving a car with the brakes lightly applied at all times. It may be possible, but it certainly is not efficient.

Angle Valves

The **angle valve** functions similar to a globe valve, but with its outlet port rotated 90 degrees relative to its inlet port. This configuration is useful in place of a standard globe valve combined with an elbow. Angle valves are frequently used as flow regulation valves on the outlet of heat emitters such as panel radiators or standing cast-iron radiators. Because the flow path through an angle valve is not as convoluted as that through a globe valve, it creates slightly less pressure drop. However, *angle valves should still not be used solely for component isolation*. As with a globe valve, *be sure to install angle valves so that entering fluid flows toward the bottom of the disc*. A example of an angle valve is shown in Figure 5–38.

A variation on the angle valve, known as a **boiler drain**, is shown in Figure 5–39. This valve allows a standard garden hose to be connected to its outlet port. Boiler

drain valves can be placed wherever fluid might need to be drained from the system.

Ball Valves

A **ball valve** uses a machined spherical plug, known as the ball, as its flow control element. The ball has a large hole through its center. As the ball is rotated through 90 degrees or arc, the hole moves from being parallel

Figure 5–39 Boiler drain valve. Photo courtesy of NIBCO, Inc., Elkhart, IN.

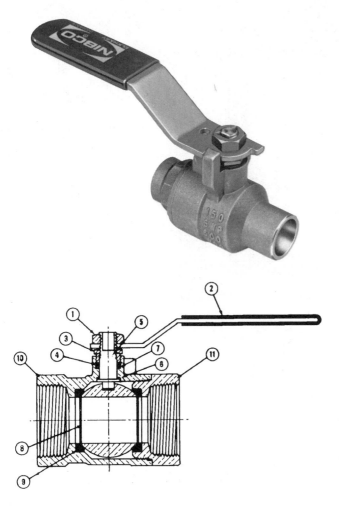

Callouts:

1 Handle Nut   7 Packing
2 Handle   8 Ball
3 Packing Nut   9 Seat Ring
4 Packing Gland   10 Body
5 Stem   11 Body End Piece
6 Thrust Washer

Figure 5–40 A conventional ball valve (a) external appearance, (b) internal construction. Photo and illustration courtesy of NIBCO, Inc., Elkhart, IN.

with the valve ports (its fully open position), to being perpendicular to the ports (its fully closed position). Ball valves are equipped with lever-type handles rather than handwheels. The position of the lever relative to the centerline of the valve indicates the orientation of the hole through the ball. An example of a ball valve is shown in Figure 5–40.

When the size of the hole in the ball is approximately the same as the pipe size of the valve, the valve is appropriately called a **full port ball valve**. If the size of the hole is smaller than the pipe size of the valve, it is called a conventional or reduced port ball valve.

Full port ball valves are well suited for component isolation purposes. In its fully open position, the valve creates relatively little flow interference, and thus minimal pressure drop.

It is generally agreed that ball valves can also be used for flow regulation when the fluid flow rate will be reduced no more than 25% from its unregulated flow rate. This limitation prevents the valve from operating in an almost closed position where high fluid velocities could erode internal surfaces.

Check Valves

In many hydronic systems it is necessary to prevent fluid from flowing backwards through a pipe. The **check valve**, in one of its many forms, is the common solution.

Check valves differ in how they prevent reverse flow. The **swing check**, shown in Figure 5–41, contains a disc that is hinged along its upper edge. When fluid moves through the valves in the allowed direction (as indicated by the arrow on the side of the valve's body), the disc swings up into a chamber out of the direct path of the fluid. The moving fluid stream holds it there. When the fluid stops, or attempts to reverse itself, the disc swings down due to its own weight and seals across the opening of the valve. The greater the back pressure, the tighter the seal.

For proper operation, swing check valves must be installed in horizontal piping with the bonnet of the valve in an upright position. Installation in other orientations can cause erratic operation creating dangerous water hammer effects.

A **spring-loaded check valve** is designed to eliminate the orientation restrictions associated with a swing check. These valves rely on a small internal spring to close the valve's disc whenever fluid is not moving in the intended direction. *The spring action allows the valve to be installed in any orientation.* The force required to compress the spring does, however, create slightly more pressure drop compared to a swing check. As with a swing check, *be sure the valve is installed with the arrow on the side of the body pointing in the direction of flow.* An example of a spring-loaded check valve is shown in Figure 5–42.

5.7 SPECIALTY VALVES FOR HYDRONIC APPLICATIONS

Several valves have very specific functions in hydronic systems. Some provide safety protection and are required by plumbing and mechanical codes in all areas of the U.S. Others automatically regulate the temperature and pressure at various points in the system. All these valves will be incorporated into system designs in later chapters.

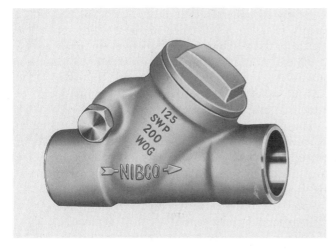

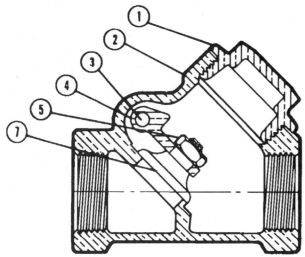

Callouts:

1 Bonnet   4 Disc Hanger
2 Body   5 Hanger Nut
3 Hinge Pin   7 Seat Disc

Figure 5–41 A swing check valve (a) external appearance, (b) internal construction. Notice flow direction arrow on body. Photo and illustration courtesy of NIBCO, Inc., Elkhart, IN.

Boiler Feed Water Valves

A **boiler feed water valve**, or pressure reducing valve, is used to lower the pressure of water from a domestic water distribution pipe before it enters a hydronic system. This valve is necessary because most buildings have domestic water pressure higher than the relief valve settings of small hydronic systems. The feed water valve allows water to pass through whenever the pressure at its outlet side drops below its pressure setting. In this way small amounts of water are automatically fed into the system as air is vented out, or when small amounts of fluid are lost through valve packings, etc. In effect, the

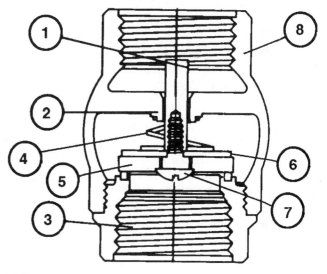

Callouts:

1 Stem   5 Disc
2 Stem Guide   6 Disc Holder
3 Body End   7 Seat Screen
4 Spring   8 Body

Figure 5–42 A spring-loaded check valve (a) external appearance, (b) internal construction. Notice flow direction arrow on body. Photo and illustration courtesy of NIBCO, Inc., Elkhart, IN.

feed water valve maintains a constant preset pressure on the system. Hydronic systems without this type of automatic feed water system tend to lose pressure as a result of air venting, and very small water losses through valves packing, pump gaskets, etc. The resulting low pressure operation can create problems with noise and pump cavitation. In multi-story buildings the reduced pressure can also prevent circulation in upper parts of the system.

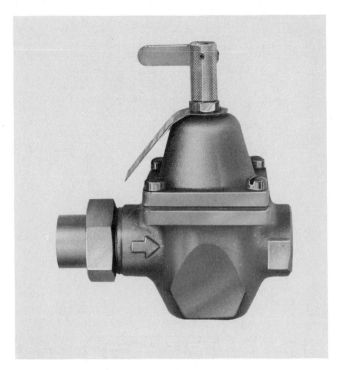

Figure 5–43 Example of a boiler feed water valve. Notice flow direction arrow on body. Courtesy of Watts Regulator Co.

The pressure setting of feed water valves can be varied by rotating its center shaft with a screwdriver. Once the desired pressure setting is attained, the shaft position can be secured by tightening the locknut surrounding the shaft. It is often convenient to have a pressure gauge mounted near the outlet of the feed water valve to indicate the current pressure setting. Most feed water valves also have a lever at the top to allow the valve to be manually opened to quickly fill the system during startup. *Be sure the valve is installed with the flow direction arrow pointing toward the hydronic system.* A typical boiler feed water valve is shown in Figure 5–43.

Pressure Relief Valves

A **pressure relief valve** *is a code requirement for any type of closed-loop hydronic heating system.* It functions by opening just below a preset pressure rating allowing system fluid to be safely released from the system before dangerously high pressures can develop. In a situation where all other system controls may have failed to limit heat production, the pressure relief valve is the final means of safety protection. In its absence, some component within the system could potentially explode with devastating results.

Most mechanical codes require pressure relief valves to be installed on any piping assembly that contains a heat source and is capable of being isolated by valves from the rest of the system.

Nearly all boilers sold in the U.S. are shipped with a factory installed ASME-rated pressure relief valve. The typical pressure rating for relief valves on small boilers is 30 psi. Other hydronic heat sources, such as heat pumps, are usually not equipped with factory installed pressure relief valves, but nonetheless require one to be present in any closed system they are a part of.

The operation of a pressure relief valve is simple. When the force exerted on the internal disc equals or exceeds the force generated by the internal spring, the disc lifts off its seat and allows fluid to pass through the valve. The spring force is calibrated for the desired opening pressure of the valve. This rated pressure, along with the maximum heating capacity of the equipment the valve is rated to protect, is stamped onto a permanent tag or plate attached to the valve. A typical relief valve is shown in Figure 5–44.

All pressure relief valves should have a waste pipe attached to their outlet port. This pipe routes any expelled fluid safely to a drain, or at least down near floor level. The waste pipe must be the same size as the valve's outlet port, with a minimal number of turns and no valves. The waste pipe should end 6 inches above the floor or drain to allow for unrestricted flow if necessary.

Backflow Preventer

Consider a hydronic heating system filled with an antifreeze solution connected to a domestic water supply

Figure 5–44 Example of a pressure relief valve. Courtesy of Watts Regulator Co.

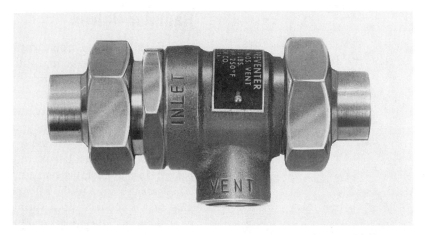

Figure 5–45   Example of a small backflow preventer. Note inlet port marking. Courtesy of Watts Regulator Co.

by way of a feed water valve. If the pressure of the domestic water system should suddenly drop, due to a rupture of the water main or other reason, the anti-freeze solution could be pushed backward by expansion tank pressure into the domestic water system, and ultimately contaminate a private or public water system. A **backflow preventer** eliminates the possibility of this happening. It consists of a pair of check valve assemblies in series, with an intermediate vent port that drains any backflow that migrates between the valve assemblies. *Be sure the valve is installed with the inlet side connected to the domestic water supply.* An example of a small backflow preventer is shown in Figure 5–45.

Most municipal plumbing codes require backflow preventers on all hydronic heating systems that are connected to domestic water piping. Even if local codes do not require this valve, it is good practice to use it,

especially considering that antifreeze or other chemical treatments might be added to the system at some later date. *A single check valve is not an acceptable replacement for a backflow preventer, and should not be relied on for this function.*

Flow Check Valves

Another variation on the basic check valve is the **flow check valve.** These valves are designed with a weighted internal plug that is sufficiently heavy to stop thermosiphoning or gravity flow of hot water while the system's circulator is off. They also prevent reverse flow in multizone hydronic systems.

Flow checks are available with either two or three ports as shown in Figure 5–46. The three-port valve allows the valve to be installed with either a horizontal

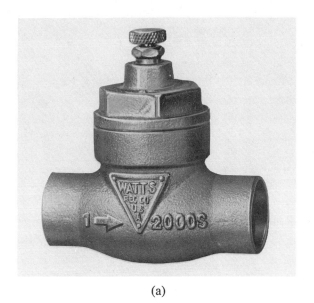

(a)

(b)

Figure 5–46   Example of flow check valves (a) two-port design. Courtesy of Watts Regulator Co. (b) three-port design. Courtesy of Armstrong Pumps. Notice flow direction arrows on valve bodies.

Figure 5–47   A manual operator knob attached to a radiator valve

or vertical inlet pipe. The inlet port that is not used is sealed with a plug. The small lever or screw at the top of these valves allows the valve to be manually opened in the event of a circulator failure. *The lever does not allow the valve to adjust flow rate.* Some manufacturers even offer a version of the flow check valve without this lever. Be sure to observe flow direction when installing these valves.

Radiator Valves

Originally from Europe, **radiator valves** provide the potential for precise room-by-room temperature control in hydronic heating systems. These valves are usually installed in the supply pipe or directly to the supply connection of individual heat emitters. They consist of two separate parts, the valve body and the operator.

The nickel-plated brass valve body is available in pipe sizes from $1/2$ inch to 1 inch, in either a straight or angle pattern. In the latter, the outlet port of the valve is rotated 90 degrees relative to the inlet port. Inside the valve is a plug that is mounted to a spring-loaded shaft. The plug is held in its fully open position by the internal spring. Fluid enters the lower body chamber and flows through the orifice toward the plug. To close the plug, the shaft must be *pushed* inward against the spring force. No rotation of the shaft is necessary. The shaft movement can be achieved manually using a threaded cap or knob assembly as shown in Figure 5–47, or automatically using a non-electric **thermostatic operator**. When a thermostatic operator is used, the valve is appropriately called a **thermostatic radiator valve (TRV)**.

Several types of thermostatic operators can be matched to radiator valve bodies. The most common configuration attaches the thermostatic operator directly to the valve body as shown in Figure 5–48. A cut-away view of such a combination is shown in Figure 5–49.

The thermostatic operator contains either a fluid or wax compound in a sealed chamber. As the air temperature surrounding the operator approaches the setpoint temperature, the fluid or wax expands with sufficient force to slowly close the plug in the valve, reducing and eventually stopping the heat output from the heat emitter being controlled. As the room air temperature begins to decrease,

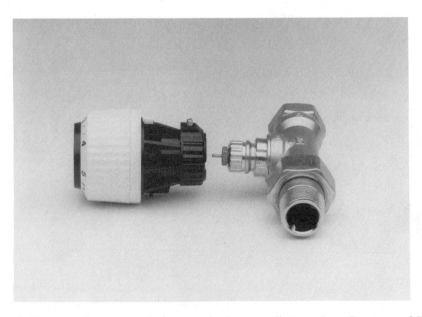

Figure 5–48   A thermostatic operator being attached to a radiator valve. Courtesy of Danfoss, Inc.

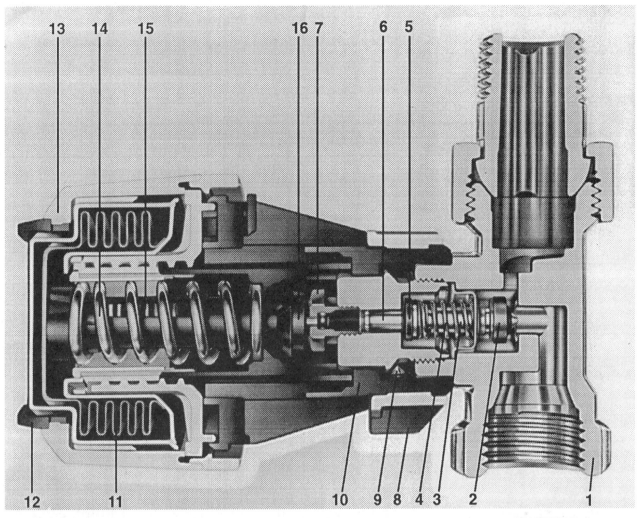

PART	MATERIAL		PART	MATERIAL
1. Valve	Nickel plated brass		10. Socket	POM
2. Valve disc	EPDM		11. Bellows	Phosphor-bronze
3. Spindle guide	Phosphor-bronze		12. Reference ring	Polyacetal
4. Spring	Stainless steel		13. Handle	ABS
5. Back seat washer	EPDM		14. Adjustment spring	Steel
6. Valve spindle	Brass		15. Safety spring	Steel
7. Pressure pin	Stainless steel		16. Pressure spindle	Polyamide No. 6
8. Clamping band	Al. alloy		Capillary tube	Steel
9. Allen screw	Steel			

Figure 5–49   Internal construction of a thermostatic operator attached to a straight radiator valve body. Courtesy of Danfoss, Inc.

the fluid or wax contracts, allowing the spring force to slowly reopen the valve plug, thus increasing the heat output from the heat emitter. The combination of the valve and thermostatic operator represents a fully **modulating** temperature control system that continually adjusts the flow rate through the heat emitter in an attempt to maintain a constant room temperature.

Because the thermostatic operator is often located in close proximity to the heat emitter, it is critical that it be mounted so as not to be affected by convective air currents created by the heat emitter. The preferred mounting position is with the valve stem in a horizontal position with the thermostatic operator facing away from the heat emitter as shown in Figure 5–50. *In no case should the valve be installed so the thermostatic operator is directly above the heat emitter.* This will cause the valve to close almost as soon as it begins to open due to warm air rising from the heat emitter and deceiving the thermostatic operator. Be sure to observe the proper flow direction indicated by the arrow on the side of the valve.

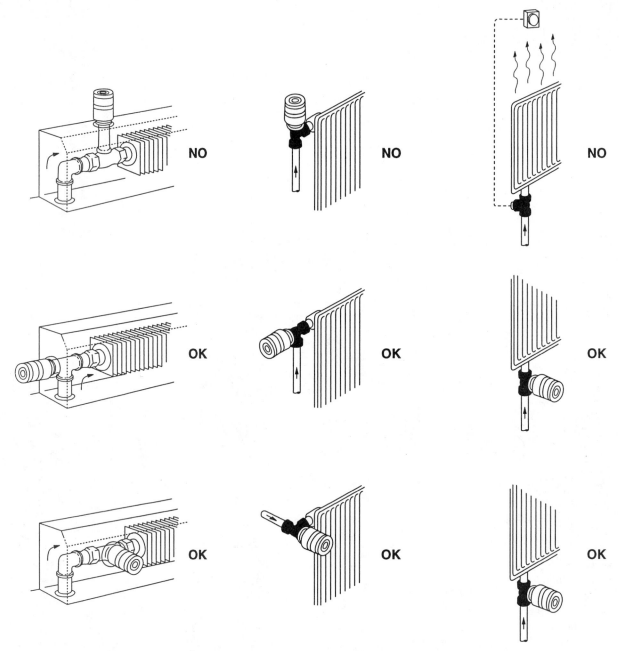

Figure 5–50 Preferred mounting orientation for thermostatic operator directly attached to a radiator valve. Courtesy of ISTA, Inc. and Danfoss, Inc.

For increased adaptability in different applications, most manufacturers of radiator valves offer thermostatic operator assemblies with a remote setpoint dial. This allows temperature adjustments to be made at normal thermostat height above floor level. A copper capillary tube runs from the setpoint dial assembly to the operator assembly mounted on the radiator valve. These capillary tubes can be ordered in lengths up to 12 feet. Figure 5–51 shows an example of a radiator valve fitted with a remote setpoint adjustment assembly. Care must be taken not to kink or break the capillary tube during installation.

An excellent application for a TRV is in conjunction with a diverter tee in what is called a 1-pipe distribution system. This combination allows the option of independently controlling the output of each heat emitter served by the distribution circuit. The accuracy and responsiveness of the TRV can quickly adjust the output of a room's heat emitter based on air temperature changes from solar or other internal heat gains. This combination is illustrated for a wall panel radiator in Figure 5–52. With proper design this concept offers tremendous opportunities for precise heat distribution. The design of 1-pipe systems using TRVs is covered in detail in Chapter 11.

Mixing Valves

Some hydronic distribution systems require lower temperature water than may be present in the heat source. For example, a radiant floor heating system may require water at 110 °F, while the water in a thermal storage tank supplying this load is at 150 °F. The lower water temperature can be obtained by blending return water from the distribution system with hot water from the heat source. The flow rate of each fluid stream needed to attain a specific outlet temperature is controlled by a **mixing valve**.

Mixing valves are available in either three-port or four-port configurations. *Three-port mixing valves are suitable when the heat source can accept the cooler return water from the distribution system without problems such as flue gas condensation.* A thermal energy storage tank would be an example of such a heat source. A typical three-port mixing valve is shown in Figure 5–53. A simple piping schematic showing how it functions is shown in Figure 5–54.

The position of the valve's handle or knob determines the amount of each fluid stream that enters the valve. The

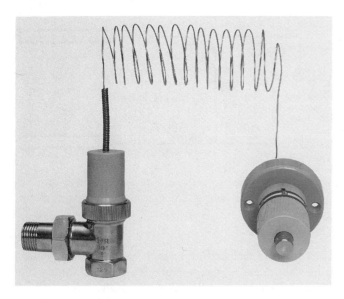

Figure 5–51   A radiator valve fitted with a remote setpoint dial and operator assembly. Courtesy of Ammark Corporation.

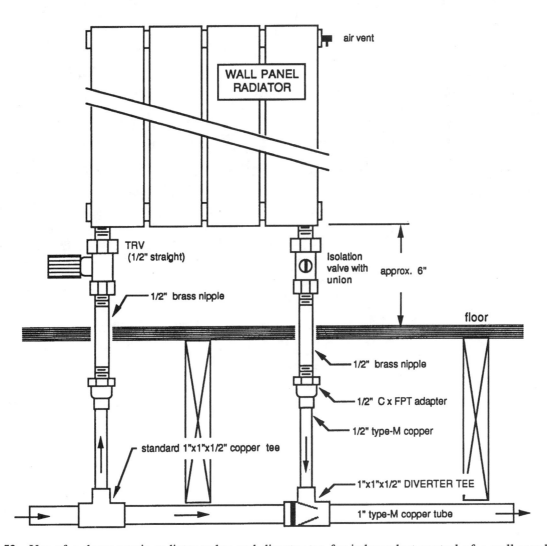

Figure 5–52   Use of a thermostatic radiator valve and diverter tee for independent control of a wall panel radiator

Figure 5–53 A three-port mixing valve with a manual adjustment knob. Note setting number on dial face. Courtesy of Danfoss, Inc.

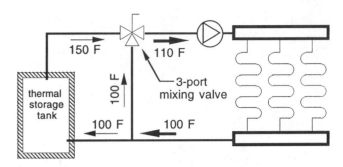

Figure 5–54 Use of a three-port mixing valve to achieve a reduced water temperature in the distribution system

flow rate and temperature of each entering stream determines the blended temperature leaving the valve. The blended temperature will always be in the range formed by the hot and cold streams entering the valve. Most three-port valves have an indicator scale on the valve body. The pointed end of the handle or knob indicates the setting of the valve between zero and ten. The full range of the handle movement is about 90 degrees. See Figure 5–55.

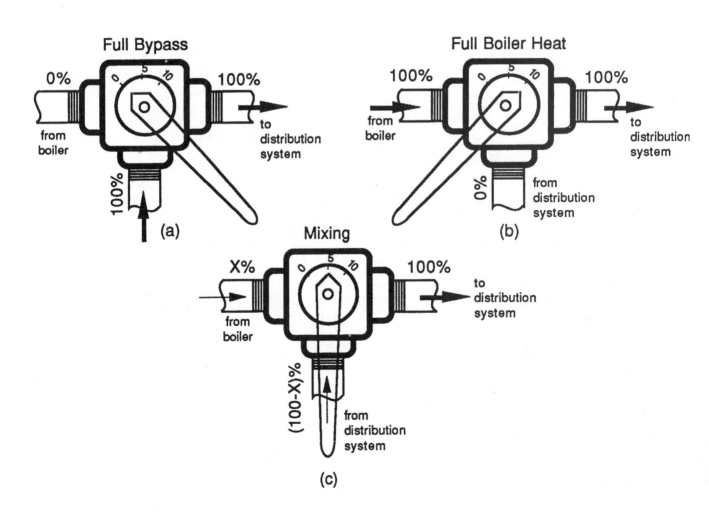

Figure 5–55 Variation in flow created by the handle setting of a three-port mixing valve

A basic three-port mixing valve does not contain an internal thermostatic element. As such it does not adjust itself to compensate for changes in the entering fluid streams. These valves can, however, be fitted with a motorized **actuator** controlled by an electronic setpoint control. This control/motor combination automatically adjusts the valve's shaft position to maintain a given outlet temperature.

Three-port mixing valves are not linear valves. The position of the handle across its range of travel is not necessarily proportional to the amount of flow through one of the inlet ports. In some cases, small changes in the position of the handle can make relatively large changes in the blended temperature.

Most oil and gas-fired boilers cannot accept cool return water without continuous condensation of the water vapor in the flue gases. *In these situations, three-way mixing valves are limited to systems where the coolest return water from the distribution system is always above the dewpoint temperature of the exhaust gases.* When this is not possible, a four-port mixing valve should be used.

In a four-port mixing valve, some of the hot fluid from the heat source is blended with the water headed for the distribution system, and the water returning to the boiler. *This maintains the boiler above dewpoint while also allowing the distribution system to operate at a lower water temperature.* This makes the four-port design the preferred type of mixing valve for use with conventional gas or oil-fired boilers. Examples of a four-port mixing valve are shown in Figure 5–56. A simple schematic showing its location in the system is shown in Figure 5–57.

Flow through a four-port mixing valve is controlled by a simple rotating vane. This vane is aligned parallel with the handle of the valve. As the handle is rotated from one extreme to the other, the vane rotates through an arc of about 90 degrees. This range corresponds to a situation where all water returning from the distribution system is routed through the boiler (handle pointer indicates a setting of ten), to the opposite extreme where none of the water circulating in the distribution loop goes through the boiler (handle pointer indicates a setting of zero). In most

cases the valve is set in an intermediate position where a blended water temperature is supplied to the distribution loop. These setting are shown in Figure 5–58.

Both three-port and four-port mixing valves are available in cast-iron or brass. Cast iron is the more common material for closed loop systems. Like the three-port valve, four-port mixing valves do not contain a thermostatic element. They too can be equipped with motorized actuators for automatic temperature control of the outlet stream. Such applications will be discussed in detail in later chapters.

Thermostatic Mixing Valves

The three-port and four-port mixing valves just discussed are set to achieve a certain outlet temperature based on two *steady* inlet temperatures and flow rates. If the temperature or flow rate of either incoming stream changes, the outlet temperature will also change unless the valve is readjusted. If the inlet conditions vary frequently, the constant manual readjustments becomes impractical. An alternative to equipping such mixing valves with motorized operators is to use a valve with an internal thermostatic element. Such a valve is appropriately called a **thermostatic mixing valve**. An example of such a valve is shown in Figure 5–59.

This type of valve reacts to temperature changes of the incoming fluids by opening or closing its inlet ports as necessary to maintain a nearly fixed outlet temperature. A knob on the valve allows this outlet temperature to be adjusted within a certain range. Such valves are useful in radiant floor heating applications where the temperature of the heat source is higher than the required supply temperature of the floor, and where this higher temperature

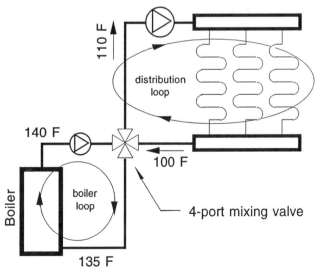

Figure 5–57 Use of a four-port mixing valve to achieve a reduced water temperature in the distribution system. Note the higher return water temperature entering the boiler.

Figure 5–56 Examples of a four-port mixing valve. Courtesy of Tekmar Controls.

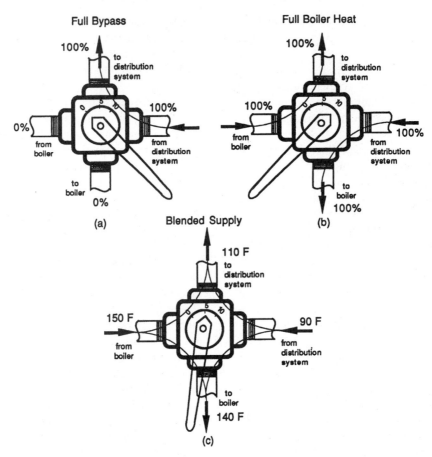

Figure 5–58 Variation in flow created by the handle setting of a four-port mixing valve

Figure 5–59 Example of a non-electric thermostatic mixing valve. Courtesy of Spargo, Inc.

is likely to fluctuate. These valves also can be used to lower the temperature of domestic hot water supplied from variable temperature sources such as an active solar energy system.

Zone Valves

Hydronic heating systems often use two or more separate piping circuits to supply heat to specific areas in a building. Such areas are called zones. The device most often used to control flow in a particular zone piping circuit is called a **zone valve**.

A variety of zone valves are available. They all consist of a valve body combined with an actuator. The actuator is the device that produces movement of the valve shaft when supplied with electricity. Some actuators use small electric motors combined with gears to produce a rotary motion of the valve shaft. Others use **heat motors** that act like miniature hydraulic cylinders to produce a linear push/pull motion. A zone valve equipped with a gear motor actuator is shown in Figure 5–60a. A zone valve using a heat motor actuator is shown in Figure 5–60b.

Zone valves operate in either a fully open or fully closed position. They do not modulate to partially open

(a)

(b)

Figure 5–60   Two examples of hydronic zone valves (a) valve equipped with a gear motor actuator. Courtesy of Erie Manufacturing. (b) valve equipped with a heat motor actuator. Courtesy of Spargo, Inc.

positions to regulate flow rate in a piping circuit. Those with gear motors can move from fully closed to fully open in about three seconds. Those with heat motors have a much slower opening time of between two and three minutes. Since the response time of a heating system is relatively slow, the slower opening of zone valves using heat motors does not make a noticeable difference in system performance.

Both two-way and three-way zone valves are available. The two-way valve either allows full flow through the valve or totally stops flow. This is the most commonly used zone valve in residential applications. A three-way zone valve has an additional port called the bypass. There will always be flow through a three-way zone valve whenever the system's circulator is on. The flow that enters the common port must either pass straight through the valve or out the bypass port. Three-way zone valves are useful when flow must be maintained in a certain portion of the piping where temperature sensors are located. They can also be used to direct incoming flow into one of two different outlet pipes, depending on the operating mode of the system. A schematic for a typical three-zone system using two-way zone valves is shown in Figure 5–61. An example of a three-way zone valve, along with a piping schematic showing its use is given in Figure 5–62 and 5-63 respectively.

In a typical residential system, zone valves are located in the supply pipe of each zone circuit, usually in the vicinity of the heat source. In some cases it is possible to locate zone valves near, or even inside the enclosure of the heat emitters they control. This is more typical of commercial systems.

Most zone valves for residential and light commercial systems are designed to operate on a 24 VAC control voltage. This is a common transformer secondary voltage used in many types of the heating and cooling equipment. This voltage allows the zone valves to be wired with standard thermostat cable. For certain applications, however,

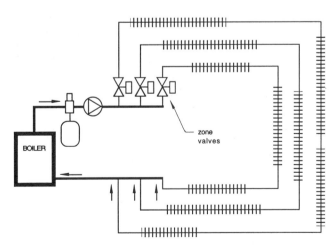

Figure 5–61   A typical piping schematic for a three-zone system using standard two-way zone valves

Figure 5–62 An example of a three-way zone valve. Courtesy of Erie Manufacturing.

it may be desirable to operate zone valves using standard 120 VAC line voltage. Some manufacturers offer actuators set up for this voltage.

Zone valves are available in two-wire, three-wire, or four-wire configurations. Two wires are always required to operate the valve's actuator. The third, and in some cases fourth, wire is connected to an **end switch** in the valve. When the valve reaches its fully open position, the end switch closes. This switch closure is used to signal the heat source and circulator to operate whenever one or more zones is calling for heat. The details of control wiring for zone valves is covered in Chapter 9.

Zone Headers

An innovative approach to multi-zone systems is the use of a **zone header**. This device saves installation time by reducing the need to mount several individual zone valves and their associated fittings in the field. One particular product consists of a cast bronze header with two, three, or four zone circuit outlets. Each outlet has an associated control valve operated by an individual heat motor actuator. The unit is prewired requiring only connection of line voltage and thermostats. An example of a four-circuit zone header and associated return header is shown in Figure 5–64.

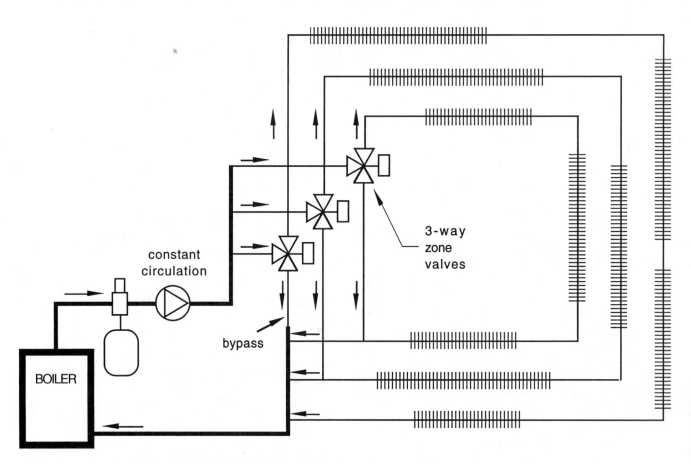

Figure 5–63 Piping schematic showing the use of three-way zone valves

Figure 5–64 Example of a four-circuit zone header system. Supply header at top, return header at bottom. Courtesy of Spargo, Inc.

Differential Pressure Bypass Valves

In certain types of multi-zone hydronic systems, particularly those with a large number of zones controlled by two-way zone valves, it is possible that a single large circulator is used to supply flow through all active zone circuits. If most of the zones are shut off, the circulator will tend to build up a large pressure differential across the active zones. This pressure differential can cause excessively high flow rates and associated flow noises in the remaining (active) zone circuits. To prevent this from occurring, a special valve called a **differential pressure bypass valve** can be installed in the system, as shown in Figure 5–66. This valve contains an internal spring assembly that can be set to maintain an almost constant upper limit on the pressure differential across the circulator. The valve functions by allowing a portion of the flow from the system's circulator to bypass the distribution system, thus limiting the pressure differential the pump can develop. An example of a differential pressure bypass valve is shown in Figure 5–65.

It is important to realize that a differential pressure bypass valve dissipates a portion of the mechanical energy added to the fluid by the circulator. In effect the valve acts like an automatically applied brake in the system.

Figure 5–65 Example of a differential pressure bypass valve. Courtesy of Honeywell Braukmann.

Good system design should rely on differential pressure bypass valves for only unusually low load conditions, or to protect a circulator from no-flow operation in the event of a control malfunction. The proper use of this valve is further discussed in Chapter 11.

Metered Balancing Valves

Multi-zone hydronic systems containing two or more parallel piping paths often require adjustments of the flow rates to properly balance the heating or cooling capacity of each branch. This is especially true in a so-called **parallel direct return piping** system. The piping system shown in Figure 5–67a would have higher flow rates in the piping paths closer to the heat source compared to those farther away from the heat source. The proper use of **metered balancing valves** can yield the desired flow rate in each branch as shown in Figure 5–67b.

Although standard globe valves can also be used to regulate flow rate, a metered balancing valve differs in three important ways from a standard globe valve. First, it uses a plug rather than a disc as the flow control element.

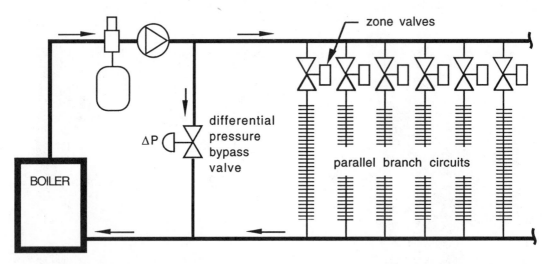

Figure 5–66 Schematic showing operation of a differential pressure bypass valve

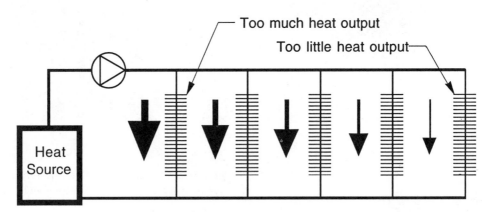

Unbalanced parallel direct return piping system
Greater flow rates in branches near circulator

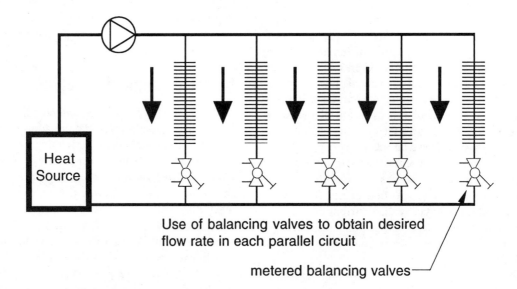

Use of balancing valves to obtain desired
flow rate in each parallel circuit

metered balancing valves

Figure 5–67 Use of metered balancing valves to adjust flow rate in parallel piping paths (a) situation without balancing valves, (b) situation after balancing valves are installed and adjusted

Figure 5–68   Cut-away view of a metered balance valve. Courtesy of Armstrong Pumps.

This allows the flow rate to vary in closer proportion to stem position. Second, the stem movement per turn of the shaft is less, allowing for more precise adjustments. Finally, metered balancing valves have two pressure metering ports that can be used to verify the flow rate through the valve based on the pressure drop across it. A cut-away view of a typical metered balancing valve is shown in Figure 5–68.

The pressure differential across the valve is measured using a digital **manometer** attached to both pressure taps on the valve body. This device calculates the flow rate through the valve based on this differential pressure drop

across it. It can be easily moved from one valve to the next in order to balance the entire system. An example of a digital manometer is shown in Figure 5–69.

Metered balancing valves are more commonly used in commercial rather than residential hydronic systems. Medium and large scale commercial systems often have many more parallel piping circuits than do residential systems. The greater complexity of these systems inherently leads to conditions that require careful balancing to achieve uniform comfort. Simpler residential systems are more forgiving of flow rate variations.

Metered balancing valves are more intricate and thus more expensive than globe valves. To use them effectively, the digital manometer must also be purchased. If the valves are used only occasionally, or in small numbers, the cost of the digital manometer is harder to justify.

Lockshield/Balancing Valves

Another specialty valve created for the hydronics industry provides the ability to isolate, balance, and even drain individual heat emitters. This valve, available in both straight and angle patterns as shown in Figure 5–70, is called a **lockshield/balancing valve**. Like many innovative hydronic components, it is a product of European design.

To adjust the flow setting of the valve, the cap of the valve is removed, and an internal brass plug is closed using an allen key. A retaining ring that limits the travel of this plug is then adjusted using a screwdriver. The plug is then rotated open until it contacts the bottom edge of the retaining ring. This arrangement allows the plug to be closed and opened repeatedly without affecting its flow rate setting when fully opened.

To drain the valve, a specialized tool is first screwed onto the valve body as shown in Figure 5–71. The shaft of this tool then engages the entire plug/retainer ring assembly and allows it to be unscrewed from the body and temporarily stored in a cavity in the drainage tool.

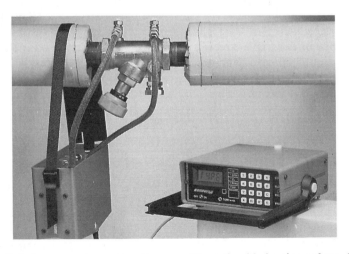

Figure 5–69   Example of a digital manometer system for use with metered balancing valves. Courtesy of Armstrong Pumps.

straight   **angle**

Figure 5–70 Straight and angle versions of a lockshield/balancing valve. Courtesy of Regin HVAC Products, Inc.

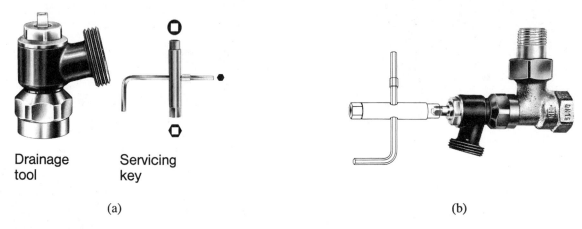

Drainage   Servicing
tool   key

(a)   (b)

Figure 5–71 Special tool for draining a heat emitter through a lockshield/balancing valve, (a) drainage tool and key, (b) drainage tool mounted on lockshield/balancing valve. Courtesy of Regin HVAC Products, Inc.

The fluid exits the drainage tool through a standard garden hose. The heat emitter can also be refilled by pumping fluid back in through this tool. When finished, the plug/retainer ring assembly is screwed back into the valve body using the shaft of the drainage tool. The assembly seats back into its original position such that the flow setting of the valve is unchanged. The drainage tool is then removed, and the valve cover replaced. These valves are equipped with a union that allows the valve and heat emitter to be quickly separated.

The simple exterior appearance of this valve conceals its versatility. It is truly an elegant solution in situations where individual heat emitters occasionally have to be temporarily removed for wall painting, etc.

5.8 SCHEMATIC SYMBOLS FOR PIPING COMPONENTS

Piping schematics are the road maps of hydronic system design. They describe the relative placement of dozens of piping components that must be put together to form a smooth operating system. Like the many small markings on a map, pumps, valves, and other piping components can all be represented by symbols.

Unfortunately there is no universal standard for piping symbols to which all people subscribe. Variations among designers, drafters, etc., are quite common. *The hydronic system designer should establish a consistent set of symbols and always identify each symbol on installation drawings using a symbol legend.* Without this information, a designer's symbol for a globe valve may be interpreted as a gate valve by the installer, etc. This needlessly wastes time and materials, and creates misunderstandings when the job is bid or installed.

Within this book, all piping schematics will use symbols based on the legend shown in Figure 5–72. Some of the components shown on this legend have not been discussed at this point, but will be used in later chapters. This legend is also shown in Appendix A.

5.9 TIPS ON PIPING INSTALLATION

As with any trade, the installation of piping requires an attitude of craftsmanship. Well planned and installed

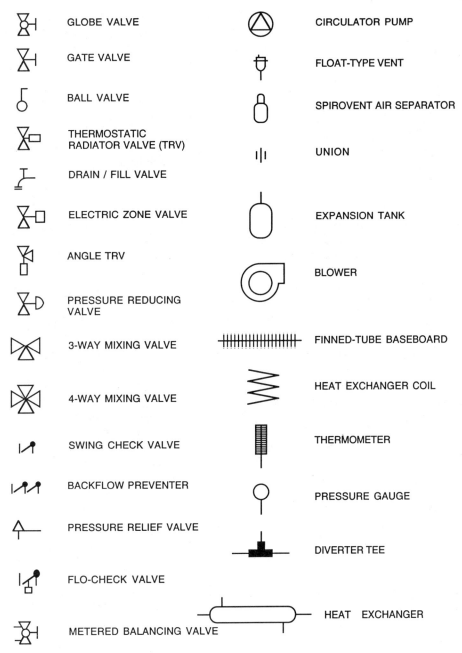

GLOBE VALVE

GATE VALVE

BALL VALVE

THERMOSTATIC
RADIATOR VALVE (TRV)

DRAIN / FILL VALVE

ELECTRIC ZONE VALVE

ANGLE TRV

PRESSURE REDUCING
VALVE

3-WAY MIXING VALVE

4-WAY MIXING VALVE

SWING CHECK VALVE

BACKFLOW PREVENTER

PRESSURE RELIEF VALVE

FLO-CHECK VALVE

METERED BALANCING VALVE

CIRCULATOR PUMP

FLOAT-TYPE VENT

SPIROVENT AIR SEPARATOR

UNION

EXPANSION TANK

BLOWER

FINNED-TUBE BASEBOARD

HEAT EXCHANGER COIL

THERMOMETER

PRESSURE GAUGE

DIVERTER TEE

HEAT EXCHANGER

Figure 5–72   Legend of schematic piping symbols used in this book

piping systems convey a message of professionalism that does not go unnoticed by the customer. The planning of efficient piping layouts is a skill acquired through experience. It is part art and part science. The following guidelines will help in planning and installing good piping layouts.

- As a general rule, piping should always be kept vertical (plumb) or horizontal (level). This ensures proper alignment of fitting joints and other components that join into the piping path. An exception is when a pipe must be pitched for proper drainage. In this case the pipe should have a minimum drop of $1/4$ inch per foot of

horizontal run. In either case use levels and/or plumb bobs to keep the piping runs accurate.
- Plan piping to minimize joints and fittings. If possible, use 20 foot sections of straight pipe on long runs to minimize couplings.
- Minimize changes in direction of piping. Do not zig-zag around other components. This adds to fitting count and makes the system look more complicated than necessary. It also wastes pumping energy.
- Try to align similar components in parallel piping branches. For example, when installing three zone valves side by side, make sure they are aligned with each other.

- Always consider the need to remove components for servicing. Do not locate serviceable components in small, hard to get to areas, or behind other equipment.
- Always provide support adjacent to components supported by the piping such as circulators, expansion tanks, or other relatively heavy devices.
- Plan piping assemblies that can be put together at a vise or flat working surface to minimize awkward working positions.
- Always allow for the takeup distances of threaded or soldered fittings when measuring and cutting a length of tubing.
- In situations where both piping and electrical wiring are in close proximity, try to keep the piping below the electrical wiring. This reduces the chance of water dripping onto electrical components during service or if a leak should develop.
- Lay out piping runs using the centerline of the piping rather than its edges. This ensures alignment between piping and fixed connection points, even when different pipe sizes are used.
- Start the piping layout by locating major pieces of equipment such as the boiler, circulator, mixing valve, etc. The positions of these devices serve as benchmarks to which interconnecting piping can be referenced.
- Secure all piping to structural surfaces to prevent sagging and to support the weight of components supported by the piping.
- Where insulation will cover the piping, be sure to allow sufficient space behind the pipe and the surface it is attached to.
- When piping is run through floor joists or other framing members, always drill holes at least 1/4 inch larger than the outside diameter of the piping to avoid binding that can cause expansion noises. Use a chalk line during layout to keep the holes precisely aligned. Plan ahead on how long piping lengths can be routed through joists.
- The optimum location for a hole through a framing member such as a floor joist is halfway between the bottom edge and top edge. This location minimizes loss of structural strength. *Never notch into the bottom of floor joists, ceiling joists, or trusses to accommodate piping.* Whenever possible, keep the edge of the hole at least two inches in from the edge of framing members to avoid nails or screws.

5.10 SUMMARY

This chapter has described both basic and specialized piping hardware used for piping hydronic heating systems. The many combinations of piping, fittings, and valves provide almost unlimited flexibility for the creative system designer. Standard piping approaches that might prove cumbersome in certain situations can be replaced by novel combinations of these components.

Unavoidably, not every valve available has been shown or discussed. New variations of the basic designs appear regularly. The designer should make a practice of scanning trade publications for new developments on a regular basis. Keep a file on both basic and specialized valves, and a variety of piping materials. These are the building blocks of both standard and innovative hydronic systems.

KEY TERMS

1-pipe systems
actuator
angle valve
backflow preventer
ball valve
baseboard tee
boiler feed water valve
check valve
copper water tube
cross-linked polyethylene (PEX) tubing
dielectric union
differential pressure bypass valve
diverter tee
domestic water
end switch
expansion compensator
female pipe thread (FMT)
fittings
flow check valve
flux
galvanic corrosion
gate valve
globe valve
hard drawn tubing
heat motor
lockshield/balancing valve
male pipe thread (MPT)
metered balancing valve
mixing valve
Monoflo® tee
national pipe thread (NPT)
nominal inside diameter
oxygen diffusion barrier
parallel direct return piping
pipe size
piping offset
polybutylene tubing
polymer
pressure relief valve
radiator valve
socket fusion

soft temper tubing
soldering
thermal expansion
thermostatic mixing valve
thermostatic radiator valve
working pressure
zone valve

CHAPTER 5 QUESTIONS AND EXERCISES

1. Describe the differences between type K, L, and M copper tubing.
2. Describe the difference between copper water tube and ACR tubing.
3. What type of solder should be used for piping conveying domestic water? Why?
4. A straight copper pipe 65 ft long changes temperature from 65 °F to 180 °F. What is the change in length of the tube? Describe what would happen if the tube was cooled from 65 °F to 40 °F.
5. A hydronic system contains 50 3/4-in, and 25 1-in soldered copper joints. Estimate the number of inches of 1/8-in wire solder required for these joints.
6. Describe a situation where two diverter tee fittings would be preferred to a single diverter tee.
7. Write a standard description for the fittings shown in Figure 5–73.
8. What should be done prior to soldering a valve with a non-metallic disc or washer?
9. What is the advantage of a four-port mixing valve over a three-port mixing valve when used with a conventional gas or oil-fired boiler?
10. Describe a situation where the high thermal conductivity of copper tubing is undesirable.
11. Describe a situation where a differential pressure bypass valve installed in a multi-zone hydronic distribution system begins to open.
12. Describe three separate functions of a lockshield/balancing valve.
13. Why should the thermostatic actuator of a radiator valve not be mounted above the heat emitter?
14. Why should a globe valve not be used for component isolation purposes? What type(s) of valves are better suited for this application?
15. What is an advantage of a spring-loaded check valve in comparison to a swing check valve? What is a disadvantage?
16. What is the difference between a full port ball valve and conventional port ball valve?
17. Sketch a system piping schematic showing how a thermostatic mixing valve would be used to deliver water at 110 °F to a radiant floor heating system, assuming the heat source is a thermal storage tank at 150 °F.
18. Name two applications for a three-way zone valve.
19. Define nominal pipe size.
20. Does the presence of an oxygen diffusion barrier on polymer tubing totally prevent oxygen entry into the system? Justify your answer.
21. What is the function of the end switch in a zone valve?
22. When can a swing check valve offer the same protection as a backflow preventer in the makeup water line of a hydronic system?

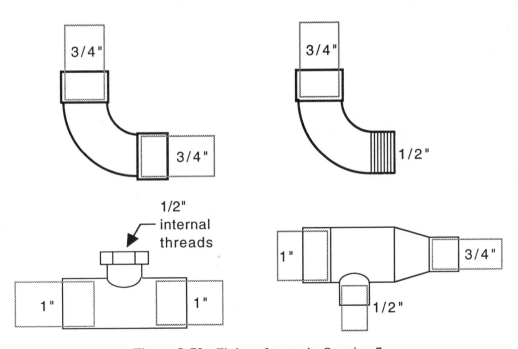

Figure 5–73 Fittings for use in Question 7

6 FLUID FLOW IN PIPING

OBJECTIVES

After studying this chapter you should be able to:

- Describe the difference between flow rate and flow velocity
- Explain how the pressure of a liquid varies with its depth
- Define the term *head* and relate it to pressure
- Calculate or look up the fluid properties factor (α)
- Explain what hydraulic resistance is and how it is used
- Calculate the hydraulic resistance of piping, fittings, and valves
- Sketch hydraulic resistance diagrams for a given piping layout
- Convert two or more parallel hydraulic resistances into an equivalent resistance
- Reduce complex piping systems into simpler equivalent diagrams
- Plot the system resistance curve of a given piping system
- Explain several considerations that relate to pipe sizing
- Estimate the operating cost of a given piping system

6.1 INTRODUCTION

An understanding of how fluids behave while moving through piping is essential to designing good hydronic systems. This chapter lays the groundwork for describing fluid flow in piping. It introduces new analytical methods for predicting flow rates in all parts of a piping system, even those with complex branches and many components. These methods will become a routine part of overall system design that will be used many times in later chapters.

Unfortunately, experience has shown that many of the people responsible for design and installation of hydronic systems don't understand the principles of fluid flow in piping well enough to scrutinize designs before committing time and materials to their installation. This is especially true of nonstandard projects for which there is no previous example to fall back on as a reference. This often leads to hesitation to try anything beyond what is already familiar. Many existing hydronic systems could have been better matched to their buildings, installed for

less money, or operated at a lower cost if their designer had had some way of confidently evaluating how a slightly different design would have operated. This chapter give the designer the tools for simulating the behavior of simple or complex piping systems *before* committing time and money to their installation.

Many of the concepts in this chapter are drawn from classical fluid mechanics, a subject not unknown for its share of mathematics. However, much effort has gone into distilling this theory into concise and easy to work with design tools suitable for day-to-day work.

To further simplify matters, the Hydronics Design Toolkit contains several programs that can quickly evaluate the equations and procedures presented in this chapter. This allows the designer to experiment with the various factors involved, and develop a good feel for how piping systems behave under a wide variety of conditions.

6.2 BASIC CONCEPTS OF FLUID MECHANICS

What Is a Fluid?

In a purely technical sense the word fluid can mean *either* a **liquid**, such as water, or a **gas**, such as air. The science of fluid mechanics includes the study of both liquids and gases. However, from the standpoint of hydronic heating, the word fluid almost always refers to a specific *liquid* that is being circulated through the system. *In the context of this book, the word fluid will always refer to a liquid unless otherwise stated.*

Most liquids, including water, are **incompressible**. This means that any given amount of liquid always occupies the same volume, with the exception of minor changes in volume due to thermal expansion. A simple example would be a long pipe completely filled with water. If another gallon of water were pushed into one end of the pipe, the exact same amount would have to flow out the other end. The additional water cannot be squeezed into the pipe.

By comparison, if the same pipe were filled with a gas such as air, it would be possible to push another gallon of air into the pipe without allowing any air out the other end. The result would be an increase in the pressure of the air within the pipe. Because of this, gases are said to be **compressible fluids**.

The fact that liquids are incompressible makes it easy to describe their motion through closed containers such as piping systems. Two basic terms associated with this motion are **flow rate** and **flow velocity**.

Flow Rate

Flow rate is a measurement of the *volume* of fluid to pass a given point in the piping in a given period of time. In the U.S., the customary units for expressing flow rate are *gallons per minute* (gpm). Within this book, flow rate is represented in equations by the symbol f. The flow rate of a fluid within a hydronic system will have a major effect on the system's performance.

Flow Velocity

The motion of a fluid flowing through a pipe is more complex than that of a solid object moving along a straight path. The speed, or flow velocity of the fluid, is different at different points across the diameter of the pipe. To visualize the flow velocity of a fluid, think about a single drop of fluid that could be colored red and remain intact as it flowed through the surrounding fluid. The drop would move faster near the centerline of the pipe compared to near the edges. If the drop's speed could be measured at several points between the edge and the centerline of the pipe, and the speeds represented by arrows of varying length, a **velocity profile**, as shown in Figure 6–1a, could be drawn.

The term flow velocity is commonly understood to mean the *average* **flow velocity** of the fluid as it moves through a pipe. This average velocity would be the velocity that, if present across the entire cross section of the pipe, would result in the same flow rate as that created by the actual velocity profile. The concept of average flow velocity is illustrated in Figure 6–1b. Unless otherwise stated, when the term flow velocity is referred to in this book, it will mean the *average flow velocity* in the pipe. The common units for flow velocity in the U.S. are feet per second, abbreviated either as ft/sec or FPS. Within equations, the average flow velocity is represented by the symbol, v.

A simple equation relates the flow rate of a liquid to its average flow velocity.

(Equation 6.1)

$$f = 3.12 \ vA$$

where:

f = flow rate through the pipe (gpm)
3.12 = constant associated with the units of the equation
v = average flow velocity in the pipe (ft/sec)
A = inside cross-sectional area of the pipe (in^2)

Example 6.1: What is the flow rate of water moving at an average velocity of 8 ft/sec through a ¾-in type M copper tube?

Solution: From Figure 5–1, the inside diameter of a ¾-in type M copper tube is 0.811 inches. Its cross-sectional area can be found using the equation for the area of a circle:

$$A = \frac{\pi d^2}{4} = \frac{\pi (0.811 \ \text{in})^2}{4} = 0.517 \ \text{in}^2$$

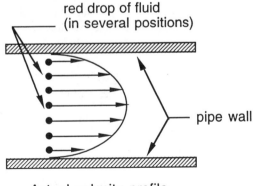

Actual velocity profile
within a pipe

(a)

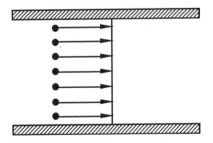

Concept of an "average" velocity profile resulting in the same flowrate as the actual velocity profile

(b)

Figure 6–1   (a) Representation of actual velocity profile, (b) equivalent "average" velocity profile.

This is now substituted into Equation 6.1 to obtain the flow rate:

$$f = 3.12\ (8)\ (0.517) = 12.9\ \text{gpm}$$

Equation 6.1 can be rearranged to express flow velocity as a function of flow rate:

(Equation 6.2)

$$v = \frac{f}{3.12\,A}$$

where:

v = average flow velocity in the pipe (ft/sec)
f = flow rate through the pipe (gpm)
3.12 = constant associated with the units of the equation
A = inside cross-sectional area of the pipe (in²)

When selecting a pipe size for a given flow rate, the resulting average flow velocity should be calculated. Average flow velocities in excess of 4 feet per second can cause flow noise within the system and thus should be avoided. Figure 6–2 lists the relationship between flow rate and average flow velocity for type M copper tube in sizes from ½ inch to 2 inches. In these equations, v represents average flow velocity in feet per second and f represents flow rate in gallons per minute. The Hydronics Design Toolkit contains a program named PIPESIZE that can also be used to compute the relationship between flow rate and average flow velocity for these same pipes.

Static Pressure of a Liquid

Consider a pipe with a cross-sectional area of 1 square inch, filled with water to a height of 10 feet, as shown in Figure 6–3.

Assuming the water's temperature is 60 °F, its density is about 62.4 lb/ft³. The weight of the water in the pipe could be calculated by multiplying its volume by its density.

Copper tube size	Flow velocity
1/2"	$v = 1.262\ f$
3/4"	$v = 0.621\ f$
1"	$v = 0.367\ f$
1-1/4"	$v = 0.245\ f$
1-1/2"	$v = 0.175\ f$
2"	$v = 0.101\ f$

v = flow velocity (in feet per second)
f = flow rate (in GPM)

Figure 6–2 Relationship between flow rate and flow velocity for type M copper tube

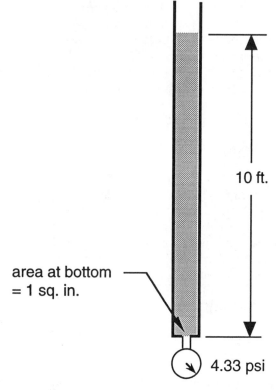

Figure 6–3 Column of water 10 feet high in a pipe with a cross-sectional area of 1 in²

$$\text{weight} = (\text{volume})(\text{density})$$

$$= (\text{cross-sectional area})(\text{height})(\text{density})$$

$$\text{weight} = (1\ \text{in}^2)(10\ \text{ft})\left(62.4\ \frac{\text{lb}}{\text{ft}^3}\right)\left(\frac{1\ \text{ft}^2}{144\ \text{in}^2}\right) = 4.33\ \text{lb}$$

The water exerts a pressure on the bottom of the pipe due to its weight. This pressure is equal to the weight of the water column divided by the area of the pipe. In this case:

$$\text{Static pressure} = \frac{\text{weight of water}}{\text{area of pipe}} = \frac{4.33\ \text{lb}}{1\ \text{in}^2}$$

$$= 4.33\ \frac{\text{lb}}{\text{in}^2} = 4.33\ \text{psi}$$

This pressure is called the **static pressure** exerted by the water. The word static means the fluid is not moving. In this case the static pressure at the bottom of the pipe is due solely to the weight of the water. If an accurate pressure gauge were mounted into the bottom of the pipe as shown in Figure 6–3, it would read exactly 4.33 psi.

Next consider a pipe with a cross-sectional area of 1 *square foot* filled with water to a height of 10 feet. The weight of this column can be calculated as:

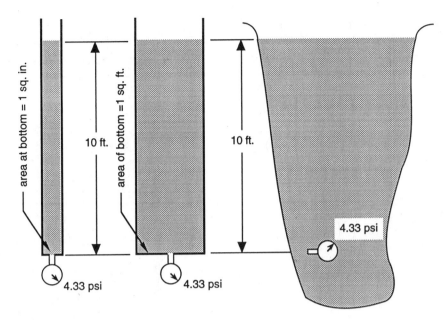

Figure 6-4   The static pressure of a fluid at a given point depends only on depth below the surface, and not on the size or shape of the container.

$$\text{weight} = (1 \text{ ft}^2)(10 \text{ ft})\left(62.4\ \frac{\text{lb}}{\text{ft}^3}\right) = 624 \text{ lb}$$

The static pressure at the bottom of the column would be:

$$\text{static pressure} = \frac{\text{weight of water}}{\text{area of pipe}} = \frac{624 \text{ lb}}{144 \text{ in}^2} = 4.33\frac{\text{lb}}{\text{in}^2}$$

$$= 4.33 \text{ psi}$$

Notice the static pressure at the bottom of the column did not change even with a large change in the diameter of the pipe. This is explained by the fact that *the larger column of water has a proportionally larger area across which to distribute its weight.* Thus the static pressure at a given depth below the surface of a liquid remains exactly the same regardless of the diameter of the pipe. This same principle remains true for liquids within containers of any size and shape, as shown in Figure 6-4. Scuba divers, for example, feel a static pressure that depends only on their depth below the surface of the water, not on how wide the body of water is.

The static pressure of a liquid increases proportionally from some value at the surface to larger values at greater depths below the surface as shown in Figure 6-5. Again the size or shape of the container makes no difference.

If the top of the system shown in Figure 6-5 were closed off, and a tank containing pressurized air at 5 psi were attached to the top of the system, the static pressure of the liquid at all locations within the container would simply increase by 5 psi as shown in Figure 6-6.

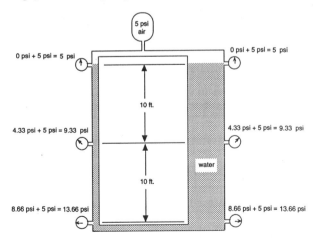

Figure 6-5   Static pressure for various depths below the fluid surface. Notice that the size of the column, or shape of the overall container, has no effect on static pressure.

Figure 6-6   The effect of increasing the pressure at top of the piping circuit. All pressures below the surface increase by the same amount.

From these observations it is possible to develop a relationship between the depth of water within a container and the static pressure it exerts at any point below its surface.

For *water only* the following approximation is useful:

(Equation 6.3)

$$P_{static} = \frac{h}{2.31} + P_{surface}$$

where:

P_{static} = static pressure at a given depth, *h*, below the surface of the water (psi)
 h = depth below the water's surface (ft)
 2.31 = unit conversion factor for water at 60 °F
$P_{surface}$ = any pressure applied at the surface of the water (psi)

If the water is open to the atmosphere at the top of the container, even through a pin hole size opening, then $P_{surface} = 0$. If the container is completely closed and filled with water, $P_{surface}$ is the pressure of the water at the top of the container.

The reason Equation 6.3 is an approximation is that the density of water, and thus the static pressure it creates within a container, changes with temperature. To account for such changes, and to allow for calculations with liquids other than water, Equation 6.3 can be modified as follows:

(Equation 6.4)

$$P_{static} = \frac{(h)(D)}{144} + P_{surface}$$

where:

P_{static} = static pressure at a given depth, *h*, below the surface of the liquid (psi)
 h = depth below the liquid's surface (ft)
 D = density of the liquid (lb/ft³)
$P_{surface}$ = any pressure applied at the surface of the liquid (psi)

To use Equation 6.4, the density of the liquid must be determined. The density of water over the temperature range of 50 °F to 250 °F can be found in Figure 4–5. The density of several other liquids commonly used in hydronic heating systems can be determined over this same temperature range using the FLUIDS program in the Hydronics Design Toolkit.

Example 6.2: Find the approximate static pressure at the bottom of a hydronic system with an overall height of 65 ft, filled with water, and having a pressure of 10 psi at its uppermost point. Refer to Figure 6–7.

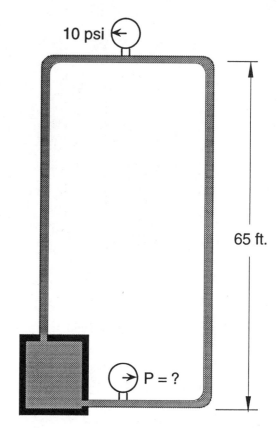

Figure 6–7   Illustration for Example 6.2

Solution: Use Equation 6.3 to calculate the static pressure:

$$P_{static} = \frac{65}{2.31} + 10 = 38.1 \text{ psi}$$

Discussion: The calculated pressure of 38.1 psi would be correct if the water temperature were about 60 °F. If, however, the same system contained water at 250 °F (having a density of 58.75 lb/ft³), Equation 6.4 would have to be used, and would yield the following:

$$P_{static} = \frac{(65)(58.75)}{144} + 10 = 36.5 \text{ psi}$$

Notice the higher temperature water would exert less pressure at the base of the system due to its lower density.

The Head of a Fluid

Fluids can contain both thermal and mechanical energy. Thermal energy content depends on the temperature of the fluid. Mechanical energy content depends on the pressure, density, and velocity of the fluid. *The term **head** refers to the total mechanical energy contained by a fluid at some pressure and flow velocity.* This is a very important concept that will be used throughout the remainder of this book.

A mathematical expression that precisely describes the head of a liquid at a given point in a piping system is given as Equation 6.5.

(Equation 6.5)

$$\text{Total Head} = H_{\text{total}} = \frac{v^2}{64.4} + \frac{P_{\text{static}}(144)}{D}$$

where:

H_{total} = total head of the liquid (ft • lb/lb, or ft)
v = flow velocity of the liquid (ft/sec)
p_{static} = static pressure of the liquid at some point in the system (psi)
D = density of the liquid (dependent on temperature) (lb/ft³)

The first term in this equation, $v^2/64.4$, is called the **velocity head** of the liquid. It indicates the mechanical energy the liquid has because of its motion. The second term in the equation, $P_{\text{static}}(144)/D$, is called the **pressure head** of the liquid. It indicates the mechanical energy the liquid has due to its pressure. This second term is derived from Equation 6.4, where the term P_{static} represents the combined pressure due to liquid depth and surface pressure. *The **total head**, or total mechanical energy content, of the liquid is the sum of the velocity head and pressure head.*

At the flow velocities present in typical hydronic systems, the amount of energy represented by the velocity head is quite small compared to that represented by the pressure head. To illustrate this, the total head of 140 °F water at 15 psi static pressure and moving at 4 ft/sec can be calculated using Equation 6.5:

$$H_{\text{total}} = \frac{4^2}{64.4} + \frac{15(144)}{61.35} = 0.248 + 35.2 = 35.45 \text{ ft of head}$$

Notice that the velocity head is only (0.248/35.45)(100) = 0.7% of the total head.

Liquids have the ability to exchange head energy back and forth between velocity and pressure. For example, when a liquid flows through a transition from a smaller to larger pipe size, its velocity decreases, while its pressure increases slightly because some of the velocity head is instantly converted into pressure head. This change in pressure, however, is so small that it would likely be unnoticed.

The units for head are (ft • lb/lb). Those having studied physics will recognize the unit of ft • lb as a unit of energy. As such it can be converted to any other unit of energy such as a Btu. However, engineers long ago chose to cancel the units of lb in the numerator and denominator of this ratio, and express head in the sole remaining unit of feet. To make a distinction between feet as a unit of distance and feet as a unit of fluid energy, the latter can be stated as **feet of head**.

When head is added to or removed from a liquid in a closed-loop piping system, there will always be an

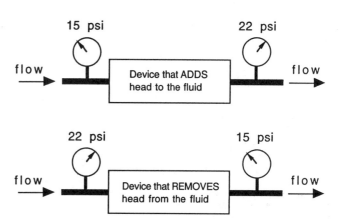

Figure 6–8 Pressure changes associated with head that is added or removed from a flowing fluid

associated pressure change in the fluid. In the case where head is lost, a drop in pressure will occur. When head is added, an increase in pressure will occur. This concept is illustrated in Figure 6–8.

Using pressure gauges to detect changes in the head of a liquid is like using thermometers to detect changes in its thermal energy content. Since energy, whether in mechanical or thermal form cannot be seen directly, its addition or removal from a liquid must be calculated based on changes in other physical quantities that can be directly measured.

Equation 6.6 can be used to convert the head that is either added to or lost from a liquid, to the associated change in pressure of a liquid. Note that this equation requires the density of the liquid which depends on its temperature with the system.

(Equation 6.6)

$$\Delta P = \frac{HD}{144}$$

where:

ΔP = pressure change corresponding to the head added or lost (psi)
H = head added or lost from the liquid (feet of head)
D = density of the liquid at its current temperature (lb/ft³)

This equation can also be rearranged to calculate the head associated with a given change in pressure:

(Equation 6.7)

$$H = \frac{144 \Delta P}{D}$$

Head Loss due to Viscous Friction

Whenever a fluid flows, an energy dissipating effect called viscous friction develops both within the fluid

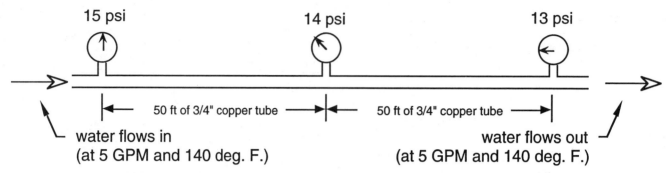

Figure 6-9 has labels: 15 psi, 14 psi, 13 psi; 50 ft of 3/4" copper tube, 50 ft of 3/4" copper tube; water flows in (at 5 GPM and 140 deg. F.); water flows out (at 5 GPM and 140 deg. F.)

Figure 6-9 Pressure drop in downstream direction associated with head loss created by viscous friction.

stream and along any surfaces it contacts. Since the mechanical energy content of the fluid is called head, it is appropriate to say that viscous friction results in a head loss from the fluid. Viscous friction develops within all fluids, but is greater for fluids with higher viscosities, or when a fluid flows over relatively rough surfaces.

The drag forces associated with viscous friction cause fluid "particles" closer to a surface to have lower velocities than those farther away from the surface. A familiar example is the slower speed of water near the banks of a stream compared to near the center of the stream. Another is lower wind speeds near the surface of the earth. Within pipes, viscous friction gives rise to a velocity profile similar to that shown in Figure 6-1a.

Because of its incompressibility, a liquid moving at a constant flow rate through a given size pipe cannot slow down as head is dissipated due to viscous friction. Therefore the *head loss reveals itself as a drop in pressure in the downstream direction of flow* as shown in Figure 6-9.

To maintain the fluid flow, it is necessary to replace this lost head. In a hydronic system this is done using a circulator pump to add head back to the fluid. Methods for calculating head loss will be detailed in Section 6.4.

Fluid-Filled Piping Loops

Nearly all smaller hydronic heating systems consist of closed piping loops or circuits. After all piping work is completed, these circuits are *completely* filled with fluid. During normal operation no additional fluid enters or leaves the system.

Consider such a piping loop as shown in Figure 6-10. Using procedures from Example 6.2, the pressure at point A due to the weight of the fluid in the left side of the loop could be calculated. It might appear that a pump placed at point A would have to overcome this pressure in order to lift the water upward to establish and maintain circulation in the loop. However, this is not true. The reason is the weight of the fluid on the right side of the

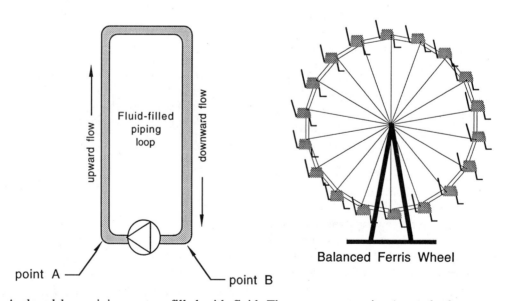

Labels: upward flow, downward flow, Fluid-filled piping loop, point A, point B, Balanced Ferris Wheel

Figure 6-10 A closed loop piping system filled with fluid. The pressure at point A equals the pressure at point B regardless of the height or shape of the loop. To maintain circulation, the pump need only overcome the head loss due to viscous friction.

loop pushes downward at point B and exactly cancels out the need to lift the fluid on the left side.

The fluid in the loop acts like a Ferris wheel with passengers of the same weight in each seat. The weight of the up-going passengers exactly balances the weight of the down-going passengers. If it weren't for friction in the bearings and air resistance, this balanced Ferris wheel would continue to rotate indefinitely once started. Likewise, if it weren't for the viscous friction of the fluid, it too would continue to circulate indefinitely within a piping loop.

The following principle can thus be stated: *To maintain a constant rate of circulation in any filled piping loop, the pump need only replace the head loss created by viscous friction in the piping system. This remains true regardless of the shape, height, or pipe size(s) used in the loop.*

This principle explains why a small pump can establish and maintain water circulation around a filled piping loop, even if the top of the loop is several stories above the pump, and contains hundreds, or even thousands of gallons of water. *Unfortunately, many pumps are needlessly oversized because this principle is not understood.*

Flow Classifications

Fluids exhibit two characteristic types of flow as they move through pipes. These are called **laminar flow** and **turbulent flow**. Both types can either enhance or detract from the performance of hydronic systems depending on where they occur.

To discuss the differences between these types of flow, it is helpful to use the concept of streamlines. A streamline indicates the path taken by an imaginary fluid particle as it moves along a pipe. Streamlines allow one to visualize flow as if the fluid were actually composed of millions of small particles, each with its own unique path.

Laminar flow is characterized by smooth, straight streamlines as the fluid moves through the pipe, as shown in Figure 6–11a. This type of flow is likely to exist when flow velocities are very low. To visualize laminar flow, think of the fluid as if it were separated into many layers that resemble concentric cylinders inside a pipe as shown in Figure 6–11b. The closer a given fluid layer is to the inside surface of the pipe, the slower it moves. The faster moving fluid layer tends to slide over the slower moving layers in such a way that mixing between layers does not occur.

Laminar flow creates minimal amounts of viscous friction. This is desirable because the lower head loss means less pumping power will be required to maintain a given flow rate. This is why large pipelines that carry heated or chilled water over hundred or thousands of feet between buildings are often designed to maintain laminar flow.

Another characteristic of laminar flow is that it allows a **boundary layer** to develop near the pipe wall. The boundary layer is a very slow moving layer of fluid that slides along the inner wall of the tube with very little mixing effect. Such a layer creates a thermal resistance between the pipe wall and the bulk of the fluid stream, reducing the rate of heat transfer between them. Because of this, *laminar flow is undesirable if the objective is to move heat between the pipe wall and fluid stream.*

The conditions present in most small hydronic piping systems almost always result in turbulent flow rather than laminar flow in most parts of the system. Turbulent flow is characterized by a vigorous mixing of the fluid as it travels down the pipe. The streamline of a single fluid particle shows it moves in an erratic path that often sweeps in toward the pipe wall, and then back out into the bulk of the fluid stream as shown in Figure 6–12. This mixing effect improves heat transfer because it drastically reduces the thickness and thermal resistance of the fluid's boundary layer. Because of this, heat exchangers, radiators, or any other device that is intended to transfer heat between a fluid and a solid surface should be designed or applied so as to maintain turbulent flow.

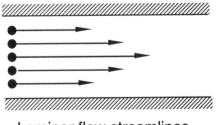

Laminar flow streamlines

(a)

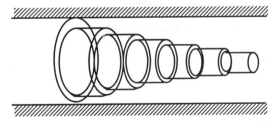

Concept of "Layers" of fluid
sliding over each other
during laminar flow.

(b)

Figure 6–11   The concept of laminar flow. (a) velocity profile within a pipe, (b) concentric layers of fluid sliding over each other.

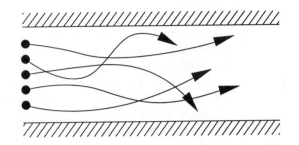

Figure 6–12 Representation of streamlines during turbulent flow

An undesirable aspect of turbulent flow is that it creates more viscous friction compared to laminar flow, and therefore requires more pumping power to maintain a given flow rate. However, the increased heat transfer capabilities of turbulent flow are usually far more important than the penalty associated with the increased pumping power.

Reynold's Number

A method exists for predicting if the flow within a pipe will be laminar or turbulent. This method, derived from experimental observations, is based on calculating the **Reynold's number** of the fluid. The calculated value is then compared to a threshold value of 4,000. *If the calculated Reynold's number is over 4,000 the flow is turbulent. If it is below 4,000, the flow may be either laminar or turbulent. If it is below 2,300 the flow will be laminar.*

The Reynold's number for flow in a pipe can be calculated using Equation 6.8:

(Equation 6.8)

$$Re\# = \frac{vdD}{\mu}$$

where:

v = average flow velocity of the fluid (ft/sec)
d = internal diameter of pipe *(ft)*
D = fluid's density (lb/ft³)
μ = fluid's dynamic viscosity (lb/ft/sec)

Example 6.3: Determine the Reynold's number of water at 140 °F flowing at 5 gpm through a ¾-in type M copper tube. Is this flow laminar or turbulent?

Solution: To calculate the Reynold's number, the density and dynamic viscosity of the water must first be determined. Using the graph in Figure 4-5, the water's density is found to be 61.35 lb/ft³. Using the graph in Figure 4–6, the water's dynamic viscosity is determined as 0.00032 lb/ft/sec. (*Note:* The density and dynamic viscosity of

water could also have been found using the FLUIDS program in the Hydronics Design Toolkit.)

The inside diameter of a ¾-in type M copper tube is found in Figure 5–1 to be 0.811 inches. This must be converted to feet to match the stated units for Equation 6.8:

$$d = (0.811 \text{ in})\left(\frac{1 \text{ ft}}{12 \text{ in}}\right) = 0.06758 \text{ ft}$$

To get the average flow velocity corresponding to a flow rate of 5 gpm, use the equation for a ¾-in copper tube from Figure 6–2:

$$v = (0.6211)(f) = (0.6211)(5) = 3.106 \text{ ft/sec}$$

These values can now be substituted into Equation 6.8:

$$Re\# = \frac{(3.106 \text{ ft/sec})(0.06758 \text{ ft})(61.35 \text{ / ft}^3)}{0.00032 \text{ lb/ft/sec}}$$

$$= 40,242$$

This value is well above the threshold of 4,000, and therefore the flow is turbulent.

Discussion: Note when the quantities with the stated units are substituted into Equation 6.8, they cancel each other out. Thus the Reynold's number is a unitless quantity. This must always be true, and serves as a check that the proper units are being used in Equation 6.8. The Reynold's number for several different fluids and pipe sizes can also be calculated using the PIPEPATH program in the Hydronics Design Toolkit.

6.3 ANALYZING FLUID FLOW IN SMOOTH PIPES

This section introduces a new method for predicting the head loss associated with fluid flow in hydronic systems assembled from **"smooth"** pipe such as copper tubing, PEX, PB, or other materials of comparable smoothness.

Most residential and light commercial hydronic systems are currently being piped with such materials. Their smooth inside surfaces create less disturbance of the fluid stream, and hence lesser amounts of head loss. *If major portions of a system will be piped with steel or iron piping, this method will somewhat underestimate head loss, and should not be used without suitable corrections.*

This new method is based on a combination of established engineering models for fluid flow in pipes, as well as methods borrowed from electrical circuit analysis. It is based on the premise that each component in a piping assembly or circuit has an associated **hydraulic resistance** that defines its ability to remove head from the fluid at any given flow rate.

The Similarity between Fluid Flow and Electricity

To most people the concepts of electricity and fluid flow seem totally different. Electrical circuits involve concepts such as voltage, current, and resistance. Piping systems deal with flow rate, head loss, and pressure drop. What could these systems possibly have in common? The answer lies in the mathematical models that engineers have developed for describing the behavior of both types of systems.

The Electrical Device Model

The classical model for describing the operating characteristics of an electrical device is Ohm's law. It is among the first concepts taught in a course on electrical circuits. Mathematically, Ohm's law can be stated as follows:

(Equation 6.9)

$$v = ir$$

where:

v = voltage difference *across* the device (volts)
i = current flowing *through* the device (amperes)
r = electrical resistance of the device (ohms)

Voltage can be though of as the *driving force* that tries to push electrons through the device. It is measured with a voltmeter across the device as shown in Figure 6–13. Current is a measurement of the *rate of flow* of the electrons through the device. Resistance is the ability of the device to *oppose* the flow of electrons.

The Piping Device Model

For the following discussion, the term *piping device* refers to any component such as a pipe, fitting, or valve through which fluid flows. The relationship between head loss, flow rate, and hydraulic resistance for such a piping device can be described by Equation 6.10:

(Equation 6.10)

$$H_L = r(f)^{1.75}$$

where:

H_L = head loss across a piping device due to viscous friction (ft of head)
r = hydraulic resistance of the piping device
f = flow rate of fluid through the piping device (gpm)
1.75 = exponent (or power) of f

The head loss can be thought of as the energy necessary to drive the fluid through the device at a certain flow rate. Recall that head loss reveals itself as a pressure drop across the device. The flow rate, f, is a measurement of the *rate of flow* of the fluid *through* the device. The hydraulic resistance is a measurement of the ability of the device to *oppose* this flow. These concepts are illustrated in Figure 6–14.

It is interesting to compare the mathematical equation for electrical devices (Ohm's law) with that for piping devices (Equation 6.10). Figure 6–15 shows a one to one correspondence between the physical quantities in these equations. Head loss *across* a piping device corresponds to voltage drop *across* an electrical device. Electrical current, the indicator of flow rate of electrons *through* an electrical device, corresponds to the flow rate of fluid *through* the piping device. Finally, electrical resistance corresponds to what we now call hydraulic resistance.

A major difference between these equations quickly becomes apparent. In the electrical device equation, the

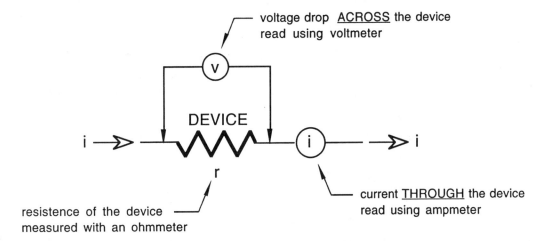

Figure 6–13 The concept of voltage, current, and resistance for an electrical device

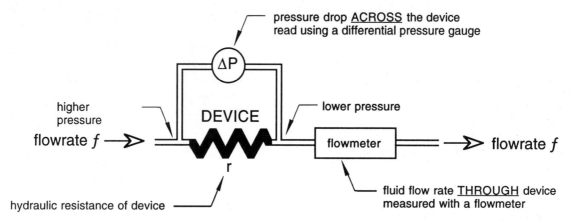

Figure 6–14 The concept of pressure drop (associated with head loss), flow rate, and hydraulic resistance for a piping device

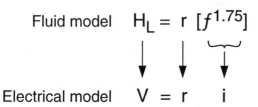

Figure 6–15 Correspondence of terms between equations for piping devices and electrical devices

electrical current through the device is directly proportional to the voltage across the device. Thus, if the voltage across the device were doubled, the current through the device would also double. In the case of the piping device model, the head loss across the device is not directly proportional to the flow rate through the device. This is because the flow rate term in Equation 6.10 is raised to the 1.75 power. If the flow rate through the piping device were doubled, the head loss across it would increase by about 3.4 times! This difference makes the fluid device equation more mathematically complex, but still manageable. Figure 6–16 shows a comparison between representative graphs of these two equations.

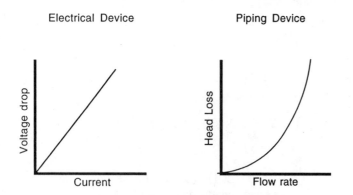

Figure 6–16 Graphical comparison between the equation describing the electrical device (Equation 6.9), and the equation describing the piping device (Equation 6.10)

Determining Hydraulic Resistance

Almost any component used in a hydronic piping circuit, be it a length of pipe, fitting, valve, heat emitter, or otherwise, can be represented by a hydraulic resistance. Devices that create relatively little interference with the fluid stream will have low hydraulic resistance. Those which have major restrictions, or twisting paths through which the fluid must flow, will have higher hydraulic resistances. After the hydraulic resistances of all devices in a piping assembly are determined, they can be combined in an **equivalent resistance** similar to how electrical resistances are combined in circuits analysis.

The hydraulic resistance term, r, in Equation 6.10 can be expanded as shown in Equation 6.11:

(Equation 6.11)
$$r = (\alpha c L)$$

therefore:

(Equation 6.12)
$$H_L = r(f)^{1.75} = (\alpha c L)(f)^{1.75}$$

where:

H_L = head loss across a device due to viscous friction (ft of head)

α = **fluid properties factor** based on the fluid's density and viscosity (Equation 6.13)

c = constant based on the size of the tube (see table in Figure 6–18)

L = length of pipe, or **equivalent length** of a piping device (ft)

f = flow rate of fluid through the component (gpm)

Equation 6.12 is a specialized relationship derived from the widely used Darcy-Weisbach equation. It incorporates an empirical relationship for the Moody

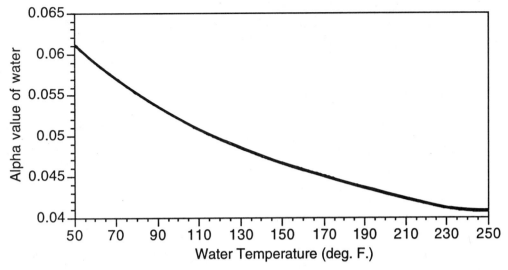

Figure 6–17   Graph of α-value of water for temperatures of 50 °F to 250 °F

friction factor based on turbulent flow in smooth piping. Because of this empirical factor, *Equation 6.12 is only valid for turbulent flow with Reynold's numbers in the range of 4,000 to about 200,000.* Fortunately, this covers most of the operating conditions found in residential and light commercial hydronic systems. At Reynold's numbers above 200,000 the equation gradually begins to underestimate head loss. At a Reynold's number of 300,000 it underestimates head loss by about 6%.

The Reynold's number for fluid flow in pipes can be found using Equation 6.8. It can also be quickly calculated for a range of pipe sizes, fluids, and temperatures using the PIPEPATH program in the Hydronics Design Toolkit. It is prudent to verify the Reynold's number is between 4,000 and 200,000 when using Equation 6.12 for system design.

The Fluid Properties Factor (α)

The fluid properties factor, α (pronounced alpha), which is a factor in Equation 6.11 and 6.12, is defined as follows:

(Equation 6.13)

$$\alpha = \left(\frac{D}{\mu}\right)^{-0.25}$$

where:

α = fluid properties factor
D = density of the fluid (lb/ft³)
μ = dynamic viscosity of the fluid (lb/ft/sec)

The α value is a total mathematical description of the fluid's properties for purposes of determining head loss in smooth piping systems.

Chapter 4 described how the density and dynamic viscosity of fluids depend on temperature. Because it is derived from these properties, the α value is also dependent on temperature. Figure 6–17 is a plot of the α value for water over a temperature range of 50 °F to 250 °F. The FLUIDS program in the Hydronics Design Toolkit can also be used to quickly find density, viscosity, and the α value for several fluids commonly used in hydronic heating systems.

The *c* in Equation 6.11 and 6.12 is a constant for a given size of copper tubing. It incorporates dimensional information such as interior diameter, cross-sectional area, and appropriate unit conversion factors into a single value. It depends only on pipe size and can be found in the table of Figure 6–18.

Example 6.4: Determine the pressure drop created when water at 140 °F flows through 100 ft of ¾-in type M copper tube at a rate of 5 gpm.

Solution: This situation requires the use of Equation 6.12. But first the value of α for water at 140 °F, as well as the *c*-value for ¾-in copper tube, must be determined.

copper tube size	c-value
1/2"	0.33352
3/4"	0.061957
1"	0.01776
1-1/4"	0.0068082
1-1/2"	0.0030668
2"	0.00083317

Figure 6–18   Values of *c* for type M copper tube for use in Equations 6.11 and 6.12

The value of α for water can be found using either Figure 6–17 or Equation 6.13. In this case the latter will be used. The density of water at 140 °F is found from Figure 4-5 to be 61.35 lb/ft^3. The dynamic viscosity of water at 140 °F is found from Figure 4–6 to be 0.00032 lb/ft/sec. The α value can now be calculated using Equation 6.13:

$$\alpha = \left(\frac{D}{\mu}\right)^{-0.25} = \left(\frac{61.35}{0.00032}\right)^{-0.25} = (191719)^{-0.25}$$

$$= 0.04779$$

The value of c is found from the table of Figure 6–18. For 3/4-in copper tube, $c = 0.061957$

All data can now be substituted into Equation 6.12:

$$H_L = (\alpha c L)(f)^{1.75} = [(0.04779)(0.061957)(100)](5)^{1.75}$$

$$= 4.95 \text{ ft of head}$$

It should be stressed that the result of Equation 6.12 is the head loss along the pipe, *not the pressure drop*. This result can be converted to a corresponding pressure drop using Equation 6.6:

$$\Delta P = \frac{(4.95)(61.35)}{144} = 2.11 \text{ psi}$$

Thus a pressure gauge at the outlet of the 100 ft of 3/4-in copper tube should have a reading 2.11 psi lower than the pressure at the inlet.

The Flow Coefficient C_v

Another method for predicting pressure drop due to head loss has been used within the valve industry for many years. It is based on a parameter called the **flow coefficient** (C_v) of the valve.

The C_v is defined as the flow rate of 60 °F water that will create a pressure drop of 1.0 psi through the valve. For example, a valve with a C_v rating of 5.0 would require a flow rate of 5.0 gpm of 60 °F water to create a pressure drop of 1.0 psi across it. Many valve manufacturers list the C_v values of their products in their technical literature. Occasionally C_v values will be listed for devices other than valves.

The C_v value can be used in an equation that estimates pressure drop based on the *square* of the flow rate of fluid through a device. This relationship, given as Equation 6.14, is often used for determining pressure drops through valves.

(Equation 6.14)

$$\Delta P = \left(\frac{D}{62.4}\right)\left(\frac{f}{C_v}\right)^2$$

where:

ΔP = pressure drop across the device (psi)
D = density of the fluid at its operating temperature (lb/ft^3)
62.4 = density of water at 60 °F (lb/ft^3)
f = flow rate of fluid through the device (gpm)
C_v = known C_v rating of the device (gpm)

Example 6.5: Estimate the pressure drop across a radiator valve having a C_v value of 2.8, when 140 °F water flows through at 4.0 gpm.

Solution: The density of 140 °F water is 61.35 lb/ft^3

Using Equation 6.14 the pressure drop is:

$$\Delta P = \left(\frac{61.35}{62.4}\right)\left(\frac{4.0}{2.8}\right)^2 = 2.0 \text{ psi}$$

For smooth piping such as copper tubing or PEX, the square relationship of Equation 6.14 will over-predict the pressure drop. A program called HYRES in the Hydronics Design Toolkit can determine an approximate value of hydraulic resistance using the C_v value of a device. This hydraulic resistance can then be used with Equation 6.12.

Use of Pressure Drop Charts

Many references on piping design contain charts such as the one shown in Figure 6–19. These charts shows the relationship between flow rate, flow velocity, and pressure drop due to head loss, for several sizes of copper tubing. They are usually based on water at a temperature of 60 °F, which is typical in domestic (cold) water distribution systems. The reference temperature of 60 °F is, however, not typical in hydronic heating systems. The higher temperature water in such systems will have lower density and lower viscosity, resulting in less head loss, and thus smaller pressure drops. Use of pressure drop charts based on 60 °F water will consistently overestimate the pressure drop in hydronic piping circuits because the water will almost always be at higher temperatures. This can result in unnecessary oversizing of the system's pumps. Because of this, *these charts should not be used for estimating pressure drop due to head loss in hydronic heating systems.* However, they can still be used for finding the relationship between flow rate and flow velocity.

Example 6.6: Use Equations 6.12 and 6.13 to recalculate the pressure drop for 100 ft of 3/4-in copper tubing transporting 60 °F water at 5 gpm. Compare the results with those of Example 6.4, and with those obtained from the chart in Figure 6–19.

Tube Type and Nominal or Standard Size, Inches

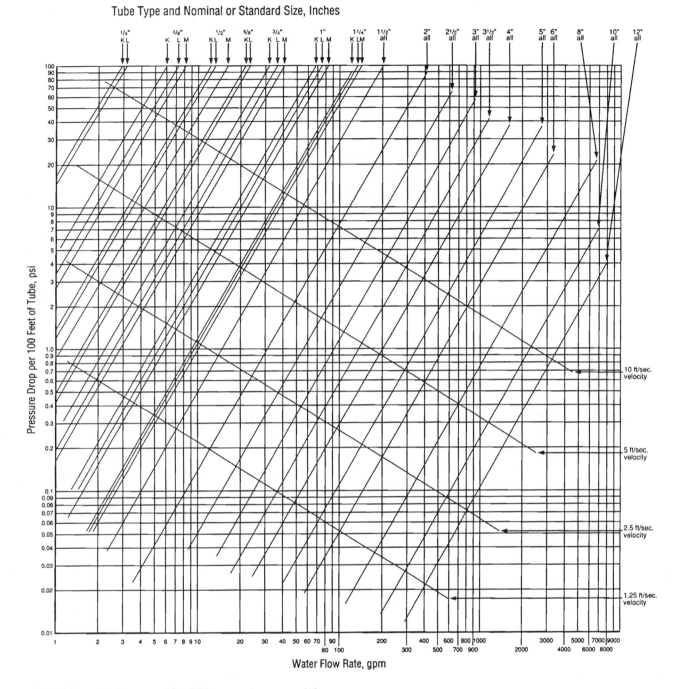

Figure 6–19   Pressure drop chart for copper tubing carrying 60 °F water. Courtesy of the Copper Development Association.

Solution: Using the graph in Figure 4-5, the density of water at 60 °F is estimated to be 62.3 lb/ft³. Using the graph in Figure 4-6, the dynamic viscosity of water at 60 °F is estimated at 0.00075 lb/ft/sec. The α value can be calculated using Equation 6.13:

$$\alpha = \left(\frac{D}{\mu} \right)^{-0.25} = \left(\frac{62.3}{0.00075} \right)^{-0.25} = (83067)^{-0.25}$$

$$= 0.0589$$

The value of c for ³/₄-in tubing is found in the table of Figure 6–18: $c = 0.061957$.

Substituting these values into Equation 6.12 along with the pipe length and flow rate:

$$H_{\text{L}} = (\alpha c L)(f)^{1.75} = [(0.0589)(0.061957)(100)](5)^{1.75}$$

$$= 6.10 \text{ ft of head}$$

The corresponding pressure drop is found using Equation 6.6:

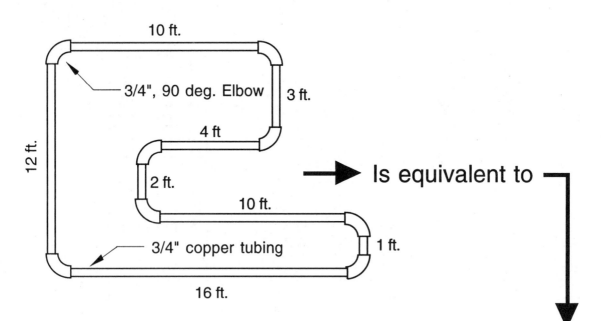

58 ft. of pipe + 8 (2 equivalent feet per elbow) = 74 ft. total equivalent length

Figure 6–20   Converting a piping circuit containing pipe and fittings to an equivalent length of straight pipe

$$\Delta P = \frac{(6.10)(62.3)}{144} = 2.64 \text{ psi}$$

The pressure drop for the same piping using 140 °F water in Example 6.4 was 2.11 psi. Comparing these results shows the pressure drop created by 60 °F water to be about 25% higher than that created by 140 °F water. This is a significant difference.

The chart in Figure 6–19 indicates a pressure drop of about 2.7 psi (for water at 60 °F). A result comparable to that found using Equation 6.12 for water at 60 °F.

6.4 HYDRAULIC RESISTANCE OF FITTINGS, VALVES, AND OTHER DEVICES

Before the total hydraulic resistance of a piping circuit can be found, the individual hydraulic resistances of all fittings, valves, or other such components must be determined. One approach is to consider each fitting, valve, or other device as an equivalent length of copper tube of the same pipe size. *The equivalent length of a component is defined as the amount of tubing, of the same pipe size, that would produce the same head loss (or pressure drop), as the actual component, at the same flow rate.* By replacing all components in the circuit with their equivalent length of piping, the circuit can be treated as if it were a single piece of pipe having a length equal to the sum of the actual pipe length, plus the total equivalent lengths of all fittings, valves, or other devices.

The piping circuit shown in Figure 6–20 contains a total of 58 feet of 3/4-inch copper pipe and eight 90 degree elbows. Assuming that the equivalent length of each elbow is 2 feet of 3/4-inch pipe, the entire circuit could be thought of as if it were 58 + 8(2) = 74 feet of 3/4-inch pipe.

This method can be used for many different fittings and valves. The table in Figure 6–21 lists representative equivalent lengths for most commonly used fittings and valves.

The data in Figure 6–21 is representative of typical fittings and valves. The exact equivalent length of a specific fitting or valve will depend on its internal shape, surface roughness, and other factors. Take note that the numbers in the table are for soldered fittings. *The equivalent lengths (of fittings) given in Figure 6–21 should be doubled for threaded fittings.* The values in Figure 6–21 can be used with Equation 6.12 to determine the head loss through any type of hydronic piping circuit using smooth piping.

Example 6.7: Water at 140 °F and 6 gpm flows through the 3/4-in piping circuit shown in Figure 6–22. Determine the head loss and associated pressure drop around the circuit.

COPPER TUBE SIZE						
Fitting or Valve₁	1/2"	3/4"	1"	1.25"	1.5"	2"
90 degree elbow	1.0	2.0	2.5	3.0	4.0	5.5
45 degree elbow	0.5	0.75	1.0	1.2	1.5	2.0
Tee (straight run)	0.3	0.4	0.45	0.6	0.8	1.0
Tee (side port)	2.0	3.0	4.5	5.5	7.0	9.0
B&G Monoflo® tee₂	n/a	70	23.5	25	23	23
Reducer coupling	0.4	0.5	0.6	0.8	1.0	1.3
Gate valve	0.2	0.25	0.3	0.4	0.5	0.7
Globe valve	15.0	20	25	36	46	56
Angle valve	3.1	4.7	5.3	7.8	9.4	12.5
Ball valve₃	1.9	2.2	4.3	7.0	6.6	14
Swing check valve	2.0	3.0	4.5	5.5	6.5	9.0
Flow-check valve₄	n/a	83	54	74	57	177
Butterfly valve	1.1	2.0	2.7	2.0	2.7	4.5

Footnotes:
1. Data for soldered fittings and valves. For threaded fittings double the listed value
2. Derived from Cv values based on no flow through side port of tee
3. Based on a standard port ball valve. Full port valves would have lower equivalent lengths
4. Based on B&G brand "flow control" valves

Figure 6–21   Representative equivalent lengths of common fittings and valves (all values expressed as feet of copper tube of the same nominal size)

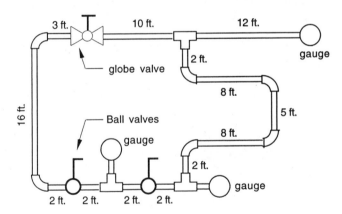

Total length of pipe <u>within circuit</u> = 62 ft.

Figure 6–22   Piping assembly for use in Example 6.7

3/4" Tubing ------------------------	62 ft.
6, 3/4" x 90 deg. elbows -------- 6(2) =	12 ft.
2, 3/4" side port tees ---------- 2(3) =	6 ft.
1, 3/4" straight through tee ------ 1(0.4) =	0.4 ft.
1, 3/4" globe valve ------------- 1(20) =	20 ft.
2, 3/4" ball valves ------------- 2(2.2) =	4.4 ft.
	TOTAL = 104.8 ft.

Figure 6–23   Adding the equivalent lengths of the fittings and valves in Figure 6–22

Solution: Start by finding the *total equivalent length* of the circuit by adding the equivalent lengths of the fittings and valves to the length of tubing as shown in Figure 6–23.

Notice that only tubing lengths *within the flow path* are included since flow does not occur in dead end pipe branches.

The circuit can now be treated as if it were simply 104.8 ft of 3/4-in copper pipe. Equation 6.12 can be used with this total equivalent length to determine the head loss at 6 gpm.

The α value for water at 140 °F can be determined from either Figure 6–17 or Equation 6.13 to be 0.0478.

The value of c for 3/4-in copper tube is found in Figure 6–18: $c = 0.061957$

This data can now be substituted into Equation 6.12 to determine the head loss:

$$H_{L} = (\alpha c L)(f)^{1.75} = [(0.0478)(0.061957)(104.8)](6)^{1.75}$$

$$= 7.14 \text{ ft of head}$$

The corresponding pressure drop around the circuit can be found using Equation 6.6. Note that this equation requires the density of water from Figure 4–5.

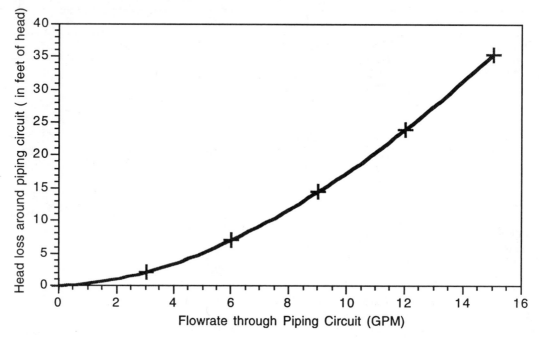

Figure 6–24 Example of a system resistance curve for a piping circuit

$$\Delta P = \frac{(7.14)(61.35)}{144} = 3.04 \text{ psi}$$

The only additional work required in this example compared to Example 6.4 was looking up and adding together the equivalent lengths of the fittings and valves.

6.5 THE SYSTEM RESISTANCE CURVE

An examination of Equation 6.12 shows that *for a given piping circuit, operating with a given fluid, at a constant temperature,* the hydraulic resistance term, αcL, remains constant:

(Equation 6.15)
$$H_{\mathrm{L}} = (\alpha cL)(f)^{1.75} = (\text{a constant})(f)^{1.75}$$

Under these conditions the head loss around the piping circuit depends only on flow rate. If the head loss through the circuit were calculated at several different flow rates, the resulting points plotted, and a smooth curve drawn through them, a graph similar to that shown in Figure 6–24 would result.

This type of graph is called a **system resistance curve**, or just system curve. *The system resistance curve is a total description of the relationship between flow rate and head loss for a given piping circuit using a specific fluid at a given temperature.* Such a curve is very useful in selecting a pump for the system as shown in Chapter 7.

Example 6.8: Using the piping circuit and data from Example 6.7, draw the system resistance curve assuming the piping carries water at 140 °F.

Solution: Start by substituting the appropriate data into Equation 6.12:

$$H_{\mathrm{L}} = (\alpha cL)(f)^{1.75} = [(0.0478)(0.061957)(104.8)](f)^{1.75}$$

Multiply the values of α, c, and L together to obtain a *single number*. This value, 0.31, is the hydraulic resistance of the piping circuit:

(Equation 6.16)
$$H_{\mathrm{L}} = (0.31)(f)^{1.75}$$

Use this equation to generate data for plotting the system resistance curve. See Figure 6–25.

Circuit Flowrate	head loss
(GPM)	(ft. of head)
0	0
3	2.12
6	7.14
9	14.5
12	24.0
15	35.5

Figure 6–25 Values of head loss calculated at several flow rates using Equation 6.16

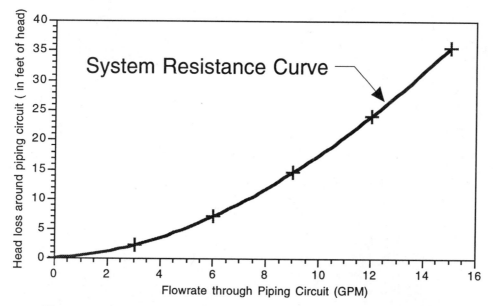

Figure 6–26   A plot of the system resistance curve using Equation 6.16

This data can now be plotted, and a smooth curve drawn through the points as shown in Figure 6–26.

If additional piping, fittings, or other components were added to the piping circuit, its total equivalent length would increase, as would its hydraulic resistance. This would result in a steeper system resistance curve. Figure 6–27a shows several system resistance curves plotted on the same axis. These curves were produced assuming the total equivalent length of the circuit described in Example 6.8 varied from a low of 50 feet to a high of 500 feet.

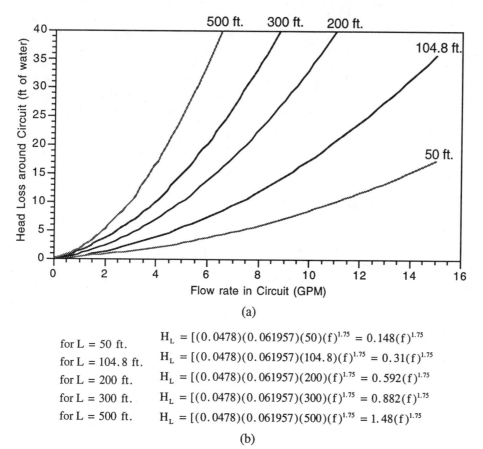

(a)

for L = 50 ft. $H_L = [(0.0478)(0.061957)(50)(f)^{1.75} = 0.148(f)^{1.75}$
for L = 104.8 ft. $H_L = [(0.0478)(0.061957)(104.8)(f)^{1.75} = 0.31(f)^{1.75}$
for L = 200 ft. $H_L = [(0.0478)(0.061957)(200)(f)^{1.75} = 0.592(f)^{1.75}$
for L = 300 ft. $H_L = [(0.0478)(0.061957)(300)(f)^{1.75} = 0.882(f)^{1.75}$
for L = 500 ft. $H_L = [(0.0478)(0.061957)(500)(f)^{1.75} = 1.48(f)^{1.75}$

(b)

Figure 6–27   (a) System resistance curves for pipe circuits of various equivalent lengths, (b) calculation of hydraulic resistance for each curve

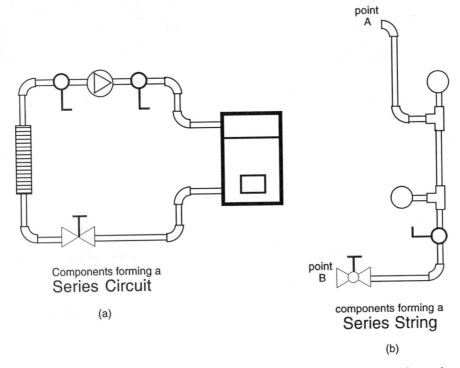

Figure 6–28 Piping components forming (a) a series circuit, (b) a series string.

The corresponding hydraulic resistances are also tabulated in Figure 6–27b.

The numbers preceding the $(f)^{1.75}$ term in the equations of Figure 6.27b are the hydraulic resistance values of each circuit. Note that hydraulic resistance increases as the equivalent length of the circuit increases. Later sections of this chapter will show how even complex piping circuits can be described by a system resistance curve.

6.6 PIPING COMPONENTS REPRESENTED AS SERIES RESISTORS

The simplest piping assemblies are formed by connecting piping components end to end. In some cases the train of components closes on itself forming a **series circuit**. In other cases only a **series piping path** is formed. Examples of these assemblies are illustrated in Figure 6–28.

To visualize the overall hydraulic resistance created by various components of a piping system, each pipe, fitting, valve, or other component can be thought of as a **hydraulic resistor**, and represented with a symbol like that used in electrical schematics. An example of how a series piping path consisting of several components would be represented is shown in Figure 6–29.

The size of a particular resistor symbol does not indicate the amount of hydraulic resistance it represents.

Only numerical values can do this. The orientation of the symbol or the shape of the series piping path also makes no difference. However, *the sequence of hydraulic resistors forming the series piping path or circuit, should always match the actual piping component layout they represent.*

A similar representation of piping components using hydraulic resistor symbols can be made for a complete series circuit, as shown in Figure 6–30.

Equivalent Resistance of Series-connected Components

The analogy of hydraulic resistance and electrical resistance also applies to how complex resistor diagrams can be reduced to simpler equivalent diagrams. In either system, *any number of series-connected resistors can be replaced, in effect, by a single **equivalent resistor** that has a resistance equal to the mathematical sum of all the individual resistances.* Equation 6.17 expresses this concept in mathematical form.

(Equation 6.17)

$$r_{e\,series} = \sum_{i=1}^{n} r_i = (r_1 + r_2 + r_3 + r_4 + \ldots + r_n)$$

This concept allows **hydraulic resistance diagrams** such as shown in Figure 6–30 to be greatly simplified. An equivalent diagram can be drawn that combines the individual hydraulic resistances of every pipe, fitting,

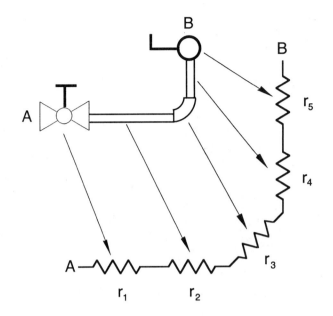

r_1 = hydraulic resistance of globe valve
r_2 = hydraulic resistance of pipe segment
r_3 = hydraulic resistance of elbow
r_4 = hydraulic resistance of pipe segment
r_5 = hydraulic resistance of ball valve

alternate way to draw resistors

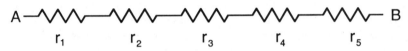

Figure 6–29 Series string of piping components represented as series string of hydraulic resistors

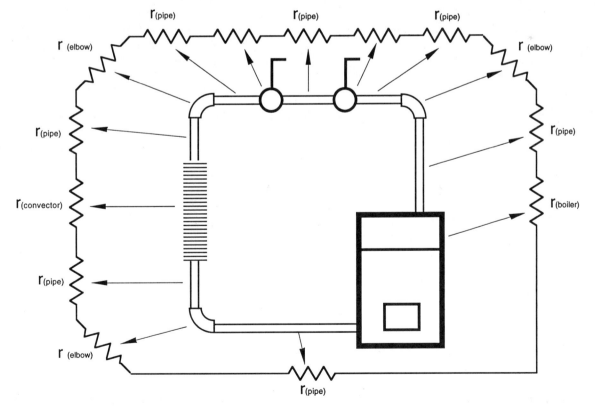

Figure 6–30 Example of a series piping circuit represented using hydraulic resistor symbols

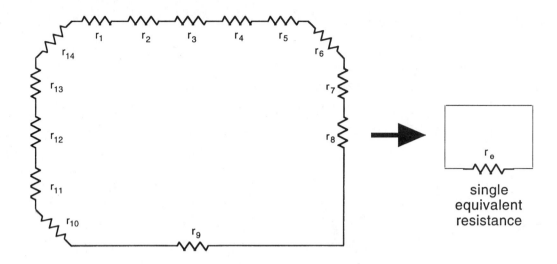

Single equivalent resistance $= r_e = r_1 + r_2 + r_3 + \ldots + r_{14}$

Figure 6–31 Concept of reducing a series circuit of resistors into a single equivalent resistance. This concept is valid for both electrical resistors and hydraulic resistors.

valve, and other component, into a single equivalent hydraulic resistance as shown in Figure 6–31. The simplified diagram totally represents the hydraulic resistance of *all* the original components, but in a much more compact and easily managed form.

In electrical circuits, the resistance of each resistor is known, and the values are simply added together. In piping circuits, however, the hydraulic resistance of each component has to be calculated using Equation 6.11, before the values can be added. Although it is certainly possible to do this, it can become quite time consuming for complex systems.

A more efficient method can be used in the common situation where the piping circuit contains a single size of pipe, fittings, and valves. This method recognizes that the fluid properties factor, α, and the pipe size factor, c, will be the same for all components, and thus need only be determined once. The only remaining value required for each component is its equivalent length (found in Figure 6–21). Finding the equivalent resistance of the circuit thus becomes a matter of looking up the equivalent lengths of the fittings and valves, combining these with the total piping length, and multiplying this **total equivalent length** by the common values of α and c as shown in Figure 6–32.

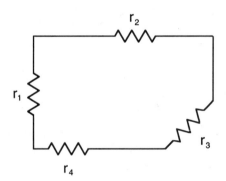

$r_{equivalent} = r_1 + r_2 + r_3 + r_4$

$r_{equivalent} = (a)\,(c)\,(L_{equivalent})_1 + (a)\,(c)\,(L_{equivalent})_2 + (a)\,(c)\,(L_{equivalent})_3 + (a)\,(c)\,(L_{equivalent})_4$

$r_{equivalent} = (a)\,(c)\,[\,(L_{equivalent})_1 + (L_{equivalent})_2 + (L_{equivalent})_3 + (L_{equivalent})_4\,]$

$r_{equivalent} = (a)\,(c)\,[\text{ total equivalent length of circuit }]$

Figure 6–32 Simplified method of adding hydraulic resistances when the piping components are all the same pipe size. Note the value of α and c are the same for all components.

Example 6.9: The series piping path shown in Figure 6–33 will transport water at 140 °F. Determine the hydraulic resistance of the entire piping path between points A and B. Use this hydraulic resistance to sketch a system resistance curve.

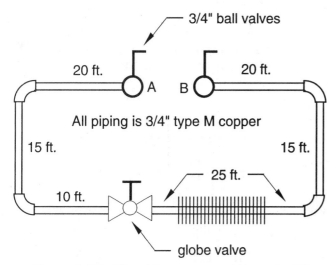

Figure 6–33   The piping system of Example 6.9

Solution: Figures 4–5 and 4–6 can be used to obtain the density and viscosity of water at 140 °F. These values are then substituted into Equation 6.13 to find α:

$$\alpha = \left(\frac{D}{\mu}\right)^{-0.25} = \left(\frac{61.35}{0.00032}\right)^{-0.25} = (191,719)^{-0.25}$$

$$= 0.0478$$

The α value for water could also have been read directly from Figure 6–17.

The value of c is found in Figure 6–18. For $^3/_4$-in copper tube $c = 0.061957$

The only remaining task is to look up the equivalent lengths of all fittings and valves and add those lengths to that of the piping. The equivalent lengths of fittings and valves is found in Figure 6–21 and Figure 6–34.

The values of α, c, and L can now be substituted directly into Equation 6.11 to find the total hydraulic resistance of the piping string:

$$r = (\alpha cL) = (0.04779)(0.061957)(137.4) = 0.407$$

Piping Component			Equivalent length
105 feet 3/4" copper tubing	=		105 ft.
4, 90 degree × 3/4" elbows	=	4(2) =	8 ft.
2, 3/4" ball valves	=	2(2.2) =	4.4 ft.
1, 3/4" globe valve	=	1(20) =	20 ft.
Total equivalent length	=	L =	137.4 ft.

Figure 6–34   Adding the equivalent lengths of piping, fittings, and valves in the system shown in Figure 6–33

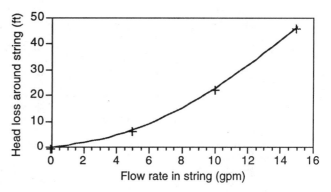

flow rate (f), (GPM)	Head loss H_L, (ft of head)
0	0
5	6.8
10	22.9
15	46.5

Figure 6–35   Using the equation of the system resistance curve to generate head loss versus flow rate data for the piping system of Figure 6–33.

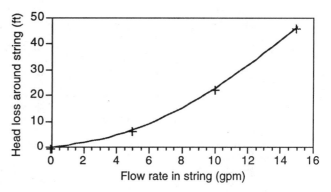

Figure 6–36   A plot of the system resistance curve for Example 6.9

This total hydraulic resistance can now be substituted into Equation 6.12 and used to generate a few points for plotting the system resistance curve as shown in Figure 6–35. The resulting data is plotted and a smooth curve drawn through the points to represent the system resistance curve as shown in Figure 6–36.

Piping Paths Containing Multiple Pipe Sizes

The piping paths shown in previous examples were relatively simple. They contained only a few fittings and valves, and all such components were of the same pipe size. A more typical hydronic system can contain dozens of fittings and valves, some of which may be of different pipe sizes. The total hydraulic resistance of such circuits can still be determined using the basic principles. However, when more than one pipe size is involved, the hydraulic resistance of all piping, fittings, valves, or other components, *of a given pipe size,* should be determined separately, and then added together. This is because each pipe size has a different c value. These calculations can organized as shown in Equation 6.18.

(Equation 6.18)

$$r_{total} = \alpha\{[c(L_{pipe} + L_{components})]_{size\,1} + [c(L_{pipe} + L_{components})]_{size\,2}$$

$$+ \ldots + [c(L_{pipe} + L_{components})]_{last\,size}\}$$

The PIPEPATH Program

The calculations represented by Equation 6.18 can become quite time consuming for large or complex piping systems. The PIPEPATH program in the Hydronics Design Toolkit can quickly evaluate such calculations to find the total hydraulic resistance of any reasonable user-specified piping path. This piping path can contain pipe, fittings, and valves in sizes from ½ inch through 2 inches. PIPEPATH contains a database of the equivalent lengths of all common fittings and valves. It automatically determines the α values of several different selectable fluids. In addition to hydraulic resistance, the program also calculates the resulting head loss of the piping path at a specified flow rate, as well as the Reynold's number of the flow. The PIPEPATH program forms the core of the Hydronics Design Toolkit. It will be referenced extensively when designing hydronic distribution circuits in later chapters. Figure 6–37a shows the main screen of the PIPEPATH program. Figure 6–37b shows the "edit table" from the program in which the user inputs piping, fitting, and valve quantities.

```
- PIPEPATH - F1-HELP  F2-FILE  F4-GRAPH    F10-EXIT TO MAIN MENU File: Samplpip
  -INPUTS-
  ENTER Average fluid temperature,(80-250)...................  180     deg.F.
  ENTER Flow rate (if known) otherwise ignore input,(0.25-90). 1       gpm
  ENTER Additional Hydraulic Resistance,(0-5.0)..............  0
  SELECT Fluid type (press F3)...............................  water
  EDIT TABLE OF PIPE, FITTINGS, AND VALVES...................  ▓Go To▓


  -RESULTS-
  Head loss of piping path   =     0.28    ft. head
  Pressure drop of piping path =   0.12    psi
  Hydraulic Resistance       =     0.275
  Reynolds No. largest pipe  =     10,806
  Reynolds No. Smallest pipe =     10,806




  MESSAGE
  Use the UP or DOWN keys to move the highlighting bar over an input. For
ENTER inputs type desired value. SELECT inputs press F3.
```

Figure 6–37a Example of the main screen of the PIPEPATH program in the Hydronics Design Toolkit

```
- PIPEPATH - F1-HELP   F5-ZERO ALL      F10-RETURN TO INPUTS  File: Samplpip
  -EDIT TABLE-
```

	1/2"	3/4"	1"	1 1/4"	1 1/2"	2"	5/8"PEX
Straight pipe(lin. ft.)	0	100	0	0	0	0	0
90 degree ells	0	0	0	0	0	0	
45 degree ells	0	0	0	0	0	0	
Tees (straight through)	0	0	0	0	0	0	
Tees (side port)	0	0	0	0	0	0	
Monoflo tees		0	0	0	0	0	
Gate valves	0	0	0	0	0	0	
Globe valves	0	0	0	0	0	0	
Angle valves	0	0	0	0	0	0	
Ball valves	0	0	0	0	0	0	
Swing check valves	0	0	0	0	0	0	
Flow control valves		0	0	0	0	0	
Butterfly valves	0	0	0	0	0	0	
Deareator		0	0		0	0	
Reducer Coupling	0	0	0	0	0	0	
Boiler (C.I. sectional)					0		

```
  MESSAGE
  Use the UP, DOWN, LEFT, RIGHT ARROW keys to move the highlighting bar
over an input. Type desired value. F10 to return to input screen.
```

Figure 6–37b Example of the "edit table" screen of the PIPEPATH program in the Hydronics Design Toolkit

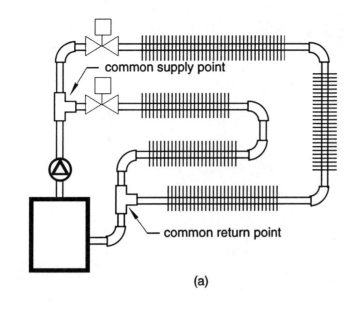

(a)

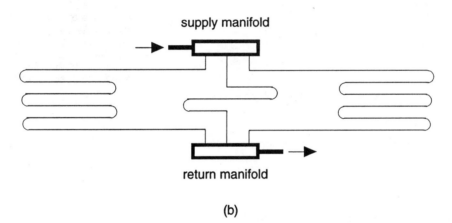

(b)

Figure 6–38 Example of parallel piping paths (a) a two-zone distribution system having two parallel piping paths, (b) a radiant floor heating distribution system with three parallel piping paths.

6.7   PARALLEL HYDRAULIC RESISTANCES

Piping systems often contain two or more piping paths connected at common points. Examples include multiple zone distribution circuits that begin and end at the same boiler as shown in Figure 6–38a, or radiant floor heating circuits that originate at a common supply manifold, and terminate at a common return manifold, as shown in Figure 6–38b. The piping paths that share common points of connection are said to be **piped in parallel**. In such arrangements the total system flow must divide up among the two or more paths at the common supply point, and then recombine at a common return point.

The analogy of electrical resistance again proves useful for analyzing how the flow divides up among two or more parallel piping paths. Figure 6–39 shows a parallel

piping assembly along with its associated hydraulic resistor diagram.

In the resistor diagram of Figure 6–39, the combined resistance of the two elbows in the upper piping branch is represented by a single resistor symbol, r_2. This is done as a matter of convenience because it reduces the number of resistor symbols that have to be drawn. Because this resistor now represents both elbows, its numerical value would be doubled. The three pipe segments in the upper branch are also represented by a single resistor symbol, r_3, as are the three in the lower branch, r_8. This simplifying method can be used whenever two or more identical components are configured in series. With experience, it easy to draw simplified resistor diagrams by lumping such resistances together.

The equivalent resistance of the entire piping assembly would be a single resistor connected between points A

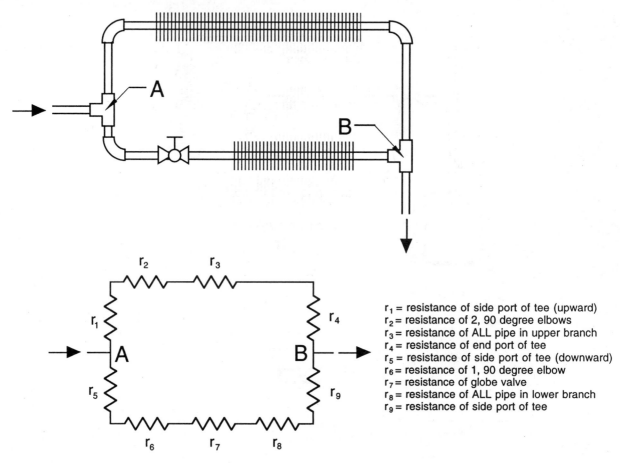

r_1 = resistance of side port of tee (upward)
r_2 = resistance of 2, 90 degree elbows
r_3 = resistance of ALL pipe in upper branch
r_4 = resistance of end port of tee
r_5 = resistance of side port of tee (downward)
r_6 = resistance of 1, 90 degree elbow
r_7 = resistance of globe valve
r_8 = resistance of ALL pipe in lower branch
r_9 = resistance of side port of tee

Figure 6–39   Parallel piping assembly with associated hydraulic resistor diagram

and B representing the combined effect of all piping components between these points. To find this equivalent resistance, start by reducing each of the brach piping paths into a single equivalent resistance as shown in Figure 6–40. Since the hydraulic resistors within each

path are in series with each other, their resistances can be added. This is represented by Equations 6.19a and 6.19b:

(Equation 6.19a)
$$r_{e1} = \alpha c(L_1 + L_2 + L_3 + L_4)$$

(Equation 6.19b)
$$r_{e2} = \alpha c(L_5 + L_6 + L_7 + L_{8} + L_9)$$

Once this step is completed, reconnect the two equivalent resistances at points A and B as shown in Figure 6–41.

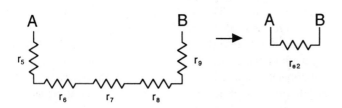

Figure 6–40   Reduce each parallel path to a single equivalent hydraulic resistance

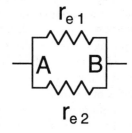

Figure 6–41   A group of two parallel hydraulic resistors

CHAPTER 6   FLUID FLOW IN PIPING   **153**

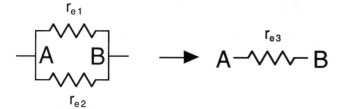

Figure 6–42   Reducing two parallel hydraulic resistances into a single equivalent hydraulic resistance

The arrangement will now be reduced to a single equivalent resistance as shown in Figure 6–42.

The equivalent resistance of two parallel hydraulic resistors is found using Equation 6.20.

(Equation 6.20)

$$R_{equivalent_{parallel}} = \left[\left(\frac{1}{r_1} \right)^{0.5714} + \left(\frac{1}{r_2} \right)^{0.5714} \right]^{-1.75}$$

The values of r_1 and r_2 in Equation 6.20 are the hydraulic resistances of the parallel resistors that are to be combined into a single equivalent resistance. This equivalent resistance then represents the entire piping assembly shown in Figure 6–39. If this assembly were to be combined with other components to form a more complex piping system, the assembly could still be represented by this single resistor within the resistor diagram of the more complex system.

In some piping systems there are more than two parallel piping strings. An example would be a radiant floor distribution system in which several piping paths originate from a common supply header, and terminate in a common return header as shown in Figure 6–43.

To find the single equivalent hydraulic resistance of this configuration, Equation 6.20 may be extended by one term as follows:

(Equation 6.21)

$$R_{equivalent_{parallel}} = \left[\left(\frac{1}{r_1} \right)^{0.5714} + \left(\frac{1}{r_2} \right)^{0.5714} + \left(\frac{1}{r_3} \right)^{0.5714} \right]^{-1.75}$$

This same principle can be extended to *any number* of parallel resistors connected in parallel. Simply add more terms to Equation 6.20 to accommodate all parallel resistances. This can be expressed mathematically for a system with *n* parallel hydraulic resistances beginning and ending at common points using Equation 6.22:

(Equation 6.22)

$$R_{equivalent_{parallel}} = \left[\left(\frac{1}{r_1} \right)^{0.5714} + \left(\frac{1}{r_2} \right)^{0.5714} + \left(\frac{1}{r_3} \right)^{0.5714} \right. $$
$$\left. + ... + \left(\frac{1}{r_n} \right)^{0.5714} \right]^{-1.75}$$

Regardless of the number of parallel hydraulic resistances one begins with, the equivalent hydraulic resistor will always be a *single* resistor connected between the common points. This single resistor represents the combined effect of *all* original resistors.

Once the equivalent resistance of a group of parallel hydraulic resistors has been found, the flow rate through each parallel piping path can also be determined using Equation 6.23.

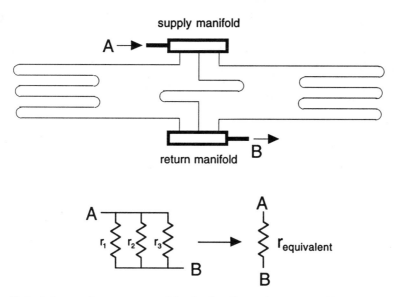

Figure 6–43   Three parallel piping paths represented by hydraulic resistances, which are then combined into a single equivalent hydraulic resistance

(Equation 6.23)

$$f_i = f_{total} \left(\frac{R_e}{r_i} \right)^{0.5714}$$

where:

f_i = flow rate through parallel path i (gpm)
f_{total} = *total* flow rate entering the common point of the parallel paths (gpm)
R_e = equivalent hydraulic resistance of all the parallel piping paths
r_i = hydraulic resistance of parallel piping path i

Example 6.10: Using the concept of parallel hydraulic resistors, find the equivalent resistance of the radiant floor distribution system shown in Figure 6–44. All four piping paths consist of $\frac{1}{2}$-in smooth tubing. The average water temperature in the system is 100 °F. To simplify the example, assume there are no valves in the manifolds, and the hydraulic resistances at the points of connection of the piping paths to the manifolds are insignificant. Assuming that water enters the supply manifold at 10 gpm, determine the flow rate in each of the branch piping paths.

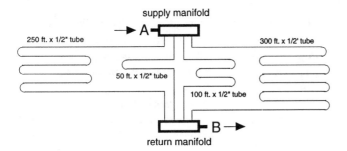

Figure 6–44 Radiant floor distribution system for Example 6.10

Solution: The resistor diagram shown in Figure 6–45 represents the assembly as four hydraulic resistors in parallel. After finding the values of these resistors, we can use Equation 6.22 to combine them into a single equivalent resistance.

Start by finding the density and dynamic viscosity of water at 100 °F. This can be done using Figures 4-5

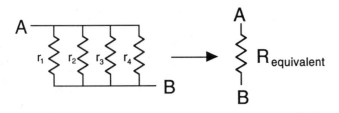

Figure 6–45 Hydraulic resistor diagram for Figure 6–44

and 4-6, or by using the FLUIDS program in the Hydronics Design Toolkit. The density of 100 °F water is 61.97 lb/ft³. The dynamic viscosity of 100 °F water is 0.0004573 lb/ ft/ sec.

From these the value of α can be calculated using Equation 6.13:

$$\alpha = \left(\frac{D}{\mu} \right)^{-0.25} = \left(\frac{61.97}{0.0004573} \right)^{-0.25} = 0.05212$$

The value of c is found in the table of Figure 6–18. For $\frac{1}{2}$-in smooth tube $c = 0.33352$.

The values of α and c will be the same for each piping path. The only remaining quantity needed is the length of each string. The lengths along with the values for α and c are now substituted into Equation 6.11:

200 ft string: $r_1 = \alpha c L_1 = (0.05212)(0.33352)(200) = 3.477$
50 ft string: $r_2 = \alpha c L_2 = (0.05212)(0.33352)(50) = 0.869$
100 ft string: $r_3 = \alpha c L_3 = (0.05212)(0.33352)(100) = 1.738$
300 ft string: $r_4 = \alpha c L_4 = (0.05212)(0.33352)(300) = 5.215$

These resulting values are the hydraulic resistances of each piping path. These resistances can now be substituted into Equation 6.22 and reduced to a single equivalent resistance:

$$R_{equivalent} = \left[\left(\frac{1}{3.477} \right)^{0.5714} + \left(\frac{1}{0.869} \right)^{0.5714} + \left(\frac{1}{1.738} \right)^{0.5714} \right.$$
$$\left. + \left(\frac{1}{5.215} \right)^{0.5714} \right]^{-1.75}$$

$$R_{equivalent} = [(0.2876)^{0.5714} + (1.1508)^{0.5714} + (0.5754)^{0.5714}$$
$$+ (0.1918)^{0.5714}]^{-1.75}$$

$$R_{equivalent} = [0.4906 + 1.0836 + 0.7292 + 0.3892]^{-1.75}$$

$$R_{equivalent} = [2.6926]^{-1.75}$$

$$R_{equivalent} = 0.1767$$

The flow rates in each piping string can now be determined by repeated use of Equation 6.23:

$$f_1 = 10 \left(\frac{0.1767}{3.477} \right)^{0.5714} = 1.82 \text{ gpm}$$

$$f_2 = 10 \left(\frac{0.1767}{0.869} \right)^{0.5714} = 4.02 \text{ gpm}$$

$$f_3 = 10\left(\frac{0.1767}{1.738}\right)^{0.5714} = 2.71 \text{ gpm}$$

$$f_4 = 10\left(\frac{0.1767}{5.215}\right)^{0.5714} = 1.45 \text{ gpm}$$

Discussion: It is worth noting that the equivalent hydraulic resistance of the entire assembly (0.1767), is smaller than even the smallest individual hydraulic resistance (0.869). This will always be true. In fact it provides a way of partially checking the results of the calculations. Those familiar with parallel electrical circuits also know this to be true in every case.

Another way of checking the results is to add the individual branch flow rates, and verify the total equals the flow rate entering the common point of the parallel paths:

$$f_{\text{total}} = f_1 + f_2 + f_3 + f_4 = 1.82 + 4.02 + 2.71 + 1.45$$

$$= 10.0 \text{ gpm } \checkmark$$

This example also shows that the larger the hydraulic resistance of a piping path, the lower its flow rate.

Using the PARALLEL Program

Example 6.10 shows the calculations for determining the equivalent resistance of several parallel piping strings can become quite involved, especially if changes in piping components or fluid temperature require the repeated use of Equation 6.22 and 6.23. To expedite such calculations, the Hydronics Design Toolkit contains a program called PARALLEL that can quickly combine up to six parallel hydraulic resistances using Equation 6.22. The individual hydraulic resistances of the piping paths can be manually calculated or, for faster results, determined using the PIPEPATH program. The PARALLEL program also calculates the flow rate in each branch piping path based on Equation 6.23. A sample screen from the PARALLEL program, using the same data as Example 6.10, is shown in Figure 6–46.

6.8 REDUCING COMPLEX PIPING SYSTEMS

The concept of using series and parallel resistors to represent piping assemblies is a powerful tool when properly applied. *A key concept is that of alternating the use of the series and parallel equivalent resistor procedures to systematically reduce a complex piping system down to a single equivalent resistance.* A system resistance curve can then be produced that will assist in choosing a circulator for the system. The single equivalent resistance can then be unwound back to the individual parallel path resistances to find the flow rates in each path. The best way of illustrating this concept is to go through an example using a typical hydronic distribution system.

```
- PARALLEL -  F1-HELP                                 F10-EXIT TO MAIN MENU
 ╔═ -INPUTS- ═══════════════════════════════════════════════════════════╗
 ║  ENTER Number of parallel piping circuits,(2 to 6).............. 4     ║
 ║  ENTER Flowrate in common piping,(.5 to 50)..................... 10  gpm║
 ║  ENTER Hydraulic resistance of circuit #1 (.001 to 10)......... 3.477  ║
 ║  ENTER Hydraulic resistance of circuit #2 (.001 to 10)......... .869   ║
 ║  ENTER Hydraulic resistance of circuit #3 (.001 to 10)......... 1.738  ║
 ║  ENTER Hydraulic resistance of circuit #4 (.001 to 10)......... 5.215  ║
 ║                                                                        ║
 ╚════════════════════════════════════════════════════════════════════════╝

 ╔═ -RESULTS- ══════════════════════════════════════════════════════════╗
 ║  Equivalent hydraulic resistance of all parallel circuits,..0.1767    ║
 ║  Flowrate in circuit # 1 ................................... 1.82gpm  ║
 ║  Flowrate in circuit # 2 ................................... 4.02gpm  ║
 ║  Flowrate in circuit # 3 ................................... 2.71gpm  ║
 ║  Flowrate in circuit # 4 ................................... 1.45gpm  ║
 ║                                                                        ║
 ╚════════════════════════════════════════════════════════════════════════╝

 ╔═  MESSAGE ═══════════════════════════════════════════════════════════╗
 ║  Use the UP, DOWN, LEFT, RIGHT ARROW keys to move the highlighting bar ║
 ║  over an input. Type desired value. F10 to return to MAIN MENU.        ║
 ╚════════════════════════════════════════════════════════════════════════╝
```

Figure 6–46   A sample screen from the PARALLEL program in the Hydronics Design Toolkit, using the data from Example 6.10

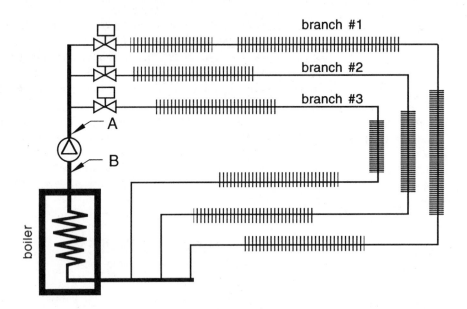

System Description:

Three zone residential system using individual zone valves. Each circuit contains a variety of fittings as follows:

Branch #1 contains 120 ft. of 3/4" copper tube and 20, 3/4"x 90 deg. elbows
Branch #2 contains 78 ft. of tube and 17, 3/4"x 90 deg. elbows
Branch #3 contains 165 ft. of tube and 26, 3/4"x 90 deg. elbows

Each zone valve is assumed to be equivalent to 30 ft. of 3/4" tube
The copper tube boiler is equivalent to 20 ft. of 3/4" copper tube
The headers are short lengths of 1.25" copper tube and thus have negligible resistance
The system contains water at 160 degree F.

Figure 6–47 Piping schematic for system used in Example 6.11

Example 6.11: Beginning with the system described in Figure 6–47, use the concepts of series and parallel resistances to reduce the system to a single equivalent resistance between points A and B. Using an assumed system flow rate of 10 gpm, find the flow rate in each branch circuit.

Solution: Start by sketching a resistor diagram of the system between points A and B. Remember that the hydraulic resistance of two or more identical components can be lumped together and represented by a single resistor symbol. The hydraulic resistor diagram shown in Figure 6–48 uses one resistor symbol to represent the combined resistance of all the tubing segments within a given branch, and another to represent the total resistance of all elbows in the branch. The third resistor symbol in each branch represents the resistance of the zone valve.

The objective is to reduce the resistor diagram of Figure 6–48 down to a single equivalent hydraulic

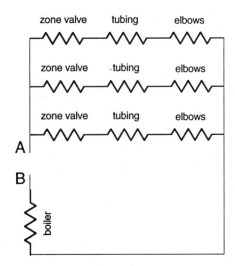

Figure 6–48 Resistor diagram for three-zone system of Example 6.11. Note that only one resistor is shown to represent all the elbows in a branch, and one resistor to represent all tubing segments in a branch.

resistance that represents the entire piping system. The first step is to calculate the hydraulic resistances of the tubing, fittings, and zone valves for each parallel piping path. Since all the tubing, fittings, and valves are the same pipe size, this becomes a matter of totaling the equivalent length of all components within a branch, and then multiplying by α and c.

branch #1: 120 ft tubing + 20(2 ft) elbows
 + 30 ft zone valve equivalent length = 190 ft

branch #2: 78 ft tubing + 17(2 ft) elbows
 + 30 ft zone valve equivalent length = 142 ft

branch #3: 165 ft tubing + 26(2 ft) elbows
 + 30 ft zone valve equivalent length = 247 ft

The density of water at 160 °F is determined as 61.02 lb/ft³. The dynamic viscosity of water at 160 °F is determined as 0.00027 lb/ft/sec. These values are then used to find the α value using Equation 6.13:

$$\alpha = \left(\frac{D}{\mu}\right)^{-0.25} = \left(\frac{61.02}{0.00027}\right)^{-0.25} = 0.04586$$

The value of c is found in Figure 6.18. For ³/₄-in smooth tube $c = 0.061957$

The hydraulic resistances of each branch can now be calculated:

branch #1 $R_1 = \alpha c L_1 = (0.04586)(0.061957)(190) = 0.5399$

branch #2 $R_2 = \alpha c L_2 = (0.04586)(0.061957)(142) = 0.4035$

branch #3 $R_3 = \alpha c L_3 = (0.04586)(0.061957)(247) = 0.7018$

The resistor diagram can now be reduced as shown in Figure 6–49:

The three parallel resistors can now be reduced to a single equivalent resistance using Equation 6.22 or the PARALLEL program in the Hydronics Design Toolkit.

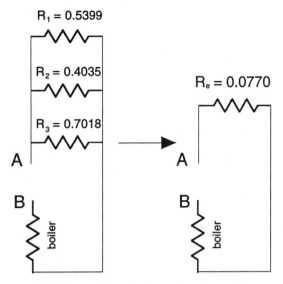

Figure 6–50   Second reduction of resistor diagram for Example 6.11

$$R_{equivalent} = \left[\left(\frac{1}{0.5399}\right)^{0.5714} + \left(\frac{1}{0.4035}\right)^{0.5714} + \left(\frac{1}{0.7018}\right)^{0.5714}\right]^{-1.75} = 0.07706$$

The resistor diagram can now be further reduced as shown in Figure 6–50.

The equivalent length of the boiler was stated as 20 ft of ³/₄-in copper tube. Its hydraulic resistance can therefore be calculated as:

$$r_{boiler} = (\alpha c L) = (0.04586)(0.061957)(20) = 0.05683$$

The hydraulic resistance of the boiler is in series with the equivalent resistance of the parallel piping paths. Therefore the hydraulic resistance of the boiler will be added to the equivalent resistance of the parallel paths to get the final overall equivalent resistance of the circuit as depicted in Figure 6–51.

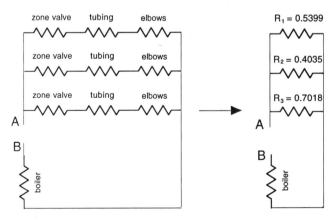

Figure 6–49   First reduction of resistor diagram for Example 6.11

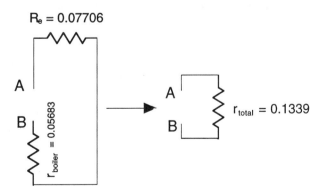

Figure 6–51   Final simplification of resistor diagram into a single hydraulic resistor that represents the entire system between points A and B

$$r = 0.05683 + 0.07706 = 0.1339$$

Assuming the circulator produces a flow rate of 10 gpm within the system, and all the zone valves are open, we can now find the flow rate in each parallel branch by repeated use of Equation 6.23:

$$f_1 = 10\left(\frac{0.07706}{0.5399}\right) = 3.29 \text{ gpm}$$

$$f_2 = 10\left(\frac{0.07706}{0.4035}\right)^{0.5714} = 3.88 \text{ gpm}$$

$$f_3 = 10\left(\frac{0.07706}{0.7018}\right)^{0.5714} = 2.83 \text{ gpm}$$

Total flow rate = 10.0 gpm ✓

Notice the sum of the branch flow rates equals the total system flow rate as a check of the calculations.

Discussion: A point worth noting is that the flow rate in any one circuit could be found assuming one or both of the other circuits were closed off at the zone valves. To do this, go back to Equation 6.22 and treat the hydraulic resistance of the closed branch (or branches) as infinite. This has the same effect as simply ignoring the closed zones as parallel resistors. For example, assuming the zone valve on branch #1 was closed, one would find the equivalent resistance of parallel resistors R_2 and R_3 only, and ignore the presence of R_1. The remainder of the procedure would be similar to that shown previously. Knowing the branch flow rates will eventually help in properly sizing the heat emitters in each branch path.

Example 6.12: The piping layout shown in Figure 6–52 represents a portion of a hydronic radiant floor heating system consisting of two manifold stations connected to a common supply and return pipe. The fluid in the system is water at a temperature of 100 °F. All pipe sizes and lengths are indicated. Using the concepts of parallel and series hydraulic resistance:

a. Reduce the piping layout to a single equivalent resistance between points A and B.
b. Sketch a system resistance curve.
c. Assuming the flow rate entering at point A is 10 gpm, determine the flow in each parallel piping path.

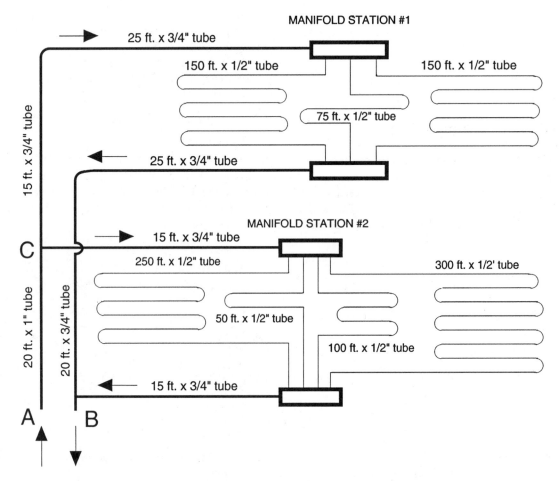

Figure 6–52 Radiant floor heating distribution system for Example 6.12

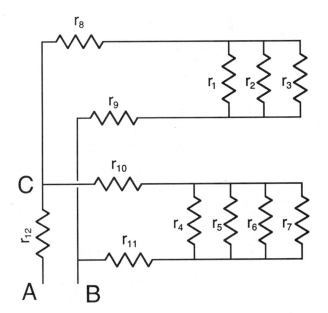

Figure 6–53   Resistor diagram for piping shown in Figure 6–52

Solution: Start by sketching a resistor diagram to represent the piping layout, as shown in Figure 6–53. A single resistor is used to represent each radiant floor branch, and each segment of supply or return pipe connecting the manifolds. After finding the values of all the hydraulic resistors, they can be combined using procedures for series and parallel resistances. Note that each resistor symbol in Figure 6–53 is numbered for reference.

By now the procedure of finding hydraulic resistances should be familiar.

Begin by getting the fluid properties factor α and pipe size constants c.

The density of water at 100 °F is 61.97 lb/ft³. The dynamic viscosity of water at 100 °F water is 0.0004573 lb/ft/sec.

From these we can calculate the value of α using Equation 6.13:

$$\alpha = \left(\frac{D}{\mu}\right)^{-0.25} = \left(\frac{61.97}{0.0004573}\right)^{-0.25} = 0.05212$$

The values of c for each pipe size used is found from Figure 6–18:

For ¹/₂-in tube   c = 0.33352
For ³/₄-in tube   c = 0.061957
For 1-in tube   c = 0.01776

The values of the hydraulic resistances can now be calculated using Equation 6.11:

r_1 150 ft × ¹/₂-in pipe:
$$r_1 = \alpha c L_1 = (0.05212)(0.33352)(150) = 2.607$$

r_2 75 ft × ¹/₂-in pipe:
$$r_2 = \alpha c L_2 = (0.05212)(0.33352)(75) = 1.304$$

r_3 150 ft × ¹/₂-in pipe:
$$r_3 = \alpha c L_3 = (0.05212)(0.33352)(150) = 2.607$$

r_4 250 ft × ¹/₂-in pipe:
$$r_4 = \alpha c L_4 = (0.05212)(0.33352)(250) = 4.346$$

r_5 50 ft × ¹/₂-in pipe:
$$r_5 = \alpha c L_5 = (0.05212)(0.33352)(50) = 0.869$$

r_6 100 ft × ¹/₂-in pipe:
$$r_6 = \alpha c L_6 = (0.05212)(0.33352)(100) = 1.738$$

r_7 300 ft × ¹/₂-in pipe:
$$r_7 = \alpha c L_7 = (0.05212)(0.33352)(300) = 5.215$$

r_8 40 ft × ³/₄-in pipe:
$$r_8 = \alpha c L_8 = (0.05212)(0.061957)(40) = 0.129$$

r_9 45 ft × ³/₄-in pipe:
$$r_9 = \alpha c L_9 = (0.05212)(0.061957)(45) = 0.145$$

r_{10} 15 ft × ³/₄-in pipe:
$$r_{10} = \alpha c L_{10} = (0.05212)(0.061957)(15) = 0.0484$$

r_{11} 15 ft × ³/₄-in pipe:
$$r_{11} = \alpha c L_{11} = (0.05212)(0.061957)(15) = 0.0484$$

r_{12} 20 ft × 1-in pipe:
$$r_{12} = \alpha c L_{12} = (0.05212)(0.01776)(20) = 0.0185$$

These values could also have been quickly determined using the PIPEPATH program in the Hydronics Design Toolkit.

Begin reducing the diagram by combining the parallel resistances (r_1, r_2, r_3) and (r_4, r_5, r_6, r_7) using Equation 6.22:

$$R_{e1} = \left[\left(\frac{1}{2.607}\right)^{0.5714} + \left(\frac{1}{1.304}\right)^{0.5714} + \left(\frac{1}{2.607}\right)^{0.5714}\right]^{-1.75}$$

$$= 0.2932$$

$$R_{e1} = \left[\left(\frac{1}{4.346}\right)^{0.5714} + \left(\frac{1}{0.869}\right)^{0.5714} + \left(\frac{1}{1.738}\right)^{0.5714} + \left(\frac{1}{5.215}\right)^{0.5714}\right]^{-1.75} = 0.2932$$

The diagram can now been reduced as shown in Figure 6–54.

The resistors r_8, R_{e1}, and r_9 are in series, and thus can be added:

$$R_{e3} = 0.129 + 0.2936 + 0.145 = 0.5676$$

The resistors r_{10}, R_{e2}, and r_{11} are also in series and can be added:

$$R_{e4} = 0.0484 + 0.1836 + 0.0484 = 0.2804$$

The resistor diagram can now be further reduced as shown in Figure 6–55.

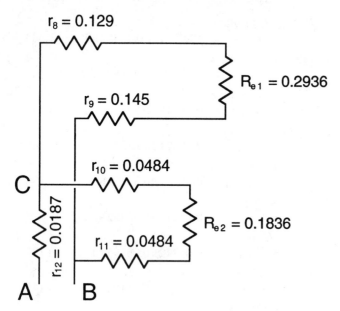

Figure 6–54 First reduction to resistor diagram for Example 6.12

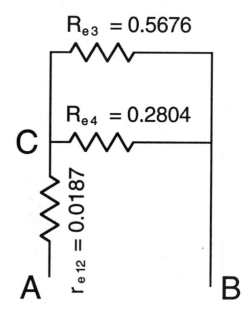

Figure 6–55 Second reduction to resistor diagram for Example 6.12

The next step is to combine resistors R_{e3} and R_{e4}, which are in parallel. Again Equation 6.22 or the PARALLEL program in the Hydronics Design Toolkit can be used:

$$R_{e5} = \left[\left(\frac{1}{0.5676} \right)^{0.5714} + \left(\frac{1}{0.2804} \right)^{0.5714} \right]^{-1.75} = 0.1145$$

The resistor diagram now appears as shown in Figure 6–56.

The final reduction of the resistor diagram is to add the two series resistances r_{12} and R_{e5}:

$$r_{12} + R_{e5} = 0.0187 + 0.1145 = 0.1332$$

This single equivalent hydraulic resistance representing the entire distribution system is shown in Figure 6–57.

This single equivalent hydraulic resistance can now be used to sketch the system curve by substituting it into Equation 6.10 and generating a few data points that can be plotted as shown in Figure 6–58. The resulting curve is plotted in Figure 6–59.

The final part of the example involves unwinding the equivalent resistance to find the flow rate within each branch. An assumed flow rate of 10 gpm enters the piping assembly at point A. All 10 gpm will pass through resistor r_{12}, and then divide between parallel resistors R_{e3} and R_{e4}. Equation 6.23 can be used to find the flow rate in each branch as shown in Figure 6–60.

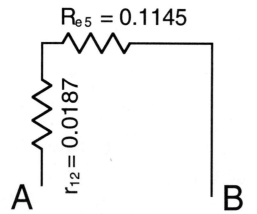

Figure 6–56 Third reduction of resistor diagram for Example 6.12

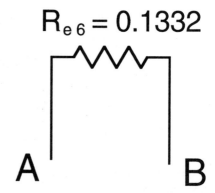

Figure 6–57 Single equivalent hydraulic resistor representing entire piping system of Figure 6–52

$H_L = rf^{1.75} = (0.1332)f^{1.75}$	
flow rate f (GPM)	head loss, H_L(ft.)
0	0
5	2.23
10	7.49
15	15.23
20	25.19

Figure 6–58   Use Equation 6.10 with the total hydraulic resistance of the system to generate data for plotting the system resistance curve

Notice the parallel resistances that are needed have already been determined as part of finding the equivalent resistance.

The values of f_1 and f_2 are the flow rates into manifold stations #1 and #2 respectively. Equation 6.23 can be applied again to find the flow rate in each piping path attached to the manifolds:

Figure 6–61 shows the process for manifold station #1.

Figure 6–62 shows the process for manifold station #2.

The branch flow rates are thus all determined. As a check of the calculations, the sum of the branch flow rates must equal the flow rate entering each manifold.

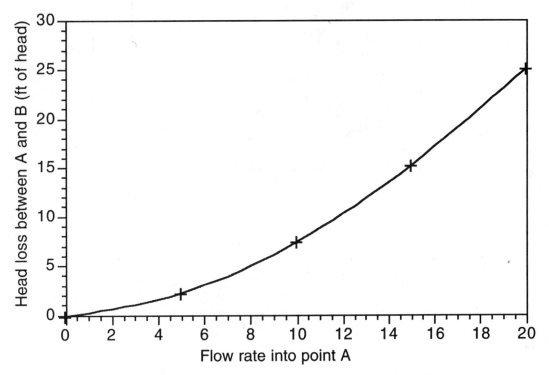

Figure 6–59   The system resistance curve for Example 6.12

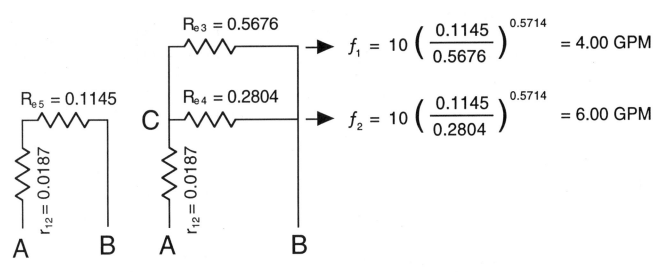

Figure 6–60   Finding the flow rate to each manifold station for Example 6.12

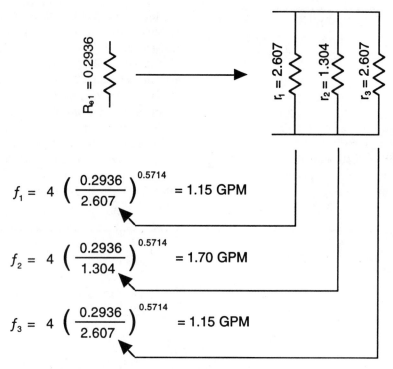

$$f_1 = 4 \left(\frac{0.2936}{2.607} \right)^{0.5714} = 1.15 \text{ GPM}$$

$$f_2 = 4 \left(\frac{0.2936}{1.304} \right)^{0.5714} = 1.70 \text{ GPM}$$

$$f_3 = 4 \left(\frac{0.2936}{2.607} \right)^{0.5714} = 1.15 \text{ GPM}$$

Figure 6-61 Flow rate in each branch of manifold station #1

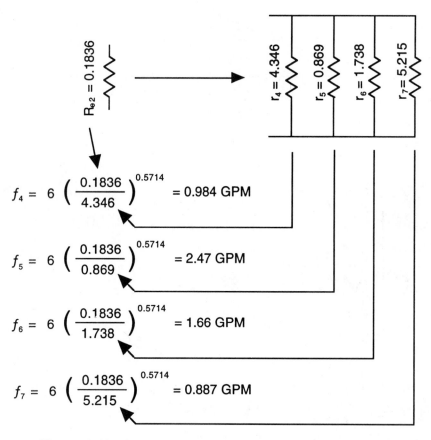

$$f_4 = 6 \left(\frac{0.1836}{4.346} \right)^{0.5714} = 0.984 \text{ GPM}$$

$$f_5 = 6 \left(\frac{0.1836}{0.869} \right)^{0.5714} = 2.47 \text{ GPM}$$

$$f_6 = 6 \left(\frac{0.1836}{1.738} \right)^{0.5714} = 1.66 \text{ GPM}$$

$$f_7 = 6 \left(\frac{0.1836}{5.215} \right)^{0.5714} = 0.887 \text{ GPM}$$

Figure 6-62 Flow rate in each branch of manifold station #2

These two flow rates in turn should equal the total flow rate entering the system at point A.

For manifold station #1:

$$f_1 + f_2 + f_3 = 1.15 + 1.70 + 1.15 = 4.0 \text{ gpm}$$

For manifold station #2:

$$f_1 + f_2 + f_3 + f_4 = 0.984 + 2.47 + 1.66 + 0.887$$

$$= 6.0 \text{ gpm}$$

For both manifolds: 4.0 gpm + 6.0 gpm = 10 gpm = total entering flow rate

Knowing the branch flow rates will allow accurate determination of the heat transfer and temperature drop along each branch. Methods for doing this will be covered in Chapter 10.

6.9   PIPE SIZING CONSIDERATIONS

Several considerations enter into the selection of a pipe size for a given application. Among these are pressure loss, flow velocity, potential for erosion damage, operating noise, installation cost, and operating cost.

Flow Velocity

Piping for hydronic heating systems should be sized so flow velocities do not exceed 4 feet per second. This minimizes flow noises created by entrained air and turbulence within the pipe. The table in Figure 6–63 lists the flow rates associated with a flow velocity of 4 ft/sec for type M copper tubing:

The equations given in Figure 6–2 or the PIPESIZE program in the Hydronics Design Toolkit can be used to determine the flow velocities associated with other flow rates for copper tube sizes of 1/2 inch through 2 inches.

Erosion Corrosion

Copper tubing is also subject to another problem if flow velocities are above the 8 to 10 feet per second

range. A condition called **erosion corrosion** can actually scrub metal off the inside wall of a tube or fittings. This effect tends to be localized at tight turns in fittings such as elbows and tees, and has not been known to be a widespread problem except in extreme cases. Still it is prudent to avoid any potential of its occurrence by sizing pipe for lower flow velocities.

Operating Cost

As the flow through a pipe increases, the head loss associated with the flow increases rapidly. Equation 6.10 can be used to show that when the flow rate through a smooth tube is doubled, the head loss increases by a factor of 3.4. The greater the head loss, the more pumping power is required to maintain a given flow. This in turn can require a larger circulator with a greater electrical power demand, and higher operating cost. *This is why a piping system should be thought of as having an operating cost as well as an installation cost.*

The determination of an optimal pipe size based on minimizing total owning and operating cost can be a complex process. It involves the use of specific installation costs, utility rates, pump efficiency, estimates of hours of operation, rate of inflation of electrical rates, and more. However, since other considerations such as flow velocity often narrow the choice down to two sizes, it is prudent to estimate their operating cost and take the results into consideration when making the final selection.

Equation 6.24 can be used to estimate the theoretical annual operating cost of any piping system transporting a fluid at a certain flow rate and head loss:

(Equation 6.24)

$$E = \frac{(3 \bullet 10^{-6})(D)(H_L)(f)(T)(k)}{\eta_p}$$

where:

E = annual operating cost of the piping system ($/yr)
D = density of the fluid at its typical operating temperature (lb/ft³)
H_L = head loss of the piping system at flow rate f (feet of head)
f = flow rate associated with the above head loss (gpm)
T = number of hours per year during which the circulator operates (hr/yr)
k = cost of electrical energy ($/kwhr)
η_p = efficiency of the motor/pump assembly (decimal percent)

Example 6.13: Determine the operating cost of a piping system that transports 160 °F water at a flow rate of 10 gpm with a head loss of 9 ft for 4,000 hours per year. Assume that electrical energy costs $0.10/kwhr, and that the pump/motor combination has an efficiency of 25%.

Tube size	flow rate at 4 ft/sec
1/2"	3.2 gpm
3/4"	6.4 gpm
1"	10.9 gpm
1-1/4"	16.3 gpm
1-1/2"	22.8 gpm
2"	39.5 gpm

Figure 6–63 Flow rates associated with flow velocity of 4 ft/sec for type M copper tube

Solution: The density of water at 160 °F is 61.02 lb/ft³. Substituting the data into Equation 6.24:

$$E = \frac{(3 \bullet 10^{-6})(61.04)(9)(10)(4,000)(0.10)}{0.25}$$

$$= \$26.37/\text{year}$$

Discussion: Keep in mind that over the life of the system, this annual operating cost will likely add up to several hundreds of dollars.

If Equation 6.24 is combined with Equation 6.12, the head loss term, H_L, can be eliminated, and the following relationship results:

(Equation 6.25)

$$E = \frac{(3 \bullet 10^{-6})(D)(\alpha c L)(f^{2.75})(T)(k)}{\eta_p}$$

where:

E = annual operating cost of the piping system ($/yr)
D = density of the fluid at its typical operating temperature (lb/ft³)
α = fluid properties factor for water at a given temperature
c = pipe size factor for the size of copper tube used
L = total equivalent length of a piping path (ft)
f = flow rate associated with the above head loss (gpm)
T = number of hours per year during which the circulator operates (hr/yr)
k = cost of electrical energy ($/kwhr)
η_p = efficiency of the motor/pump assembly (decimal percent)

This equation can be used to compute the annual operating cost of any piping system using a pump to convey fluid through copper pipe at a given temperature.

Example 6.14: Assume a hydronic heating system has a total equivalent length of 250 ft of ³/₄-in copper tube. It must convey 140 °F water at 6 gpm for 4,000 hours per year. Electrical energy costs $0.10/kwhr, and the circulator will have an efficiency of 25%. Determine the annual operating cost. Repeat the calculation for the same system using 1-in copper tube.

Solution: Using the graph in Figure 4-5, the water's density is found to be 61.35 lb/ft³. Using the graph in Figure 4-6, the water's dynamic viscosity is found to be 0.00032 lb/ft/sec. The α value can be calculated using Equation 6.13:

$$\alpha = \left(\frac{D}{\mu}\right)^{-0.25} = \left(\frac{61.35}{0.00032}\right)^{-0.25} = 0.04779$$

The value of c is found in Figure 6.18. For ³/₄-in copper tube $c = 0.061957$

Substituting all data into Equation 6.25:

$$E = \frac{(3 \bullet 10^{-6})(61.35)(0.04779)(0.061957)(250)(6^{2.75})(4,000)(0.10)}{0.25}$$

$$= \$30.08/\text{year}$$

If the same system were built using 1-in copper tube, and operated at the same 6 gpm flow rate, only the value of c would change in the above equation. For 1-in tubing: $c = 0.01776$. The revised operating cost would be:

$$E = \frac{(3 \bullet 10^{-6})(61.35)(0.04779)(0.01776)(250)(6^{2.75})(4,000)(0.10)}{0.25}$$

$$= \$8.62/\text{year}$$

In this particular case there would be a *theoretical* savings of $21.46 per year if the 1-inch tube were used instead of the ³/₄-inch tube. Such a savings could quickly pay for the increased installation cost of the larger tubing. However, in order to obtain this theoretical energy savings, a circulator that would produce *exactly* the same flow rate and head within the larger piping would have to be available and used. Since only a finite selection of circulators is available, the actual energy savings will depend on the difference in wattage of the smaller circulator used to replace the larger circulator. This savings could be calculated using Equation 6.26.

(Equation 6.26)

$$S = \frac{(w_{PL} - w_{PS})(T)(k)}{1,000}$$

where:

S = annual *savings* from being able to use a small circulator ($/yr)
w_{PL} = wattage of the larger circulator (watts)
w_{PS} = wattage of the smaller circulator (watts)
T = number of hours per year during which the circulator operates (hr/yr)
k = cost of electrical energy ($/kwhr)

Example 6.15: By using 1-in copper tubing rather than ³/₄-in copper tubing, a designer finds that a smaller circulator that operates at 85 watts can be used instead of a larger circulator that operates at 200 watts. Estimate the annual savings in using the smaller circulator assuming it will operate for 4,000 hrs/yr in a location where electricity costs $0.10/kwhr.

Solution: Substituting the data into Equation 6.26:

$$S = \frac{(200 - 85)(4,000)(0.10)}{1,000} = \$46/\text{year}$$

Discussion: Again it should be emphasized this savings will accumulate year after year. The savings will also be increased each time the cost of electricity increases. Total savings over the life of the system could be several hundreds of dollars.

Selecting a Pipe Size

The following procedure incorporates the previously discussed concepts into a method for choosing an appropriate pipe size:

Step 1. Select a tentative pipe size based on the criteria of not exceeding a flow velocity of 4 ft/sec.

Step 2. Estimate the installed cost of the system's piping, or portion thereof, using this tentative pipe size.

Step 3. Estimate the operating cost of the system using Equation 6.25 for the tentative pipe size.

Step 4. Estimate the installed cost of the system's piping, or portion thereof, using the next larger pipe size.

Step 5. Estimate the operating cost of the system using Equation 6.25 for the next larger pipe size.

Step 6. If the *potential* savings in operating cost between Steps 3 and 5 would return the higher installation cost of the large pipe in a reasonable period of time (suggested as 10 years or less), go on to Step 7. If not, use the pipe size from Step 1.

Step 7. Using methods from Chapter 7, select two circulators, one for each pipe size being considered, that will produce the desired flow rate within the system. Assuming a smaller, less expensive, and less power consuming circulator is available for use with the larger pipe size, compare the difference in installation and operating cost of the two circulators. If the savings in installation and operating cost of the small circulator over a period of a few years exceeds the higher installation cost of the larger pipe, use the larger pipe.

Example 6.16: Assume the same pipe systems described in Example 6.14. The system's circulator produces 6 gpm, and operates 4,000 hours per year using electrical energy purchased at $0.10/kwhr. Assume ¾-in copper pipe costs $0.85/ft, and 1-in copper pipe costs $1.40/ft. Also assume the following circulators were selected for each pipe size:

Using ¾-in pipe: Circulator cost = $110
                 Operating wattage = 150 watts

Using 1-in pipe: Circulator cost = $65
                 Operating wattage = 90 watts

Determine which pipe size should be used after considering both technical and economic factors.

Solution:

Step 1. Check that the flow velocity in the smaller pipe is equal to or less than the 4 ft/sec limit. From Figure 6–63 one finds that a ¾-in copper pipe can carry up to 6.4 gpm and not exceed this flow velocity limit. Thus ¾-in pipe is technically acceptable at the stated flow rate of 6 gpm, and the potential use of the large pipe size will depend on economic considerations.

Step 2. The cost of the ¾-in pipe will be (250 ft) ($0.85/ft) = $212.50.

Step 3. The estimated operating cost using ¾-in piping was calculated in Example 6.14 to be $30.08.

Step 4. The cost of the 1-in pipe will be (250 ft) ($1.40/ft) = $350.00.

Step 5. The estimated operating cost using 1-in piping was calculated in Example 6.14 to be $8.62.

Step 6. The theoretical savings in operating cost is: $30.08 – $8.62 = $21.46 per year.

The extra installation cost of the 1-in tubing is: $350 – $212.50 = $137.50.

The payback associated with use of the larger tubing is:

$$\frac{\$137.50}{\$21.46/\text{yr}} = 6.4 \text{ years}$$

Since this is a reasonably short payback period, proceed to compare costs and savings associated with the circulators.

Step 7. The savings in purchase cost of the smaller circulator is $110 – $65 = $45.

The *net* increase in system cost using the larger pipe would be the higher cost of the larger pipe minus the savings due to the smaller circulator:

Net increase in system cost = $137.5 – $45 = $92.50

The actual savings in annual operating cost can be calculated using Equation 6.26:

$$E = \frac{(150 - 90)(4,000)(0.10)}{1,000} = \$24/\text{yr}$$

The time required for operating cost savings to recover the net cost increase would be:

$$\frac{\$92.50}{\$24/\text{yr}} = 3.9 \text{ years}$$

Discussion: This example shows a relatively fast return on the extra investment in the larger pipe. Furthermore, any increase in electrical rates would reduce this payback period. *Over a system life of perhaps 30 years, the operating cost savings associated with the larger pipe size would return the higher initial investment several times over!* The larger pipe size is well justified in this case.

SUMMARY

This chapter has laid the foundation for describing fluid flow in piping systems. Many of the fundamentals of fluid mechanics have been condensed and presented in a way especially suited for hydronic heating systems. Terms such as head loss, pressure drop, flow rate, flow velocity, and more have been described. Such terms will be routinely used in later chapters.

The analogy between hydraulic resistance and electrical resistance represents a new approach for analyzing fluid-filled closed-loop piping systems, one that can be a powerful tool when properly applied. This approach will also be used in later chapters dealing with system design.

The reader is strongly encouraged to use the Hydronics Design Toolkit software to expedite many of the calculations demonstrated in this chapter. Programs such as `PIPEPATH`, `HYRES`, `PARALLEL`, `PIPESIZE`, and `FLUIDS` were specifically designed for this purpose. They also allow the reader to try many what-if scenarios related to piping system performance. This is an excellent way to gain a solid understanding of the principles in this chapter.

KEY TERMS

Average flow velocity
Boundary layer
Compressible fluid
Equivalent length (of a component)
Equivalent resistance
Equivalent resistor
Erosion corrosion
Feet of head
Flow coefficient (C_v)
Flow rate
Flow velocity
Fluid properties factor (α)
Gas
Head
Hydraulic resistance
Hydraulic resistance diagrams
Hydraulic resistor
Incompressible fluid
Laminar flow
Liquids
Piped in parallel
Pressure head
Reynold's number
Series circuit
Series piping path
Smooth pipe
Static pressure
System resistance curve
Total equivalent length (of a piping assembly)

Total head
Turbulent flow
Velocity head
Velocity profile

CHAPTER 6 QUESTIONS AND EXERCISES

Note: Questions and exercises requiring the Hydronics Design Toolkit are indicated with marginal symbol.

1. Find the flow velocities corresponding to the following conditions:
   a. 2.5 gpm flow rate in a ½-in type M copper tube
   b. 10 gpm flow rate in a ¾-in type M copper tube
   c. 15 gpm in a tube with an inside diameter of 1.1 inches

2. Use data from Chapter 4 to determine the value of the fluid properties factor, α, for water at 120 °F. *Hint:* Use Equation 6.13. Compare this to the value obtained using Figure 6–17.

3. What water pressure would be required at street level to push water to the top of a 300-ft building? Express the answer in psi. If the water pressure at the top of the building had to be 50 psi, how would this effect the pressure at the street level?

4. Water at 150 °F moves through a pipe with an internal diameter of 1.3 ins, and at a flow velocity of 3 ft/sec. Determine the Reynold's number for this flow. Would the flow be laminar or turbulent under these conditions?

5. Determine the head loss when 200 °F water flows at 10 gpm through 400 ft of 1-in type M copper tube. Determine the pressure drop (in psi) associated with this head loss.

6. A valve has a C_v rating of 3.5. What will be the pressure drop across the valve when 140 °F water flows through it at 6 gpm?

7. Find the total equivalent length of the the piping path shown in Figure 6–64.

8. Find the hydraulic resistance of the piping assembly shown in Figure 6–64 assuming water at 120 °F will flow through it.

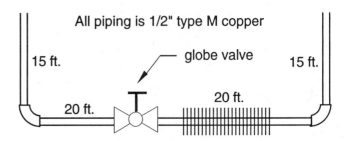

Figure 6–64 Piping assembly for Exercise 7

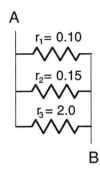

Figure 6–65 Hydraulic resistor diagram for Exercises 10 and 11

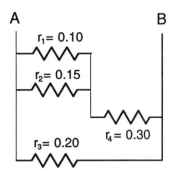

Figure 6–66 Hydraulic resistor diagram for Exercise 12

9. Calculate five data points representing head loss versus flow rate through the piping assembly shown in Figure 6–64 for 120 °F water. Plot this data and sketch a system resistance curve. *Hint:* Use Equation 6.10 to generate the necessary head loss versus flow rate data.

10. Use Equation 6.22 to find the equivalent hydraulic resistance of the three parallel hydraulic resistors shown in Figure 6–65.

11. Assuming a flow rate of 8 gpm enters at point A in Figure 6–65, determine the flow rate in each of the three branches. Check your results.

12. Determine the single equivalent resistance between points A and B for the hydraulic resistor diagram shown in Figure 6–66

13. Draw a hydraulic resistor diagram for the piping assembly shown in Figure 6–67.
Assuming all piping is 1-in copper tubing, and that the system operates with 140 °F water:
a. Using series and parallel resistance concepts, reduce the resistor diagram to a single equivalent resistance.

b. Assuming flow enters the piping assembly at 15 gpm, find the flow rate in each branch.

c. Find the flow velocity in each branch.

14. Use the FLUIDS program in the Hydronics Design Toolkit to find the answer to Exercise 2. Also find the α value for the following fluids and conditions:
a. 140 °F water
b. 50% propylene glycol solution at 140 °F
c. 30% ethylene glycol solution at 160 °F

15. Use the PIPEPATH program to determine the hydraulic resistance of the piping assembly shown in Figure 6–64.

16. Use the PIPEPATH program to determine the hydraulic resistance of both branches of the piping assembly shown in Figure 6–67.

17. Use both the PIPEPATH and PARALLEL programs to confirm the answer obtained for Example 6.12.

18. Explain the difference between head loss and pressure drop.

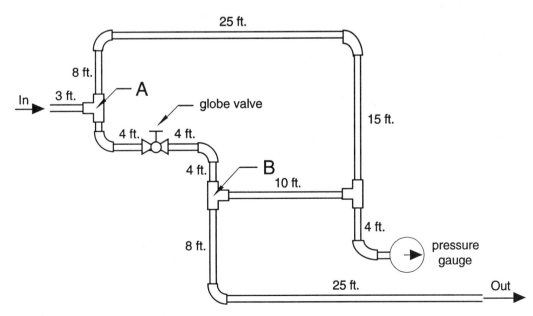

Figure 6–67 Piping assembly for Exercise 13

7 CIRCULATING PUMPS

OBJECTIVES

After studying this chapter you should be able to:

- Describe different types of circulating pumps used in hydronic systems
- List the main components of a centrifugal pump
- Determine the proper placement of the pump(s) within the system
- Calculate the flow rate a specific circulator will produce in a given piping system
- Estimate the flow rate through a pump from the pressure differential across it
- Work with both graphical and analytical descriptions of pump performance
- Predict the performance of pumps connected in series and parallel
- Explain what cavitation is and how to predict it
- Describe methods of avoiding cavitation
- Select a pump for proper performance and high efficiency
- Use the PUMPCURV and PUMP/SYS programs in the Hydronics Design Toolkit

7.1 INTRODUCTION

The pump is the heart of a hydronic system. Although smaller and less expensive than the heat source, it can prove to be just as vital to proper system performance. The wide variety of circulator pumps available in today's market allows great flexibility in how the overall system is designed and controlled.

This chapter provides an overview of the type of circulating pumps used in small and medium scale hydronic systems. It goes on to define and illustrate the concept of a pump curve. This curve is combined with the system resistance curve discussed in Chapter 6 to determine the flow rate at which the system will operate. The often overlooked concepts of series and parallel pumps are discussed from the standpoint of customizing a pumping system to the specific needs of a piping system. The proper placement of the pump in the system is also discussed. Special attention is given to the subject of cavitation and its avoidance. Finally, two additional programs from the Hydronics Design Toolkit are illustrated. These tools make short work of the mathematics involved in describing pump performance, allowing many conceivable designs to be evaluated prior to making a final selection.

7.2 PUMPS FOR HYDRONIC SYSTEMS

Pumps come in a wide variety of designs, sizes, and performance ranges. In closed-loop, fluid-filled hydronic systems, the pump's function is to circulate the system fluid around the piping. No lifting of the fluid is involved in such systems, as discussed in Chapter 6. For this reason, and especially when small fractional horsepower motors are used, pumps are often called **circulators**. *Within this book the terms circulator and pump are synonymous.*

The type of pump commonly used in hydronic systems is known as a **centrifugal pump**. It uses a rotating component called an **impeller** to add mechanical energy (head) to the fluid. Figure 7–1 illustrates the basic construction of a centrifugal pump.

As the impeller rotates, fluid within the center opening (or **eye**) of the impeller is rapidly accelerated through the passageways formed by the impeller vanes between the two impeller disks. The fluid's speed and mechanical energy content are increased as it is accelerated toward the outer edge of the impeller. As it leaves the impeller, the fluid impacts against the inside surface of the chamber surrounding the impeller. This chamber is called the **volute**. Its former speed (or velocity head) is converted to a pressure increase (pressure head). The fluid then flows around the contoured volute and exits through the discharge port.

For this process to be continuous, the rate of fluid entering the pump must be identical to that leaving. Intuition might suggest a pump continuously sucks fluid into the eye of its impeller. This is not true. *Water entering a centrifugal pump must be pushed in by system pressure upstream of the inlet port.* This is a very important point, and is often misunderstood. *If the proper conditions are not provided for fluid to be pushed into the inlet port, very undesirable operating characteristics will result.* These are discussed later in the chapter.

Centrifugal pumps can be built with differently-shaped volutes while still maintaining the same internal operation.

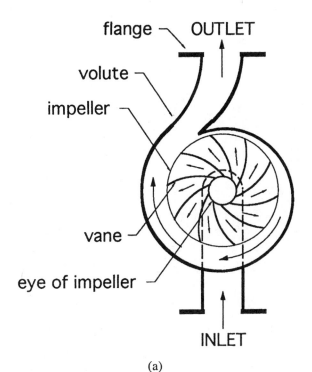

(a)

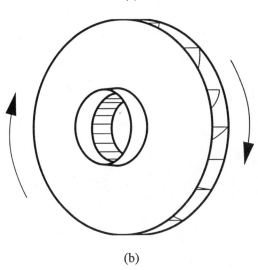

Figure 7–1 (a) Simplified cross section of a centrifugal pump, (b) an impeller.

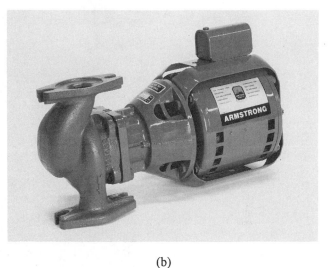

(b)

Figure 7–2 Examples of inline circulators. (a) A wet rotor circulator, (b) a mechanically-coupled three-piece circulator. Courtesy of Armstrong Pumps.

The volute's shape determines how the pump will be connected to the system's piping. Figure 7–2 shows two examples of **inline circulators**. These circulators have their inlet and discharge ports along a common centerline. An inline circulator can be placed into a piping path without need for any lateral offsets between the inlet and discharge ports. *The inline circulator design is by far the most common type used in residential and light commercial hydronic systems.*

By modifying the shape of the volute, a design called an **end suction pump** is created. An example of an end suction centrifugal pump is shown in Figure 7–3.

End suction pumps create a 90 degree turn in the system piping. In most cases an offset between the centerlines of the inlet and discharge piping is also required. End suction pumps are more common for medium- and large-scale systems.

Wet Rotor Circulators

Over the last 30 years, a specialized design for small- and medium-sized circulators has been refined specifically for use in hydronic systems by a number of pump manufacturers. This design, known as a **wet rotor circulator**, combines the motor, shaft, and impeller into a single assembly that is housed in a chamber filled with system

Figure 7–3 Example of an end suction centrifugal pump. Courtesy of Bell and Gossett.

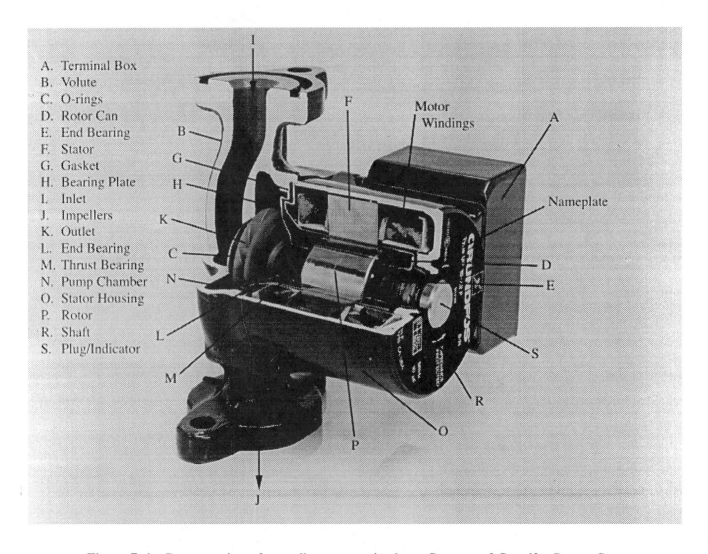

A. Terminal Box
B. Volute
C. O-rings
D. Rotor Can
E. End Bearing
F. Stator
G. Gasket
H. Bearing Plate
I. Inlet
J. Impellers
K. Outlet
L. End Bearing
M. Thrust Bearing
N. Pump Chamber
O. Stator Housing
P. Rotor
R. Shaft
S. Plug/Indicator

Figure 7–4 Cut-away view of a small wet rotor circulator. Courtesy of Grundfos Pumps Corp.

fluid. An example of a wet rotor circulator is shown in Figure 7–4.

The motor of a wet rotor circulator is totally cooled and lubricated by the system's fluid. As such it has no fan or oiling caps. The rotor assembly is supported on ceramic or graphite bushings in the rotor can. These bushings contain no oil, but ride on a film of system fluid. The rotor can is surrounded by the stator assembly of the motor.

Some advantages of wet rotor circulators are as follows:

• No oiling is required
• No leakage of system fluid due to worn pump seals
• Their small size makes them easy to locate and support
• The absence of a cooling fan and external coupling makes for quiet operation
• Several models are available with multiple speed motors
• They are relatively inexpensive due to fewer parts
• They are ideal for applications where limited flow rate and head are required
• They can be **close-coupled** for series pump applications
• Most have permanent split capacitor (PSC) motors that can be operated over a wide speed range by suitable electronic controls

Their disadvantages include:

• The low starting torque of the PSC motors may not be able to free a stuck impeller after a period of prolonged shutdown
• Servicing anything not contained in the external junction box requires opening the wetted part of the pump, resulting in some fluid loss and air entry into the system

Wet rotor circulators are the most commonly used pumps in modern residential and light commercial hydronic systems. They are available with cast iron volutes for use in closed hydronic systems, or with bronze or stainless steel volutes for direct contact with domestic water, or other open loop applications. Most wet rotor circulators have impellers constructed of stainless steel, bronze, or synthetic materials. The smaller models are specifically designed for use in multi-zone hydronic systems, where a separate circulator is used for each zone.

Three-piece Circulators

Another common pump design used in small- and medium-sized hydronic systems is called a **three-piece circulator**. It consists of the pump body assembly, coupling assembly, and motor assembly as shown in Figure 7–5.

Unlike wet rotor circulators, the motor of a three-piece circulator is totally separate from the wetted portion of the pump. This allows the motor or coupling to be serviced or replaced without needing to open the piping system. The design of the coupling assembly between the

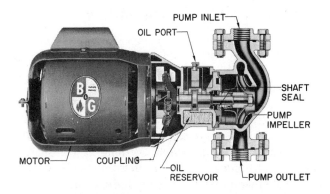

Figure 7–5   Example of a three-piece circulator. Courtesy of ITT Corporation.

motor and pump shaft varies among manufacturers. A common design employs a spring assembly that absorbs vibration or high torque between the two shafts as the motor starts. The impeller shaft penetrates the volute through **shaft seals** that must limit leakage of system fluid, even under high pressure. Although it is normal for these seals to experience minor fluid losses, the leakage rate of modern pump seals is so small the fluid usually evaporates before it is seen.

The advantages of a three-piece pump design are:

• Potential for longer life if bearings are properly lubricated
• Easy servicing of the motor without need to open wetted portion of pump
• Able to produce higher starting torque to overcome a stuck impeller condition after a prolonged shutdown

The disadvantages include:

• Heavier construction requiring strong supports
• Must be oiled periodically
• More operating noise due to external motor and coupling assembly
• Potential maintenance of mechanical seals and coupling assembly
• Defective or worn shaft seals could allow leakage of system fluid

Like wet rotor circulators, three-piece circulators are available in materials suitable for both open loop and closed loop applications.

Pump Mounting Considerations

Most circulators used in smaller hydronic systems are designed to be installed with their shafts in a horizontal position. This removes the thrust load on the bushings due to the weight of the rotor and impeller.

The direction of flow through the pump is usually indicated by an arrow on the side of the volute. *The installer should always check that the pump is installed in the correct flow direction.* As long as the shaft is horizontal, the pump can be mounted with the flow arrow pointing upward, downward, or horizontal. Of these, upward flow is slightly preferred, because it allows the circulator to rapidly clear itself of air bubbles.

As a general rule, the weight of the circulator should not be supported by system piping unless the circulator is relatively light and the piping itself is well supported immediately adjacent to the circulator. This is especially true for three-piece circulators that create a bending effect on the pipe due to the offset weight of the motor. If the circulator is supported by system piping, the piping should not be rigidly mounted to walls that can transmit vibrations to the building structure. The use of vibration-absorbing mounting brackets is recommended.

A common practice in many residential installations has been to mount several small circulators that are part of a multi-zone system on a common header assembly directly threaded into the boiler. Threaded steel or black iron pipe in the 1 in to 1.5 in range is often used to fabricate this header. Although this pipe is strong enough to handle the weight of the circulators, the bending effect created by the weight of the circulators places great stress on the boiler connection. Anytime a common header-type mounting is used, the far side of the header should be supported to remove this bending load as shown in Figure 7–6.

Connecting the Circulator to the Piping

Unlike fittings and valves, circulators need to be mounted so they can be removed for servicing if necessary. The usual method of connecting a circulator to piping is with bolted **flanges**. One side of the flanged joint is an integral part of the pump's volute. The other flange is threaded onto the piping. As the flange bolts are tightened, an O-ring or gasket is compressed between the faces of the flanges to make the seal. For small- and medium-sized circulators, a two bolt flange is common.

Pump flanges are available in cast iron for closed system applications, as well as bronze or brass for direct contact with domestic water or other open loop applications. A special type of flange, known as an **isolation flange**, contains a built-in ball valve that can be operated using a screwdriver. With the isolation flanges closed, the circulator may be unbolted and removed from the system. Only the small amount of fluid within the volute will be lost. An alternative method of isolating the pump is to use standard flanges combined with a gate valve or ball valve on each side of the pump. These options are shown in Figure 7–7.

On some very small circulators the connection to the piping is made using a half-union rather than a flange. The half-union threads directly onto the pump's volute. Half unions are available with or without an integral ball valve for isolation. In still other cases small circulators are designed to be directly soldered to copper tubing. Figure 7–8 shows the various flange and half-union options offered by one pump manufacturer.

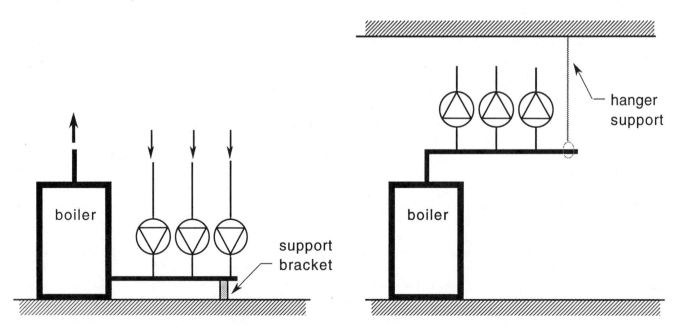

Figure 7–6 Support for piping headers that support circulators

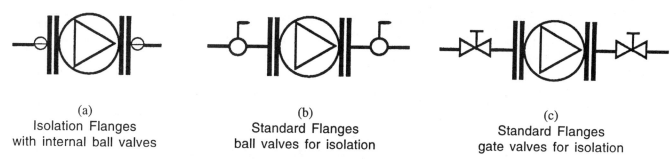

(a)
Isolation Flanges
with internal ball valves

(b)
Standard Flanges
ball valves for isolation

(c)
Standard Flanges
gate valves for isolation

Figure 7–7   Isolating a pump from system piping using (a) isolation flanges, (b) ball valves, (c) gate valves.

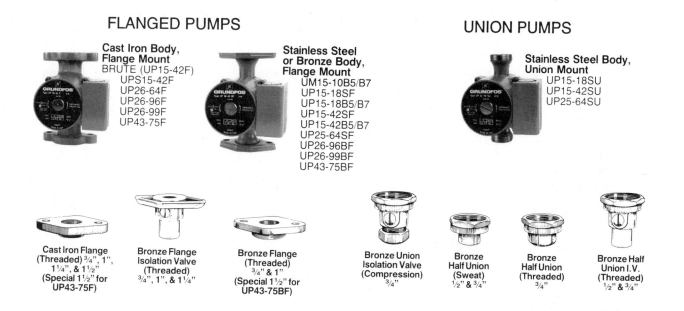

FLANGED PUMPS

**Cast Iron Body,
Flange Mount**
BRUTE (UP15-42F)
UPS15-42F
UP26-64F
UP26-96F
UP26-99F
UP43-75F

**Stainless Steel
or Bronze Body,
Flange Mount**
UM15-10B5/B7
UP15-18SF
UP15-18B5/B7
UP15-42SF
UP15-42B5/B7
UP25-64SF
UP26-96BF
UP26-99BF
UP43-75BF

UNION PUMPS

**Stainless Steel Body,
Union Mount**
UP15-18SU
UP15-42SU
UP25-64SU

Cast Iron Flange
(Threaded) ¾", 1",
1¼", & 1½"
(Special 1½" for
UP43-75F)

Bronze Flange
Isolation Valve
(Threaded)
¾", 1", & 1¼"

Bronze Flange
(Threaded)
¾" & 1"
(Special 1½" for
UP43-75BF)

Bronze Union
Isolation Valve
(Compression)
¾"

Bronze
Half Union
(Sweat)
½" & ¾"

Bronze
Half Union
(Threaded)
¾"

Bronze Half
Union I.V.
(Threaded)
½" & ¾"

Figure 7–8   Example of flange and half-union connections for connecting pumps to piping. Courtesy of Grundfos Pumps Corp.

7.3 PLACEMENT OF THE CIRCULATOR WITHIN THE SYSTEM

The location of the circulator(s) relative to the other components in a hydronic system can make the difference between quiet, reliable operation or constant problems. One guiding rule summarizes the situation: *Always place the circulator so that its inlet is close to the connection point of the system's expansion tank.* This principle has been applied for years in commercial hydronic systems, yet somehow has been largely ignored in residential installations.

To understand why this rule should be followed, one needs to consider the interaction between the circulator and expansion tank. In a closed-loop piping system the amount of fluid, including that in the expansion tank, is fixed. It does not change regardless of whether the circulator is on or off. The upper portion of the expansion tank contains a captive volume of air at some pressure. The only way to change the pressure of this air is to either push more fluid into the tank to compress the air, or to remove fluid from the tank to expand the air. This fluid would have to come from, or go to, some other location within the system. But since the system's fluid is incompressible, and the amount of fluid in the system is fixed, this cannot happen regardless of whether the circulator is on or off. *The expansion tank thus fixes the pressure of the system's fluid at its point of attachment to the piping.* This is called the **point of no pressure change** within the system.

Consider a horizontal piping circuit filled with fluid and pressurized to some pressure, say 10 psi, as shown in Figure 7–9. When the circulator is off, the pressure is the same (10 psi) throughout the piping circuit. This is indicated by the solid horizontal pressure line shown above the piping. When the circulator is turned on, it

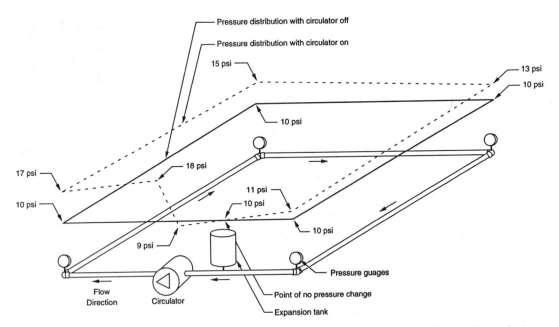

Figure 7–9 Pressure distribution in horizontal piping circuit. Solid line is pressure distribution when pump is off. Dashed line is pressure distribution when pump is on. Note expansion tank is near the inlet port of pump.

immediately creates a pressure difference between its inlet and discharge ports. However, *the expansion tank still maintains the same (10 psi) fluid pressure at its point of attachment to the system.* The combination of the pressure difference across the circulator, the pressure drop due to head loss in the piping, and the point of no pressure change, results in a new pressure distribution as shown by the dashed line in Figure 7–9.

Notice that the pressure increases in nearly all parts of the circuit when the circulator is turned on. This is desirable because it helps eject air from vents. It also minimizes the chance of cavitation. The short segment of piping between the expansion tank and the inlet port of the circulator experiences a slight drop in pressure due to head loss in the piping. The numbers used for pressure in Figure 7–9 are illustrative only. The actual numbers will of course depend on flow rates, fluid properties, and pipe sizes.

Now consider the same system with the expansion tank connected near the *discharge* port of the circulator. The dashed line in Figure 7–10 illustrates the new pressure distribution in the system when the circulator is on.

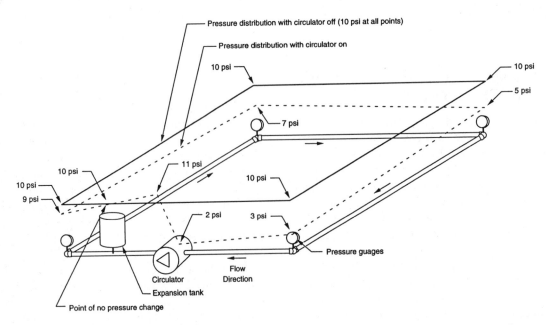

Figure 7–10 Pressure distribution in horizontal piping circuit with pump on (dashed lines). Note the expansion tank is near the discharge port of the pump.

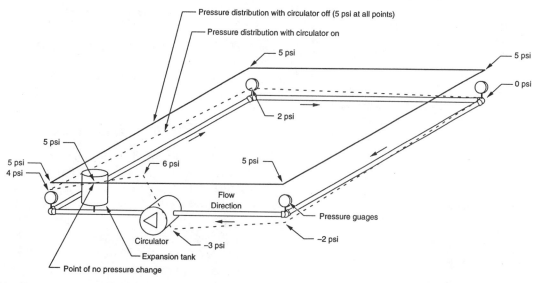

Figure 7–11   Pressure distribution in horizontal piping circuit with pump on (dashed lines). Note the expansion tank is near the discharge port of the pump.

Again the point of no pressure change remains at the expansion tank connection. This causes the pressure in most of the system to *decrease* when the circulator is turned on. The pressure at the inlet port has dropped from 10 psi to 2 psi. This situation is not desirable because it reduces the system's ability to expel air. It can, in combination with other factors, lead to pump cavitation.

To see how problems can develop, imagine the same system with a pressurization of only 5 psi. When the circulator starts, the same 9 psi differential will be established between its inlet and discharge ports, and the entire pressure profile shown with dashed lines in Figure 7–10 will be shifted downward by 5 psi (10 – 5 = 5 psi), as shown in Figure 7–11.

Notice the pressure in the piping between the upper right hand corner of the circuit and the inlet port of the circulator is now *below atmospheric pressure*. If air vents were located in this portion of the circuit, the subatmospheric pressure would actually suck air into the system every time the circulator operates. The circulator is also much more likely to cavitate under these conditions.

Unfortunately, this scenario has occurred in many residential hydronic systems. To see why, consider the piping schematic of a typical hydronic system shown in Figure 7–12.

In this arrangement the circulator is pumping *toward* the expansion tank. Although there appears to be a fair distance between the circulator and the tank, the pressure drop due to head loss through a typical sectional boiler is very low. Thus, from a pressure drop standpoint, the circulator's discharge port is very close to the expansion tank. This arrangement will cause the system pressure to drop from the expansion tank connection, around the distribution system to the inlet port of the pump, whenever the circulator is on.

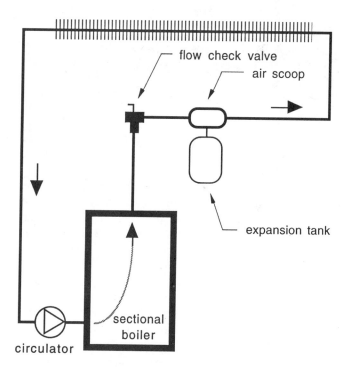

Figure 7–12   Typical piping configuration of a residential hydronic system. Note the placement of the circulator relative to the expansion tank.

One reason circulators were originally located on the return side of the boiler was to allow them to operate with cooler return water. It was believed the lower operating temperature prolonged the life of the packing, seals, and motor. This is not a concern for most currently produced wet rotor circulators that are often rated for *continuous operation* at fluid temperatures up to 230 °F.

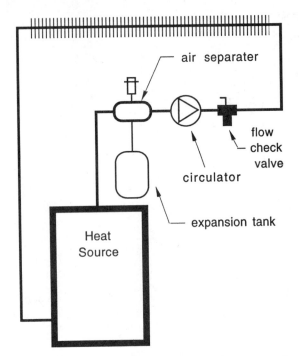

Figure 7-13 Proper arrangement of circulator and flow check valve relative to expansion tank. Note the expansion tank is located close to the inlet port of the circulator.

Some hydronic systems that are piped as shown in Figure 7-12 have worked fine for years. Others have had problems from the first day they were put into service. Why is it that some systems work and others do not? The answer lies in a number of factors that interact to determine the exact pressure distribution in any given system. These include system height, fluid temperature, pressure drop, and system pressurization. The systems most prone to problems from this type of circulator placement are those with high fluid temperature, low static pressure, low system height, and high pressure drops around the piping circuit. Rather than gamble on whether these factors will work, it is best to arrange the piping as shown in Figure 7-13.

This arrangement will *increase* the pressure in nearly all parts of the system when the circulator is on. *Many systems that have chronic problems with air in the piping can be cured with this type of component rearrangement.*

For systems employing conventional sectional boilers with low head loss characteristics, the expansion tank can also be placed at the return side of the boiler with virtually identical results. This should not be done, however, with a heat source having a high head loss characteristic, such as hydronic heat pumps or copper tube boilers. The higher pressure drop of these heat sources will cause a significant pressure drop between the expansion tank connection and the inlet of the circulator, making the circulator more prone to cavitation.

7.4 PUMP PERFORMANCE

This section presents analytical methods for describing the performance of hydronic circulators. These methods are vital in properly matching the circulator with the flow requirements of a piping system.

Explanation of Pump Head

The head produced by a pump is a commonly misunderstood concept. Some references define it as the height to which a pump can lift and maintain a column of water, others as the pressure difference the pump can produce between its inlet and discharge ports. Both definitions are partially correct but somewhat incomplete.

The head produced by a pump is actually an indication of the mechanical energy transferred to the fluid by the pump. This energy can reveal itself as an increase in flow rate or as an increase in pressure. The former is a form of kinetic energy, the latter a form of potential energy.

For incompressible fluids such as those used in hydronic systems, the flow rate and flow velocity entering a pump will always be the same as those exiting the pump. This means the added head (e.g., added mechanical energy) will reveal itself as an increase in fluid pressure between the inlet and discharge ports of the pump. The extent of this pressure increase depends on the flow rate through the pump and the density of the fluid being pumped. *As the flow rate through a pump increases, the pressure difference across it decreases* as shown in Figure 7-14.

In the U.S., the head produced by a pump is expressed in units of *feet of head*. Although this seems like an unusual unit for measuring the mechanical energy

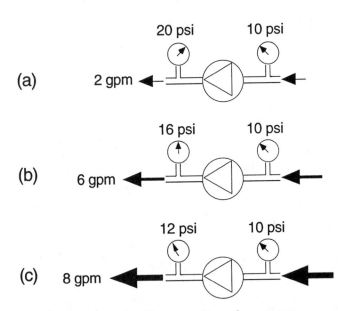

Figure 7-14 As the flow rate through a pump increases, the pressure difference between the inlet and outlet ports decreases.

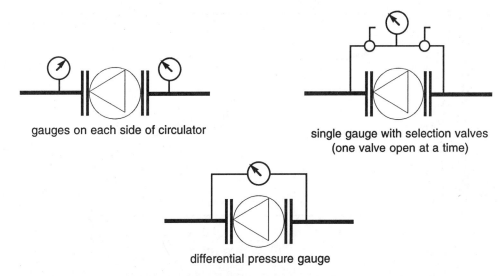

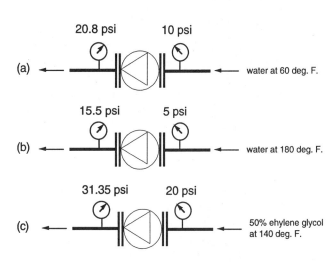

Figure 7–15 Three options for reading the pressure differential across a circulator

transferred to the fluid by the pump, it results from a simplification of the following units:

$$\text{Head} = \frac{\text{ft} \cdot \text{lb}}{\text{lb}} = \frac{\text{energy}}{\text{lb}} = \frac{\text{ft} \cdot \cancel{\text{lb}}}{\cancel{\text{lb}}} = \text{ft of head}$$

Those having studied physics will recall the unit of ft • lb is a unit of energy. Head may therefore be described as the *mechanical energy added per pound of fluid.*

The head added by a pump at some given flow rate is independent of the fluid being pumped. For example, a pump producing a head of 20 ft while pumping 60 °F water at 5 gpm will also have a head of 20 ft while pumping a solution of 50% ethylene glycol at the same 5 gpm. However, the pressure differential measured across the pump will be different because of the difference in the density of the fluids.

Converting between Head and Pressure Difference

Many times the performance of a circulator can be measured by the pressure difference between its inlet and discharge ports. Figure 7–15 shows three options for equipping a circulator with gauges for reading this pressure differential. On some medium-sized circulators, threaded openings are provided in the volute flanges for direct attachment of pressure gauges.

Equation 6.7 can be used to convert the pressure differential produced by a circulator to head. This calculation requires the determination of the density of the fluid being pumped.

(Equation 6.7)

$$H = \frac{(\Delta P)144}{D}$$

where:

H = head produced by the pump (feet of head)
ΔP = pressure differential between inlet and discharge ports of circulator (psi)
D = density of the fluid being pumped (lb/ft³)

Example 7.1: Based on the pressure gauge readings, determine the head produced by the pump in the three operating conditions shown in Figure 7–16.

Solution: In all cases the density of the fluid being pumped needs to be determined. The density of water at 60 °F and 180 °F can be found in Figure 4–5. The density of the 50% solution of ethylene glycol was found using the FLUIDS program in the Hydronics Design Toolkit.

Figure 7–16 Pressure differences across pumps for Example 7.1

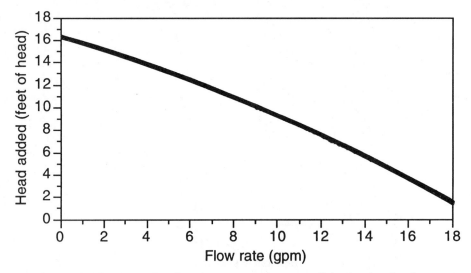

Figure 7–17 Example of a pump curve for a small hydronic circulator

a. (60 °F water): density = 62.31 lb/ft³
b. (180 °F water): density = 60.47 lb/ft³
c. (140 °F ethylene glycol): density = 65.39 lb/ft³

The head produced in each situation can be calculated using Equation 6.7:

a. $H = \dfrac{(20.8 - 10)(144)}{62.31} = 25.0 \text{ ft of head}$

b. $H = \dfrac{(15.5 - 5)(144)}{60.47} = 25.0 \text{ ft of head}$

c. $H = \dfrac{(31.35 - 20)(144)}{65.39} = 25.0 \text{ ft of head}$

Notice even though the pressure differentials across the pump are different, the pump produces the same head in all three cases.

Pump Performance Curves

A graph of a pump's ability to add head to a fluid is called a **pump curve**. An example of such a graph is shown in Figure 7–17. This curve shows the head added to the fluid by the pump as a function of the flow rate through the pump. In the U.S., head is usually expressed in units of feet, and flow rate in units of gallons per minute (gpm).

Pump curves are developed from test data based on water in the temperature range of 60 °F to 80 °F. For fluids with higher viscosities such as glycol-based antifreeze solutions, there is a very small decrease in head and flow rate capacity of the pump. However, for the fluids and temperature ranges commonly used in small hydronic heating systems this variation is so small that it can be safely ignored. Thus, *for the applications discussed in this text, pump curves may be considered to be independent of the fluid being pumped.* The reader is reminded, however, that the system curve that indicates head loss versus flow rate for a piping system is very dependent on fluid properties and temperature as discussed in Chapter 6.

Pump curves are extremely important in matching the performance of a pump to the flow requirements of a piping system. All pump manufactures publish the curves associated with their pumps. In some cases the pump curves for an entire series or family of circulators is plotted on the same set of axes so performance comparisons can be made. This is illustrated in Figure 7–18.

Operating Point of a Pump/Piping System

The reader will recall that system resistance curves were discussed in Chapter 6. Such a curve shows the head

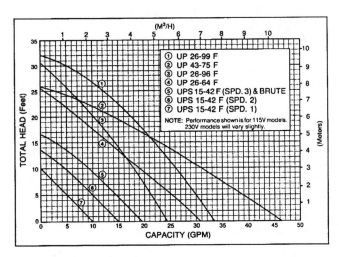

Figure 7–18 Example of several pump curves plotted on a common set of axes. Courtesy of Grundfos Pumps Corp.

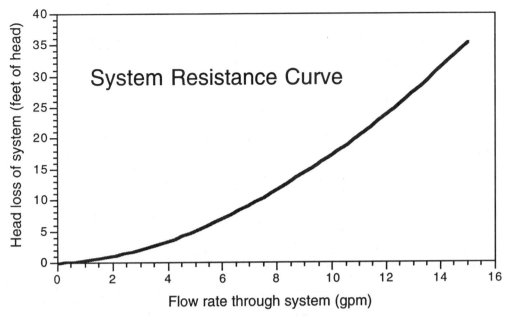

Figure 7–19   Example of a system resistance curve

loss of a given piping system as a function of flow rate for a given fluid at a given temperature. It is a unique graphical description of the hydraulic characteristics of a particular piping system. An example of a system resistance curve is shown in Figure 7–19.

Notice the pump curve in Figure 7–17 and the system resistance curve in Figure 7–19 have similar quantities on each axis. The only difference is the system curve shows head *loss* on the vertical axis, whereas the pump curve shows head *added* on its vertical axis. The pump curve can be thought of as the ability of the pump to *add* mechanical energy to a fluid over a range of flow rates. The system resistance curve can be thought of as the ability of the piping system to *remove,* or dissipate, mechanical energy from the fluid over a range of flow rates. In both cases the gain or loss of mechanical energy is expressed in feet of head.

A basic scientific principle known as the first law of thermodynamics states that when the rate at which energy is added to a system exactly equals the rate at which energy is removed from that system, the system is in equilibrium and will continue to operate at those conditions until one of the energy flows is changed. Chapter 1 discussed the concept of equilibrium for thermal energy flows. More specifically it stated that when the rate of heat input to a hydronic system exactly equals the rate of heat dissipation by the system, that system is in thermal equilibrium and temperatures will remain constant.

The concept of equilibrium also holds true for the mechanical energy added or removed from the fluid in a piping system. *When the rate at which a pump adds head to the fluid in a piping system exactly equals the rate at which the piping system dissipates this head due to viscous friction, the flow rate in that system will remain constant.*

To find this equilibrium condition, the pump curve and the system resistance curve are plotted on the same set of axes as shown in Figure 7–20. The point at which the pump curve and the system resistance curve intersect is called the **operating point** of the system. This point indicates the condition of mechanical equilibrium of the system's fluid. The flow rate and head across the circulator at the operating point can be found by extending vertical and horizontal lines from this point to each axis as shown in Figure 7–20.

A performance comparison of several circulators within a given system can be made by plotting their individual pump curves on the same set of axes as the system's resistance curve. The intersection of a given circulator's pump curve with the system resistance curve indicates the operating point for that particular circulator. By projecting vertical lines from the operating points down to the horizontal axis, the designer can determine the flow rate each circulator would produce within the system. This concept is illustrated in Figure 7–21.

Notice even though the curves for circulators 1 and 2 are markedly different, they intersect the system curve at almost the same point. These two circulators would yield almost identical flow rates of about 7.5 gpm and 7.8 gpm in this particular system. The flow rate produced by circulator 3, about 5 gpm in this case, is considerably lower.

High Head versus Low Head Pumps

Some circulators are designed to produce relatively high heads at lower flow rates. Others produce lower, but relatively stable heads over a wide range of flow rates. These characteristics are fixed by the manufacturer's design of the circulator, in particular the diameter and

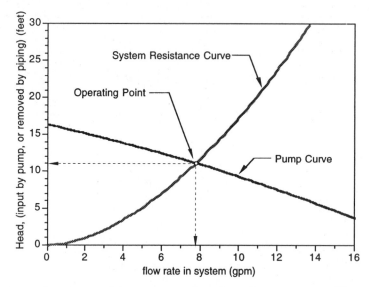

Figure 7–20 A pump curve and system resistance curve plotted on the same axes. The operating point is where the curves cross.

width of the impeller. Figure 7–22 compares the internal construction of a circulator designed for high heads at low flow rates, versus one designed for lower heads over a wider range of flow rates. Notice the impeller of the high head circulator has a relatively large diameter, but a very small separation between its disks. The low head circulator, on the other hand, has a small diameter impeller with widely separated disks.

The pump curves for each circulator shown in Figure 7–22 are plotted in Figure 7–23. The high head pump is said to have a "steep" pump curve. The other pump has a "flat" curve. Interestingly, both of these pumps have the same $1/15$th horsepower motor rating, and operate at the same rpm. The design of the impeller obviously has a major effect on how the motor's energy is converted into mechanical energy of the fluid.

Circulators with steep pump curves are intended for systems having high head losses at modest flow rates. Examples would include series piping circuits containing several components with high flow resistance characteristics, earth heat exchangers for hydronic heat pumps, or radiant floor heating systems using long circuits of small diameter tubing.

Circulators with flat pump curves should be used when it is desirable to maintain a relatively steady pressure differential across a given distribution system, over a

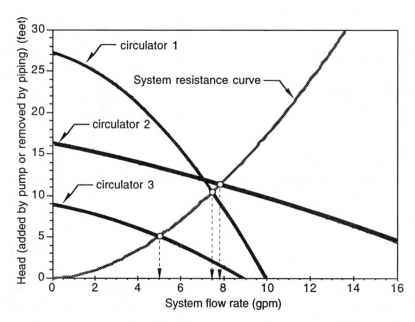

Figure 7–21 Three separate pump curves plotted with a system resistance curve. The vertical lines descending from the operating points indicate flow rates produced by each circulator.

(a)

(b)

Figure 7–22 Comparison between (a) high head impeller design, and (b) low head impeller design. Courtesy of Taco, Inc.

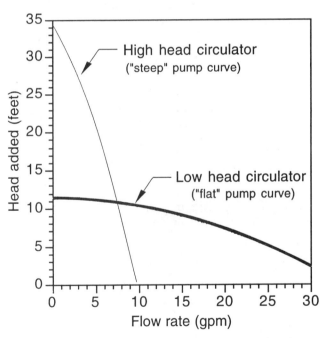

Figure 7–23 Pump curves for circulators shown in Figure 7–22

wide range of flow rates. An example of such an application is a multi-zone system in which zone valves are used to control flow through several individual distribution circuits. Depending on the number of zone valves open at a given time, and the shape of the pump curve, the flow rate through the circulator could change considerably.

A circulator with a flat pump curve will minimize such changes and thus maintain relatively stable flow rates in the active zones.

Pumps in Series

Occasionally a system design requires a larger pumping head than can be supplied by a single small circulator. The designer has the option of finding a larger circulator that can supply the necessary head, or possibly using two smaller circulators connected in **series**. The second option is often the most economical choice.

When two identical circulators are connected in series, the resulting pump curve can be found by doubling the head produced by a single circulator at each flow rate as shown in Figure 7–24. In effect the two circulators are acting as a multi-stage pump with the head added by the first being doubled by the second.

Notice that the flow rate resulting from two circulators in series (about 10.5 gpm), is not double that obtained with the single circulator (about 8 gpm). This is typical, and is the result of the curvature of both the pump and system curves. It demonstrates the fallacy of trying to double the flow rate in a system by adding a second circulator in series.

There are two ways of locating circulators so they are in series with each other. The first, known as **close coupling**, is accomplished by bolting the discharge flange of the first circulator directly to the inlet flange of the second as shown in Figure 7–25. This is practical for

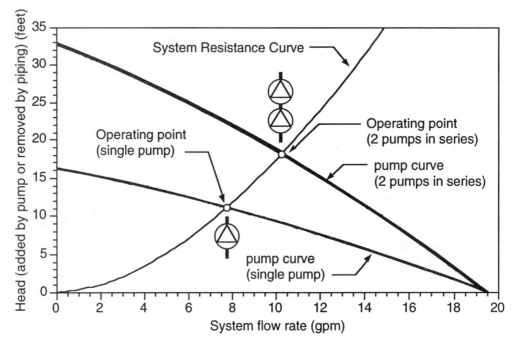

Figure 7–24 Effect of placing two identical circulators in series

(a)

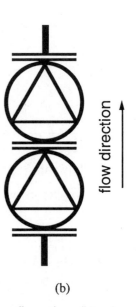

(b)

Figure 7–25 (a) Example of two small circulators in a close-coupled series configuration, (b) schematic representation of this assembly.

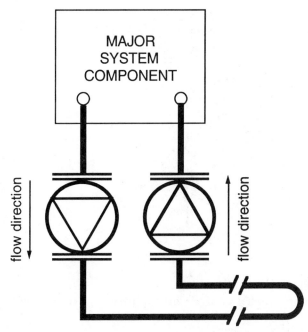

Figure 7–26   Concept of two small circulators in a push/pull series configuration

Another way of installing circulators in series is called a **push/pull** arrangement, and is shown in Figure 7–26. In this case the two circulators are separated by several feet of piping, or a major system component such as the heat exchanger coil of a heat pump. This approach will distribute the pressure increase of each circulator more evenly throughout the system. It also can reduce the chances of cavitation relative to the previously described close-coupled arrangement. This configuration is often used for the earth heat exchanger of ground source heat pump systems where a high head is required due to the combined flow resistance of both the heat pump coil and earth heat exchanger piping.

Pumps in Parallel

Another way of connecting two or more identical circulators is in parallel. The inlet ports of the circulators are connected to a common header pipe, as are the discharge ports. This arrangement is useful when a relatively high flow rate is required at a modest head. An example would be a multi-zone parallel distribution system with several zones.

The pump curve for two identical circulators connected in parallel is obtained by doubling the flow rate of a single circulator at each head value, as shown in Figure 7–27. For three identical circulators in parallel, the flow rate would be tripled at each head value, and so on.

System resistance curve #1 in Figure 7–27 is for a piping system with relatively low flow resistance. Point A is the operating point for this piping system using a single circulator. Point B is the operating point using two of the same circulators in parallel. For this particular piping system and circulator combination, using two

relatively small circulators that will always be operated at the same time. It has the advantage of requiring only one set of flanges to connect the resulting assembly to the system piping. *An obvious precaution is to make sure the flow direction arrow on each circulator is facing the same direction before bolting them together.* The high heads and associated pressure differential resulting from such an assembly makes it critical the system's expansion tank is located near the inlet port of the first circulator. If this is not done, the likelihood of cavitation is greatly increased.

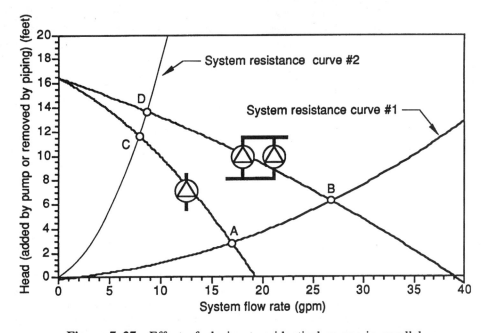

Figure 7–27   Effect of placing two identical pumps in parallel

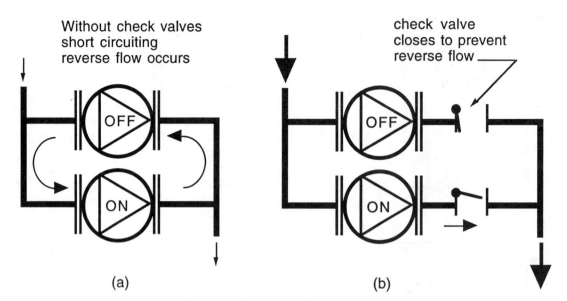

Without check valves short circuiting reverse flow occurs

check valve closes to prevent reverse flow

OFF

ON

(a)

OFF

ON

(b)

Figure 7–28 (a) Reverse flow occurs through inactive circulator, (b) check valve closes to prevent reverse flow through inactive circulator.

parallel pumps increases the flow rate from about 17 gpm to about 27 gpm, a significant gain.

Points C and D in Figure 7–27 are the operating points for the same circulator, and parallel circulator combination, in a piping system having a higher head loss characteristic as shown by system resistance curve #2. In this system, adding a second parallel circulator results in a small increase in flow rate from about 8 gpm to 9 gpm, a very small gain considering twice the pumping power would be used.

These curves demonstrate that the flow rates obtained using multiple circulators in parallel can differ considerably depending on the system resistance curve of the piping system into which they are placed. They also show that *adding a second circulator in parallel will not double the system's flow rate*, even for a system with low head loss characteristics.

When two or more identical parallel circulators operate at the same time, their discharge ports can simply be manifolded together. However, if the system is designed to operate with one of the circulators on and the other off, a check valve must be installed in the discharge pipe from *each* circulator as shown in Figure 7–28b. Without this check valve, much of the flow produced by the active circulator(s) would simply short circuit backward through the inactive circulator(s) as shown in Figure 7–28a.

7.5 ANALYTICAL METHODS FOR PUMP PERFORMANCE

The previous sections have shown graphical methods of finding the intersection of a system resistance curve and pump curve. In many situations it is convenient to

be able to find this operating point without having to draw the system and pump curves. This can be done by representing each curve in mathematical form and solving for their intersection.

The pump curves for most circulators can be accurately represented using Equation 7.1:

(Equation 7.1)
$$H_{pump} = b_0 + b_1(f) + b_2(f)^2$$

where:

H_{pump} = head produced by the circulator (feet of head)

f = flow rate through the circulator (gpm)

b_0, b_1, b_2 = constants that are *specific for a given pump*. These numbers can be positive or negative

Equation 7.2 is an example of this equation with actual numbers:

(Equation 7.2)
$$H_{pump} = 10.88 + (-0.206)(f) + (-0.00971)(f)^2$$

When plotted, this equation produces the pump curve shown in Figure 7–29.

The PUMPCURV Program

The Hydronics Design Toolkit contains a program called PUMPCURV that finds the values of the constants b_0, b_1, and b_2 in Equation 7.1 for any circulator used in residential and light commercial hydronic systems. The program requires the user to input three sets of data

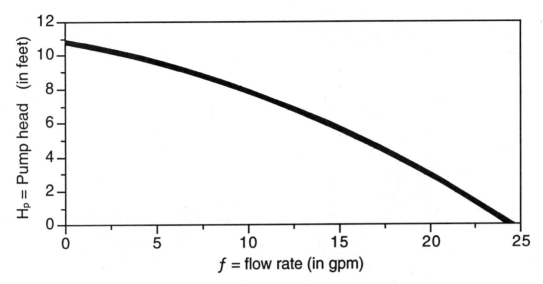

Figure 7-29   Graph of Equation 7.2 representing a pump curve

consisting of a flow rate and associated head. This data can be taken from the pump curve in the manufacturer's literature. The program then finds the values of b_0, b_1, and b_2. PUMPCURV also graphs the resulting pump curve along with the data used to generate it. The user can assign a name to the circulator, and store the data to disk under that name. This allows the circulator performance information to be read by other programs in the Hydronics Design Toolkit. It also eliminates the need to reenter information on the same circulator in subsequent runs of

the program. An example of the input screen is shown in Figure 7-30. An example of the graph generated by the program is shown in Figure 7-31.

Analytical Method for Finding the Operating Point

Chapter 6 gave several examples of how to construct the system resistance curve once the total hydraulic resistance of the system was determined. The system

```
- PUMPCURV -   F1-HELP  F2-FILE  F4-GRAPH    F10-EXIT TO MAIN MENU File: Samplpum
 -INPUTS-
  ENTER 1st Flowrate   0   gpm        ENTER 1st Head Pressure   16.4ft. water
  ENTER 2nd Flowrate   10  gpm        ENTER 2nd Head Pressure   9.3 ft. water
  ENTER 3rd Flowrate   19.5gpm        ENTER 3rd Head Pressure   0   ft. water


 -RESULTS-
 The Coefficients of the pump curve equation are:     b0=  16.4000
                                                       b1=  -0.5721
                                                       b2=  -0.0138

    Equation is:   Delta P =  16.4000   -0.5721 x gpm  -0.0138 x gpm²
```

```
 .MESSAGE
  Use the UP or DOWN keys to move the highlighting bar over an input. For
  ENTER inputs type desired value. All calculations are instant after input.
  10730
```

Figure 7-30   Example of the input screen from the PUMPCURV program

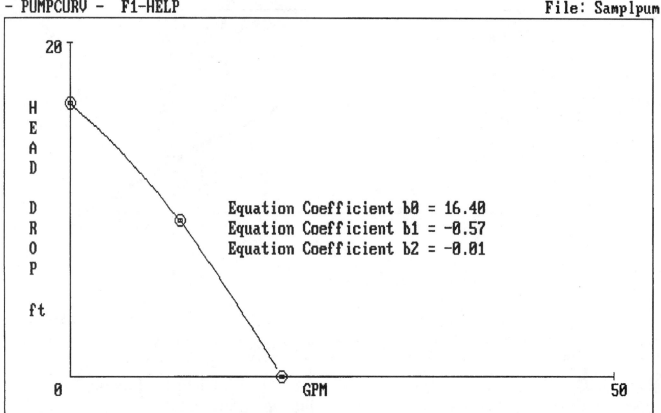

Figure 7-31 Example of the graphics screen from the PUMPCURV program showing the pump curve and data used to create it

resistance curve represents the head loss of the system as a function of flow rate, and can be represented using Equation 6.10.

(Equation 6.10)

$$H_L = r(f)^{1.75}$$

For example, the system curve determined in Example 6.8 was:

(Equation 6.16)

$$H_L = 0.31(f)^{1.75}$$

The operating point of a piping system such as that described by Equation 6.16, and a pump curve such as described by Equation 7.2, is the point where these two equations have the same value for H_{pump} and H_L. The manual method of finding this operating point involves a logical trial and error process called **iteration**. The following steps describe this process:

1. Set up a table with one column for flow rate, f, another for the head added by the circulator, H_{pump}, and a third for the head loss of the piping system, H_L.
2. Make a reasonable first estimate for the system's flow rate.
3. Use Equation 7.1 with specific values for b_0, b_1, and b_2 to calculate H_{pump}.
4. Use Equation 6.10 with a specific value of r to calculate H_L.
5. If H_{pump} is greater than H_L, the estimated flow rate is too low. Increase the value of flow rate f and go back to step 3.
6. If H_{pump} is less than H_L, the estimated flow rate is too high. Lower the value of flow rate f and go back to step 3.
7. Repeat this procedure until the values of H_{pump} and H_L are within 0.1 ft or less of each other. The value of flow rate, f, now approximates the actual system flow rate. The head across the circulator is estimated by averaging the last values of H_{pump} and H_L.

flow rate f (GPM)	Pump head H_{pump} (feet)	System head loss H_L (feet)
4.0	9.9	3.507
6.0	9.29	7.13
8.0	8.61	11.8
7.0	8.96	9.33
6.5	9.13	8.2
6.9	8.99	9.106
6.85	9.01	8.99

← close enough

Figure 7–32 Iterative solution to find operating point in Example 7.1

Example 7.2: Find the flow rate corresponding to the pump curve described by Equation 7.2 and the system curve described by Equation 6.16 using a manual iteration procedure.

Solution:

Pump curve:

$$H_{pump} = 10.88 + (-0.206)(f) + (-0.00971)(f)^2$$

System resistance curve: $H_L = 0.31(f)^{1.75}$

The table in Figure 7–32 shows successive estimates of flow rate with corresponding values of pump head, and system head loss calculated from the two equations. The last estimate of 6.85 gpm showed the pump head and system head loss to be well within 0.1 ft of each other. Hence, this was the last calculation performed.

The head across the pump could now be estimated by averaging the final values of H_{pump} and H_L:

$$\frac{9.01 + 8.99}{2} = 9.00 \text{ ft of head}$$

As a check, the graph in Figure 7–33 was plotted using Equations 7.2 and 6.16. The graph confirms the operating point is at a flow rate of about 6.85 gpm, and a head of about 9.0 ft. When this analytical method is used to find the system's flow rate, the graph is not needed and is normally not plotted.

The PUMP/SYS Program

The Hydronics Design Toolkit also contains a program called PUMP/SYS that can find the intersection of a pump curve and system curve using the iterative method shown in Example 7.2. The user calls up a disk file for the circulator, previously created using the PUMPCURV program. The user also calls up a piping file previously created using the PIPEPATH program. The PUMP/SYS program immediately calculates the flow rate and head at the operating point of this combination. It also displays a graph of the pump and system curves showing their intersection. Figure 7–34 is an example of such a graph.

The combined use of the PIPEPATH, PUMPCURV, and PUMP/SYS programs makes it possible to find the operating point of an unlimited number of circulator/piping system combinations. Example 7.3 illustrates the combined use of these programs.

Example 7.3: A piping circuit contains 275 ft of 3/4-in copper tube, 25 90 degree elbows, 6 gate valves, and 8 straight through tees. The piping circuit will operate with water at 140 °F. The proposed circulator has a pump

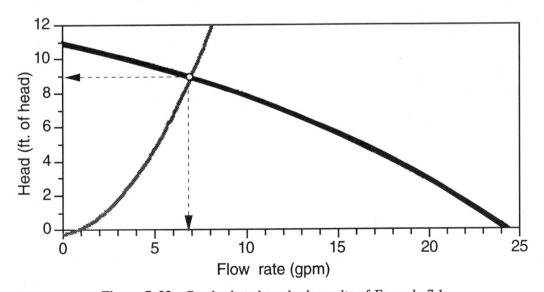

Figure 7–33 Graph plotted to check results of Example 7.1

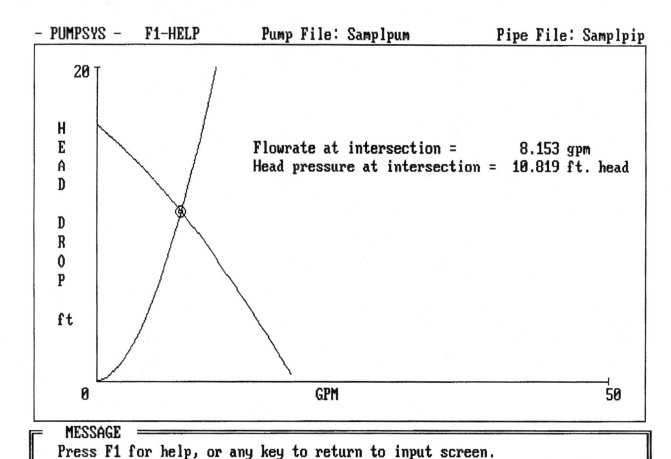

Figure 7-34 Screen shot from the PUMP/SYS program in the Hydronics Design Toolkit

curve as shown in Figure 7–35. Find the flow rate at which the system will operate:

a. Using manual calculations and the pump curve graph of Figure 7–35.
b. Using the PIPEPATH, PUMPCURV, and PUMP/SYS programs.

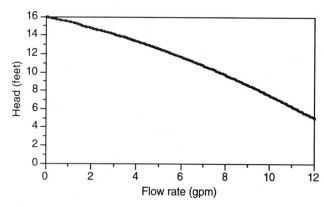

Figure 7–35 Pump curve for circulator used in Example 7.3

Solution (part a): To determine the system resistance curve, the hydraulic resistance of the piping system using the specified 140 °F water must be determined. The necessary data for fluid properties and equivalent lengths of the piping components can be found by referring to earlier chapters.

Using the graph in Figure 4–5, the water's density is found to be 61.35 lb/ft³. Using the graph in Figure 4–6, the water's dynamic viscosity is found to be 0.00032 lb/ft/sec. The α value can be calculated using Equation 6.13:

$$\alpha = \left(\frac{D}{\mu}\right)^{-0.25} = \left(\frac{61.35}{0.00032}\right)^{-0.25} = (191,719)^{-0.25}$$

$$= 0.04779$$

Alternatively, the α value for water could have been read from Figure 6–17.

The value of c for 3/4-in copper tube is found in Figure 6–18. $c = 0.061957$

The equivalent lengths of the piping components can be found from Figure 6–21:

flow rate f (GPM)	Pump head H_{pump} (feet)	System head loss ΔP (feet)
5.0	16.32	12.5
3.0	6.67	14.1
4.0	11.04	13.5
4.2	12.03	13.2
4.4	13.05	13.0

◄── close enough

Figure 7-36 Iterative solution to find operating point in Example 7.3

$3/4$-in tubing	275 ft
25 $3/4$-in × 90 degree elbows	25(2) = 50 ft
8 $3/4$-in straight through tee	8(0.4) = 3.2 ft
6 $3/4$-in gate valve	6(0.25) = 1.5 ft
	Total = 329.7 ft

This data can now be substituted into Equation 6.11 to obtain the equation for the system resistance curve:

$$H_L = [\alpha c L]f^{1.75} = [(0.04779)(0.061957)(329.7)](f)^{1.75}$$

$$= 0.976(f)^{1.75}$$

The table in Figure 7-36 shows the iterative procedure to find the operating point. Within each row, the system's head loss was calculated from the equation, and the pump head was estimated directly from the graph in Figure 7-35.

The flow rate in this system will be approximately 4.4 gpm.

Solution (part b): This example can be completed very quickly using the Hydronics Design Toolkit.

First, the PIPEPATH program is used to determine the hydraulic resistance of the specified circuit. A screen shot is shown in Figure 7-37. The information is stored as a disk file named system1.

Second, the PUMPCURV program is used with three widely separated points from the plotted pump curve to generate the equation representing the pump curve. A screen shot is shown in Figure 7-38. The pump curve is stored as a disk file named pump1.

Finally, the PUMP/SYS program is used by loading the previously created disk files system1 and pump1. The program instantly finds the operating point and graphs both the system resistance curve, and pump curve as shown in Figure 7-39.

7.6   PUMP EFFICIENCY

Like any device that converts one form of energy to another, a pump can be thought of as having an efficiency. Recall that efficiency is defined as the ratio of the rate

```
        F1-HELP                    F10-RETURN TO INPUTS  File: SYSTEM1
 ┌─ -EDIT TABLE- ═══════════════════════════════════════════════════════════
 │                         1/2"    3/4"   1"    1 1/4"  1 1/2"   2"
 │ Straight pipe (lin. ft.) 0      275    0     0       0        0
 │ 90 degree ells           0      25     0     0       0        0
 │ 45 degree ells           0      0      0     0       0        0
 │ Tees (straight through)  0      8      0     0       0        0
 │ Tees (side port)         0      0      0     0       0        0
 │ Monoflo tees                    0      0     0       0        0
 │ Gate valves              0      6      0     0       0        0
 │ Globe valves             0      0      0     0       0        0
 │ Angle valves             0      0      0     0       0        0
 │ Ball valves              0      0      0     0       0        0
 │ Swing check valves       0      0      0     0       0        0
 │ Flow control valves             0      0     0       0        0
 │ Butterfly valves         0      0      0     0       0        0
 │ Deareator                       0      0     0       0        0
 │ Reducer Coupling         0      0      0     0       0        0
 │ Boiler (C.I. sectional)                            0
 └──────────────────────────────────────────────────────────────────────────

 ┌═ MESSAGE ═════════════════════════════════════════════════════════════════
 │  Use the UP, DOWN, LEFT, RIGHT ARROW keys to move the highlighting bar
 │  over an input. Type desired value. F10 to return to input screen.
 └──────────────────────────────────────────────────────────────────────────
```

Figure 7-37   Use of the PIPEPATH program for Example 7.3

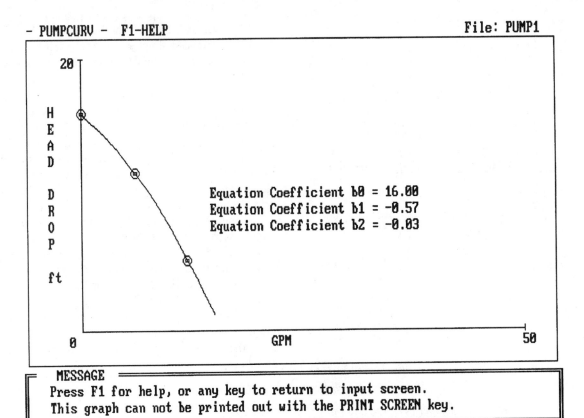

Figure 7–38   Use of the PUMPCURV program for Example 7.3

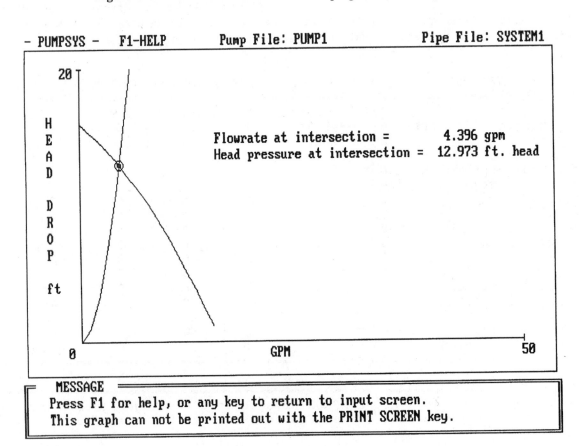

Figure 7–39   Use of the PUMP/SYS program for Example 7.3

of energy output divided by the rate of energy input. In the case of a pump, this corresponds to the rate at which the pump imparts mechanical energy to the fluid, divided by the rate at which energy is supplied to rotate the pump's shaft. Equation 7.3 can be used to calculate the **efficiency of a pump**.

(Equation 7.3)

$$\eta_{\text{pump}} = \frac{f \Delta P}{1,716(hp)}$$

where:

η_{pump} = efficiency of the pump
f = flow rate through the pump (gpm)
ΔP = pressure differential measured across the pump (psi)
1,716 = units conversion constant
hp = horsepower delivered to the pump shaft by the motor

Although Equation 7.3 is easy to derive from basic physics, it is not necessarily easy to use because measurements of input horsepower to the pump's shaft are difficult to obtain outside of a testing laboratory. Furthermore, the input horsepower to a pump increases with flow rate.

To provide efficiency information to designers, most pump manufacturers plot the efficiency of a pump versus its flow rate either as a separate graph, or as a group of contour lines overlaid on the pump curve as shown in Figure 7–40. Unfortunately, such information is often omitted from the curves of small circulators. As a rule, however, the efficiency of most circulators is maximum near the knee of the pump curve as shown in Figure 7–41. For the circulator to operate at or near its maximum efficiency, the system curve must pass through this region of maximum efficiency.

Efficiency of a Pump/Motor Combination

The efficiency obtained using Equation 7.3 is that of the pump only. It does not include any energy losses associated with the electric motor driving the pump. No electric motor is 100% efficient in converting electrical energy into mechanical energy. Motor efficiencies vary considerably for different types of construction and different loading conditions. The induction motors used on many commercial and industrial pumps have efficiencies in the range of 80% to 90% when loaded between 25% and 100% of their rated horsepower. The smaller permanent split capacitor motors used on wet rotor circulators often have efficiencies below 50%. A significant amount of the

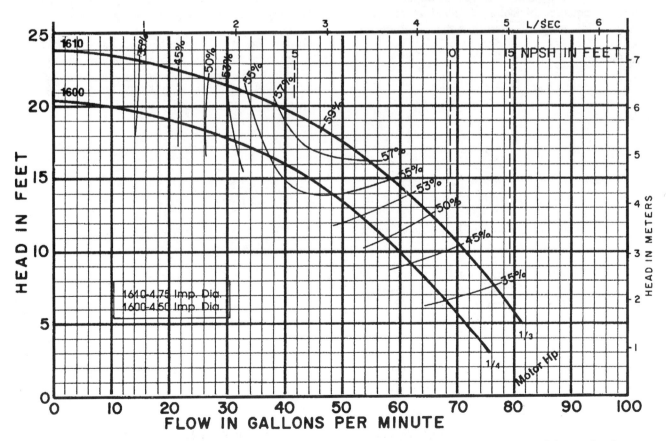

Figure 7–40   Example of efficiency contours plotted with pump curves. Typical of medium and large circulators. Courtesy of Taco, Inc.

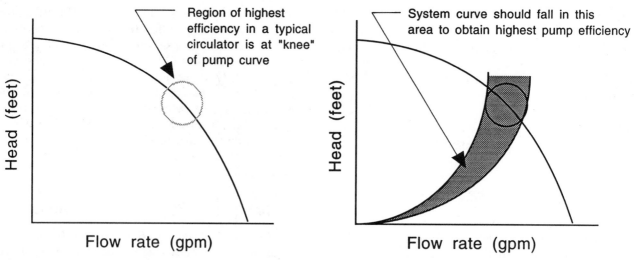

Figure 7-41 Region of highest efficiency for small circulators. For highest efficiency, the system curve should be within shaded area.

electrical energy input to the motor is therefore lost before being transferred to the pump shaft as mechanical energy. This energy loss takes the form of heat that is either dissipated to the circulating fluid or directly to the air around the motor housing.

To account for the efficiencies of *both* the pump and the electric motor, Equation 7.3 can be combined with the defining equation of motor efficiency to obtain the following relationship for a **pump/motor combination**:

(Equation 7.4)

$$\eta_{\text{pump \& motor}} = \frac{0.4344f(\Delta P)}{w}$$

where:

$\eta_{\text{pump \& motor}}$ = efficiency of the pump and motor assembly as a whole

f = flow rate through the pump (gpm)

ΔP = pressure differential measured across the pump (psi)

w = input wattage required by the motor (watts)

0.4344 = units conversion factor

The input wattage required by a pump is not a fixed value, but depends on where the pump operates on its curve. Therefore, to make use of Equation 7.4, one needs to know the input wattage *at a particular point on the pump curve*. This data may be available from the pump's manufacturer, or it can be measured using a watt meter. Figure 7-42 shows the variation in both electrical power and amperage drawn by the motor of a small wet rotor circulator as a function of flow rate.

Figure 7-43 shows the pump curve of this same circulator along with its calculated efficiency. Notice that peak efficiency occurs near the knee of the pump curve. For this particular pump, it occurs at a flow rate of about

9.4 gpm. Ideally, the intersection of the pump curve and system curve will be at or near this point.

Example 7.4: A small circulator operates at a point where it produces a flow rate of 8.0 gpm of 140 °F water against a head of 9.7 ft. A watt meter indicates the pump is drawing 85 watts of electrical input. What is the efficiency of the overall motor/pump unit?

Solution: Equation 7.4 can be used, but since it requires the pressure difference across the pump, not the head, Equation 6.6 is first used to convert the stated head to a pressure difference. The fluid density is also required, and is found from Figure 4-5:

$$\Delta P = \frac{(9.7)(61.35)}{144} = 4.133 \text{ psi}$$

This pressure difference can now be substituted into Equation 7.4 along with the other data:

$$\eta_{\text{pump \& motor}} = \frac{0.4344(8)(4.133)}{85}$$

$$= 0.169 \text{ or } 16.9\% \text{ efficient}$$

Discussion: This value seems surprisingly low, but is in fact typical of most small, wet rotor circulators currently produced. A number a factors contribute to this low efficiency. One is the unavoidable decrease in impeller efficiency as its diameter is reduced. Another is the fact that the volute chambers of small circulators do not contain diffuser vanes to guide the fluid off the impeller and out of the volute with minimal turbulence. The economics of producing small circulators simply does not justify marginal efficiency gains at significantly increased production costs.

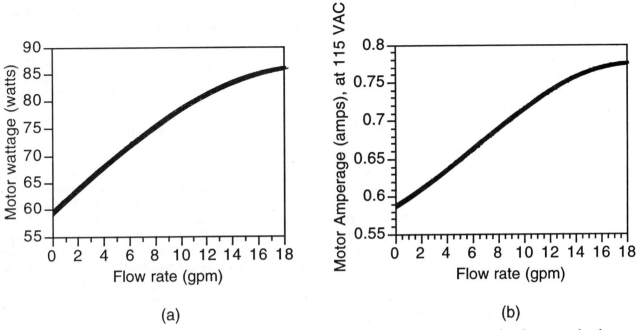

Figure 7–42 (a) Wattage drawn by a small wet rotor circulator, (b) amperage drawn by the same circulator.

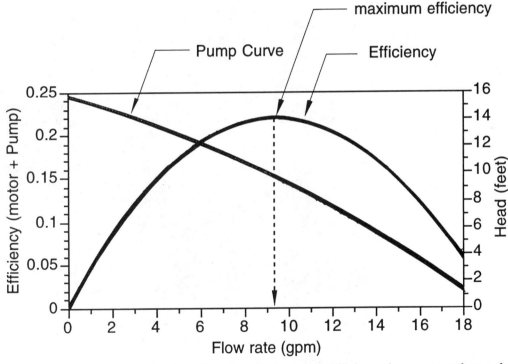

Figure 7–43 Efficiency curve and pump curve for a small circulator. Efficiency is represented as a decimal percent, i.e., 0.1 = 10%

7.7 PUMP CAVITATION

Chapter 4 discussed the fact that liquids will begin boiling at specific combinations of pressure and temperature. The minimum pressure that must be maintained on a liquid to prevent it from boiling at some temperature is called its vapor pressure. The process of boiling involves the formation of vapor pockets in the liquid. These pockets look like bubbles in the liquid, but should not be confused with air bubbles. They will form even in water that has been completely deaerated, *instantly appearing* whenever the liquid's pressure is lowered below the vapor pressure corresponding to its current temperature. This process is often described as the water "flashing" into vapor.

When vapor pockets form in water, the density inside the vapor pocket is about 1,500 times lower than the surrounding liquid. This is comparable to a single kernel of popcorn expanding to the size of a baseball. If the pressure of the liquid then increases above the vapor pressure, the vapor pockets instantly collapse inward in a process known as **implosion**. Although this sounds relatively harmless, on a microscopic level vapor pocket implosion is a very violent effect. It is so violent that it can actually rip away material from surrounding surfaces, even hardened metal surfaces.

For a liquid at a given temperature, cavitation will occur *anywhere* the liquid's pressure drops below its vapor pressure. In hydronic systems, the most likely locations for cavitation to occur are in partially closed valves or inside the circulator's impeller. The greater potential for damage is in the impeller.

As liquid flows into the eye of a circulator's impeller, it experiences a rapid drop in pressure. How low the pressure gets depends on the nature of the system and where the circulator is located relative to the expansion tank. If the pressure at the eye of the impeller drops below the vapor pressure, thousands of vapor pockets instantly form. The vapor pockets are carried outward through the impeller, and as soon as the local fluid pressure increases above the vapor pressure, they implode. This usually occurs near the outer edges of the impeller. The violent implosions create the churning or popping sounds that are characteristic of a cavitating pump. A noticeable drop in pump flow rate and head occurs because

the fluid in the pump is now partially compressible. If the pump continues to operate in this mode, the impeller will be severely eroded in as little as a few weeks of operation. Obviously, pump cavitation must be avoided if the system is to have a long, reliable life. Figure 7–44 shows the effects of cavitation damage to a pump impeller.

Net Positive Suction Head Available

The best way of preventing pump cavitation is to know how to predict its occurrence, and then design the system to avoid these conditions. Over the years, engineers have standardized a method of describing the head of a fluid as it enters the pump in a given piping system. This head is then compared to the known minimum value required for a given pump to avoid cavitation. This method requires the calculation of an index called the **Net Positive Suction Head Available** to the pump, or **NPSHA**. The following equation can be used to calculate NPSHA:

(Equation 7.5)

$$\text{NPSHA} = \frac{v^2}{64.4} + H_e - H_L + (p_s + 14.7 - p_v)\left(\frac{144}{D}\right)$$

where:

NPSHA = net positive suction head available at the pump inlet (ft of head)

v = velocity of the liquid in the pipe entering the pump (ft/sec)

H_e = distance to the liquid's surface above (+) or below (−) the pump inlet (ft)

Figure 7–44 Example of cavitation damage to a pump impeller. Courtesy of ITT Corporation.

H_L = head loss due to viscous friction in any piping components between the liquid surface and the inlet of the pump (ft of head)

P_s = *gauge pressure* at the surface of the fluid (psig)

P_v = vapor pressure of the liquid as it enters the pump (psi *absolute*)

If the pressure of the *moving fluid* can be measured *at the entrance to the pump*, it is not necessary to know the pressure at the surface of the liquid, the distance to the liquid's surface, or the head loss due to viscous friction in the piping leading to the pump. These effects will be accounted for in the pressure gauge reading itself. This allows Equation 7.5 to be simplified as follows:

(Equation 7.6)

$$\text{NPSHA} = \frac{v^2}{64.4} + (p_i + 14.7 - p_v)\left(\frac{144}{D}\right)$$

where:

NPSHA = net positive suction head available at the pump inlet (ft of head)

v = velocity of the liquid in the pipe entering the pump (ft/sec)

p_i = gauge pressure of the moving fluid measured at the pump entrance (psig)

p_v = vapor pressure of the liquid as it enters the pump (psi *absolute*)

The NPSHA indicates the difference between the total head of the fluid at the inlet port of the pump and the head at which the fluid will boil. *NPSHA is entirely determined by the system into which the pump will be placed and the properties of the fluid being pumped. It does not depend on the pump itself.*

Example 7.5: Determine the NPSHA for the piping system shown in Figure 7–45.

a. Assuming the water temperature is 140 °F.
b. Assuming the water temperature is 200 °F.

Solution (part a): A number of quantities need to be determined before using Equation 7.5:

The velocity of the water in the pipe entering the pump is found using the equation for 1-in copper tubing from Figure 6–2: $v = 0.367f = 0.367(8) = 2.94$ ft/sec.

The distance to the (top) surface of the water as shown in Figure 7–45 is 15 ft. This distance is considered positive because the surface of the water is above the inlet to the pump. Thus $H_e = 15$ ft.

The head loss from the (top) surface of the water to the inlet of the circulator will be assumed to be that created by 20 ft of 1-in type M copper tubing carrying the 140 °F water at the stated flow rate of 8 gpm. This can be determined using Equation 6.12:

(Equation 6.12)

$$H_L = (\alpha c L)(f)^{1.75}$$

The value of α for water at 140 °F is found from Figure 6–17: $\alpha = 0.0475$

The value of c for 1-in type M copper tube is found in Figure 6–18: $c = 0.01776$

Substituting into Equation 6.12:

$$H_L = [(0.0475)(0.01776)(20)](8)^{1.75} = 0.642 \text{ ft of head}$$

The pressure on the surface of the water would, in the case of a closed fluid-filled piping system, be the pressure at the top of the piping loop. Therefore $p_s = 5$ psig.

The vapor pressure of 140 °F water can be read from Figure 4–4: $p_v = 2.9$ psia

The density of water at 140 °F is found from Figure 4–5: $D = 61.35$ lb/ft³

Substituting all data into Equation 7.5 yields the following:

$$\text{NPSHA} = \frac{2.94^2}{64.4} + 15 - 0.642 + (5 + 14.7 - 2.9)\left(\frac{144}{61.35}\right)$$

$$= 53.9 \text{ ft of head}$$

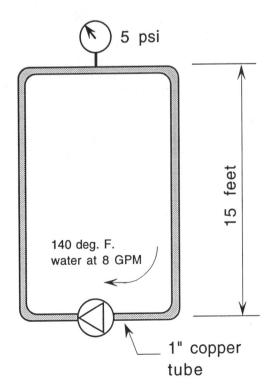

Figure 7–45 Piping system for Example 7.5

Solution (part b): Because the water's temperature has changed, its density, vapor pressure, and head loss due to viscous friction must be recalculated.

From Figure 4–5, the density of water at 200 °F is 60.15 lb/ft³.

From Figure 4–4, the vapor pressure of 200 °F water is 11.5 psia.

From Figure 6–17, the value of α for 200 °F water is 0.043.

The revised head loss is again found using Equation 6.12 with the new value of α:

$$H_L = [(0.043)(0.01776)(20)](8)^{1.75} = 0.581 \text{ ft of head}$$

Substituting these changes into Equation 7.5:

$$\text{NPSHA} = \frac{2.94^2}{64.4} + 15 - 0.581 + (5 + 14.7 - 11.5)\left(\frac{144}{60.15}\right)$$

$$= 34.2 \text{ ft of head}$$

Discussion: Note that increasing the water temperature significantly lowered the NPSHA of an otherwise unchanged system. Other factors that tend to lower the NPSHA of a system include decreasing the system's height, lowering its pressure, increasing inlet pipe head loss due to viscous friction, or slowing its flow velocity. Of these, the head associated with flow velocity is usually a very small part of the total NPSHA value. This can be seen in this example since the value of $v^2/64.4$ is only 0.134 ft of head.

Net Positive Suction Head Required

Any given pump has a minimum required value of net positive suction head in order to operate without cavitation. This value is called the **Net Positive Suction Head Required**, or **NPSHR**. Its value depends on the design of the pump, and to some extent, the flow rate at which the pump operates. Pump manufacturers determine values of NPSHR by lowering the NPSHA to a pump until cavitation occurs. This is done for several flow rates so a curve may be plotted.

Avoiding pump cavitation is simply a matter of ensuring the NPSHA to a pump always exceeds the NPSHR of the pump under all possible operating conditions. The higher the NPSHA is relative to the NPSHR, the wider the safety margin against cavitation. A common design practice is to add a safety margin of at least 2 feet of head to the NPSHR of a pump, and not allow the NPSHA to fall below this value.

Example 7.6: The circulator shown in Figure 7–46 has a NPSHR value of 3.5 ft of head. The expected operating point of the pump is at a flow rate of 5 gpm with a 13 ft head across the pump. The system is charged to a static pressure of 3 psig. Determine if cavitation will occur.

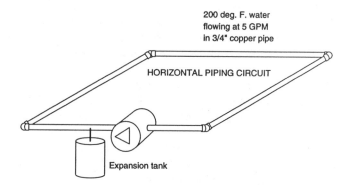

200 deg. F. water flowing at 5 GPM in 3/4" copper pipe

HORIZONTAL PIPING CIRCUIT

Expansion tank

Figure 7–46 Horizontal piping loop for Example 7.6

Solution: Because the expansion tank is located at the *outlet* of the circulator, the 13 ft of head produced by the pump will lower the pressure of the fluid entering the pump by an amount corresponding to the head loss of 13 ft. This allows the pressure of the moving water entering the pump to be determined. The pressure drop around the loop corresponding to the head loss of 13 ft can be determined using Equation 6.6. Note the density of 200 °F water is 60.15 lb/ft³.

$$\Delta P = \frac{HD}{144} = \frac{(13)(60.15)}{144} = 5.43 \text{ psi}$$

The pressure at the inlet port of the circulator would be the static pressure of the system (3 psig), *which remains constant at the point of attachment of the expansion tank,* minus the pressure drop due to the head loss around the loop (5.43 psig). The inlet pressure is thus $3 - 5.43 = -2.43$ psig. The negative sign indicates this pressure is below atmospheric pressure.

Since the pressure at the inlet of the circulator is known, Equation 7.6 can be used to calculate NPSHA. Again several pieces of information need to be gathered before substituting into the equation.

The flow velocity corresponding to a flow rate of 5 gpm in a ³/₄-in copper tube is found using the appropriate equation from Figure 6–2: $v = 0.6211$, $f = 0.6211(5) = 3.106$ ft/sec.

The vapor pressure of 200 °F water is 11.5 psia.

Substituting into Equation 7.6:

$$\text{NPSHA} = \frac{3.106^2}{64.4} + (-2.43 + 14.7 - 11.5)\left(\frac{144}{60.15}\right)$$

$$= 1.99 \text{ ft of head}$$

Discussion: *Since the NPSHA of 1.99 ft is less than the NPSHR of 3.5 ft, the pump will cavitate.* It is interesting to note the factors that contributed to the presence of cavitation. These include low system static pressurization, a low system height, and high temperature water with its associated high vapor pressure. Any of these factors could potentially be altered to prevent the pump from cavitating.

Gaseous Cavitation

Another type of cavitation can result from the presence of air in the liquid. The air may be present as small air bubbles entrained by the liquid. Air may also appear inside the vapor pockets formed by vaporous cavitation because the lowered pressure allows some dissolved air to come out of solution. This so-called **gaseous cavitation** is not as violent or damaging as vaporous cavitation. In properly designed and deaerated hydronic systems it is often only a temporary phenomenon when the system is first put into service.

Guidelines for Avoiding Cavitation

A number of guidelines for avoiding cavitation can be developed based on the previous discussions. These guidelines widen the safety margin against the onset cavitation. Most attempt to make the NPSHA of the system as high as possible. The quantitative effect of each guideline would have to be assessed using Equations 7.5 or 7.6 for a given piping system.

- Keep the static pressure on the system as high as practical
- Keep the fluid temperature as low as practical
- Always place the expansion tank near the inlet side of the circulator

- Keep the circulator low in the system to maximize static pressure at its inlet
- Do not place any components with high flow resistance (especially flow regulating valves) near the inlet of the circulator
- If the system has a static water level, such as a partially filled tank, keep the inlet of the pump as far below this level as practical
- Be especially careful in the placement of high head pumps or close-coupled series pump combinations because they create greater pressure differentials
- Provide a straight length of pipe at least 10 pipe diameters long leading to the inlet port of the pump
- Install good deareating devices in the system

The schematics in Figure 7–47 depict some of these guidelines in action.

Correcting Existing Cavitation

It is not uncommon for an installer or designer to be called upon to correct an existing noisy pump. If the pump exhibits the classic signs of cavitation (i.e., churning or popping sounds, vibration, poor performance), it probably can be corrected by applying one or more of the previous guidelines.

The pump should first be isolated and the impeller inspected for cavitation damage, especially if the problem

Design "inviting" cavitation
(what's wrong)

1. Relatively low system pressure
2. Throttling valve near pump inlet
3. Expansion tank near pump outlet
4. High temperature fluid
5. Lack of deareator

Design "discouraging" cavitation
(what's right)

1. Higher system pressure
2. Throttling valve on outlet side of pump
3. Expansion tank near pump inlet
4. Lower temperature fluid
5. Use of good deareator
6. Larger/straight pipe into pump

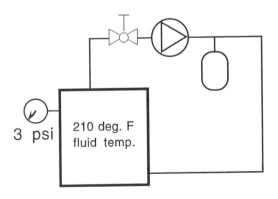

(a)

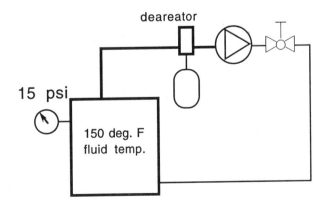

(b)

Figure 7–47 Schematics depicting good and bad design practices for controlling cavitation has been present for some time. Look for signs of abrasion or erosion of metal from the surfaces of the impeller. If evidence of cavitation damage is present, the impeller (or rotor assembly in a wet rotor circulator) should be replaced.

Because every system is different, there is no definite order in which to apply these guidelines in correcting the cavitation problem. In some cases, simple actions such as changing the system's pressure or operating temperature will totally solve the problem. In others, the layout may be so poor that extensive piping modifications may be the only way to solve the problem. Obviously, corrections that involve only setting changes are preferable to those requiring piping modifications. These can be attempted first, assuming they will not in turn create new problems.

In cases where cavitation is not severe, increasing the system's static pressure will often correct the problem. This can usually be done by increasing the pressure setting of the boiler feed water valve. Be sure the increased pressure does not cause the pressure relief valve to open each time the fluid is heated up to operating temperature.

If the system is presently operating at a relatively high water temperature (above 190 °F), try lowering the boiler aquastat setting by about 20 °F. In this temperature range, the vapor pressure of water decreases rapidly as its temperature is lowered. This may be enough to prevent vapor pockets (the cause of cavitation) from forming. If this adjustment is made, be sure the system has adequate heat output at the lower operating temperature.

Always check the location of the expansion tank relative to the circulator. If the circulator is pumping toward the expansion tank, consider moving the tank near the inlet port of the circulator. This may not only eliminate cavitation, but also improve the ability of the system to rid itself of air.

Also look for major flow restrictions near the circulator inlet. These include throttling valves, plugged strainers, or piping assemblies made up of many closely spaced fittings. If necessary, eliminate these components from the inlet side of the circulator.

7.8 SELECTING A CIRCULATOR

The ideal circulator for a given application would produce the exact flow rate desired, while operating at maximum efficiency, and drawing the least amount of electrical power. It would also be competitively priced, and built to withstand many years of service. These conditions, although possible, are not likely in all projects due to the following practical considerations:

* There is only a finite number of circulators available for any given range of flow rate and head requirements. This contrasts with the fact that an unlimited number of piping systems can be designed for which one of these circulators must be chosen.
* Most small circulators are single speed units that operate on a fixed pump curve. These circulators cannot be fine tuned to the system requirements through impeller changes, motor changes, etc., as is often possible with larger pumps. The exception is multi-speed circulators where each speed represents a different pump curve.
* The choices may be further narrowed by material compatibility. For example, cast iron circulators should never be used in open loop hydronic applications because of corrosion.
* Purchasing channels and inventories of spare parts may limit the designer's choices to the product lines of one or two manufacturers.

The selection process must consider these limitations while also attempting to find the best match between the head and flow rate requirements of the system and circulator. A suggested procedure for finding this match is as follows:

Step 1. Determine the desired flow rate in the system based on the thermal and flow requirements of the heat emitters. Flow requirement of various heat emitters will be discussed in Chapter 8.

Step 2. Use the system resistance curve or its equation to determine the head loss of the system at the required flow rate. Detailed methods for determining the system resistance curve were given in Chapter 6.

Step 3. Plot a point representing the desired flow rate and associated head loss on the same set of axes as the pump curve(s).

Step 4. The selected circulator should have 10% to 20% extra head over what the piping system will theoretically require at the desired flow rate. This is justified by the fact that the system may be installed with more fittings, valves, etc., than what the original design was based on. Using pumps that produce more than 20% extra head at the desired flow rate is wasteful due to higher input energy requirements. Excessive head might also increase flow velocities above acceptable limits creating noise or other problems.

Step 5. The curve of the selected pump or pump combination should intersect the system curve in or near the region of highest pump efficiency. Equation 7.4 can be used to check this when motor wattage is known at the desired flow rate. If no wattage data is available, remember the region of highest pump efficiency is usually near the knee of the pump curve.

The curves shown in Figure 7–48 illustrate some of the above considerations. The pump curves of circulators 1 and 2 would be capable of providing the desired flow rate, but would operate in a region of lower efficiency. Their motors would also likely require more electrical energy than necessary. The curve for circulator 3 intersects

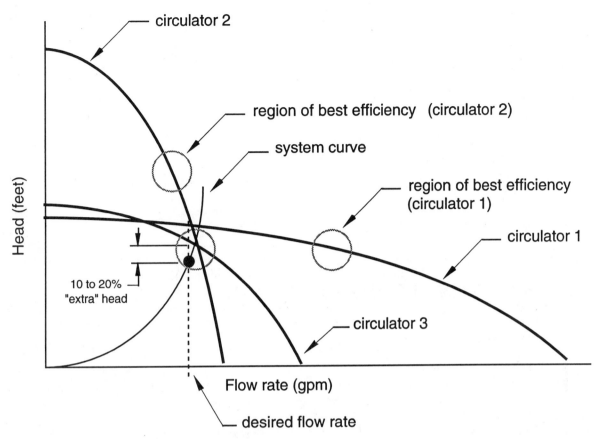

Figure 7–48   Circulator 3 provides the best match of head, flow, and efficiency for the system curve shown.

the system curve in a region where pump efficiency will be near maximum. Its head is also slightly greater than theoretically required, allowing for some inevitable variations in the installed system. Circulator 3 would be the best choice in this application.

SUMMARY

This chapter has discussed the types of circulators used in hydronic heating systems, as well as methods for describing their performance in a given piping system. The fundamental concept of determining the intersection of the pump curve and system resistance curve will be used in several later chapters as an inherent part of system design. The placement of the circulator so its inlet is near the connection point of the system's expansion tank is also a vital concept to be repeatedly applied in subsequent system design.

By applying the concepts presented in this chapter, the designer should be able to select a circulator that properly supplies the flow rate and head requirements of a piping system, while operating at or near maximum efficiency and without the destructive effects of cavitation.

The reader is encouraged to experiment with the PUMPCURV and PUMP/SYS programs introduced

in this chapter. Within a short time several dozen pump curves can be saved as disk files ready for routine use in system design.

KEY TERMS

Centrifugal pump
Circulator
Close-coupling (of pumps)
End suction pump
Eye (of an impeller)
Feet of head
Flanges
Gaseous cavitation
Impeller
Implosion
Inline circulator
Isolation flanges
Iteration
Mechanical energy
Net Positive Suction Head Available (NPSHA)
Net Positive Suction Head Required (NPSHR)
Operating point
Parallel pumps
Point of no pressure change

Pump
Pump cavitation
Pump curve
PUMPCURV program
Pump efficiency
Pump head
Pump/motor combination
PUMP/SYS program
Push/pull
Series pumps
Shaft seals
Three-piece circulator
Volute
Wet rotor circulator

CHAPTER 7 QUESTIONS AND EXERCISES

Note: Questions and exercises requiring the Hydronics Design Toolkit are indicated with marginal symbol.

1. Describe the difference between a circulator with a steep pump curve and one with a flat pump curve. Which would be better for use in a multi-zone system involving zone valves? Why?
2. Describe the difference between the head of a pump and the difference in pressure measured across the pump using pressure gauges. How is this pressure difference converted to a head value?
3. A circulator with a pump curve as shown in Figure 7–49 is installed in a system operating with 140 °F water. Pressure gauges on the inlet and discharge ports have readings as shown. What is the flow rate through the pump?

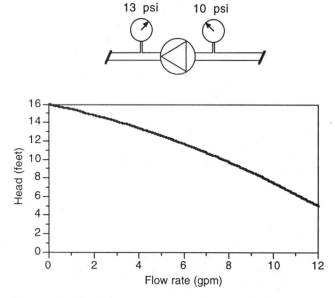

Figure 7–49   Pressure gauge readings and pump curve for Exercise 3

4. Will using two pumps in parallel double the flow rate in a piping system? What about using two pumps in series? Justify your answer by drawing a pump curve as well as the curves for two such circulators in parallel and in series. Show the intersection of these curves with a system resistance curve sketched on the same set of axes. Estimate the flow rates at each operating point.
5. A piping system has a system resistance curve described by the following equation:

$$H_{loss} = 0.85(f)^{1.75}$$

where the head loss is in ft, and the flow rate (f) is in gpm.

   A circulator with the pump curve shown in Figure 7–50 is being considered for the system. Using iterative calculations, find the flow rate and head at the operating point of this piping system and circulator.
6. A circulator has a pump curve described by the equation:

$$H_{pump} = 7.668 + 0.1086(f) - 0.009857(f)^2$$

Head is in ft, and flow rate f is in gpm. Assuming this pump is used with the same piping system described in Exercise 5, find the flow rate and head at the new operating point using manual iterative calculations.
7. The static pressure in the circulator of a given system is 5 psig when the circulator is off. When the circulator operates, a head of 20 ft is established across it. The fluid being pumped is 140 °F water. Determine the readings of pressure gauges tapped into its inlet and discharge ports when it is running assuming:
   a. The expansion tank is connected at the inlet port of the circulator.
   b. The expansion tank is connected to the discharge port of the circulator.
8. A circulator has a pump curve as shown in Figure 7–50. On a sheet of graph paper, sketch the resulting pump curve for the following situations:
   a. Two of these circulators connected in series.
   b. Three of these circulators connected in parallel.

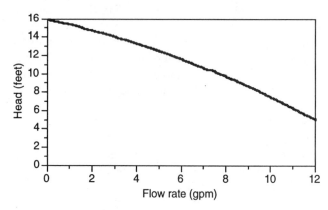

Figure 7–50   Pump curve for Exercise 5

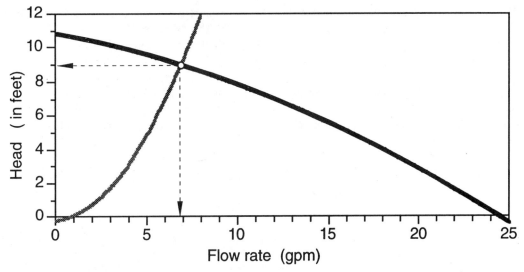

Figure 7–51 Pump curve and system resistance curve for Exercise 9

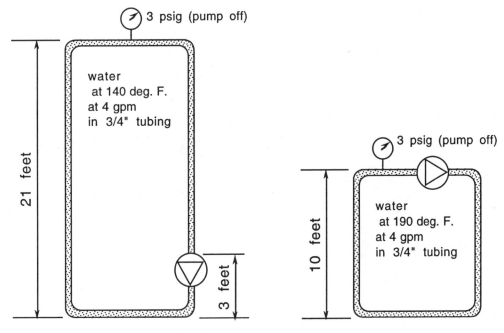

Figure 7–52 Piping diagrams for Exercise 10

9. A piping system and wet rotor circulator have the head versus flow rate curves shown in Figure 7–51. The measured power consumption of the circulator in this application is 75 watts. What is the over-all efficiency of the motor/pump assembly in this application?

10. Determine the NPSHA at the inlet of the pump for the two systems shown in Figure 7–52.

11. A pump has the following performance data:

flow rate (gpm)	head (ft)
0	15
5	11
10	6
12.5	2.5

Use the PUMPCURV program in the Hydronics Design Toolkit to generate a pump curve for this pump. Write down the values of the constants b_0, b_1, and b_2. Give the pump a name and save it to disk.

12. A piping system contains 315 ft of 1-in copper tubing, 30 90 degree elbows, 13 45 degree elbows, 6 side port tees, and 8 gate valves. The system fluid is a 30% mixture of propylene glycol and water operating at 150 °F. Use the PIPEPATH program in the Hydronics Design Toolkit to determine the system resistance curve. Give the piping circuit a name and save it to disk.

13. Use the PUMP/SYS program in the Hydronics Design Toolkit to load the pump file created in

Exercise 11 and the piping system file created in Exercise 12. Find the operating point of the circulator in this system.

14. Use the PUMPCURV program to create a pump curve for the following arrangements:
    a. Two pumps as described by the data in Exercise 11 connected in series.
    b. Three pumps as described by the data in Exercise 11 connected in parallel.

15. Repeat Exercises 12 and 13 for the following fluids:
    a. 150 °F water
    b. 20% propylene glycol at 150 °F
    c. 40% propylene glycol at 150 °F
    d. 50% propylene glycol at 150 °F

    Describe what happens to the flow rate in the system as the percentage of propylene glycol in the system fluid is increased. Make a graph showing these results.

HEAT EMITTERS 8

OBJECTIVES

After studying the chapter you should be able to:

- Describe several different types of heat emitters
- Identify good applications for various types of heat emitters
- Explain the difference between a radiator and a convector
- Properly size finned-tube baseboard convectors
- Work with the SERIESBB program from the Hydronics Design Toolkit
- Explain the installation methods used for various heat emitters
- Estimate the performance of heat emitters over a wide range of conditions
- Explain which kinds of heat emitters are best suited for low temperature systems
- Calculate the rate of heat loss from distribution piping

8.1 INTRODUCTION

Hydronic **heat emitters** release heat from the system's fluid into the space to be heated. One distinct advantage of hydronic heating is the wide variety of heat emitters available to suit almost any job requirement. These range from small "kick-space heaters" to underfloor systems that use the entire floor area of a building as the heat transfer surface. The latter type of system is specialized enough to justify a separate chapter.

The system designer must select and size heat emitters for a given project, so comfort and control are achieved in all areas of the building. Technical, architectural, and economic issues must be considered. The technical issues involve temperature, flow rate, and heat output characteristics of the heat emitter options. Architectural issues concern the appearance of the heat emitters in the building, as well as any interference they may create with furniture, door swings, etc. Economic considerations can greatly expand or restrict heat emitter options. The designer must provide a workable, if not optimal, selection of heat emitters within a given budget. Since most heating systems are designed after the floor plan of the building

has been established, the heating designer is often faced with the challenge of making the system fit the plan. A good knowledge of the available options can lead to creative and efficient solutions.

8.2 CLASSIFICATION OF HEAT EMITTERS

Some heat emitters directly heat the air in a room. They are properly classified as **convectors** because convective heat transfer is the primary means by which heat is released. Other types of heat emitters release the majority of their heat as thermal radiation, and are properly called **radiators**. The balance between convective heat transfer and radiative heat transfer can significantly affect the comfort achieved by various heat emitters. Each type has advantages and disadvantages that need to be understood before making a selection. Often a combination of two or more types provides an ideal match for the requirements or restraints imposed by a given project.

In this chapter the following types of heat emitters will be discussed:

- Finned-tube baseboard convectors
- Fan-coil convectors
- Panel radiators
- Radiant baseboards

These generic categories encompass most of the available types of heat emitters for residential or small commercial hydronic systems. Sometimes a specialized name is used to describe a specific heat emitter. A **kick-space heater**, for example, is a small horizontal **fan-coil** designed to mount under a cabinet or stair tread. A **towel warmer** is a name used for certain styles of panel radiators. These specialties will be discussed within the above groupings.

8.3 FINNED-TUBE BASEBOARD CONVECTORS

In the U.S., the dominant type of hydronic heat emitter used in residential and small commercial buildings is the

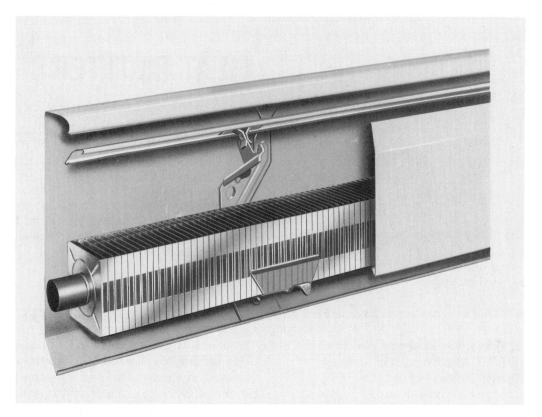

(a)

(b)

Figure 8–1   (a) Cut-away view of a typical residential finned-tube baseboard convector. Courtesy of Weil McLain Division of the Marley Co. (b) Installed finned-tube baseboard with closed damper.

finned-tube baseboard convector. It was developed during the 1940s as a flexible, modular, and quickly-installed alternative to standing cast-iron radiators. Its widespread use in smaller hydronic systems is largely due to its price advantage.

Figure 8–1 shows a cut-away view of a typical finned-tube baseboard convector. It consists of a sheet metal enclosure containing a copper tube fitted with aluminum fins. The finned tube is often called the **element** of the baseboard, since it is the component from which heat is released.

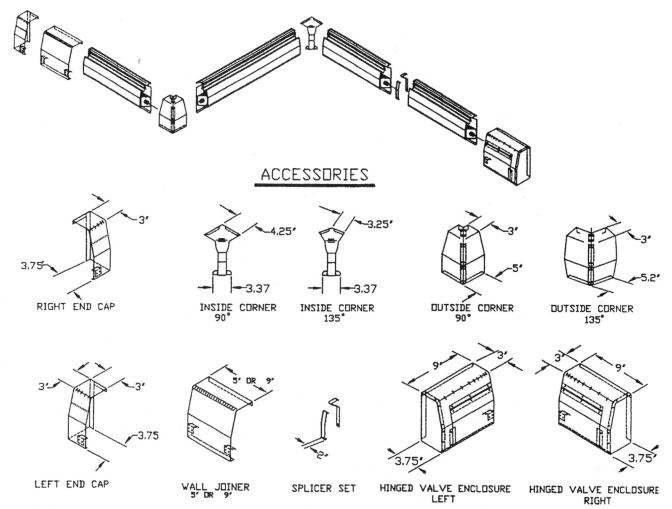

ACCESSORIES

RIGHT END CAP

INSIDE CORNER
90°

INSIDE CORNER
135°

OUTSIDE CORNER
90°

OUTSIDE CORNER
135°

LEFT END CAP

WALL JOINER
5' OR 9'

SPLICER SET

HINGED VALVE ENCLOSURE
LEFT

HINGED VALVE ENCLOSURE
RIGHT

Figure 8–2 Examples of trim parts for residential finned-tube baseboard. Courtesy of Embassy Industries.

There are many styles of finned-tube baseboard available. Those intended for residential use usually have elements consisting of either 1/2-inch or 3/4-inch copper tubing with rectangular aluminum fins about 2.25 inches to 2.5 inches wide and high. The enclosures are fabricated from steel sheet, and vary from about 6 inches to 8 inches in height.

Most manufacturers sell finned-tube baseboard in lengths from 2 feet to 10 feet long in increments of 1 foot. For long runs, combinations of these lengths are joined together.

Larger styles of finned-tube convectors are available, intended for use in commercial buildings. They have elements consisting of either copper or steel pipe in sizes from 3/4 inch up to 1.25 inches. Their fins are larger (3 inches to 5 inches on a side), and are available in both aluminum and steel. These larger convectors can release considerably more heat per foot of length than the smaller residential baseboards. They are also built to withstand the greater physical punishment associated with commercial or institutional usage. Because of their size and cost, however, they are often eliminated from consideration in residential systems.

In addition to straight lengths of finned-tube baseboard, most manufacturers offer sheet metal enclosure fittings such as end caps, inside and outside corner trim, and joint covers. These allow the baseboard to be fitted to the shape of the room when necessary. They are usually designed to snap into place over the straight length enclosure without the use of fasteners. Figure 8–2 depicts some of the available enclosure trim pieces.

Finned-tube baseboard convectors operate on the principle that heated air rises due to its lower density. This is the same effect that lifts a hot air balloon or makes smoke rise up a chimney. Air in contact with the finned-tube element is heated by natural convection. It rises through the fins, and up through the slot at the top of the enclosure. Cooler air near the floor level is drawn in through a slot at the bottom of the enclosure. As long as the finned-tube element is warmer than the surrounding air, natural convection will continue to pump air up through the enclosure. Most baseboard enclosures are equipped with a hinged damper over their outlet slot. This can be manually set to limit air flow through the baseboard.

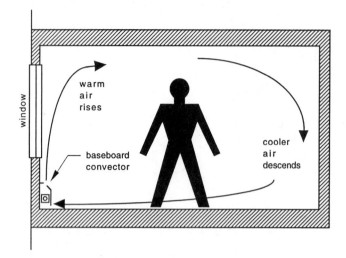

Figure 8–3 Room air circulation established by operation of finned-tube baseboard convector

When fully closed, heat output is reduced 40% to 50%.

The upward flow of warm air can negate the effects of downward drafts from exterior walls, and especially windows. Because of this, *finned-tube baseboard is usually placed along exterior walls, and especially under windows.* The circulation of room air due to operation of the finned-tube baseboard is depicted in Figure 8–3.

Heating a room by natural convection of warm air creates both desirable and undesirable effects. On the plus side, the slow moving air produces no sound. The rising warm air is also an effective means of limiting condensation of water vapor on cold glass surfaces. It negates the uncomfortable drafts associated with downward movement of cool air along these surfaces. When the baseboard is arranged to blanket an exterior wall with warm air, the wall's surface temperature is increased, improving the comfort in the room.

An undesirable result of heating by natural convection is that warm air tends to stay near the ceiling of the room, while cooler air remains near floor level. This effect, often referred to as stratification, is much more noticeable in rooms with high ceilings. It is worsened by convectors operating at high water temperatures. The higher the element temperature, the faster the air travels as it leaves the convector. This allows it to rise toward the ceiling without proper mixing with room air, increasing temperature stratification. As a rule, *convectors are not well suited for rooms with high vaulted ceilings.*

Anytime air is used to transport heat, dust movement can become a problem. Finned-tube convectors that operate at high temperatures in dusty or smoky environments have been known to produce noticeable streaks on wall surfaces above their enclosures. This effect is by no means limited to finned-tube baseboard. It should, however, be considered when job conditions could make it a problem.

Placement Considerations

It has already been stated that baseboard convectors should be placed, whenever possible, along exterior walls and especially under windows. A number of other factors also enter into placement decisions. Among these are available wall space, furniture placement, and door swings.

Any room that is to be heated by baseboard convectors must have sufficient unobstructed wall space to mount the necessary linear footage of heating element. This requirement often presents a problem, especially in smaller houses. Kitchens and bathrooms in particular often lack the necessary wall space because of cabinets and fixture placement. However, an estimate of baseboard length must be made before determining if available wall space is adequate. Methods for doing so are presented in the next section.

Heat emitters often compete with furniture for wall space. Most homeowners believe the heating system should be fit around furniture placement. While this may not always be possible, it should be reasonably attempted by the system designer. When interference is likely, it should be brought to the attention of the owner so a compromise can be worked out. The degree to which heat emitters allow flexibility of interior design can have a big impact on customer satisfaction with the overall system. It is unfortunate how often furniture placement is unnecessarily restricted by a careless attitude toward placement of heat emitters.

As an absolute minimum, 6 inches of free space should be available in front of any finned-tube baseboard. Without it, heat output will be significantly lowered due to inadequate air circulation. Closer spacing can also lead to degradation of fabrics, woods, and finishes due to prolonged exposure to relatively hot, low humidity air rising from the baseboards.

Door swings should also be checked before placing baseboard convectors. Baseboard convectors should not be installed within the arc of the door swing. Their thickness will prevent the door from making its normal swing back to the wall. The closer the baseboard is to the door opening, the worse the problem. Wooden doors that are left open against baseboard can also be discolored, warped, or cracked by prolonged exposure to the warm air.

Experience proves that every new job can bring its own set of unusual circumstances. It would be impossible to address all situations affecting the placement of baseboard convectors. However, a sampling of some of the most common placement considerations is shown in Figure 8–4.

Installing Baseboard Convectors

When measuring walls for baseboard convectors it is important to account for the space required by end caps and other trim accessories. These dimensions will vary

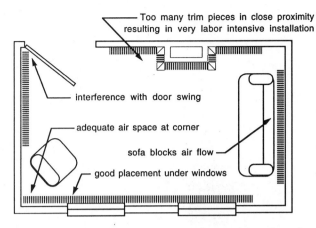

Figure 8–4 Good and bad placement of baseboard convectors

from one manufacturer to another. In some cases trim accessories slide over the enclosure for simple adjustment of the overall assembly length. This is very helpful in situations where the enclosure runs wall-to-wall. In situations where stock enclosure lengths combined with sliding end caps will not allow a wall-to-wall fit, the enclosure can be carefully cut to length. The finned-tube element can be shortened as necessary by sliding some of the fins off the end of the tube, and cutting the tube. *To avoid sliced fingers, always wear gloves when handling the finned-tube element.*

Sometimes the required length of baseboard will end 1 or 2 feet short of a corner of a room. When this happens the enclosure can be capped with a standard end cap, or the enclosure assembly without the element can be extended to the wall. This dummy enclosure may or may not contain a standard copper tube soldered to the end of the finned-tube element. Its only real purpose is to continue the aesthetics of the baseboard enclosure from wall to wall.

The four basic steps of installing a baseboard convector are:

1. Drill holes for the riser pipes
2. Attach the enclosure to the wall
3. Install the finned-tube element
4. Install the front cover and trim accessories

The holes for the riser pipes should be located so as to avoid drilling into floor joists, girders, wiring, or other plumbing beneath the floor. One way to do this is to locate a nearby object such as a plumbing stack *that penetrates through the floor*, measure the location of the joists or other objects to this reference object under the floor, then transfer these measurements above the floor to locate the joists. Another method is to drill a small 1/8-inch pilot hole and insert a rigid wire to help locate the hole from under the floor. The distance from this wire to the joists can be measured under the floor and again

transferred to the top of the floor. Since most floor joists are located either 16 inches or 24 inches on center, the position of adjacent joists can be measured easily. The location of the floor joist, or other obstructions, may require the enclosure, and/or the finned-tube element to be moved one way or the other. Always check this before fastening the enclosure to the wall.

When the centerline of one riser pipe is located, obtain the centerline of the other by measuring the *assembled length* of the heating element, along with any fittings or valves that are attached to it. Locate the other hole, and again *check that it does not fall over a floor joist or other obstruction before boring through*. This method of assembling the finned-tube element, with the required fittings and valves, is simpler than trying to calculate the exact position of the holes while accounting for lengths of the finned-tube element, valves, fittings, etc.

The holes for the riser pipes to the finned-tube element should have a diameter about 1/2 inch greater than the nominal pipe size used. This provides a space for the tubing to expand when heated without stressing the pipe. It also allows some tolerance for errors in measurement or alignment when soldering. Since the assembly is installed at room temperature, it is close to its minimum length. Because of this, most of the 1/2-inch space created by the larger hole should be located on the side of the tube *away* from the baseboard as shown in Figure 8–5. This allows maximum room for piping expansion.

The installer should be careful to mount the enclosure high enough to clear the finish flooring materials. As a rule, the bottom of the enclosure should be at least 1 inch above the subflooring. This allows sufficient space for medium thickness carpeting and pad. The enclosure can be nailed or screwed to the wall. On wood frame walls the enclosure should be fastened at intervals not exceeding 32 inches (to every other stud in a typical wall). Allow a slight gap with the wall surface at the end of the enclosure to allow trim pieces to be snapped in place.

After the enclosure is mounted, the finned-tube assembly can be placed on the cradles of the enclosure and all soldered joints made up. In some cases the riser pipes

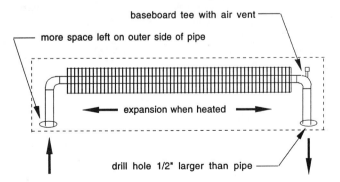

Figure 8–5 Installation of riser piping to a finned-tube baseboard. Note oversized holes and space allowed for expansion of the finned-tube element.

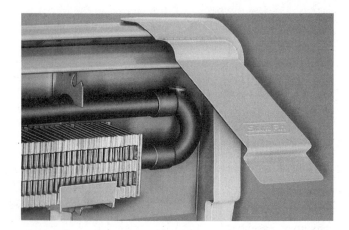

Figure 8–6 Example of a 180° return bend when a baseboard is supplied and returned from the same end. Courtesy of Slant/Fin Corp.

can be soldered into the assembly with the finned-tube element and fittings. This assembly can then be placed into the enclosure by sliding the riser pipes down through the holes in the floor. *Be sure the element rests on the plastic glides supplied with the enclosure.* These glides allow the element to expand and contract without creating squeaking sounds as its changes temperature.

In some situations, the supply and return piping must exit the same end of a baseboard. This approach is often used when one end of the baseboard cannot be accessed through the floor. This type of installation requires a 180 degree return bend at the end of the finned-tube element. The return pipe then runs back over the top of the finned-tube element. A detail of this return bend, with a tapping for an air vent, is shown in Figure 8–6. The return pipe also requires support within the enclosure. It is usually supported by a loop of copper wire wrapped around the brackets in the enclosure.

To complete the baseboard installation, snap the front cover in place, then install the necessary trim pieces. Be sure the damper blade on the enclosure is left in its fully open position.

8.4 THERMAL RATINGS AND PERFORMANCE OF FINNED-TUBE BASEBOARD

The heat output of finned-tube baseboard is expressed in Btu/hr *per foot of active element length.* The active length of the element is the length of pipe covered with fins. This is usually about 4 inches less than the length of the straight enclosure, not including the end caps.

The heat output of the baseboard is dependent on the fluid temperature in the finned-tube element. Manufacturers publish tables showing the heat output of their baseboard units at several different water temperatures. An example of such a ratings table is shown in Figure 8–7.

These tables are based on testing conducted by the Hydronics Institute using the IBR Testing and Rating Code for Baseboard Radiation. Output ratings are usually listed for average water temperatures ranging from 220 °F to 150 °F. Any output rating at water temperatures below 150 °F must be estimated using correction factors established by the testing standard. These are listed in Figure 8–8.

RATINGS: Fine/Line 15 Series
(Hot water ratings, BTU/HR per linear ft. with 65°F (18.3°C) entering air)

ELEMENT	WATER FLOW	PRESSURE DROP *	140°F †	150°F	160°F	170°F	180°F	190°F	200°F	210°F	215°F	220°F
No. 15-75E Baseboard with ³/₄" element	1 GPM	47	290	350	420	480	550	620	680	750	780	820
	4 GPM	525	310	370	440	510	580	660	720	790	820	870
No. 15-50 Baseboard with ¹/₂" element	1 GPM	260	310	370	430	490	550	610	680	740	770	800
	4 GPM	2880	330	390	450	520	580	640	720	780	810	850

* Millinches per foot.
† Ratings at 140°F determined by multiplying 150°F rating by the I=B=R conversion multiplier of .84.

NOTE: Ratings are for element installed with damper open, with expansion cradles. Ratings are based on active finned length (5" to 6" less than overall length) and include 15% heating effect factor. Use 4 gpm ratings only when flow is known to be equal to or greater than 4 gpm; otherwise, 1 gpm ratings must be used.

Figure 8–7 Example of a thermal rating table for finned-tube baseboard. Courtesy of Slant/Fin Corp.

Average water Temperature (deg. F.)	correction factor to heat output rating at 150 deg. F.
150	1.0
140	0.84
130	0.69
120	0.55
110	0.41
100	0.28

Figure 8-8 Correction factors for heat output ratings of copper tube/aluminum fin baseboard at water temperatures lower than 150 °F. Based on IBR Testing and Rating Code for Baseboard Radiation from the Hydronics Institute.

Note that these factors multiply the output rating of the baseboard at 150 °F water temperature. They can be used to show how the heat output of finned-tube baseboard decreases rapidly at lower water temperatures. For example, the heat output of a baseboard operating with 100 °F average water temperature is only 28% of the heat output while operating at 150 °F. It soon becomes evident that finned-tube baseboard convectors are impractical in systems operating at low water temperatures.

The ratings in Figure 8-7 are based on air entering the baseboard at 65 °F and flow rates of 1 gpm and 4 gpm. The heat output at 4 gpm flow rate is obtained by multiplying the heat output at 1 gpm by 1.057. The slightly improved performance at higher flow rates is due to increased convection between the water and the inside surface of the tube. A footnote below the ratings table indicates the 1 gpm rating should be used unless the flow rate is known to be equal to or greater than 4 gpm. This in turn requires a knowledge of the hydraulic performance of the system the baseboard is a part of, using methods from Chapter 6. Analytical methods will soon be presented to allow the baseboard performance to be adjusted for any reasonable flow rate.

Note the footnotes accompanying the thermal ratings indicate that a 15% **heating effect factor** has been included in the thermal ratings. *This means the listed values for heat output are actually 15% higher than the measured heat output established by laboratory testing.* The current rating standard for finned-tube baseboard allows manufacturers to add 15% to the tested performance provided a note indicating such is listed along with the rating data. The origin of the heating effect factor goes back several decades. At that time it was used to account for the higher output of baseboard convectors placed near floor level rather than at a height typical of standing cast-iron radiators. Unfortunately, it is a factor that has outlived its usefulness. Since the 15% higher heat output

cannot be documented by laboratory testing, it is inaccurate to select baseboard lengths that assume its presence. *It is the author's recommendation that baseboard **heat output** ratings that include a 15% heating effect factor be corrected by dividing them by 1.15 before selecting baseboard lengths.* This bases selection on actual tested performance.

Analytical Model for Heat Output per Unit Length

When the rating data shown in Figure 8-7 is combined with the low temperature correction factors given in Figure 8-8, it is possible to plot the heat output of a finned-tube baseboard over a wide range of temperature. This data, and a smooth curve drawn through it, are shown in Figure 8-9.

All types of finned-tube baseboard exhibit a similar curve of heat output versus water temperature. The values on the vertical axis will vary for baseboards with different size elements, but the shape of the curve over the same range of temperature will be very similar.

A generalized way of expressing heat output for a finned-tube baseboard involves two modifications to the data shown in Figure 8-8. First, the heat output data is normalized by dividing each heat output value by the heat output at 200 °F. Second, the heat output is expressed as a function of *temperature difference* between the air temperature entering the baseboard and the water temperature in the baseboard. This allows heat output to be determined for situations where the baseboard is heating rooms not kept at normal comfort temperature. A factor is also included to account for heat output over a range of flow rates. This factor is based on methods used in the IBR Testing and Rating Code. The result is an empirical equation that gives the heat output of a generic finned-

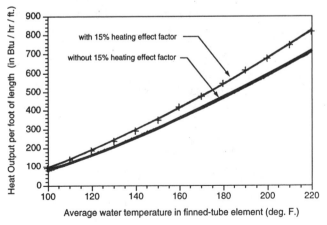

Figure 8-9 Heat output of a finned-tube baseboard versus water temperature for 65 °F entering air and water flow rate of 1 gpm. The upper curve includes the 15% heating effect factor. The lower curve has the heating effect factor removed.

tube baseboard (operating with water) over a wide range of conditions:

(Equation 8.1)
$$q' = \beta(f^{0.04})[0.00096865(T_w - T_{air})^{1.4172}]$$

where:

q' = heat output in Btu/hr/ft of finned-tube element length

β = heat output of the baseboard at 200 °F average water temperature, 1 gpm from manufacturer's literature.* (Btu/hr/ft)

f = flow rate of water through the baseboard (gpm)

T_w = water temperature in the baseboard (°F)

T_{air} = air temperature entering the baseboard (°F)

The coefficients and exponents are the result of the empirical derivation.

Example 8.1: A baseboard has an output of 650 Btu/hr/ft at an average water temperature of 200 °F, entering air temperature of 65 °F, and a flow rate of 1 gpm. Find the output of this baseboard while operating at an average water temperature of 135 °F, flow rate of 3 gpm, and having an entering air temperature of 50 °F. The heat output rating was taken from manufacturer's literature where a 15% heating effect factor was included in the ratings data.

Solution: The first step is to remove the heating effect factor by dividing the heat output rating of 650 Btu/hr/ft by 1.15.

$$\beta = \frac{650}{1.15} = 565.2 \text{ Btu/hr/ft}$$

This value can now be substituted along with the other data in Equation 8.1.

$$q' = 565.2(3^{0.04})[0.00096865(135 - 50)^{1.4172}]$$

$$= 565.2(1.0449)[0.5254] = 310.3 \text{ Btu/hr/ft}$$

8.5 SIZING FINNED-TUBE BASEBOARD

This section describes two methods for selecting the lengths of finned-tube baseboards connected in a series circuit. These range from a simple but less accurate

---

* If the literature indicates a 15% heating effect factor is included in the data, divide the value by 1.15 to remove it.

method suitable for preliminary sizing, to a more detailed method suitable for final design. After discussing manual sizing procedures, a program called 𝕊𝔼ℝ𝕀𝔼𝕊𝔹𝔹 from the Hydronics Design Toolkit will be discussed.

Sizing Method #1 (IBR Method)

Preliminary lengths of finned-tube baseboard can be selected by basing the heat output of each baseboard on the *average* temperature of the fluid in the piping circuit. The IBR method as it is often called, customarily takes this to be 10 degrees lower than the outlet temperature of the heat source. The procedure is as follows:

Step 1. Determine the average temperature of the fluid in the circuit:

(Equation 8.2)
$$T_{average} = T_{boiler\ outlet} - 10$$

Step 2. Look up the heat output rating of the baseboard at this temperature from the manufacturer's literature. Interpolate between the listed values if necessary. Again it is the author's recommendation that the 15% heating effect factor be eliminated when present in the published ratings.

(Equation 8.3)
$$\text{True heat output rating} = \frac{\text{IBR heat output rating}}{1.15}$$

Step 3. Divide the design heating load of each room by the value obtained in step 2. Round off the number to the next higher whole foot length of baseboard.

(Equation 8.4)
$$\text{Baseboard length} = \frac{\text{Room design heating load (Btu/hr)}}{\text{True heat output rating (Btu/hr/ft)}}$$

Example 8.2: A series piping circuit of finned-tube baseboard will be used to heat the rooms having the design heating loads indicated, in the order they are listed. The heat output ratings of the baseboard are given in Figure 8–7. The heating effect factor should be removed from these ratings for purposes of sizing. The boiler outlet temperature is 170 °F. Use sizing method #1 to select a baseboard length for each room.

Design heating load of rooms:

1. Living room	8,000 Btu/hr
2. Dining room	5,000 Btu/hr
3. Bedroom #1	3,500 Btu/hr
4. Bedroom #2	4,500 Btu/hr
5. Bathroom	1,500 Btu/hr

Solution:

Step 1. Using Equation 8.2, the average water temperature in the circuit is assumed to be 10 °F less than the boiler outlet temperature:

$$T_{average} = 170 - 10 = 160 \text{ °F}$$

Step 2. The IBR heat output rating of the baseboard at 160 °F is read from Figure 8–7 as 420 Btu/hr/ft. The heating effect factor, which is noted in the footnotes as being present in these ratings, is removed using Equation 8.3:

$$\text{True heat output} = \frac{420}{1.15} = 365.2 \text{ Btu/hr/ft}$$

Step 3. The length of each baseboard is found by dividing the true heat output rating value into the design heat load of each room using Equation 8.4, and rounding the result up to the next whole foot length:

1. Living room   8,000/365.2   =   21.9 or 22 ft
2. Dining room   5,000/365.2   =   13.7 or 14 ft
3. Bedroom #1   3,500/365.2   =   9.6 or 10 ft
4. Bedroom #2   4,500/365.2   =   12.3 or 13 ft
5. Bathroom   1,500/365.2   =   4.1 or 5 ft

Discussion: Although this method is very simple to use, it can lead to problems if used as the final means of sizing the baseboards. Its weakness is the fact that all baseboard lengths are based on an assumed average water temperature. In reality, each successive baseboard in the circuit will operate at a lower temperature than the preceding baseboard. *Because this method is based on average temperature, it tends to oversize baseboards near the beginning of the circuit and undersize baseboards near the end. This could lead to overheating some rooms and underheating others.* Another point worth noting is the average temperature of the fluid is dependent on the flow rate through the system. It may be 10 degrees less than the boiler outlet temperature, or it may be 5 or even 15 degrees less. This method does not account for this possibility at all. *Because of these limitations, sizing method #1 should only be used to make initial estimates of baseboard length for purposes of planning piping layouts.*

Sizing Method #2

The accuracy of baseboard sizing can be greatly improved by accounting for the temperature drop of the system fluid as it moves from one baseboard to the next. This allows the length of each baseboard to be based on the fluid temperature at its installed position in the circuit, rather than an overall average temperature.

This method is based on the use of water as the heat transfer fluid. It is also based on the use of standardized heat output ratings established at an entering air temperature of 65 °F. It requires a reasonable estimate of the flow rate through the baseboard. This can be made using methods from Chapters 6 and 7 or with the aid of the PIPEPATH, PUMPCURV, and PUMP/SYS programs from the Hydronics Design Toolkit.

Step 1. Using the rating data from the baseboard manufacturer, make a graph of the heat output of the baseboard versus water temperature, similar to that shown in Figure 8–10. If the ratings data contains the 15% heating effect factor (most do), remove it from each data point using Equation 8.3 *before* plotting the points on the graph. If water temperatures below 150 °F are expected in any of the baseboards, use the data from Figure 8–8 to calculate and plot estimated performance data as shown in Figure 8–10. Draw a smooth curve through all the data points. Save this graph for use on future projects using the same baseboard.

Step 2. Calculate the average water temperature in the first baseboard on the circuit using Equation 8.5:

(Equation 8.5)

$$T_{aveBB} = T_{in} - \frac{\text{Room Load}}{1,000(f)}$$

where:

T_{aveBB} = average water temperature in the baseboard (°F)

T_{in} = water temperature entering the baseboard (°F)

Room Load = design heating load of the room (Btu/hr)

f = flow rate of water through the baseboard (gpm)

Step 3. Using the graph prepared in step 1, estimate the heat output of the baseboard at the average water

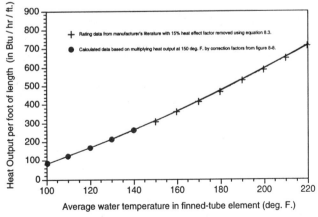

Figure 8–10 Graph of heat output rating data prepared from manufacturer's data. Note: The heating effect factor has been removed from all data points using Equation 8.3.

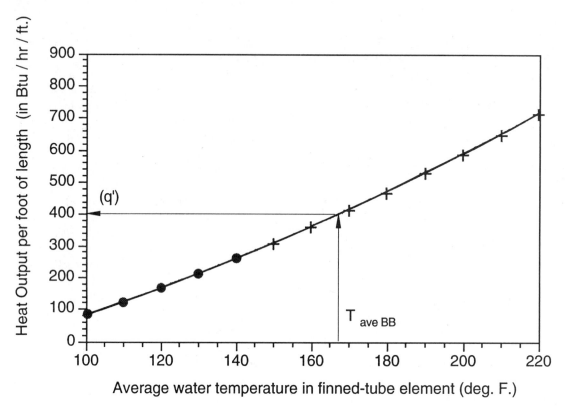

Figure 8–11 Use of graph prepared in step 1

temperature found in step 2. This is illustrated in Figure 8–11. The heat output value read from the graph is designated as q'.

Step 4. Calculate the length of the baseboard by dividing the design heating load of the room by the heat output per foot found in step 3:

(Equation 8.6)

$$L = \frac{\text{Room Load}}{q'}$$

Round this length off to the next larger whole foot. This length is designated $L_{rounded.}$

Step 5. Calculate the outlet temperature of the baseboard based on its rounded length using the following equation:

(Equation 8.7)

$$T_{out} = T_{in} - \frac{(L_{rounded})(q')}{500(f)}$$

Step 6. Using the outlet temperature of the first baseboard (calculated in Step 5) as the inlet temperature for the second baseboard, repeat this procedure beginning with Step 2. In a similar manner, use the outlet temperature of each baseboard as the inlet temperature for the next baseboard in the series circuit. Eventually all baseboard lengths will be determined, as will the outlet temperature of the last baseboard.

A good way to organize these calculations is to make a table as shown in Figure 8–12. This table is filled in by working across the top row to obtain the outlet temperature of the first baseboard. This value is then copied to the first column of the second row, and the process is repeated. The last value to be obtained is the outlet temperature of the last baseboard. This table can be extended, if necessary, for circuits with more than ten series-connected baseboards.

Example 8.3: Using sizing method #2, repeat the baseboard selection process for the same rooms described in Example 8.2. Assume a water flow rate of 2.25 gpm. Use the graph in Figure 8–13 for heat output ratings.

Solution: The table shown in Figure 8–14 shows the results of applying the procedure for sizing method #2. Note the outlet temperature of the first baseboard becomes the inlet temperature to the second baseboard, and so on.

Discussion: It is interesting to compare the results obtained using sizing method #1 in Example 8.2, with those of sizing method #2 in this example. The best comparison can be made by looking at the *unrounded lengths of baseboard* calculated for each room. The unrounded lengths show the greatest differences between the calculated results, and are given in the table of Figure 8–15.

Room Name	Design Heat Load	Inlet Temp.	Average water temp (equ. 8.5)	Heat Output of baseboard (from graph)	Calculated Baseboard Length (equ. 8.6)	Rounded Baseboard Length	Outlet Temp. (equ. 8.7)
1.							
2.							
3.							
4.							
5.							
6.							
7.							
8.							
9.							
10.							

Figure 8–12   Table used to organize calculations for sizing method #2

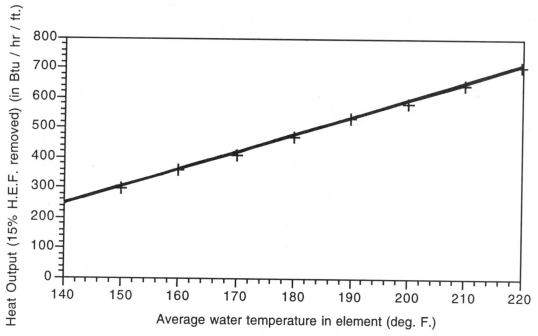

Figure 8–13   Graph of baseboard heat output ratings for Example 8.3. Fifteen percent heating effect factor has been removed from all data.

Note the length of the first baseboard obtained using method #1 is longer than the length obtained using method #2. Conversely, method #1 calculates a shorter length for the last baseboard on the circuit compared to method #2. These results are due to the use of a single average water temperature in method #1, versus the average

temperature *at each baseboard* in method #2. They reveal the tendency of method #1 to oversize baseboards at the beginning of the circuit and to undersize them near the end.

The greater accuracy of method #2 comes at the expense of more extensive calculations. Furthermore, since the results of the first baseboard affect the input

Room Name	Design Heat Load	Inlet Temp.	Average water temp (equ. 8.5)	Heat Output of baseboard (from graph)	Calculated Baseboard Length (equ. 8.6)	Rounded Baseboard Length	Outlet Temp. (equ. 8.7)
1. Living	8000	170	166.4	399.5	20.02	20	162.9
2. Dining	5000	162.9	160.7	366.6	13.6	14	158.3
3. Bedroom #1	3500	158.3	156.7	343.2	10.2	11	154.9
4. Bedroom #2	4500	154.9	152.9	321.5	13.99	14	150.9
5. Bathroom	1500	150.9	150.2	306.6	4.89	5	149.5
6.							
7.							
8.							
9.							
10.							

Figure 8–14   Table for sizing method #2 filled with results for Example 8.3

Lengths using method #1	Lengths using method #2
21.9 ft.	20.02 ft.
13.7 ft.	13.6 ft.
9.6 ft.	10.2 ft.
12.3 ft.	13.99 ft.
4.1 ft.	4.89 ft.

Figure 8–15   Comparison of unrounded baseboard lengths obtained using sizing methods #1 and #2 for the same room heating loads

data to the second, and so on, any error in the calculations will invalidate all remaining numbers from that point to the last value obtained. If, for example, the flow rate through the baseboards is changed, the entire table has to be recomputed. Still, the greater accuracy obtained using method #2 justifies its use for final baseboard selection.

An ideal way to automate these calculations is to build the procedure into a spreadsheet program on a PC. This allows long series strings to be calculated error free, and the results of changes in flow rates, inlet temperatures, etc., to be obtained almost instantly.

The SERIESBB Program

The Hydronics Design Toolkit contains a program named SERIESBB that executes a sizing procedure slightly more precise than even sizing method #2. It contains data that allows baseboard performance to be adjusted based on use of different fluid temperature and a wide range of entering air temperatures, both of which cannot be handled by sizing method #2. The program can work with up to 15 baseboards connected in series. The heat output rating of each baseboard at 200 °F can be specified independently, as can the entering air temperature and the flow rate. The latter allows baseboards connected using Monoflo® tees to be included anywhere in the series string.

SERIESBB determines the rounded length, heat output, inlet temperature, and outlet temperature of each baseboard specified. It is a powerful program suitable for use in day-to-day design work. The reader is encouraged to experiment with the program to determine the effect of various system design changes on baseboard length. Examples of the input screen and edit table are shown in Figure 8–16 and 8–17 respectively.

8.6   HYDRONIC FAN-COILS

When rooms lack the wall space needed to mount finned-tube baseboard, one alternative is to use a hydronic

```
- SERIESBB -  F1-HELP       F4 FIND RESULTS              F10-EXIT TO MAIN MENU
  -INPUTS-
 ENTER Number of baseboards in series, (1-15)................ 5
 ENTER Default IBR baseboard output with 200F water,(100-1000) 600    Btu/hr ft
 ENTER Default air temp. adjcent to baseboard,(35-90)........ 65      deg.F.
 ENTER Inlet fluid temperature to first baseboard,(90-250).... 180    deg.F.
 ENTER Default flowrate through baseboards,(0.25-20).......... 1      gpm
 SELECT Fluid type (press F3)................................. water
 Question Remove IBR heating effect factor(Y/N).............. YES
 EDIT TABLE OF BASEBOARD DATA.................................Go To
```

```
  MESSAGE
   Use the UP or DOWN keys to move the highlighting bar over an input. For
 ENTER inputs type desired value. SELECT inputs press F3. F4 to calculate.
```

Figure 8-16   Screen shot of the main screen of the SERIESBB program in the Hydronics Design Toolkit

```
- SERIESBB -  F1-HELP                              F10-RETURN TO INPUTS
  -EDIT TABLE-
   BASEBOARD     ASSIGNED LOAD    IBR RATING @200F   AIR TEMP.    FLOW RATE
   NUMBER          (Btu/hr)        (Btu/hr/F)        (DEG. F)      (GPM)
                  (500-20000)      (100-1000)        (35-90)      (.25- 1)
   Baseboard 1      5000             600               65            1
   Baseboard 2      5000             600               65            1
   Baseboard 3      5000             600               65            1
   Baseboard 4      5000             600               65            1
   Baseboard 5      5000             600               65            1
```

```
  MESSAGE
   Use the UP, DOWN, LEFT, RIGHT ARROW keys to move the highlighting bar
 over an input. Type desired value. F10 to return to input screen.
```

Figure 8-17   Screen shot of the edit table from the SERIESBB program in the Hydronics Design Toolkit

fan-coil. These devices come in all different sizes, shapes, and heating capacities. They all have two common components, a finned-tube coil heat exchanger through which the system fluid is routed, and a blower that forces air to flow across the coil. The basic concept of a fan coil is shown in Figure 8–18.

The size and physical orientation of the coil and the blower are largely what distinguish one type of fan-coil from another. Most fan-coils used in residential or small commercial hydronic systems are designed to blow air directly into the heated space. Some are designed to be mounted on a wall, or recessed into it, others to

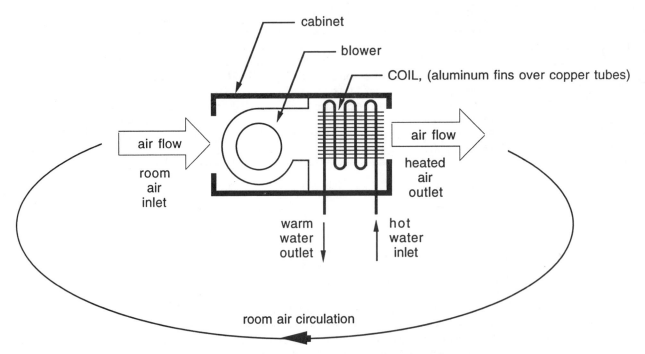

Figure 8–18 Basic concept of a hydronic fan-coil

be hung from the ceiling. A limited number are designed to be mounted flush with the floor. Another type of fan-coil called an **air handler** is designed for concealed mounting either above the ceiling or below the floor of a room. Air is delivered to the room through a duct or ducting system. Each of these configurations will be discussed individually.

Advantages of Fan-coils

Unlike finned-tube baseboard convectors, fan-coils rely on forced-convection to transfer heat from the coil surface to the room air. Force-convection is much more effective than natural convection in transferring heat from a surface to air. This allows the surface area of the coil to be many times smaller than that of a finned-tube baseboard element of equivalent capacity. A fan-coil unit can often provide equivalent heating capacity using only a fraction of the wall space required by finned-tube baseboard.

Because of the efficiency of forced convection heat transfer, many fan-coils can operate with significantly lower fluid temperatures compared to finned-tube baseboard. They can often be used with low temperature hydronic heat sources such as hydronic heat pumps, or active solar systems, where the use of finned-tube baseboard would be marginal at best.

Disadvantages of Fan-coils

The space saving and low fluid temperature advantages of fan-coils do not come without some compromises. First, any type of fan-coil will create some operating noise due to the blower operation and relatively high rate of air flow. Noises can be worsened by vibration, loose fitting grills, trim, etc. Manufacturers attempt to limit operating noise through design, but it will always be present to some degree. *The likelihood of some operating noise should be discussed with the owners to ensure it is acceptable in certain areas of the building.*

Another factor associated with fan-coils is dust and dust movement. Regardless of how clean a room is kept, dust is always present in the air and will be carried into the intake of any type of fan-coil. Quality fan coils will have filters capable of removing most of the dust before it is deposited on the coil surfaces or blown back into the room. When filters are not present, or if the unit is not periodically cleaned, dust accumulation on coil surfaces can reduce the heating capacity of the unit, especially if it operates at low fluid temperatures.

Also to be considered is the fact that fan-coils require connection to line voltage electrical wiring to operate the blower. Although the amount of power consumed by the blowers in residential size fan-coils is quite small, the designer must ensure a source of electricity is available at or near the proposed location of a fan-coil.

Wall-mounted Fan-coils

A common type of fan-coil for residential and light commercial applications is designed to be mounted on, or recessed into, a wall or partition. Figure 8–19 illustrates both surface-mounted and recessed fan-coils for wall mounting.

Wall-mounted fan coils usually draw room air in near the bottom of the unit and discharge it horizontally near

(a)

(b)

Figure 8-19   (a) example of a surface mounted fan-coil convector, (b) example of a recessed fan-coil convector. Courtesy of Beacon Morris Division of Mestek, Inc.

the top. Some are designed for installation in short wall spaces such as under windows. Others are designed to fit in tall but narrow wall spaces.

Often the fan-coils are equipped with controls to adjust the speed of the blower. This allows some degree of control over heat output. On some units, two or three specific speeds can be selected. In others, a solid state speed control allows a wide range of blower speed settings.

Some fan-coils come with an integral thermostat that can turn the blower off once the room reaches its temperature setting. The usefulness of the thermostat depends on the other controls used in the system. For example, most fan-coil thermostats are not wired back to the heat source so that a call for heat from the fan-coil can also start the heat source and circulator. More typically, fan-coil thermostats function as an upper limit control for room temperature. Sometimes a low limit thermostat is used to prevent the blower from operating until the proper water temperature is reached within the coil. By studying the control wiring supplied with the fan-coil, it may be possible to connect its internal controls with those of the heat source and circulator. Control wiring is covered in detail in Chapter 9.

Piping connections to fan-coils are usually made to $1/2$-inch copper tube stubs or FPT connections. In some situations the piping enters the back of the cabinet, and is thus concealed from view. If this is not possible, piping risers are routed up from the floor into the bottom of the cabinet.

Recessed fan-coils generally provide better aesthetics than those mounted on the surface of a wall. Their use requires good planning during the construction of a building. A properly sized wall cavity must be framed, and the piping and wiring routed to the cavity. The sooner this planning takes place the better. It is not uncommon to find a lack of space for recessed mounting, especially in bathroom areas where plumbing and electrical systems can present interference. Recessed fan-coils are usually designed so that a base-pan assembly can be mounted into the wall cavity before the walls are closed in. The blower/coil assembly then slides into position over the finished

wall. Installation of recessed fan-coils in retrofit situations is usually harder than in new construction because the walls are already closed-in. *Be sure to take a careful look at retrofit jobs before committing to recessed fan-coil unit(s).*

The location of the fan-coil should be such that air flow is not blocked by furniture. Since a fan coil often has to heat an entire room it is very important the air stream be allowed to penetrate into the space. *Discuss the positioning with the owners so that interference problems with furniture or other furnishings can be avoided.* The units should also not blow directly on the occupants of a room. For example, imagine trying to read a newspaper in a reclining chair that is only two or three feet away from, and directly in line with, the discharge grill of a wall-mounted fan-coil. The idea is to diffuse the air stream into the room air before it contacts the occupants.

Under-cabinet Fan-coils

Another type of hydronic fan-coil is designed to be mounted in the 4-inch space beneath a kitchen or bathroom cabinet, or even under the lower tread of a set of stairs. These units, often called kick space heaters, are designed for situations where wall space is not available, particularly kitchens and bathrooms. An example of a typical under-cabinet heater is shown in Figure 8–20. A typical one foot wide under-cabinet heater can yield the heat output of 8 to 10 linear feet of finned-tube baseboard. Wider units are available with greater heating capacities.

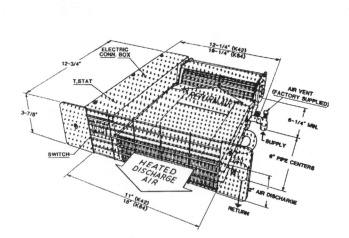

Figure 8–20   Example of a small kick-space heater. Courtesy of Beacon Morris Division of Mestek, Inc.

Air flows into the unit through the upper part of the front grill. It is drawn in by a transverse blower mounted at the rear of the unit. The blower reverses the air flow, and sends it out through the horizontal coil in the lower part of the unit where it is heated. These types of fan-coils do not have any type of filter, and are somewhat prone to dust accumulation. They should be periodically cleaned with a vacuum cleaner.

Kick space heaters are usually supported on rubber grommets that reduce vibration transfer to the floor or cabinet. Connections are typically ½-inch copper tube stubs. Integral junction boxes are used for wiring connections. Most under-cabinet heaters are equipped with a low limit aquastat that prevents the blower from operating until the water temperature in the coil reaches a certain temperature. This setting is determined by the manufacturer and is usually not adjustable. Typical temperatures are 110 °F to 140 °F. *This thermostat may have to be bypassed if the unit will be used with a lower temperature heat source.*

When installing an under-cabinet heater it is convenient to provide an access hatch through the bottom shelf of the cabinet. This can be done by carefully sawing out the bottom shelf over the area where the unit will be mounted

and installing wooden ledgers to support the shelf when replaced. Be sure to allow sufficient space to fully access the side-mounted piping and wiring connections. This arrangement permits easy access for cleaning and to the air vent. It also allows the unit to be lifted out if servicing is necessary.

When installing under-cabinet heaters be careful not to accidentally bend the exposed blades of the blower wheel. The resulting imbalance can cause noticeable vibration and associated noise.

Overhead Fan-coils

Commercial buildings are often designed with overhead piping for mechanical systems. One method of delivering heat from overhead piping without routing it down near floor level is the use an overhead fan-coil, or **unit heater**.

Unit heaters are often used in commercial spaces such as service garages, machine shops, supermarkets, etc., where overhead mounting allows them to be out of the way of heavy traffic or rough usage. They are usually supported by threaded steel rods or perforated angle struts attached to structural members such as roof trusses or beams. Figure 8–21 shows a typical unit heater designed

Figure 8–21   Example of a horizontal air flow unit heater. Note adjustable louvers on front of unit. Courtesy of Sterling Radiator Division of Mestek, Inc.

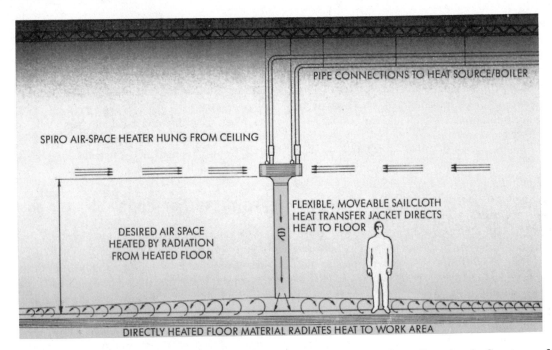

Figure 8–22 Overhead fan-coil with downward air flow and sock to carry air to floor level. Courtesy of Spirotherm, Corp.

for angled horizontal air flow. Louvers on the front of the unit allow the heated air stream to be directed toward the floor level at various angles.

Another type of unit heater is designed for downward air flow. This style is particularly effective in reducing temperature stratification in areas with high ceilings. One manufacturer equips their vertical unit heaters with a flexible synthetic duct that delivers the heated air to just above floor level. The air stream is deflected by the floor and travels horizontally, warming the floors as it goes. An example of such a heater is shown in Figure 8–22.

The mounting height of a unit heater also affects the height of the comfort zone in the space. In buildings with high ceilings, heated air tends to stratify. By mounting vertical unit heaters at a lower level it is possible to maintain the comfort zone in the lower, occupied areas. This improves comfort and saves energy.

8.7 THERMAL PERFORMANCE OF FAN-COILS

The heat output of a fan-coil is dependent on several factors. Some are fixed by the construction of the fan-coil and cannot be changed. These include:

- The surface area of the coil
- The size, spacing, and thickness of the fins
- The number of tube passes through the fins
- The air-moving ability of the blower

Other performance factors are dependent on the system into which the fan-coil is installed. These include:

- The entering water temperature
- The water flow rate through the coil
- The entering air temperature
- The air flow rate through the coil

When selecting a fan-coil, the designer must match the characteristics of the available units with the temperatures and flow rates the system can provide, while also obtaining the required heating capacity. This usually involves a search of manufacturer's performance data in an attempt to find a good match among several simultaneous operating conditions. This data is often provided in the form of tables that list the heat output at several combinations of temperature and flow rate for both the water-side and air side of the unit. An example of a thermal performance table for a wall-mounted fan-coil is shown in Figure 8–23.

This table lists the Btu/hr heating capacity of two units over a wide range of entering water temperatures, two different air flow rates, and two water flow rates. Each heat output value can be thought of as a snapshot of heating performance at specific conditions. Since the heat output of the fan-coil varies continuously over a range of temperatures and flow rates, there is no guarantee it will operate at one of the conditions listed in the table.

When it is necessary to estimate the heat output of the unit at conditions other than those listed in the table, a process called **interpolation** can be used. This process is

MODEL	FLOW RATE (GPM)	FAN SPEED	AIR DELIVERY (CFM)	ENTERING WATER TEMPERATURE (°F)										
				110*	120*	130*	140	150	160	170	180	190	200	210
WH 50	1.0	High	89	1690	2030	2354	2783	3205	3580	3887	4263	4638	5079	5354
		Low	50	1390	1610	2010	2400	2825	3200	3605	4020	4380	4750	5040
	3.0	High	89	1996	2456	2866	3309	3717	4094	4470	4913	5323	5664	6142
		Low	50	1450	1910	2217	2590	2900	3410	3790	4299	4570	4950	5300
WH 90	1.0	High	121	3700	4350	5025	6000	6780	7640	8400	9290	10050	10800	11500
		Low	75	3250	3900	4505	5350	6000	6750	7420	8190	8830	9550	10200
	3.0	High	121	3800	4520	5320	6130	7160	7940	8710	9600	10420	11250	12090
		Low	75	3350	4020	4800	5580	6380	7050	7730	8500	9200	9980	10790

WH SERIES BTU/HR OUTPUT FOR MYSON FAN CONVECTORS — ENTERING AIR AT 65° F

*For Output At These Water Temperatures, Optional 107° Low Temperature Cutoff Required.

Figure 8–23   Table of thermal performance data for wall-mounted fan-coil. Courtesy of Myson, Inc.

usually not as simple as averaging between the values that are above and below the desired value. Rather it sets up *proportions* to arrive at a more accurate estimate. The best way to illustrate interpolation is through an example.

Example 8.4: The heat output of the fan-coil represented by the data in Figure 8–23 is 3,205 Btu/hr at 150 °F entering water temperature, 1 gpm water flow rate, and high fan speed. At 160 °F and the same flow rate the output is listed as 3,580 Btu/hr. Estimate the heat output of the fan-coil at 152 °F water temperature.

Solution: Set up the known and unknown quantities as shown in Figure 8–24, and subtract the quantities connected by the brackets. Form the ratios of these differences, and equate these ratios to each other. Solve the resulting equation for the answer.

Notice the algebraic signs (+ or –) take care of themselves so the answer is positive. Another way of thinking about the problem is that the temperature of 152 is one-fifth of the way between the numbers 150 and 160. This implies the desired value of heat output, *x,* will be one-fifth of the way between 3,205 and 3,580. One-fifth of this difference is:

$$(1/5)(3,580 - 3,205) = 75$$

Adding 75 to the lower value of 3,205 gives the same answer of 3,280 Btu/hr.

In some cases **double interpolation** is required to obtain an estimate of heat output that falls between both the column and row rating points of the table. Again, an example is the easiest way to show how this is done.

Example 8.5: Make an estimate of the heat output of the fan-coil represented by the data in Figure 8–23, at an entering water temperature of 152 °F and a water flow rate of 2.5 gpm.

Solution: Begin by interpolating between the temperatures of 150 and 160 two times, first for the heat output values listed at a flow rate of 1 gpm, and again for the heat outputs listed at a flow rate of 3 gpm. To finish the calculation, take the results of the first two (temperature) interpolations and interpolate them between the lower flow rate of 1 gpm and the upper flow rate of 3 gpm.

The first temperature interpolation is the same as shown in Example 8.4, with a result of 3,280 Btu/hr at 152 °F.

The second interpolation is shown in Figure 8–25.

The final interpolation will be between the flow rates of 1 and 3 gpm, and will use both previously calculated capacities at 152 °F entering water temperature. The procedure is shown in Figure 8–26.

The estimated capacity of this fan-coil at 152 °F entering water temperature and 2.5 gpm water flow rate is thus 3,664 Btu/hr.

$$\frac{150-152}{150-160} = \frac{3205-x}{3205-3580}$$

$$\frac{-2}{-10} = \frac{3205-x}{-375}$$

$$0.2\,(-375) = 3205 - x$$

$$x = 3205 + 75 = 3280 \text{ Btu / hr}$$

Figure 8–24   Interpolation procedure for Example 8.4

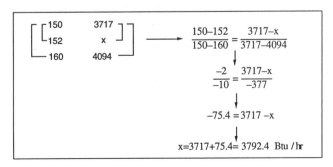

$$\frac{150-152}{150-160} = \frac{3717-x}{3717-4094}$$

$$\frac{-2}{-10} = \frac{3717-x}{-377}$$

$$-75.4 = 3717 - x$$

$$x = 3717 + 75.4 = 3792.4 \text{ Btu / hr}$$

Figure 8–25   First interpolation procedure for Example 8.5

$$\begin{bmatrix} \begin{bmatrix} 1 & 3280 \\ 2.5 & x \end{bmatrix} \\ 3 & 3792.4 \end{bmatrix} \longrightarrow \frac{1-2.5}{1-3} = \frac{3280-x}{3280-3792.4}$$

$$\frac{-1.5}{-2} = \frac{3280-x}{-512.4}$$

$$-384.3 = 3280-x$$

$$x = 3280+384.3 = 3664 \text{ Btu/hr}$$

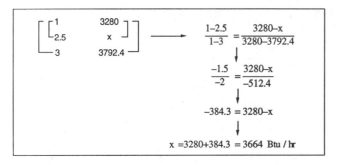

Figure 8–26 Final interpolation procedure for Example 8.5

Discussion: The final value should always be checked to see if it is reasonable. It should fall between the values listed in the tables both vertically and horizontally.

Another way to perform interpolation is the use of a spreadsheet program on a PC. Another useful method, especially if the data will be used repeatedly, is to plot the data on a graph and draw a smooth curve through the points. The estimated performance between data points is easily read from this curve.

Sometimes manufacturers list the heat output of a fan-coil unit at a single reference condition. This reference

Model No.	Output BTU/ HR*	GPM	Final Air °F	Prssr. Drop FT./H_2O	Motor HP	RPM	Nominal CFM	Outlet FPM	Nom. Amps @ 115VAC
HS-108 A	8,030	.80	91	.80	9 Watt	1550	245	250	.8
	6,800		90			1350	210	215	.8
HS-118A	18,400	1.9	94	2.2	9 Watt	1550	500	500	.8
	15,650		96			1350	420	420	.8
HS-125A	24,800	2.5	102	2.2	1/47	1550	580	590	1.1
	21,230		106			1350	460	450	1.1
HS-136A	35,900	3.6	99	3.0	1/30	1070	850	550	1.1
	32,300		100			900	750	480	1.1
HS-18	13,050	1.3	95	.005	9 Watt	1550	395	395	.8
	11,725		99			1350	350	350	.8
HS-24	17,400	1.8	96	.014	9 Watt	1550	450	450	.8
	15,600		98			1350	380	380	.8
HS-36	26,100	2.7	103	.09	1/47	1550	550	550	1.1
	23,500		103			1350	480	480	1.1
HS-48	34,800	3.5	103	.12	1/30	1070	750	550	1.3*
	31,300		111			900	630	460	1.3*
HS-60	43,600	4.4	105	.17	1/30	1070	900	650	1.3*
	39,200		112			900	700	510	1.3*

(a)

*Hot Water Conversion Factors Based on 200° Entering Water 60° Entering Air 20° Temperature Drop

Entering Air Temperature	Entering Water Temperature—20° Water Temperature Drop										
	100°	120°	140°	160°	180°	200°	220°	240°	260°	280°	300°
30°	0.518	0.666	0.814	0.963	1.12	1.26	1.408	1.555	1.702	1.85	1.997
40°	0.439	0.585	0.731	0.878	1.025	1.172	1.317	1.464	1.609	1.755	1.908
50°	0.361	0.506	0.651	0.796	0.941	1.085	1.231	1.375	1.518	1.663	1.824
60°	0.286	0.429	0.571	0.715	0.857	1.000	1.143	1.286	1.429	1.571	1.717
70°	0.212	0.353	0.494	0.636	0.777	0.918	1.06	1.201	1.342	1.483	1.63
80°	0.140	0.279	0.419	0.558	0.698	0.837	0.977	1.117	1.257	1.397	1.545
90°	0.069	0.207	0.345	0.483	0.621	0.759	0.897	1.035	1.173	1.311	1.462
100°	0	0.137	0.273	0.409	0.546	0.682	0.818	0.955	1.094	1.23	1.371

(b)

Figure 8–27 Example of reference rating conditions (a) with correction factors, (b) for the heating capacity of a unit heater. Courtesy of Sterling Radiator Division of Mestek, Inc.

condition would consist of a specified value for the entering temperatures and flow rates of both the water and air. Accompanying this will be several tables or graphs of correction factors that can be used to adjust the reference performance up or down for variation in temperature and flow rate of both the entering water and entering air. An example of tables listing these performance correction factors for a unit heater is shown in Figure 8–27.

General Principles of Fan-coil Performance

The heat output of fan-coils can be described in the form of simple principles that can be verified by examining the rating data supplied by fan-coil manufacturers. By varying one operating condition such as entering temperature or flow rate, while all others remain fixed, the designer can get a feel for which factors have the greatest impact on heat output. This can be very helpful in evaluating the feasibility of various design options.

Principle #1: The heat output of a fan-coil is approximately proportional to the temperature *difference* between the entering air and entering water.

If, for example, a particular fan-coil can deliver 5,000 Btu/hr with a water temperature of 140 °F and room air entering at 65 °F, its output using 180 °F water can be estimated as follows:

$$\text{estimated output at } 180\ °F = \frac{(180 - 65)}{(140 - 65)}(5,000)$$

$$= 7,670\ \text{Btu/hr}$$

A proportional relationship between any two quantities will result in a straight line that passes through or very close to the origin when the data is plotted on a graph. Figure 8–28 shows this to be the case when the heat output of the fan-coil from Figure 8–23 is plotted against the difference between the entering water and entering air temperatures.

This concept is useful for estimating heat output when a fan-coil is used in a room kept above or below normal comfort temperatures. For example, if a fan-coil can deliver 5,000 Btu/hr with a water temperature of 140 °F and room air entering at 65 °F, its output while heating a garage maintained at 50 °F could be estimated using the same proportionality method described above:

$$\text{estimated output at } \atop 50\ °F\ \text{entering air} = \frac{(140 - 50)}{(140 - 65)}(5,000)$$

$$= 6,000\ \text{Btu/hr}$$

Although this principle is very helpful, it is still an approximating technique and should not replace the use of thermal rating data supplied by the manufacturer, if available at the desired conditions.

Principle #2: Increasing the flow rate of water through the coil will increase the heating capacity of the fan-coil.

Interestingly, many heating professionals instinctively disagree with this statement. They argue that because the water moves through the coil at a faster speed, it has less time in which to release its heat. However, the time a given particle of water stays inside the fan-coil is irrelevant in a system with continuous circulation. The faster moving fluid improves convective heat transfer between the fluid and the inner surface of the tubes, resulting in greater heating capacity.

Another way of justifying this principle is to think of the temperature difference between the in-going and out-going fluid. As the flow rate through the fan-coil is increased, this temperature difference decreases. This means that the average temperature inside the coil is increasing, and thus the coil should output more heat. Those who claim a smaller temperature difference

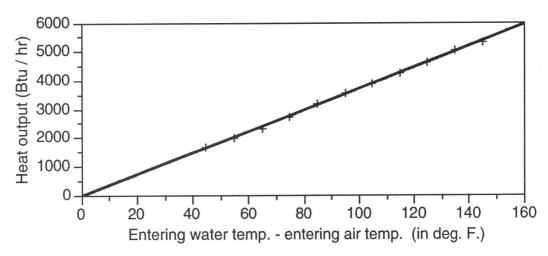

Figure 8–28   Plot of fan-coil heat output versus temperature difference between entering water and entering air based on data from Figure 8–23

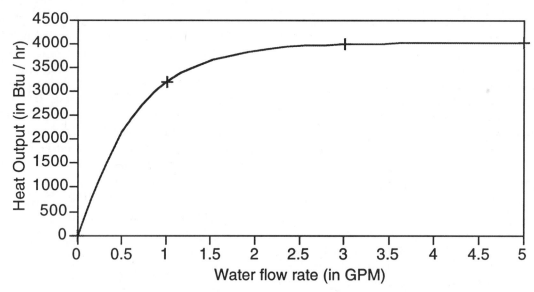

Figure 8–29 The effect of water flow rate on heat output of a small hydronic fan-coil

between the ingoing and outgoing fluid means less heat is being released, are ignoring the fact that the rate of heat transfer depends on both the temperature difference and flow rate, as described by the sensible heat rate equation of Chapter 4.

The rate at which heat output increases with flow rate is very dependent on the nature of convective heat transfer. The rate of gain is very fast at low flow rates, then becomes very marginal at flow rates above about 50% of the maximum flow rate the coil is rated for. This can be seen in Figure 8–29 where the manufacturer's data for heat output versus water flow rate of a fan-coil has been plotted.

Notice the fan-coil has about 50% of its maximum heat output capacity at only about 10% of its maximum flow rate. The strong curvature of this relationship means that *attempting to control the heat output of a fan-coil by adjusting flow rate can be very tricky.* Very minor valve adjustments can create major changes in heat output at low flow rates, and almost no change at all at higher flow rates. This curve also points out that attempting to boost heat output from a fan-coil by driving it at an unusually high flow rate will yield very minor gains. The argument against doing this is further supported when one considers the rapid gain in head loss through the coil at these higher flow rates.

Principle #3: Increasing the flow rate of air through the coil will marginally increase the heat output of the fan-coil.

This principle is again the result of improved forced-convection heat transfer between the coil surfaces and the air. Faster moving air "scrubs" heat off the coil at a faster rate. As with water flow rates, however, the gain in heat output is very minor above the nominal air flow rate the unit is designed for.

Principle #4: Fan-coils with large coil surfaces and/or **multiple tube passes** through the coil will be capable of operating at lower entering water temperatures.

Convective heat transfer is proportional to the contact area between the surface and the fluid, as well as the temperature difference between them. If this contact area is increased, the temperature difference can be made smaller while the fan-coil retains the same heating capacity. This principle can be well applied in hydronic systems that must operate with low temperature heat sources such as solar collectors or heat pumps. In these applications, the designer may need to use fan-coils with larger coils and/or more tube passes through the fins of the coil, to drive heat from the coil at the required rate. On larger air-handlers, manufacturers may offer higher performance coils with up to six rows of tubes as an option. This is usually not true on smaller residential fan-coils. For these products the only way to increase coil surface area is to install more fan-coils. This may become cost prohibitive, forcing the designer to look at other options.

Using Fan-coils for Cooling

Larger commercial buildings frequently use hydronic distribution systems for supplying both heated and chilled water to fan-coils or air-handler units. This same approach can be used in residential or small commercial systems if a water chilling device such as a hydronic heat pump is used as the heating/cooling source.

Fan-coils can be used as heat emitters for chilled water cooling if and only if they are equipped with condensate drip pans. These pans are located beneath the coil. They act as a catch basin for the water droplets that continually form on the coil during cooling operation. The drip pans are usually connected to a drain pipe or floor

Figure 8–30 Example of a small fan-coil equipped with a drip pan for chilled water cooling applications. Courtesy of Burnham Corp.

drain so the condensate can be disposed of at the same rate it is formed. Many of the hydronic fan-coils described in this chapter are designed only for heating. They do not have condensate pans, and should not be used for chilled water cooling applications. The condensate formed would simply run out the bottom of the unit causing damage to the structure. An example of a small wall-mounted fan-coil equipped with a drip pan, and thus suitable for chilled water cooling, is shown in Figure 8–30.

8.8   PANEL RADIATORS

A relatively recent arrival to the U.S. market, **panel radiators** have been used for many years in Europe. They are available in hundreds of sizes, shapes, and capacities, to fit different job requirements. They can be used throughout a building or in combination with other types of heat emitters.

Most panel radiators are made of steel. Some are built of preformed steel sheets welded together at their perimeter. Others are constructed of tubular steel components. At least one model available in the U.S. is constructed of aluminum extrusions. Since both steel and aluminum are subject to oxidation, radiators constructed of either

material are limited to use in closed-loop hydronic systems. Many panel radiators feature a high quality enamel finish, in several different colors to coordinate with interior design.

Some panel radiators are designed to release most of their heat in the form of thermal radiation. These panels have a flat front and usually only project about 2 inches out from the wall surface they are mounted to. They are truly radiators, tending to warm the objects in a room rather than the air itself. This often produces a noticeable increase in comfort. Furthermore, because the air is not heated to the temperature it would be if finned-tube baseboard convectors were used, there is less temperature stratification within the room.

Figure 8–31 shows an example of a **vertical panel radiator**. These panels are designed to be used in tall but narrow wall spaces. They are especially useful in kitchens, bathrooms, or offices where horizontal wall space is often limited. Lengths in excess of 20 feet are available.

Such radiators are normally mounted with their bottom 6 to 8 inches above the floor. Since the panels are available in many standard heights, finding a suitable height is seldom a problem. Most panel radiators hang from mounting brackets fastened to the wall. The brackets, shown in Figure 8–32, should be fastened directly to wood framing members or into solid masonry. In new

Figure 8–31 Example of vertical panel radiator. Courtesy of Runtal USA.

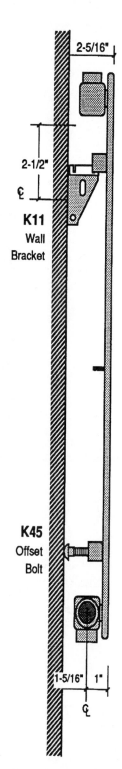

Figure 8–32 Side view of vertical panel radiator showing mounting brackets. Courtesy of Runtal USA.

construction with studded walls, wooden backer blocks should be positioned at the proper height and location to accept the fasteners for the brackets. It is important that the heating system designer specifies the position of these brackets so they are installed before the wall is closed up.

Piping connections can be made by running ½-inch or ¾-inch risers up through the floor, and connecting to threaded tappings at the bottom of the radiator. This is one location where brass pipe nipples rather than standard copper tubing provide better protection against physical damage of the exposed pipe. They also eliminate any visible soldered joints above the floor. Holes for the risers should be accurately located and neatly drilled about ¼ inch larger than the outside diameter of the riser pipe. This prevents noise from thermal expansion.

To provide for individual control of the panel radiator, a manual or thermostatic radiator valve is often mounted in the *upgoing* riser directly to the inlet connection of the radiator. To facilitate removal of the radiator for wall painting, it is common to mount a shut-off valve in the downgoing riser. A lockshield valve with integral union,

such as discussed in Chapter 5, is a good choice for this location. The use of these valves also shortens the length of piping exposed below the radiator. The combination of a radiator valve with integral union in one riser, and a drainable lockshield valve with integral union in the other, allows the panel to be isolated from the system, neatly

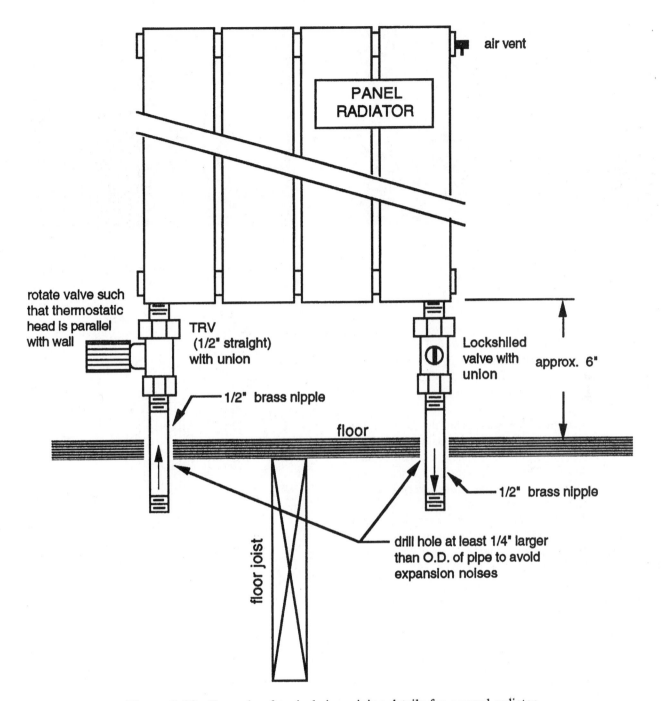

air vent

PANEL
RADIATOR

rotate valve such
that thermostatic
head is parallel
with wall

TRV
(1/2" straight)
with union

Lockshiled
valve with
union

approx. 6"

1/2" brass nipple

floor

1/2" brass nipple

floor joist

drill hole at least 1/4" larger
than O.D. of pipe to avoid
expansion noises

Figure 8–33   Example of typical riser piping details for a panel radiator

drained through a hose, and easily disconnected from the piping. It then can be lifted off its brackets for full access to the wall behind. A representation of these details is shown in Figure 8–33.

Panel radiators are also available as horizontal panels. This design is intended for wider but shorter wall spaces such as beneath windows. Again, several widths and lengths are available to match job requirements. Figure 8–34 shows two types of **horizontal panel radiators**. The mounting and riser piping details are similar to those shown in Figure 8–33.

Most manufacturers also offer panels with steel fins welded to the back. These increase convective heat output, creating a stronger current of rising warm air above the panel. High output panel radiators consist of several finned panels stacked together like a sandwich. The outer panel surface provides radiant heat while the air rising through the fins also provides a strong convective output. Such radiators will typically project 4 inches to 8 inches out from the wall depending on the number of stacked panels in the assembly. The more fins and panels in the assembly, the greater its heat output.

(a)

(b)

Figure 8–34 Examples of horizontal panel radiators (a) four-tube high panel, (b) six-tube high panel. Courtesy of Runtal USA.

Narrower units can be mounted directly to the wall using brackets supplied by the manufacturer. The wider and heavier models are best supported by floor pedestals, again supplied by the manufacturer. Examples of finned panel radiators and their associated mounting brackets are shown in Figure 8–35.

Another variation of the basic panel design is known as a towel warmer. These panels are designed for bathrooms or kitchens and provide a space in which to hang one or more towels. The simple appeal of a warm, dry towel after a shower or bath has proven very effective in marketing these panels. One manufacturer even coordinates paint colors with a major manufacturer of plumbing fixtures. When budgets allow, towel warmers are available with chrome, brass, and even gold plating. An example of a towel warmer is shown in Figure 8–36.

Some panel radiators are available with thermostatic valves already built in. A special chamber in the radiator is designed so that water can bypass the radiator when heat is not needed, or flow through the radiator when the thermostatic element opens up. This integral bypass design allows multiple radiators to be connected in a simple series arrangement. This design gives the same functionality as if the panel were connected using diverter tees, and a separate external radiator valve as shown in Figure 5–52. An example of a panel radiator with integral control valve is shown in Figure 8–37.

Thermal Performance of Panel Radiators

Thermal ratings for panel radiators are usually given in the form of Btu/hr *per foot of panel length* for several panel widths and entering water temperatures as shown in Figure 8–38a. Correction factors for other entering water temperatures and room air temperatures usually accompany these ratings, as shown in Figure 8–38b.

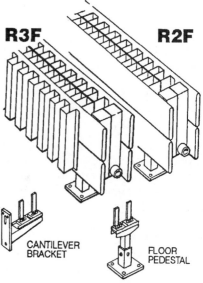

Figure 8–35 Examples of finned panel radiators and mounting brackets. Courtesy of Runtal USA.

Figure 8–36   Example of a towel warmer radiator.
Courtesy of Myson, Inc.

When a system requires panel radiators to be piped in series, it is important to account for the drop in fluid temperature as it moves from one panel to the next. A procedure similar to method #2 for sizing baseboard should be used. The size of each panel should be based on the water temperature at its location in the series circuit.

In some situations it is undesirable to change the height or width of the panels, such as when panel radiators will be placed under several identical windows in a room. In such cases a single panel size can be selected based on the average temperature among all the panels. The average water temperature can be estimated using the following equation:

(Equation 8. 9)

$$T_{average} = T_{in} - \frac{Q_{total}}{1,000(f)}$$

where:

$T_{average}$ = average temperature at which to select panel size (°F)

T_{in} = temperature of water entering the first panel (°F)

(a)

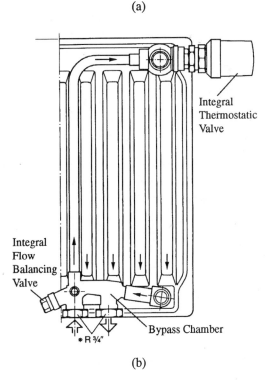

(b)

Figure 8–37   (a) Top view of a panel radiator with an integral thermostatic valve. Courtesy of Wisconsin Intertrade, Inc. (b) Cross section of radiator with integral bypass chamber and thermostatic valve. Courtesy of Buderus Hydronic Systems, Inc.

Q_{total} = total required heat output of all identical panels (Btu/hr)

f = water flow rate through the series string of panels (gpm)

Note Equation 8.9 is almost identical to Equation 8.5 used for finding the average temperature in a baseboard.

BTUH/ft Ratings @ 65°F EAT						
	MODEL	HEIGHT inches	DEPTH inches	215°F	180°F	140°F
PERIMETER STYLE	R-1	2.8	1.6	230	160	90
	R-2	5.7	1.6	420	300	170
	R-3	8.6	1.6	620	440	250
	R-4	11.5	1.6	820	580	330
	R-5	14.4	1.6	1020	720	410
	R-6	17.3	1.6	1220	860	500
WALL PANEL	R-7	20.2	1.6	1430	1010	580
	R-8	23.1	1.6	1640	1160	660
	R-9	26.0	1.6	1850	1300	750
	R-10	29.0	1.6	2060	1450	830

(a)

EAT

		45°F	50°F	55°F	60°F	65°F	70°F	75°F	80°F	85°F	90°F	95°F
	240°F	1.365	1.350	1.304	1.266	1.220	1.171	1.124	1.086	1.039	1	0.953
	235°F	1.343	1.305	1.267	1.219	1.171	1.124	1.086	1.038	1	0.952	0.910
	230°F	1.305	1.267	1.219	1.171	1.124	1.086	1.038	1	0.952	0.910	0.868
	225°F	1.267	1.219	1.171	1.124	1.086	1.038	1	0.952	0.910	0.868	0.826
	220°F	1.219	1.171	1.124	1.086	1.038	1	0.952	0.910	0.868	0.826	0.785
	215°F	1.171	1.124	1.086	1.038	1	0.952	0.910	0.868	0.826	0.785	0.744
	210°F	1.124	1.086	1.038	1	0.952	0.910	0.868	0.826	0.785	0.744	0.704
	205°F	1.086	1.038	1	0.952	0.910	0.868	0.826	0.785	0.744	0.704	0.664
	200°F	1.038	1	0.952	0.910	0.868	0.826	0.785	0.744	0.704	0.664	0.625
	195°F	1	0.952	0.910	0.868	0.826	0.785	0.744	0.704	0.664	0.625	0.587
	190°F	0.952	0.910	0.868	0.826	0.785	0.744	0.704	0.664	0.625	0.587	0.549
	185°F	0.910	0.868	0.826	0.785	0.744	0.704	0.664	0.625	0.587	0.549	0.511
	180°F	0.868	0.826	0.785	0.744	0.704	0.664	0.625	0.587	0.549	0.511	0.474
	175°F	0.826	0.785	0.744	0.704	0.664	0.625	0.587	0.549	0.511	0.474	0.438
AWT	170°F	0.785	0.744	0.704	0.664	0.625	0.587	0.549	0.511	0.474	0.438	0.403
	165°F	0.744	0.704	0.664	0.625	0.587	0.549	0.511	0.474	0.438	0.403	0.369
	160°F	0.704	0.664	0.625	0.587	0.549	0.511	0.474	0.438	0.403	0.369	0.334
	155°F	0.664	0.625	0.587	0.549	0.511	0.474	0.438	0.403	0.369	0.334	0.301
	150°F	0.625	0.587	0.549	0.511	0.474	0.438	0.403	0.369	0.334	0.301	0.269
	145°F	0.587	0.549	0.511	0.474	0.438	0.403	0.369	0.334	0.301	0.269	0.237
	140°F	0.549	0.511	0.474	0.438	0.403	0.369	0.334	0.301	0.269	0.237	0.207
	135°F	0.511	0.474	0.438	0.403	0.369	0.334	0.301	0.269	0.237	0.207	0.177
	130°F	0.474	0.438	0.403	0.369	0.334	0.301	0.269	0.237	0.207	0.177	0.149
	125°F	0.438	0.403	0.369	0.334	0.301	0.269	0.237	0.207	0.177	0.149	0.122
	120°F	0.403	0.369	0.334	0.301	0.269	0.237	0.207	0.177	0.149	0.122	0.096
	115°F	0.369	0.334	0.301	0.269	0.237	0.207	0.177	0.149	0.122	0.096	0.071
	110°F	0.334	0.301	0.269	0.237	0.207	0.177	0.149	0.122	0.096	0.071	0.050
	105°F	0.301	0.269	0.237	0.207	0.177	0.149	0.122	0.096	0.071	0.050	0.030
	100°F	0.269	0.237	0.207	0.177	0.149	0.122	0.096	0.071	0.050	0.030	0.011

EXAMPLE: To find the BTUH/ft Rating for an R-6 Panel at 155°F AWT and 65°F EAT,
Multiply the Correction Factor (0.511) by the BTUH/ft Rating at 215°F (1219),
e.g. (0.511) X (1219) = 623 BTUH/ft

(b)

Figure 8–38   (a) Thermal performance ratings given as heat output per foot of panel length for several panel widths (R_1 = 1 tube wide, R_2 = 2 tubes wide, etc.). (b) Correction factors to estimate heat output at other operating conditions. EAT = entering air temp, AWT = average water temp. Courtesy of Runtal USA.

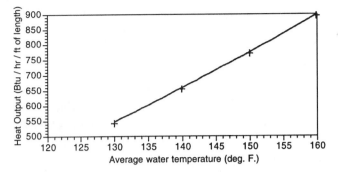

Figure 8–39   Heat output data for the panel radiator used in Example 8.6.

Example 8.6: A group of four panel radiators will be connected in a series. Each of the panels will be mounted beneath an identical window within a single room, and should be the same size to maintain consistent aesthetics. A nominal height of 24 ins has been selected. Determine the width of each panel knowing the total heat output of the four panels must be 9,000 Btu/hr and the water enters the first panel at 2 gpm and 150 °F. A portion of the manufacturer's data for a 24-in high panel operating in 65 °F room air has been plotted on the graph in Figure 8–39.

Solution: The average water temperature for the four panels can be found using Equation 8.9.

$$T_{average} = 150 - \frac{9,000}{1,000(1.5)} = 144 \; °F$$

From the graph of Figure 8–39, the heat output at 144 °F is approximately 710 Btu/hr/ft. The total linear footage can be determined as:

$$\frac{9,000 \; \text{Btu/hr}}{710 \, \text{Btu/hr/ft}} = 12.67 \; \text{ft}$$

The width of each panel would be one-fourth of this total or 38 inches. The manufacturer may not have a panel exactly 38 inches wide. In this case select the next larger width.

Discussion: In effect, this example sizes a hypothetical single large panel to the total load of the room and divides it up in four equal pieces. When sizing panel radiators it is common to calculate a panel size or length that may not be exactly the same as that produced by the manufacturer. In such cases select the next larger panel size.

8.9   RADIANT BASEBOARD

Another relatively new type of hydronic heat emitter on the U.S. market is referred to as **radiant baseboard**. The active portion of the baseboard consists of an aluminum extrusion that forms a very slim profile about

1 inch wide and 5 inches high. The extrusion has either two integral aluminum tubes, or two channels that tightly hold lengths of PEX tubing. In some cases copper tubing is inserted into the aluminum channels and mechanically expanded to fit tightly against the aluminum. A cut-away view of a radiant baseboard is shown in Figure 8–40a. A complete panel with compression-type fittings is shown in Figure 8-40b.

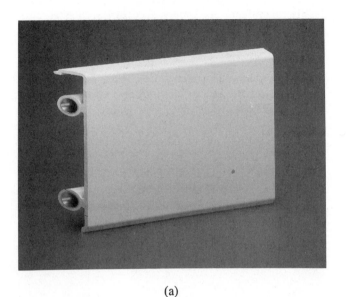

(a)

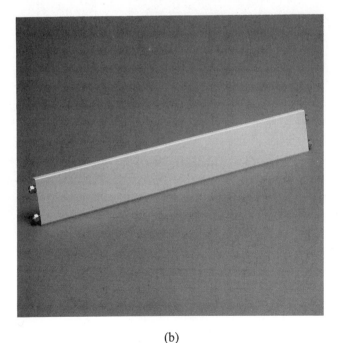

(b)

Figure 8–40   (a) Cut-away view of a radiant baseboard consisting of an aluminum extrusion with copper-lined fluid passageways. (b) A complete panel with fittings. Courtesy of Radiant Technology, Inc.

Unlike finned-tube baseboard, radiant baseboards have no fins. Their output is approximately 85% thermal radiation and 15% convection. The radiant output warms the floors and objects near the floor. This enhances thermal comfort, while also reducing temperature stratification in a typical room to about 2 °F from floor to ceiling.

Radiant baseboard is sold in straight lengths of 1 to 10 feet in increments of 6 inches. To join these lengths together, as well as to accommodate corners and other details, manufacturers supply accessories such as compression fittings, mounting brackets, corner assemblies, and filler panels for covering the joints.

Radiant baseboards are designed to replace the conventional wooden baseboards used at the base of a wall. *They are typically run all the way around a room.* Their narrow and clean lines, and high quality enamel finish, blend well with most interiors. They are less obtrusive than conventional finned-tube baseboard, and can also withstand more physical punishment without bending or denting. The baseboard panels are held to the wall by metal clips that snap to the extrusion. The clips are fastened to the wall first, and the baseboard is then pushed into place.

Connections between baseboard segments are made with brass compression fittings tightened by wrenches. No soldering is required. The joint is then covered with a filler panel that matches the active baseboard panel. At corners, a short length of PEX tubing with a tight 90 degree turn is used to connect the panels on each wall. The tubing for corners can be purchased preformed, or formed by heating with a hot air gun. Some of these joint details are shown in Figure 8–41.

The double tube design allows both supply and return connections to be made at the same end of the unit. The other end of the baseboard can be at any location in the room without regard for getting the pipe down through the floor.

A common piping design used with radiant baseboard is to supply and return each room from a central manifold station as shown in Figure 8–42. PEX tubing is often used as supply and return piping between the manifolds and radiant baseboards because it is easy to route through walls, floor joists, etc. This design allows independent control of the heat output within each room, provided suitable control hardware is used. It also allows the same water temperature to be supplied to each baseboard,

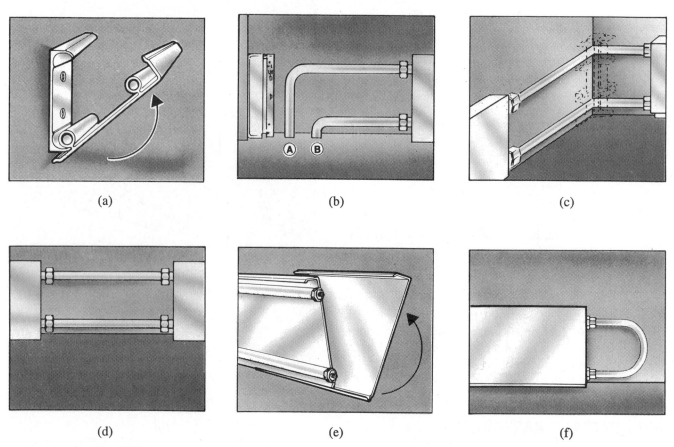

(a)   (b)   (c)

(d)   (e)   (f)

Figure 8–41 Installation details for radiant baseboards: (a) Panel section snaps into mounting bracket, (b) PEX tubing penetrations up through floor, (c) Preformed PEX tubing at corners, (d) PEX tubing between straight panel segments,(e) Cover plate snaps over joints, (f) 180 degree PEX tubing bend at end of run. Courtesy of Radiant Technology, Inc.

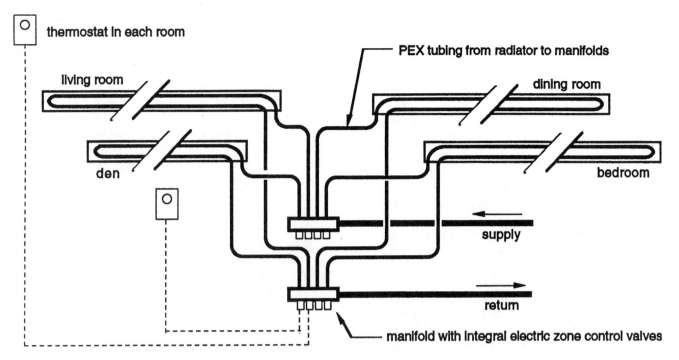

Figure 8–42 Example of a radiant baseboard system where each room is piped to a manifold system allowing for independent zone control

making for simpler sizing and lower temperature operation of the system. This type of piping design can also be used for most other types of heat emitters. It will be discussed in detail in Chapter 11.

Thermal Ratings for Radiant Baseboard

Because of its smaller size and surface area, a given length of radiant baseboard usually has a lower heat output than an equivalent length of finned-tube baseboard.

A comparison between the heat output of a typical residential finned-tube baseboard and a 5 in by 1 in radiant baseboard is shown in Figure 8–43.

The lower heat output of the radiant baseboard requires a greater length of baseboard. A rule of thumb would be to allow for twice the linear footage of radiant baseboard compared to a typical residential finned-tube baseboard. This length requirement may be a problem depending on wall space. Remember the radiant baseboard is intended to go all the way around the perimeter of a room.

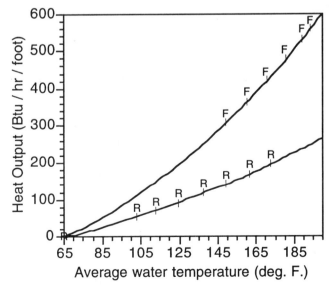

Figure 8–43 Comparison of heat output of a residential finned-tube baseboard and radiant baseboard. The finned-tube baseboard rating does not include the 15% heating effect factor.

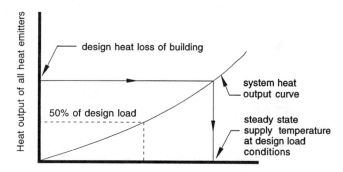

Figure 8–44 Example of a system heat output curve

8.10 SYSTEM HEAT OUTPUT CURVE

The heat output versus water temperature ratings of the heat emitters used in a hydronic system will determine the temperature the system operates at under steady state conditions. The higher the water temperature supplied to the distribution system, the greater the heat output will be. This relationship can be represented by a **system heat output curve** such as shown in Figure 8–44.

To find the steady state operating temperature of the system, find the design heating load on the vertical axis. Project a horizontal line to the left until it intersects the system heat output curve, then project down to the horizontal axis to read the supply water temperature.

If the upper limit control of the heat source is set below this temperature, the system will not be able to produce design heat output. If it is set above this temperature, the heat source would theoretically not be able to reach its upper limit setting under design load conditions, assuming its heat output is equal to the design heating load of the building.

This graph also shows when the building's heating load is less than its design load (which it is for most of the heating season), the water temperature supplied to the heating system can be reduced. Methods for doing so will be described in Chapter 9.

It should also be noted that the concept of a system heat output curve is theoretically only valid for systems that have constant circulation and control heat output by varying water temperature. Systems that turn the circulator on at each call for heat seldom operate under steady state conditions. The water temperature in such systems is largely influenced by the need to warm the thermal mass of the system components during each on cycle, and cannot be accurately predicted based on a system heat output curve.

If the heat output versus water temperature characteristics of a heat emitter can be expressed as an equation, it can be plotted as a continuous curve as shown in Figure 8–45a.

Such an equation could have the following form:

(Equation 8.10)

$$Q = a(T_{w_{ave}} - T_{room\ air})^b$$

where:

Q = heat output rate of the heat emitter (Btu/hr)
$T_{w\ ave}$ = average water temperature in the heat emitter (°F)
$T_{room\ air}$ = air temperature adjacent to the heat emitter (°F)
a, b = constants to be determined by curve fitting methods

If the output is only known at several specific temperatures, it can be plotted as a group of connected line segments as shown in Figure 8–45b.

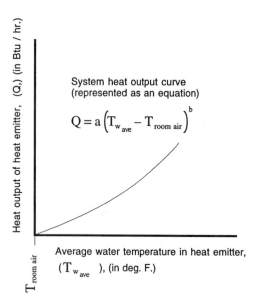

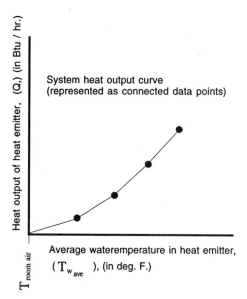

Figure 8–45 Examples of system heat output curves. (a) Heat output curve plotted from Equation 8.10, (b) heat output curve represented as connected data points.

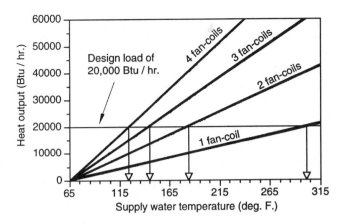

Figure 8-46 System heat output curves for one, two, three, and four parallel connected fan-coil units serving a total heating load of 20,000 Btu/hr. Downward arrows from intersection points indicate system operating temperatures.

Most hydronic systems consist of several heat emitters that share the total load of the building. In some cases two or three different types of heat emitters are used in the same system. *To construct the overall system heat output curve, the individual heat output curves of each heat emitter are graphically added on a common set of axes.*

Figure 8-46 shows the heat output curve for a single fan-coil, as well as the curves assuming two, three, and four of these fan-coils were used (in parallel) in the system. Notice when two fan-coils are used, the heat output value of a single fan-coil is simply doubled at each temperature. When three are used, the heat output value is tripled at each temperature, and so on. The intersections of the resulting system heat output curves, with the design heating line at 20,000 Btu/hr, indicates the operating temperature of the overall system based on how many fan-coils are used.

The supply temperatures at which the system would operate, under design heat output, would be as follows:

Heating load supplied	Number of fan-coils used	Required supply water temperature
20,000 Btu/hr	1	300
20,000 Btu/hr	2	183
20,000 Btu/hr	3	144
20,000 Btu/hr	4	123

This relationship between system water temperature and the number of fan-coils used is depicted in Figure 8-47. Obviously, the 300 °F supply temperature required if a single fan-coil were to be used is out of the question.

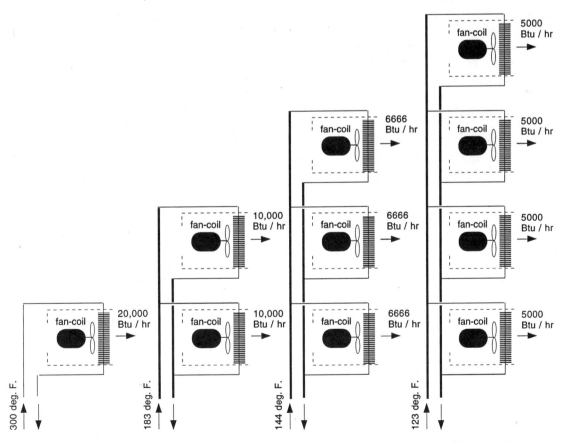

Figure 8-47 Concept of decreasing supply water temperature as an increased number of heat emitters are used to supply the same heating load

The 183 °F temperature would be more typical of a higher temperature residential system. Using three fan-coils would further lower the required supply water temperature to 144 °F. By the time the fourth fan-coil was added to the system, the water temperature is so low that condensation would likely form if a gas-fired boiler was used as the system's heat source. The use of four fan-coils would, however, be beneficial in combination with a low temperature heat source such as a hydronic heat pump or active solar collector array. The lower operating temperature would improve the efficiency of such heat sources.

8.11 HEAD LOSS OF HEAT EMITTERS

When analyzing piping systems containing heat emitters, it is important to know the head loss they create. This will vary with each type and size of unit. The best, and probably only, source of head loss data will be that provided by the manufacturer. The following guidelines describe typical head loss characteristics of each of the heat emitters discussed:

Head Loss of Finned-tube Baseboard

Since the element of a finned-tube baseboard is simply a length of copper tubing with fins attached, the pressure drop will be the same as that of a standard copper tube of the same nominal size and length. This can easily be determined using the methods from Chapter 6 or the PIPEPATH program in the Hydronics Design Toolkit.

Head Loss of Fan-Coils

Most manufacturers of fan-coils list the head loss of their products at two or three specific flow rates. If at least three points are listed, the data can be plotted and

a smooth curve drawn for use in estimating the head loss at other flow rates. An example is shown in Figure 8–48. Although usually not stated in manufacturer's literature, one data point will always be 0 head loss at a flow rate of 0 gpm. This is of course true for any piping device.

If the head loss is only given for one flow rate, it can be estimated at other flow rates using the following equation:

(Equation 8.11)

$$H_2 \cong H_1 \left(\frac{f_2}{f_1} \right)^2$$

where:

H_2 = head loss to be estimated (ft of head)
H_1 = known head loss (ft of head)
f_1 = known flow rate (gpm)
f_2 = flow rate at which head loss is to be estimated (gpm)

Example 8.7: A fan-coil has a head loss of 2 ft at a flow rate of 1.5 gpm. Estimate its head loss at a flow rate of 4 gpm.

Solution: Substituting the data in Equation 8.11:

$$\text{Head loss at 4 gpm} \cong (2) \left(\frac{4}{1.5} \right)^2 = 14.2 \text{ ft of head}$$

Notice the large increase in head loss as the flow rate is increased.

Another option when at least three head loss versus flow rate values are known is the use of the HYRES **program** in the Hydronics Design Toolkit. This program can determine the hydraulic resistance of the fan-coil that can then be used to describe its hydraulic characteristics

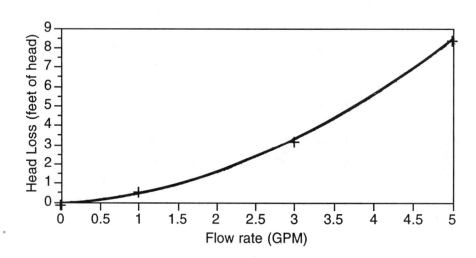

Figure 8–48 Graph of manufacturer's data for head loss versus flow rate of a small fan-coil

```
- HYRES -      F1-HELP                        F10-EXIT TO MAIN MENU
  -INPUTS-
  ENTER Maximum flow rate for the device,(2 to 100)......... 10    gpm
  ENTER Average fluid temperature,(80-250)................. 180   deg.F
  SELECT System fluid,(press F3 ).......................... water
  QUESTION Do you know the Cv value of the device?......... NO

                                  Flow rate        Pressure drop
                             (.01 to 100 gpm)  (.01 to 100 ft. head)
  ENTER First data point............... 2                2
  ENTER Second data point.............. 4                8
  ENTER Third data point............... 8                32


  -RESULTS-
  The fitting value of "a" =                                0.5000
  The fitting value of "b" =                                2.0000
  The Hydraulic resistance of the specified device  =       0.6407


    MESSAGE
  Use the UP, DOWN, LEFT, RIGHT ARROW keys to move the highlighting bar
  over an input, then type in a value. F10 to exit to MAIN MENU
```

Figure 8–49 Screen shot of the HYRES program that determines the hydraulic resistance of a component based on head loss versus flow rate data

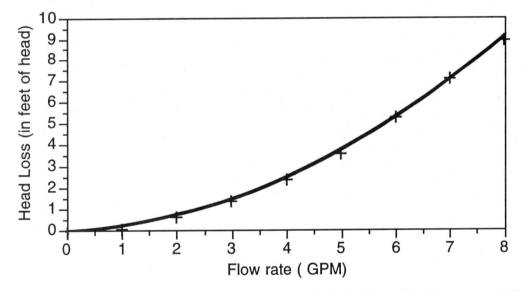

Figure 8–50 Graph of head loss versus flow rate for a 2 ft wide by 7 ft tall panel radiator prepared from manufacturer's data

in the PIPEPATH program. A screen shot of the HYRES program is shown in Figure 8–49.

Head Loss of Panel Radiators

Each size, shape, and make of panel radiator will have a different head loss characteristic. The only source of information will be the manufacturer's literature. A graph of head loss versus flow rate for a panel radiator is shown

in Figure 8–50. A smooth curve drawn through the data can be used to estimate head loss at other flow rates. The HYRES program can also be used to establish the hydraulic resistance of a panel radiator.

Head Loss of Radiant Baseboard

The head loss of radiant baseboard can be estimated by considering the baseboard to be a tube twice as long as the baseboard, and the same nominal size as the baseboard tube. An allowance for two elbows and several couplings can also be included with the tube length to account for the return bend and joints. The PIPEPATH program in the Hydronics Design Toolkit can be used to combine the head loss of these components into a single hydraulic resistance to represent the overall baseboard.

8.12   HEAT LOSS FROM COPPER PIPING

Any pipe carrying fluid at a temperature above that of the surrounding air releases heat to the air. If the pipe runs through heated space, this heat offsets a portion of the building's heating load. If the pipe runs through unheated space, the heat is lost.

Several issues related to system design can be addressed if the designer has a method for evaluating piping heat loss. For example, the designer may want to know the heat loss and temperature drop in the piping between the outlet of the heat source, and the inlet of the first heat emitter. If the system requires long piping runs through relatively cool spaces, significant heat loss and temperature drop can occur between heat emitters. This will obviously lower the heat output capability of the heat emitters. The designer may also want to know how much heat is released by pipe routed through heated space since this heat partially heats the building.

The heat output of a pipe will be affected by the pipe material, pipe size, flow rate, fluid, and surrounding air temperature. If the piping is insulated, its heat loss will be considerably reduced. The complexity of evaluating heat loss under all combinations of these conditions is beyond what can be presented in this book. However, Equation 8.12 can be used to determine the outlet temperature of a *bare copper tube carrying water* at various temperatures:

(Equation 8.12)

$$T_{out} = T_{ait} + \left[(T_{in} - T_{air})^{-0.244} + \frac{\sigma L}{f} \right]^{-4.09}$$

where:

T_{out} = temperature of the water leaving the pipe (°F)
T_{in} = temperature of the water entering the pipe (°F)
T_{air} = temperature of the air surround the pipe (°F)
L = length of the pipe (ft)
f = flow rate of water in the pipe (gpm) Note:
  $f > 0$
σ = constant based on pipe size from the table in Figure 8–51

pipe size	σ value
1/2"	5.402×10^{-5}
3/4"	7.221×10^{-5}
1"	9.026×10^{-5}
1-1/4"	1.078×10^{-4}
1-1/2"	1.244×10^{-4}
2"	1.566×10^{-4}

Figure 8–51   Table of σ values for use with Equation 8.12

Once the outlet temperature of the pipe is known, the heat loss can be calculated from the familiar sensible heat rate equation for water:

(Equation 8.13)

$$Q = 500(f)(T_{in} - T_{out})$$

where:

Q = rate of heat loss from the pipe (Btu/hr)
f = flow rate of water through the pipe (gpm)
T_{in} = temperature of water entering the pipe (°F)
T_{out} = temperature of the water leaving the pipe (°F)

Example 8.8: Water enters a $^3/_4$-in copper pipe 75 ft long at 3.5 gpm and 160 °F. The pipe runs through an area with an air temperature of 50 °F. What will be the water temperature leaving the pipe? What will be the rate of heat transfer from this pipe under these conditions?

Solution: Look up the appropriate σ value. For $^3/_4$-in pipe, σ = 7.221×10^{-5}. Substitute this and the remaining data into Equation 8.12:

$$T_{out} = 50 + \left[(160 - 50)^{-0.244} + \frac{(7.221 \bullet 10^{-5})(75)}{3.5} \right]^{-4.09}$$

$$T_{out} = 50 + \left[(110)^{-0.244} + 1.5473 \bullet 10^{-3} \right]^{-4.09}$$

$$T_{out} = 50 + 0.31916^{-4.09}$$

$$T_{out} = 156.8 \text{ °F}$$

The rate of heat output can be found using Equation 8.13:

$$Q = 500(3.5)(160 - 156.8) = 5,600 \text{ Btu/hr}$$

Discussion: This rate of heat loss is significant. It would probably be sufficient to heat a modest room in a well-insulated house. If totally wasted in an unheated space, it represents a sizable amount of wasted fuel over a heating season. Certainly this pipe would be a good candidate for insulation.

```
- PIPELOSS - F1-HELP    F3-SELECT                F10-EXIT TO MAIN MENU
  -INPUTS-
   ENTER Lenght of pipe,(1 to 500)...................... 10      ft.
   ENTER Temperature of fluid entering pipe,(80 to 250).. 150    deg. F
   ENTER Flowrate in pipe,(.5 to pipe limit)............. 1      gpm
   ENTER Air Temperature adjacent to pipe, (-20 to 80)... 55     deg. F
   SELECT Pipe size,(press F3 to select)................. 1/2" copper
   SELECT Pipe insulation,(press F3 to select)........... none
   SELECT Fluid type,(press F3 to select)............... water

  -RESULTS-
   Pipe inlet temperature..............    150.0 deg. F
   Pipe outlet temperature.............    149.3 deg. F
   Heat loss from pipe.................      322 BTU/hr




  MESSAGE
  Use the UP, DOWN, LEFT, RIGHT ARROW keys to move the highlighting bar
  over an input. Type desired value. F10 to return to MAIN MENU. F3 to SELECT
```

Figure 8–52   Screen shot of the PIPELOSS program from the Hydronics Design Toolkit

Equation 8.12 and 8.13 can be applied to any uninsulated copper pipe carrying water. They can be used to evaluate the heat loss and temperature drop of the piping segments between each of several heat emitters. To be practical such calculations require the use of a computer or programmable calculator. The PIPELOSS **program** in the Hydronics Design Toolkit is specifically designed to evaluate these equations. It also allows a comparison of the heat loss with and without several types of pipe insulation, and for piping carrying several different system fluids. A screen shot from the PIPELOSS program is shown in Figure 8–52.

SUMMARY

This chapter has discussed several types of hydronic heat emitters suitable for a wide variety of applications in residential and light commercial systems. They have ranged from units having mostly convective heat output such as finned-tube baseboard and fan-coils, to those with mostly radiant output, such as panel radiators and radiant baseboard. No one type is ideal for all applications. Attempts to use only one type for all kinds of jobs will inevitably lead to compromises and possible dissatisfaction. The designer needs to be aware of the range of products available, and use them as resources to match the system to the specific needs of the building and the budget. Knowing the strengths and weaknesses of each

type of unit can help narrow the selection during preliminary design.

Always consider the temperature and flow rate requirements of heat emitters before committing to using them. As a rule, do not match heat emitters designed around relatively high water temperatures with low temperature heat sources. If this were done, the heat emitters will represent a bottleneck for heat flow from the system into the building. The converse to this rule is also true. Do not directly couple a low temperature, low thermal mass distribution system to heat sources that are designed to operate at higher temperatures. Such a system could have very short operating cycles, felt as blasts of heat by occupants. The result will be uncomfortable conditions and dissatisfied customers. The lower temperature heat emitters could also pull the system operating temperature below the dewpoint of convention gas and oil-fired boilers.

Chapter 11 will discuss several different ways of combining the heat emitters described in this chapter into an overall distribution system.

KEY TERMS

Air handler
Convectors
Double interpolation
Element
Fan-coil convector

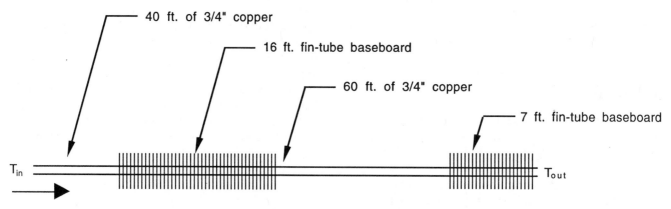

Finned-tube baseboard convector
Heat emitters
Heating effect factor
Heat output rating
Horizontal panel radiator
HYRES program
Interpolation
Kick-space heater
Multiple tube pass coil
Panel radiator
PIPELOSS program
Radiant baseboard
Radiators
SERIESBB program
System heat output curve
Towel warmer
Unit heater
Vertical panel radiator

Room	Design heating load
1. Master bedroom	4,500 Btu/hr
2. Bedroom	3,000 Btu/hr
3. Den	8,000 Btu/hr
4. Living room	11,000 Btu/hr
5. Bathroom	1,500 Btu/hr

The series baseboards will be in the same order as the listing of the rooms. Use sizing method #1, along with the following data, to select a length for each baseboard.

Baseboard thermal ratings given by lower curve in Figure 8–9:
Water flow rate = 1 gpm
Water temperature entering first baseboard = 180 °F
Air temperature at floor level = 65 °F

3. Repeat Exercise 2 using the more accurate sizing method #2. Compare the results with those obtained in Exercise 2.

4. The output of a small fan-coil is rated at 4,500 Btu/hr at an entering water temperature of 180 and flow rate of 1 gpm, in a room with 65 °F air. Estimate the heat output of the fan-coil at:
a. 160 °F entering water, 65 °F entering air, 1 gpm
b. 140 °F entering water, 55 °F entering air, 1 gpm

5. A fan-coil has the thermal performance ratings given in Figure 8–23. Interpolate between the data to estimate the heat output of the unit at 167 °F entering water temperature, 65 °F entering air temperature, and 3 gpm water flow rate.

6. Estimate the output of the fan-coil described in Exercise 5 using principle #1 discussed in Section 8.7. How does this compare to the answer obtained in Exercise 5?

CHAPTER 8 QUESTIONS AND EXERCISES

Note: Questions and exercises requiring the Hydronics Design Tookit are indicated with marginal symbol.

1. A finned-tube baseboard convector has the heat output ratings given in Figure 8–7. Estimate the heat output of a 1 ft segment of this baseboard operating at an average water temperature of 135 °F, and 3 gpm in a room with 60 °F air near floor level. Use Equation 8.1.

2. The following rooms are to be heated by a series string of finned-tube baseboards:

Figure 8–53 Piping and baseboard arrangement for Exercise 10

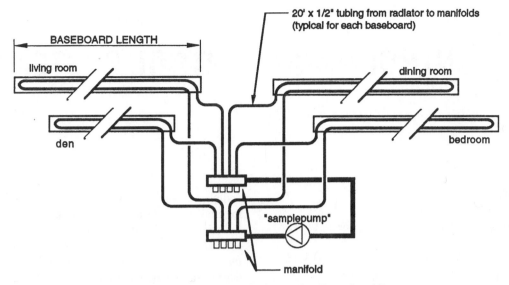

Figure 8–54   System piping for Exercise 15

7. Construct the heat output curve for the WH50 fan-coil described in Figure 8–23, assuming 65 °F room air temperature, and 1 gpm flow rate. Construct the system heat output curve assuming four of these fan-coils are used in parallel. Find the operating temperature of the four fan-coil system when supplying a total load of 18,000 Btu/hr.

8. Water enters a 1-in copper pipe at 9 gpm and 180 °F. The pipe is 100 ft long. Determine the outlet temperature and rate of heat loss under the following conditions:
   a. 70 °F surrounding air temperature
   b. 35 °F surrounding air temperature

9. Using the operating conditions from Exercise 8, part b, calculate the temperature of the water halfway through the pipe. Is the temperature drop half of the total temperature drop over the full 100 ft of pipe? Explain your findings.

10. Using the SERIESBB program and the PIPELOSS program in sequence, determine the rate of heat output from both piping segments and baseboards shown in Figure 8–53 for the following conditions:
   a. T_{in} = 180 °F
   b. T_{in} = 130 °F
   Hint: The outlet temperature of the first pipe becomes the inlet temperature to the first baseboard, and so on. Work the calculations for each component from left to right.

11. Use the SERIESBB program in the Hydronics Design Toolkit to find the required baseboard lengths for the room loads described in Exercise 2.

12. Repeat Exercise 11 assuming the order of the room loads is reversed. Are there any changes in baseboard length? If so, explain why.

13. Use the SERIESBB program and the room loads stated in Exercise 2 to find the required baseboard lengths under the following assumptions:

a. The circuit flow rate is 0.5 gpm
b. The circuit flow rate is 2.0 gpm
c. The circuit flow rate is 5.0 gpm
Explain why the baseboard lengths are changing.

14. A small hydronic fan-coil has the following data supplied by its manufacturer:

Flow rate (gpm)	Head loss (ft of head)
1	1.5
2	5.2
3	10.8

Use the HYRES program in the Hydronics Design Toolkit to find the hydraulic resistance for this fan-coil.

15. A radiant baseboard system consists of several baseboards connected to a common supply and return manifold. Each baseboard contains two fluid passageways as shown in Figure 8–54. Each passageway may be considered to have a head loss characteristic equivalent to ½-in copper tube. The lengths of each baseboard are given in the table below. Each base-board has the equivalent of 20 ft of ½-in tubing connecting it to the common manifold (10 ft supply + 10 ft return). Use the PIPEPATH program to get a hydraulic resistance for each baseboard circuit. Then use the PARALLEL program to combine these hydraulic resistances into a single equivalent resistance. Use the PUMP/SYS program to find the system flow rate assuming the circulator is described by the default samplepump file. Finally, go back to the PARALLEL program and find the flow rate through each baseboard.

Room	Baseboard length
Living room	95 ft
Den	25 ft
Dining room	55 ft
Bedroom	30 ft

9 CONTROLS AND CONTROL SYSTEMS

OBJECTIVES

After studying this chapter you should be able to:

- Compare different control strategies for heating systems
- Describe several basic control components
- Describe switches in terms of poles and throws
- Use ladder diagrams to lay out control systems
- Explain the internal design and operation of relays
- Compare electro-mechanical and electronic controls
- Explain the operation of reset controls
- Design custom control systems for hydronic heating applications
- Connect both three and four wire zone valves
- Diagnose common wiring errors in control circuits

9.1   INTRODUCTION

The control system is the brain of a hydronic heating system. It determines exactly when and for how long devices such as circulators, burners, mixing valves, etc., will operate. The comfort, efficiency, and longevity of the system are as dependent on the controls as they are on any other component or subsystem.

A well-designed control system can be a real work of art, providing sophisticated operating logic to optimize efficiency and comfort. It can also be designed for easy troubleshooting and maintenance. A carelessly designed control system can be a nightmare in terms of achieving stable and efficient system operation. Selecting the best heat source, circulators, valves, etc., will not compensate for an improperly designed or adjusted control system.

Controls for hydronic heating represent technology that spans several decades. They range from time-proven devices such as the bimetal room thermostat to microprocessor-based controls that have the intelligence to optimize system operation for minimum fuel use and maximum comfort. Both high-tech and low-tech controls have their place in modern hydronic systems. With experience, a resourceful system designer will learn to integrate them for maximum reliability and efficiency.

Many of the decisions made by the controls of heating systems are the result of the position of one or more switches at any given time. Some of these switches are manually set. Others are operated by temperature. A knowledge of what types of switch-based controls are available and how to organize them in a consistent manner, is very important for designing a control system, especially when customized system designs require a control system that is not simply an off-the-shelf item. This chapter will show how **ladder diagrams** can be used to neatly and consistently document operation of the control system.

Electronic and microprocessor-based control components have worked their way into nearly all areas of heating and cooling technology over the last decade. Hydronic system designers now have a broad range of these controls to work with. Many functions that previously required human intervention can now be performed by such controls. These include automatic adjustments of system water temperature based on weather conditions, automatic shutdown of system pumps during warm weather, and even periodic exercise of components such as mixing valves and circulators to prevent setup during non-operational periods. At the same time, the accuracy of electronic controls can save energy and improve comfort. This chapter will acquaint you with several electronic controls and show how to match them to the requirements of the job.

9.2   METHODS OF CONTROL

The operation of almost any control device used in hydronic heating falls into one of the following categories:

- On/off control
- Staged control
- Modulating control
- Outdoor reset control

These categories describe the type of output signal(s) from the control. Other system components respond to these signals in different ways. For example, a circulator may start, stop, or run at a certain speed, while a mixing valve actuator may rotate very slightly. The combined status of all controlled devices within a system place it into a predetermined **operating mode**. Depending on its

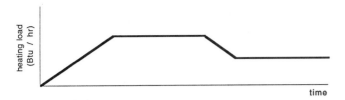

Figure 9-1   Hypothetical building heating load profile

complexity, a given system may have several different operating modes.

All four methods of control can be used to regulate the same process with varying degrees of accuracy. A good example is controlling the output of a heating system so it matches the current heating load of a building. This familiar requirement will be used to illustrate the first three types of control operation.

On/off Control

Control devices that operate by simply opening or closing electrical contacts are by far the most common type used in heating systems. Devices such as **room thermostats**, burner relays, and **setpoint controls** all function by either allowing or preventing an electrical signal to pass through a set of contacts. *When on/off control devices are used to regulate the heat output of a system, they can only do so by turning the heat source on and off. They cannot vary the rate of heat output of the device they control.* For example, turning a room thermostat to a high setting does not make the boiler burn fuel at a faster rate. It only keeps it on for a longer time while the room temperature increases toward the higher setpoint.

A hypothetical building load is shown in Figure 9-1. The load begins at zero, rises at a constant rate to its maximum design value, then decreases at a steady rate back to half its design value. The time over which this occurs is somewhat arbitrary, and can be assumed to be a few hours.

In Figure 9-2, this load profile is overlapped with a series of rectangles that represent the on-time of the heat source attempting to match this heat requirement using an on/off control device such as a room thermostat. For simplicity, assume that the heating capacity of the heat

source is exactly the same as the design heating load of the building.

Notice the height of all the rectangles is the same. This indicates the heat source puts out a constant rate of heat flow when it is on. Also notice the width of the rectangles increases as the load approaches its maximum value. This means the on-time of the heat source is increasing. While design load occurs, the rectangle remains uninterrupted. Since the boiler is assumed to be sized to exactly meet the design load of the building, it must operate continuously whenever design load conditions occur. As the load decreases the rectangles become progressively narrower. When a steady load of half the design load is established and maintained, the width of the rectangles remain constant.

The shaded area of each rectangle represents the *quantity of heat* delivery by the heat source during its on cycle. Ideally the area contained under the load profile, out to some point in time, should exactly equal the area in the shaded rectangles, from the beginning of the graph to the same point in time. This would indicate the total heat output from the heat source exactly matches the total heat lost from the building over the period of time.

The match between heat output and heating load at low and medium load conditions is certainly not ideal. For example, consider the period of time when the load is half the design load. Under these conditions the heat source outputs pulses of heat. The height of these pulses corresponds to the total heating capacity of the heat source, which during this period of time is twice the heating load of the building. To compensate for the high rate of heat output relative to the building load, the heat source must operate only about half the elapsed time.

In heating systems having low thermal mass, on/off cycling of the heat source is very noticeable and undesirable. A forced air system with a constant speed blower and fixed firing rate is a good example. Under partial load conditions the furnace still operates at its maximum rate of heat output. Since the air in the building has very little thermal mass for absorbing this heat, its temperature increases very quickly. The system will operate only until the room thermostat is satisfied and shut off. This could be as little as two or three minutes, depending on where the thermostat is located. The occupants feel this surge in air temperature and usually do not like it, although many will tolerate it. It tends to make the room stuffy or make people feel drowsy. When the furnace is shut off, drafts from windows and doors quickly reestablish themselves and decrease comfort.

Hydronic systems have an advantage in this respect. A typical boiler and piping system has a significantly greater thermal mass than a forced air system. This allows the distribution system to temporarily absorb some of the extra heat output and spread out its delivery to the building over a greater period of time. This effect is further enhanced if the circulator runs continuously. The greater the thermal

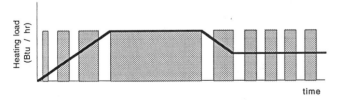

Figure 9-2   Rectangles representing on-time of heat source attempting to match the heating load profile of Figure 9-1

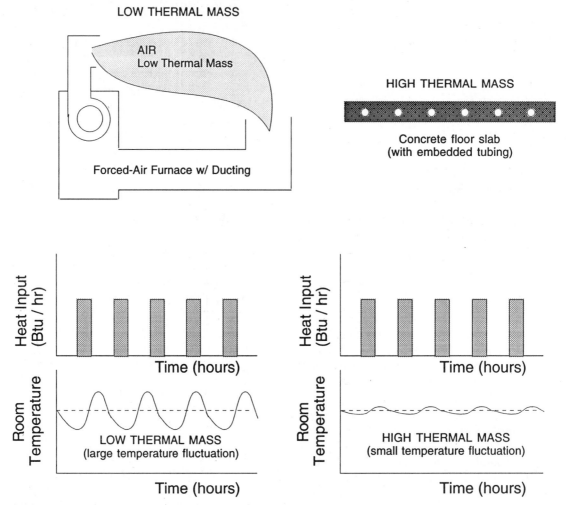

Figure 9–3 Room temperature fluctuations associated for heating system having low and high thermal mass

mass of the system, the more stable the room temperatures will remain during partial load conditions.

Radiant floor heating using a 4-in to 6-in thick concrete slab is perhaps the ultimate hydronic system as far as thermal mass is concerned. A 2,000 ft² floor, 4 ins thick contains 93,000 pounds of concrete. It could absorb about 20,000 Btu while only increasing its temperature 1 °F. By contrast, the air in the same building (assuming an 8 ft ceiling height) could absorb only about 290 Btu with a temperature rise of 1 °F. The thermal mass of the floor slab is almost 70 times greater than the air. This allows the slab to accept pulses of heat during partial load conditions while delivering a relatively steady heat output to the building. A controls engineer would say the thermal mass of the system "dampens" the temperature response of the system. A comparison of systems with high and low thermal mass is shown in Figure 9–3.

Staged Control

Another way of trying to match the output of the heat source with the building load is with multi-stage controls.

This approach increases the heat capacity of the heat source in fixed increments as the load increases. The number of increments, or **stages**, between zero and maximum heat output can vary from 2 to 12 or more. The higher number of stages is more common in larger commercial systems.

Stages come on and go off in sequence. The first stage represents the lowest steady heat output of the system. When this is insufficient to meet the load, the second stage is started to provide additional heat output. Likewise, any additional stages continue to come on until heat output matches or slightly exceeds the heating load. When all stages are on at the same time, the system is at maximum steady state heat output. Figure 9–4 illustrates the use of a two-stage control system applied to the load profile of Figure 9–1.

Notice during low load conditions the first stage of heat output operates intermittently. The width of the rectangles becomes wider as the load increases toward the half design load condition. As the load continues to increase above the half design load, the first stage output remains on continuously, while the second stage operates

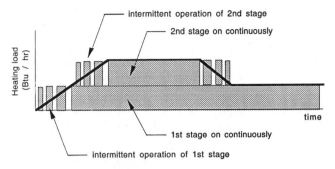

Figure 9-4 Use of a two-stage heating system to meet the heating load profile of Figure 9-1

intermittently to make up the difference between the load and the output of the first stage. Maximum heat output occurs when both stages remain on continuously. This diagram assumes the total heat output of the heat source(s) exactly matches the design load of the building. Any oversizing would necessitate some off-time for the second stage. As the load decreases, the second stage resumes intermittent operation. Since the load profile ends at half of design load, the first stage continues to stay on.

Again, the objective is that the area under the load profile equals the total area of the shaded rectangles up to any point in time. A properly designed staged system will always provide a closer match between heat output and the current heating load. The more stages the better the match. Multiple-stage heat sources also help reduce the temperature fluctuations associated with low thermal mass distribution systems. An example of a four-stage system applied to the same load profile is shown in Figure 9-5.

The practicality of using a multi-stage heat source depends on the total heating load of the building and the type of heat source used. Small- and medium-sized houses usually have loads low enough that only single-stage heat sources are available. This is especially true with presently available oil-fired boilers that often have minimum heat outputs of about 60,000 Btu/hr because of nozzle size limitations. Two-stage operation can be considered for gas-boilers or hydronic heat pumps. In such cases a modular boiler (or modular heat pump) system using two identical units could be used. Electric boilers come standard with staged heating elements. Larger residential and light commercial buildings are very

adaptable to multi-stage controls. In making the decision, the designer should weigh the extra cost associated with a multi-stage system against the potential fuel savings and more accurate system control.

Modulating Control

The best match between the output of a heating system, and the heating load of a building, is obtained when the heat output is *continuously variable over a range from zero to full output*. Controls that allow continuous variations in heat output are called **modulating controls**. Figure 9-6 shows the ideal case where the heat output of the system exactly matches the load profile.

There are several ways modulating controls can adjust the heat output of a system. Most are based on *controlling the temperature of the water supplied to the heat emitters*. As the supply temperature decreases, so does heat output. One method of doing this uses a motorized mixing valve to create a specific outlet temperature based on the proportions of supply and return water entering it. Another way is the use of a variable speed pump to inject hot water in a constantly circulating distribution circuit. Still another method, used with electric resistance boilers, is to control the electrical current through the element(s) using solid state power regulating devices. Some non-electric hydronic system components also operate as modulating devices. One example is the thermostatic radiator valve that moves between fully open and fully closed as room temperatures slowly increase.

Outdoor Reset Control

One of the best methods for matching the heat output of a hydronic system to the heating load of a building at any given time is to make small but frequent adjustments to the water temperature supplied to the distribution system. As the outdoor temperature decreases and building heating load increases, the supply water temperature needs to be raised. When outdoor temperature increases, the supply water temperature can be decreased. This method of control is called **outdoor reset control**.

When a distribution system is designed, its heat output at a certain water temperature is established. For example, a finned-tube baseboard system may be designed to

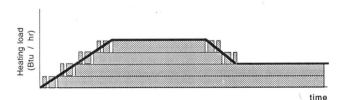

Figure 9-5 Use of a four-stage heating system to meet the the heating load profile of Figure 9-1

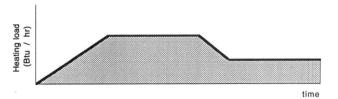

Figure 9-6 Modulating control achieves near ideal match between output of heat source and heating load of building

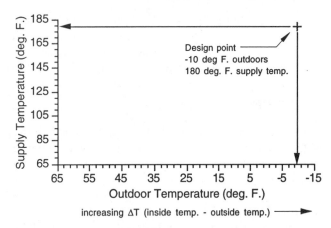

Figure 9–7 Correspondence between outdoor temperature and supply water temperature for a specific distribution system

supply the design heating load of a building at –10 °F outdoor temperature using supply water at 180 °F. This operating point can be plotted on a graph as shown in Figure 9–7.

The heat output of this distribution system will decrease if the supply water temperature is lowered. In Chapter 8 the relationship between heat output and supply water temperature was plotted as a system heat output curve. Recall that the relationship between heat output and the difference between supply temperature and room temperature was approximately proportional, especially when fan-coils were used as heat emitters. Also recall from Chapter 2 that the relationship between heating load and the difference between inside and outside temperature was proportional—the greater the difference, the greater the heating load. These two relationships imply a third, approximately proportional relationship between supply water temperature and outdoor temperature as shown in Figure 9–8.

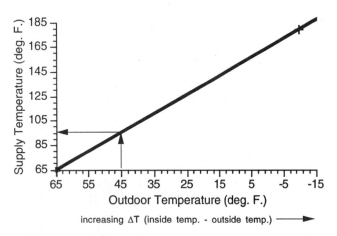

Figure 9–8 Approximately proportional relationship between supply water temperature and the difference between inside and outside air temperature

The sloping line in Figure 9–8 represents (theoretically) all combinations of outdoor temperature and supply water temperature where the heat output of the system would exactly match the heating load of the building. This is called a **reset line** or **heating curve** for the system. The supply water temperature that corresponds to any given outdoor temperature between 65 °F and –10 °F can be found by projecting a line up from the horizontal axis to the reset line, then horizontally across the vertical axis. For example, at an outdoor temperature of 45 °F, the necessary supply water temperature is about 96 °F. When the outside air temperature has increased to 65 °F the supply water temperature is also 65 °F, and there would be no heat output from the system, even with constant water circulation.

The slope of the reset line is called the **reset ratio**, and can be calculated as the change in supply water temperature divided by the change in outdoor temperature between any two points on the reset line.

(Equation 9.1)

$$\text{Reset ratio} = \frac{\Delta T_{\text{supply water}}}{\Delta T_{\text{outdoor}}}$$

In the case illustrated in Figure 9–8 the reset ratio would be:

$$\text{Reset ratio} = \frac{180 - 65}{65 - (-10)} = \frac{115}{75} = 1.53$$

Different types of heat emitters such as finned-tube baseboard, fan-coils, and radiant floor heating have significantly different supply water temperature requirements at the same outdoor design temperature and heating load. Typical values would be as follows:

Heat emitter type	Supply water temperature at design load
Fin-tube baseboard	180 °F
Fan-coil	130 °F
Radiant floor	105 °F

These systems would each have a different reset line as shown in Figure 9–9.

The reset ratios for each system could be calculated as follows:

Fin-tube baseboard:

$$\text{Reset ratio} = \frac{180 - 65}{65 - (-10)} = \frac{115}{75} = 1.53$$

Fan-coil:

$$\text{Reset ratio} = \frac{130 - 65}{65 - (-10)} = \frac{65}{75} = 0.87$$

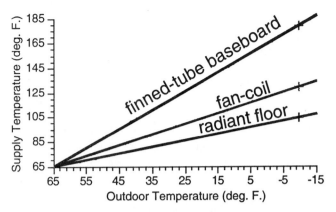

Figure 9–9 Typical reset lines for system using various heat emitters

Radiant floor:

$$\text{Reset ratio} = \frac{105-65}{65-(-10)} = \frac{40}{75} = 0.53$$

It should be emphasized these reset ratios are representative only. *Any given hydronic heating system would have its own unique reset ratio depending on the type(s) of heat emitters used and heat loss characteristics of the building.* In most cases this inherent reset ratio is found by making minor adjustments above or below a calculated starting value until the heating system tracks the building heating load while maintaining a consistent interior temperature.

Reset control can be accomplished in several different ways using both electro-mechanical and electronic controls. Later sections will detail methods of implementing reset control.

9.3   SWITCHES AND RELAYS

Many of the operating modes of heating and cooling control systems are set using specific arrangements of electrical contacts. These contacts must be either open or closed. An open contact prevents an electrical signal (e.g., a voltage) from passing a given point in the circuit. A closed contact allows the signal to pass through. Such contacts may be part of a manually operated switch or an electrically operated switch known as a **relay** or **contactor**. When the settings of the various contacts allow a complete circuit to be formed, a current will flow, and some predetermined control action will take place.

Switches

Switches and relays are classified according to the number of **poles** and **throws** they have. *The number of poles is the number of independent and simultaneous electrical paths through the switch.* Most of the switches

Figure 9–10 Example of a DPDT toggle switch

used in heating control system have one, two, or three poles. They are often designated as single pole (SP), double pole (DP), or triple pole (3P).

The number of throws a switch or relay has corresponds to the number of position settings where a current can pass through the switch. The majority of switches and relays used in heating control systems have either one or two throws, and are called single throw (ST), or double throw (DT), switches. Figure 9–10 illustrates a typical double pole double throw (DPDT) toggle switch, as might be used in a heating control system.

Notice the printing on the side of the switch indicates it is rated for a maximum of 10 amps of current at 250 volts (AC), and 15 amps of current at 125 VAC. *When selecting a switch for a given application be sure its current and voltage ratings equal or exceed the current and voltage it will operate with.*

Figure 9–11 shows the schematic symbols used to designate switches having different numbers of poles and throws. On the double and triple pole switches, a dashed line indicates a non-conducting mechanical coupling between the metal blades of the switch. This coupling ensures that all contacts open or close at the same instant.

Relays

Relays are simply electrically operated switches. They can be operated from a remote location by a low power electrical signal. They consist of two basic subassemblies: the coil and the contacts. Other parts include a spring, pendulum, and terminals. These components and basic operation of a relay are illustrated in Figure 9–12.

The spring holds the contacts in their "normal" position. When the proper voltage is applied to the coil, a magnetic force is created that pulls the common contact attached to the pendulum to the other position. In most

	Single Throw	Double Throw
Single Pole	SPST	SPDT
Double Pole	DPST	DPDT
Triple Pole	3PST	3PDT

Figure 9–11 Schematic symbols used to represent switches having different numbers of poles and throws

relays this action takes about 10 to 15 thousandths of a second. A light click can be heard as the contacts move to their other position. As soon as power is removed from the coil, the spring snaps the contacts back to their deenergized position.

Relay contacts are designated as **normally open** (N.O.), or **normally closed** (N.C.). In this context, the word normally means when the coil of the relay is *not energized* (e.g., no voltage is applied to the coil). A normally open contact will not allow an electrical signal to pass through while the coil of the relay is off. A normally closed contact will allow the signal to pass. Like switches, relays are classified according to the number of poles they have. Figure 9–13 shows the most common configurations used in heating control systems.

Unlike the terminals of switches, the contacts of a relay do not necessarily appear in the same vicinity on the schematic drawing of the control system. One or more of the contacts may appear in the **line voltage** section,

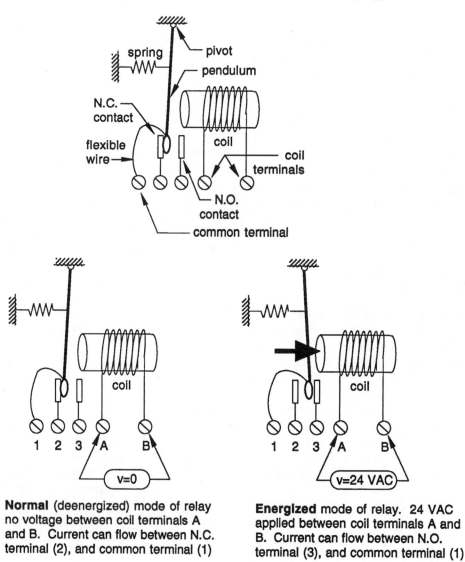

Normal (deenergized) mode of relay no voltage between coil terminals A and B. Current can flow between N.C. terminal (2), and common terminal (1)

Energized mode of relay. 24 VAC applied between coil terminals A and B. Current can flow between N.O. terminal (3), and common terminal (1)

Figure 9–12 Basic parts and operation of a relay

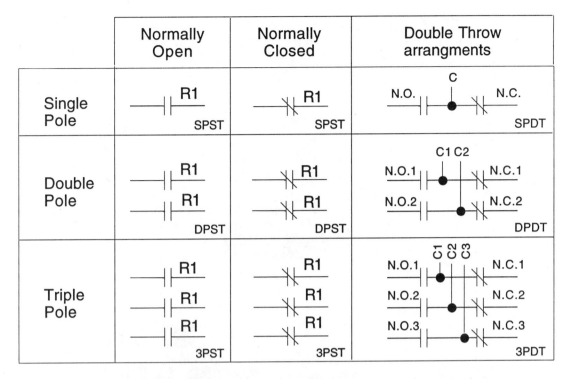

Figure 9–13   Common relay contact configurations used in heating control systems

while another appears in the **low voltage** section. This is done to simplify the drawing. Physically, all the contacts are located next to each other in the enclosure of the relay. Because the contacts of a single relay can appear in several locations on the same drawing, an alphanumerical designation such as R1 or RR should appear next to *every contact* of the relay, so that it is associated with the proper coil.

The coil of a relay is rated to operate at a specific voltage. In heating control systems the most common coil voltage is 24 VAC. Relays with coil operating voltages of 12, 120, and 240 VAC are also available when needed. The coil consists of several hundred windings of very fine wire around an iron core. When the rated coil voltage is applied, sufficient magnetic force is created to pull the pendulum to its inward position and hold it there against the force of the spring, changing the position of the contacts. There are two common symbols used to designate the coil of a relay on electrical schematics. These are shown in Figure 9–14. Notice an alphanumerical designation is always shown. This identifies the coil with its associated contacts.

The electrical power consumed by an energized relay coil is usually expressed in VAs (volt-amps). The coil of

a typical small relay capable of switching line voltage loads with contact currents up to about 12 amps will draw about 2 VA when energized. This represents an almost insignificant power demand. However, the total VAs of all relays, and other control voltage components that could operate simultaneously, should be added up to help determine the minimum **VA rating** of the transformer supplying the control voltage.

General Purpose Relays

Some control systems require relays to be installed as separate components. So-called general purpose relays are often used in these applications. They are built as plug-in modules with clear plastic enclosures. Their external terminals are designed to plug into a **relay socket**. These sockets have either screw terminal or quick-connect tabs for connecting to external wires. Each terminal on the socket is numbered the same as the connecting pin on the relay that connects to that terminal. *This numbering is very important because it allows the sockets to be wired into the control system without the relay being present.* Relay manufacturers provide wiring diagrams that indicate which terminal numbers correspond to the normally closed, normally open, common, and coil terminals. Figure 9–15 shows a typical general purpose 3PDT relay, a relay socket into which it mounts, and its wiring diagram.

Relay sockets can be fastened directly to a panel or chassis, or snapped into a **DIN rail**. The latter type of mounting allows sockets to be added or removed without fasteners. DIN rail mounting has become a standard

Figure 9–14   Two schematic symbols for the coil of a relay

(a)

(b)

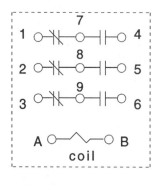

(c)

Figure 9–15   (a) Bottom side of a general purpose, triple pole double throw (3PDT) relay, (b) a suitable mounting socket (c) a wiring diagram for the relay.

Figure 9–16   Example of relays and a small modular control mounted on a DIN rail

modular mounting system in the controls industry. Many small modular controls other than relay sockets are designed for it. By mounting a generous length of DIN rail in a control cabinet, relays and other components can be easily added, removed, or moved to a different location. An example of a DIN rail with two relay sockets and a small modular reset control is shown in Figure 9–16.

Time Delay Relays

Quite often it is necessary to incorporate a delay between two control events. Examples include restarting the compressor of a heat pump following a shutdown, or waiting to see if first stage heat output will satisfy the load before turning on the second stage. These operations are accomplished using **time delay relays**.

Time delay relays incorporate adjustable timing circuits in their plastic housings. They operate in one of two modes: They can prevent the coil from attaining its required operating voltage until a certain period of time has elapsed, or they can sustain the coil voltage at operating levels for a designated time after the input voltage has been interrupted. The first type is called a delay on make relay. The second type is called a delay on break relay. Of the two, a delay on make relay has more application in heating control circuits.

Time delay relays are manufactured in several different forms. The type likely to be used in a custom control system is designed to plug into the same type of socket used for general purpose relays. It is distinguished by the knob at the top of the plastic case which is used to adjust

the time delay period. An example of a time delay relay is shown in Figure 9–17.

Like other relays, time delay relays can be obtained with several configurations of poles, throws, and operating voltages. Delay periods ranging from a fraction of a

Figure 9–17   Example of a plug-in time delay relay. Note the contact wiring diagram on the case. Courtesy of Magnecraft Electric Co.

second to several minutes are available. Because of their complexity, time delay relays cost more than general purpose relays. Their ability to protect compressors against unequalized pressure starts or prevent short cycle operation of second stage heating more than justifies their use. In the long run they may prevent mechanical breakdowns that could cost hundreds of dollars to repair.

9.4 LADDER DIAGRAMS OF CONTROL SYSTEMS

It is often necessary to combine several control components to build an overall control system. The way these components are connected to each other, and to devices such as pumps, oil burners, etc., will obviously determine how the system will operate. A ladder diagram provides a standardized method for developing and documenting the necessary control action. The finished diagram can then be used for installation and troubleshooting. Such a diagram should always be part of the documentation of a hydronic heating system.

Ladder diagrams have two basic sections, the line voltage section and the control voltage section. A **transformer** separates the two sections. Its **primary side** is connected to the line voltage section, its **secondary side** to the control voltage section. In North America most heating and cooling systems use transformers with a nominal secondary voltage of 24 VAC. An example of a basic ladder diagram is shown in Figure 9–18.

The vertical lines in this diagram represent the sides of the ladder. Any horizontal line connected between the two sides is called a rung of the ladder. A rung connected across the line voltage section will be exposed to full line voltage (120 VAC nominal). A rung connected across the control voltage section will be exposed to 24 VAC. The overall control circuit is constructed by adding as many rungs as necessary to create the desired operating modes required by the system.

Relays are often used to operate line voltage loads such as circulators, oil-burners, or blowers based on the action of low voltage components such as thermostats. Ladder diagrams are an ideal way to show how this is accomplished. When a circuit path is completed through a relay coil in the control voltage section of the ladder, one or more contacts of that relay are operated in the line voltage section, enabling one or more line voltage devices to operate. The way relay coils, switches, thermostats, etc., are physically wired together determines how the system will operate in any given control mode, and is often called **hard-wired control logic**.

Consider a situation where a line voltage circulator is to be operated by a low voltage switch. Since the pump needs line voltage to run, it is connected across the upper (line voltage) section of the ladder diagram as shown in Figure 9–19. To prevent it from running other than when needed, it is wired in series with a normally open (N.O.) relay contact. To close this contact, the coil of the relay must be energized. This requires a completed circuit path across the lower (control voltage) section of the ladder diagram. By wiring the relay coil in series with the switch, the coil is energized when the switch contacts are closed, and deenergized when they are open.

Although this example is relatively simple, it illustrates the basic use of both the line voltage and control voltage portions of the ladder diagram.

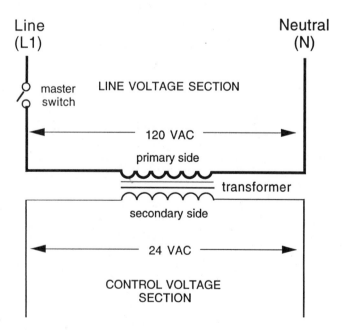

Figure 9–18 Basic layout of a ladder diagram

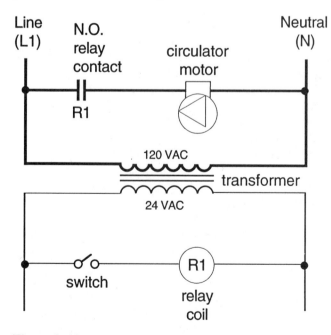

Figure 9–19 Example of a ladder diagram for operating a line voltage circulator using a low voltage switch

9.5 ELECTRO-MECHANICAL CONTROLS

Many of the controls used in hydronic systems have existed for several decades. Although these controls have been improved over time, their basic operating principles remain the same. These so-called **electro-mechanical controls** use components such as springs, bimetal strips, magnets, and fluid-filled temperature sensing bulbs to open and close electrical contacts based on sensed quantities such as temperature and pressure.

The section examines several of the more common electro-mechanical controls used in small hydronic systems. Keep in mind there are literally hundreds of electro-mechanical controls available for specific purposes and applications in hydronic heating. Often one basic control will have several available options enabling it to work in several different situations.

Single-stage Room Thermostats

No control is more familiar by sight and name than the room thermostat. For many homeowners it is the only control they know of, or want to know of, in the entire heating system. If the house is too cold they turn it up. If it is too warm, they turn it down. The room thermostat is, however, only one of several control components, even in simple hydronic systems.

The function of a single-stage room thermostat is to monitor the temperature of the room in which it is mounted, and when necessary, close a set of electrical contacts signalling the balance of the system that heat is needed. An expression often used to describe this contact closure is that the thermostat is "calling for heat."

A single-stage room thermostat is simply a temperature operated switch. As such it can have only one of two possible states at any given time (open contacts or closed contacts). The fact that the thermostat contacts are closed does not affect the *rate* at which heat is produced by the heat source. Many homeowners do not understand this. The common practice of turning a room thermostat to a high setting to *quickly* increase room temperature is proof.

Most electro-mechanical room thermostats contain a **bimetal element** consisting of two strips of metal bonded together. Each strip is made of a different metal having a different rate of thermal expansion when heated. When heated or cooled, this bimetal element will bend or rotate due to these different rates of expansion. The motion of the bimetal element can be used to pull contacts together or push them apart. In a heating thermostat the contacts are arranged so when the bimetal element is sufficiently cooled by the surrounding air, the contacts snap together. This contact closure can be used to start the heat source, open a valve, or initiate some other control action. In a cooling thermostat just the opposite takes place. When the bimetal element is sufficiently heated by the surrounding air, the contacts snap together, completing a circuit to start the cooling device. The schematic symbols used to designate room thermostats are shown in Figure 9–20.

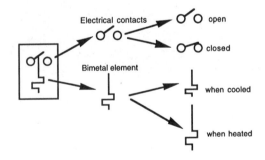

	Operating Mode	Stand by Mode	Description of operation
Heating Thermostat			Bimetal element contracts when cooled pulling contacts together. Element expands when heated pushing contacts apart.
Cooling Thermostat			Bimetal element expands when heated pushing contacts together. Element contracts when cooled pulling contacts apart

Figure 9–20   Schematic symbols for representing room thermostats in electrical drawings

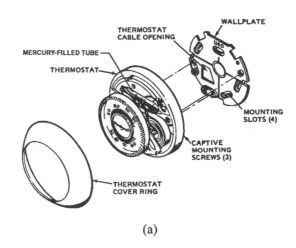

(a)

(b)

Figure 9–21   (a) Internal design of a coiled bimetal thermostat using a glass tube containing mercury. Courtesy of Honeywell. (b) External appearance.

One type of room thermostat uses a bimetal element coiled into a spiral shape. A small glass tube containing the ends of two flexible wire leads, and a small quantity of mercury, is attached to the outer portion of this element as shown in Figure 9–21. As the temperature changes, the glass tube is slowly rotated by the movement of the bimetal element. When the tube reaches a certain angle, the mercury it contains flows to the end of the tube containing the wire contacts. Being a good conductor, the mercury completes the circuit between the wire leads and thus through the thermostat. If the tube rotates in the other direction, the mercury eventually flows to the other end of the tube, and the circuit is broken. *This type of thermostat must be precisely leveled when mounted on the wall to ensure proper flow of the mercury.*

Another variation of the bimetal thermostat uses the bimetal element itself as a portion of the conducting path through the thermostat. As the room temperature slowly drops, the element brings the contact points closer together. When they are a certain small distance apart, the force created by a magnet snaps the contacts together. This prevents the erratic operation that could occur when very minor temperature fluctuations cause the contacts to bounce between open and closed. An example of this snap-action thermostat is shown in Figure 9–22.

The ideal room thermostat would operate the heating system to maintain the room precisely at a desired **setpoint temperature**. If the thermostat detected even the slightest drop below the setpoint, it would bring on the heating system. When the temperature of the room

Figure 9–22 Example of a snap-action bimetal room thermostat

increased the slightest amount above the setpoint, it would stop the system. This would of course result in extremely short and frequent operating cycles of the heat source. Although this is possible with electric resistance heating elements, it would cause extreme wear on boilers and heat pumps. To prevent such short cycling, room thermostats have a built-in range of temperature between the point where the contacts close and open. This is called the **differential** of the thermostat, and is illustrated in Figure 9–23.

Some thermostats have an adjustable differential, others are fixed by the manufacturer. The narrower the differential, the greater the number of on/off cycles the heating system will experience during a given period of time. Narrow differentials are beneficial for comfort control *if they can be obtained without creating abnormally short operation of the heat source.* Electric resistance boilers can be used with thermostats having differentials as low as 1 °F. Their heating element(s) can be switched on and off many times an hour without incurring any abnormal wear and tear. Hydronic systems with oil or gas-fired boilers typically

use thermostats with differentials of 3 °F to 5 °F to prevent short cycles, while also keeping room temperature swings tolerable.

Another complication of controlling heating systems with simple on/off devices such as thermostats is that of overshoot and undershoot, as illustrated in Figure 9–23. Overshoot is caused by the residual heat that remains in the heat emitters and distribution piping after the heat source is turned off. The air temperature continues to rise as this residual heat is released into the room. Undershoot is caused by the delay in providing heat to the room due to the warm-up time required by the system's components and fluid.

To limit overshoot, a **heat anticipator** is built into most room thermostats. This consists of a small adjustable resistor that has current flowing through it whenever the contacts of the thermostat are closed. The heat given off by the resistor makes the temperature felt by the bimetal element just slightly higher than room temperature. This fools the thermostat into opening its contacts just before room temperature reaches half the differential above setpoint. The residual heat in the heat emitters and piping then provides enough heat for the system to coast to its upper temperature limit with minimum, if any, overshoot.

Whenever an electro-mechanical thermostat is installed, its heat anticipator should be set for the electrical current that flows through the thermostat when its contacts are closed. This current depends on the device the thermostat is connected to. In some cases the anticipator setting is printed on the control the thermostat is connected to. In other cases it has to be measured with an ammeter and set as part of the startup procedure. The adjustment is made using a small screwdriver to set the anticipator dial to the proper current setting. Figure 9–24 shows the heat anticipator adjustment on a room thermostat.

Room thermostats should be mounted on interior walls away from obvious sources of localized heat such as lights, cooking equipment, heat emitters, or direct sunlight. A mounting height of about 5 feet is typical. Choosing a thermostat location should not be considered

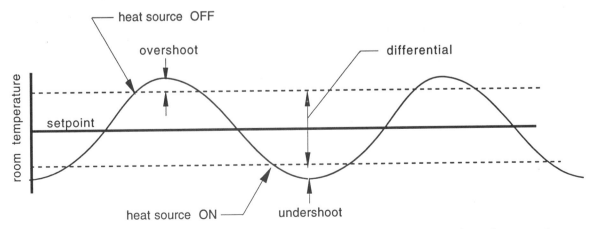

Figure 9–23 Graph showing variation in room temperature and associated operation of a room thermostat

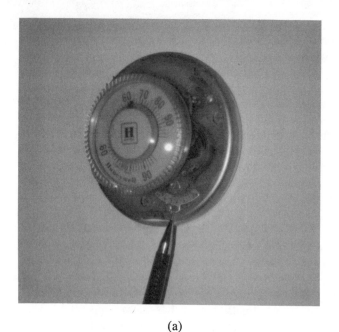

(a)

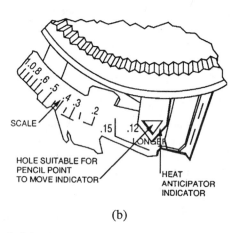

SCALE

HOLE SUITABLE FOR
PENCIL POINT
TO MOVE INDICATOR

HEAT
ANTICIPATOR
INDICATOR

(b)

Figure 9–24 (a) Heat anticipator adjustment on a low voltage room thermostat, (b) close-up of amperage values on anticipator scale. Courtesy of Honeywell.

a trivial matter. *Careless placement can significantly detract from the performance of an otherwise well-installed system.*

Room thermostats can be further classified as either low voltage or line voltage devices. The majority of those used in hydronic heating are designed for low voltage operation. They typically switch a circuit operating at a nominal 24 VAC. This voltage is a standard transformer secondary voltage used throughout the U.S. heating and cooling industry. Because low voltage thermostats operate at low voltages and low currents, they qualify as class 2 devices under the national electrical code, and can be wired with relatively small diameter conductors in the range of 18 to 24 wire gauge. Thermostat cable is widely available in this size range. It can be purchased

with two to eight color-coded conductors in a flexible plastic sheathing.

Line voltage room thermostats are designed for the nominal 120 or 240 VAC line voltage. They are intended to directly switch line voltage devices such as heating elements, pumps, or fans. All line voltage thermostats have an associated **ampacity** rating. This rating lists the maximum current the contacts can safely handle. Often two current ratings are given. The greater one is for resistive loads such as heating elements, the lower one for inductive loads such as AC induction motors. *When selecting a line voltage thermostat always check that these ratings meet or exceed the total amperage of all devices that draw their current through the thermostat.*

Two-stage Room Thermostats

Occasionally a hydronic system requires control of two stages of heat input. In all such cases, one stage is given priority over the other. For example, many heat pump systems use electric heating elements to boost heat output should the heat pump's output become too low under extreme operating conditions. This electric heating is significantly more expensive to operate than the heat pump, and thus it is preferable to operate it only when necessary to prevent the building from becoming uncomfortable. Another possible two-stage application is a combination wood/oil boiler. When the heat output of the wood fire is sufficient to maintain comfort there is no need to fire the oil burner. As the wood fire burns out, however, the room begins to cool and the oil-burner will need to be fired.

These control schemes can be handled using a two-stage thermostat. This device has two sets of independent contacts, one for each stage. The contacts are moved by either a single or dual bimetal element(s). *The first stage contacts will always close before the second stage contacts.* The heating device with operating priority is controlled by the first stage contacts. The supplemental heating device is controlled by the second stage contacts.

Two-stage thermostats usually have a fixed temperature differential of 2 °F to 4 °F between stages. If, for example, this differential was 3 °F and the first stage contacts closed at a room temperature of 70 °F, the second stage contacts would close at 67 °F. As the room warms up the second stage contacts would open first. If the room temperature continued to rise, the first stage contacts would eventually open.

Two-stage room thermostats are available for both line voltage and low voltage applications. Figure 9–25 shows a simplified wiring schematic using a two-stage thermostat to control both the compressor and supplemental electric heating elements of a heat pump system.

It is possible to obtain room thermostats with three and even four stages of heating control. Their use in smaller hydronic systems, however, is quite uncommon given the other control options available.

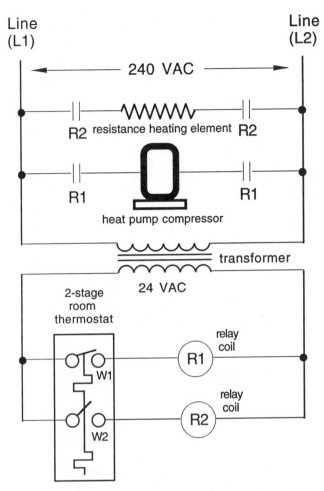

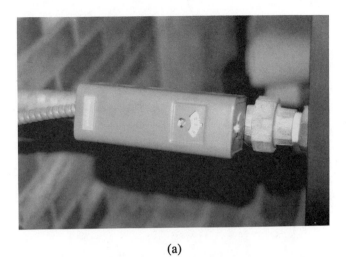

(a)

Figure 9–25   Ladder diagram of a two-stage room thermostat controlling a heat pump with electrical resistance supplemental heating. Stage 1 operates the compressor. Stage 2 operates the electric heating element.

Setpoint Controls (Aquastats)

There are often several locations in a hydronic system where fluid temperatures are measured for safety and control purposes. The opening or closing of electrical contacts is often required based on whether a particular temperature is above or below a predetermined setting. This is accomplished using what is called a setpoint control, or **aquastat**.

One example of an electro-mechanical set point control is called a strap-on aquastat. It is designed to mount directly on a pipe using a metal strap or hose clamp. A sensing bulb is squeezed tightly against the pipe as the clamp is tightened. This bulb contains a fluid that increases in pressure when heated. The pressure exerted by the fluid's vapor pushes against a diaphragm or bellows assembly in the control, causing it to expand. This movement in turn opens or closes the electrical contacts in the control. Most strap-on aquastats are designed to switch several amps of current at line voltage.

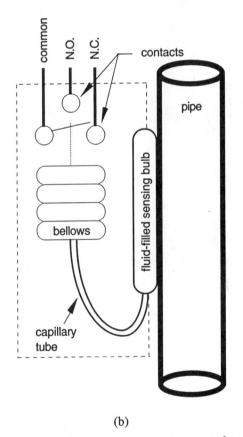

(b)

Figure 9–26   (a) A strap-on aquastat mounted to a pipe (b) simplified illustration of internal components.

Current and voltage ratings are typically listed on the cover plate. These controls can also be used in low voltage systems. An example of a strap-on aquastat is shown in Figure 9–26.

Another common type of setpoint control is the remote bulb aquastat shown in Figure 9–27. The only major difference between it and the strap-on aquastat is the length of the copper capillary tube between the sensing

Figure 9-27 A remote bulb setpoint control. Note the capillary tube that connects the control's body and sensing bulb. Courtesy of White Rogers.

bulb and the internal diaphragm or bellows assembly. Remote bulb aquastats allow the control body and electrical wiring to be located up to several feet away from the location of the sensing bulb. Controls can be ordered with capillary tubes up to 12 feet long. *The length of this capillary tube cannot be changed in the field because the working fluid would be lost.* Any extra length of capillary tubing should be neatly coiled near the control and fastened to something solid to prevent vibration. *Care should be used in routing the capillary tube to prevent kinking.*

In applications where the bulb will be sensing the temperature of a surface, it should be tightly secured using a clamp or tube strap. It should then be covered with insulation to minimize the error from heat loss to the surrounding air.

The most accurate method of sensing a fluid temperature is to immerse the sensing bulb directly in the fluid. Some controls are available with sensing bulbs having MPT threads. These bulbs can be screwed directly into a fitting such as a tee. Bear in mind, however, if the control has to be replaced, some spillage of system fluid is inevitable during removal and replacement of the sensing bulb. If such spillage is to be avoided, the sensing bulb can be mounted into a well. Such wells usually consist of a segment of copper tube slightly larger than the sensing bulb, closed at one end, and open at the other. The open end is equipped with male pipe threads so it can be screwed into a tapping on a tank or a tee. The closed end prevents fluid from entering the well. For best results, the sensing bulb should first be coated with a special heat-conducting paste often supplied with the well before being inserted into it. This paste improves heat conduction between the bulb and the inside surface of the well, resulting in a faster response to changing temperatures. An example of a sensing well is shown in Figure 9-28.

Combination High Limit/Switching Relay Control

Because many residential hydronic systems use a single boiler, single circulator, and single room thermostat, they have virtually identical control requirements. The similarity of these systems allows manufacturers to design a specialized control that can coordinate the operation of all these components. Manufacturers of packaged boilers often equip their boilers with such a

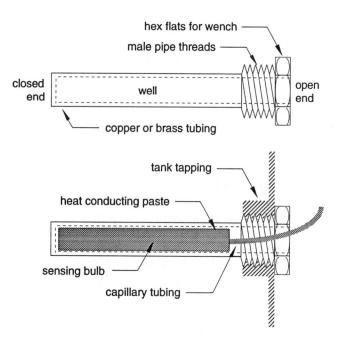

Figure 9-28 Illustration of a sensor well for use with a remote bulb setpoint control

(a)

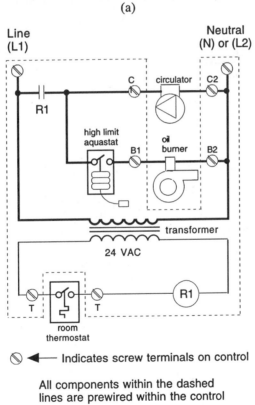

(b)

Figure 9–29 (a) Example of a boiler high limit/switching relay control. Courtesy of White Rogers. (b) Ladder diagram showing internal wiring.

control, and usually pre-wire it to both the burner and circulator. An example of a high limit/switching relay control for an oil-fired boiler is shown in Figure 9–29a. Its circuitry is shown in the form of a ladder diagram in Figure 9–29b.

This control incorporates several components such as a transformer, relay, and aquastat into a small case that mounts directly onto the boiler. The sensing bulb of the aquastat projects out the rear of the case and into a well in the boiler block.

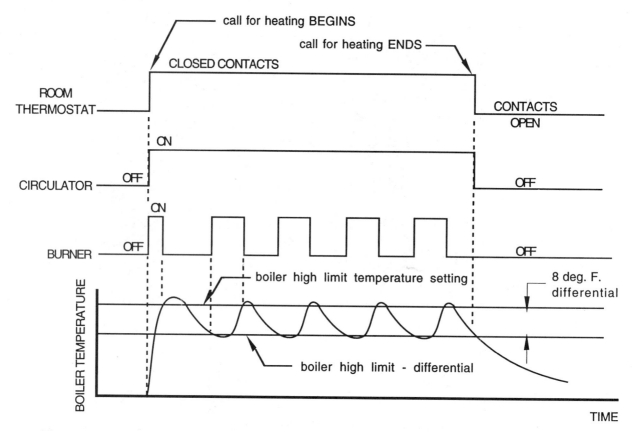

Figure 9–30 Operation of a combination high limit/switching relay control over a period of time

A typical residential system with this control will operate as follows:

1. When the room thermostat contacts close, a low voltage (24 VAC) circuit running through the coil of the internal relay is energized.
2. The relay contacts close to start both the system circulator and oil-burner.
3. If the water temperature in the boiler eventually climbs to the setting of the aquastat (i.e., if the heating load is less than the heat output of the boiler), the burner contacts open and shut it off. The circulator, however, continues to run.
4. If the boiler water temperature decreases by a certain differential (typically about 8 °F), the burner is again fired to bring the boiler temperature back up.
5. Only when the room thermostat contacts open the circuit between the "T" terminals will the circulator *and* the burner be turned off.

This boiler temperature floats between high limit, and high limit minus the differential whenever the thermostat is calling for heat. This operation is depicted in Figure 9–30. Notice the boiler temperature will eventually go down to room temperature if the boiler remains off long enough.

Triple Action Control

For many years was common practice to produce domestic hot water using a **tankless coil** mounted in the boiler block. Cold water is heated as it passes through this copper coil on its way to a faucet. Because hot water is needed year round, at any time of the day, *boilers with tankless coils must maintain a minimum water temperature at all times.* Due to the year round standby heat loss of the boiler, this method of domestic hot water heating is considered inefficient by today's standards, although many are still in use. Those individuals having to replace or repair this type of system should be familiar with its operation.

Because of the past widespread use of boilers equipped with tankless coils, several control manufacturers offer a unit control to manage the overall operation of the system. This control is often called a **triple action** (or sometime triple aquastat) **control**. An example of such a control is shown in Figure 9–31a. Its associated schematic is shown in Figure 9–31b.

The triple action control also mounts directly to the boiler. It has a single temperature probe that inserts into a well in the boiler block. The control has two dials for temperature adjustments, one to set the high limit of the boiler water, the other to set the minimum standby temperature (low limit) of the boiler.

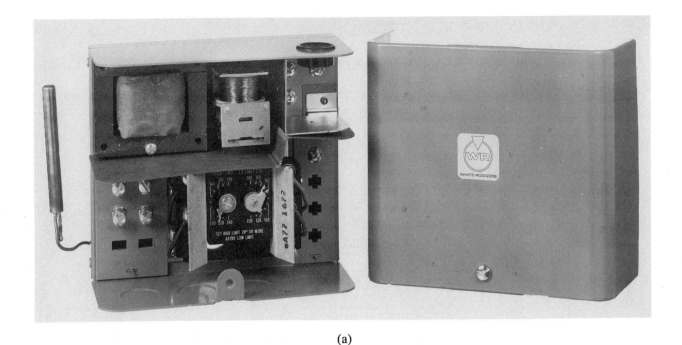

(a)

(b)

Figure 9–31 (a) Example of a triple action control. Note individual settings for high limit and low limit temperatures. Courtesy of White Rogers. (b) Ladder diagram showing internal wiring.

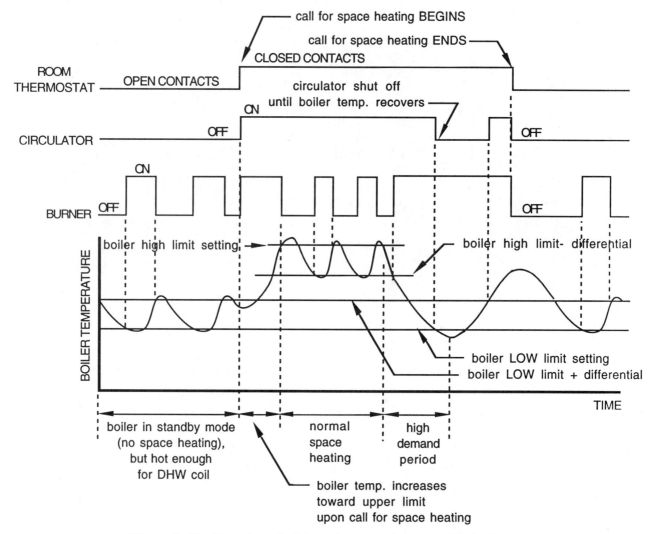

Figure 9-32 Operation of triple action control over a period of time

The triple action control fires the burner if the room thermostat calls for heat or if the boiler temperature falls below the low limit setting. If the load on the boiler is high due to simultaneous demands for space heating and domestic water heating, and the boiler temperature drops below the low limit setting, the circulator supplying water to the space heating distribution system is temporarily shut off. It restarts when the boiler temperature recovers its low limit setting plus differential. A sequence showing control operation is shown in Figure 9-32.

Multi-zone Relay Centers

The **multi-zone relay center** is a relatively new arrival in the field of hydronic controls. It was designed to centralize the wiring and control operation of up to six independent zone circulators from a single unit control. The savings in time, space, and cost relative to using separate controls for each circulator is considerable. An example of a multi-zone relay center is shown in Figure 9-33.

A common hydronic system configuration is to have several independent heating zones, each with its own circulator, along with an indirectly fired domestic water heater. A schematic for this type of system is shown in Figure 9-34.

Most multi-zone relay center controls allow the domestic water heating to take priority over space heating. This is accomplished by temporarily stopping all circulators supplying hot water to space heating zones whenever the aquastat on the domestic hot water tank is calling for heat. Since the full boiler capacity can then be directed to domestic water heating, hot water is produced very quickly. This is very desirable in applications that have large demands for domestic hot water, such as homes with large whirlpool tubs. A typical indirectly fired storage water heater, set up with priority control, can produce about ten times more hot water in a given period of time compared to a conventional residential electric water heating tank. This allows tank sizes as small as 20 and 30 gallons to be routinely used in residential applications.

Figure 9–33   A multi-zone relay center with cover removed. Note line voltage wiring to circulators at bottom, and low voltage wiring to thermostats at top.

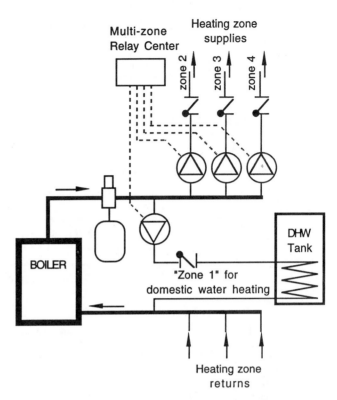

Figure 9–34   Typical piping schematic for a multi-zone heating system with an indirectly fired storage water heater

A schematic representation of a multi-zone relay center with domestic water heating as a priority load is shown in Figure 9–35. The same circuitry could of course be assembled on the job using individual components, but the extra labor and materials cost would far outweigh the cost and convenience of the preassembled relay center.

The isolated relay contact designated B in Figure 9–35 is used to fire the heat source on a call for heat from any of the zones. Because this contact is electrically isolated from the remainder of the circuit, there is no need to be concerned about matching the secondary control voltage, or phasing, of the transformers in the relay center and heat source.

Multi-zone relay centers are available in configurations for three to six zones. They are also available for operating 24 VAC zone valves rather than line voltage zone circulators. In some cases it is wise to consider purchasing a unit with one extra zone relay rather than what is immediately required. This allows easy expansion of the system in the future. If more than six zones are needed, multiple relay centers can be used.

Zone Valves

Zone valves are often used to permit or prevent flow within a branch circuit (e.g., zone) of a hydronic heating system. A variety of zone valves are available. Several types were described in Chapter 5. They all consist of a valve body combined with an actuator. The actuator is the device that produces movement of the valve shaft when an electrical voltage is applied.

Some actuators use small electric motors combined with gears to produce a rotary motion of the valve shaft. These actuators can fully open or close the valve in about three seconds. This is fast enough for rapid heat delivery, yet slow enough to prevent water hammer. Other zone valves use heat motor actuators that produce a linear motion when heated by an internal resistor. This type of actuator can take two to three minutes from when the operating voltage is applied to when the valve is fully open. An example of each type is shown in Figure 5–60.

In addition to opening the valve body, zone valves are often equipped with internal switches that are used to signal other components in the system that heating is being requested, and that the circulator and heat source should be started. These switches are called **end switches** because their contacts close when the actuator reaches the end of its travel and the valve is fully open. End switch contacts open as soon as the valve starts to close.

The most common types of zone valves have either three or four terminals or wires. The difference between them is in how the end switch is wired. In a three-wire zone valve, one contact of the switch is wired in parallel with the zone valve motor. In a four-wire zone valve the end switch is isolated (not connected to any other internal wiring). This difference determines how the zone valve interfaces with the rest of the control system. A simplified representation of both three-wire and four-wire zone valves is shown in Figure 9–36.

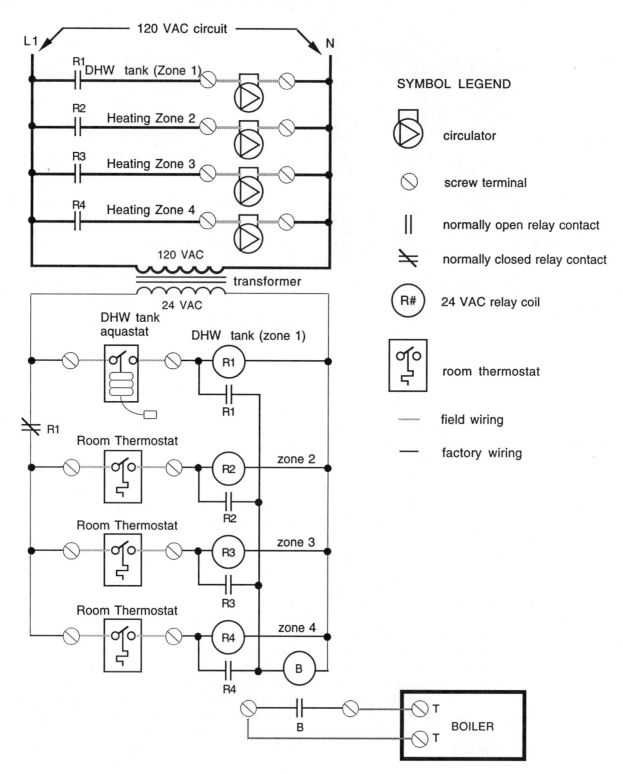

Figure 9–35 Schematic of multi-zone relay center for three heating zones and one zone of priority domestic water heating

When a three-wire zone valve is used, the signal passed through its end switch is supplied by the same voltage source that operates its actuator motor. This is usually the 24 VAC output of a transformer. The power required to operate a zone valve is expressed in VAs (volt-amps), and is usually listed on the zone valve or its data sheet. *Care must be taken that the VA rating of the transformer supplying several zone valves is greater than the total VAs drawn when all zone valves are operating at the same time.* If this is not done, the transformer will be overloaded and eventually burn out. Zone valves with heat motors typically require more VAs than those using gear motors.

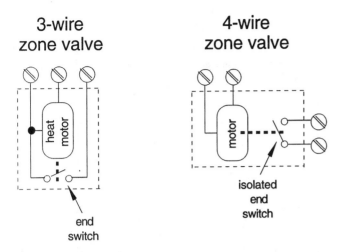

Figure 9–36   Simplified representation of three- and four-wire zone valves

One common approach is to obtain the 24 VAC operating power required by the zone valves from the transformer in the heat source's control. Before doing this be sure the transformer in the control can provide the total VAs required by all connected zone valves, plus any other controls being powered. Manufacturers usually list the maximum *external* VAs the control's transformer can safely supply. A schematic showing the general wiring diagram for this approach in conjunction with three-wire zone valves and an oil-fired boiler is shown in Figure 9–37.

The operating sequence of this control system is as follows: When either of the two room thermostats call for heat, the actuator of the associated zone valve is energized, opening the valve. When the actuator reaches the end of its travel (e.g., the valve is fully open), the end switch closes, completing a circuit across the "T" terminals of the high limit/switching relay control. This control then starts the circulator and oil burner, and operates as previously described. When the thermostat is satisfied, it opens the circuit through the zone valve motor. The burner and circulator stop and the zone valve closes. Burner and circulator operation can be initiated by *any* of the zone valves, and will continue as long as *any* of the zone thermostats are calling for heat. If the external VA rating of the transformer in the high limit/switching relay control allows, additional zone valves could be added by simply duplicating the wiring used for one of the zone valves shown.

If the external VA rating of the heat source's transformer is not high enough to supply the total VAs required by all zone valves, an external transformer must be used. This relieves the control's transformer from supplying the VA requirement of the zone valves. Typical wiring for this approach in combination with three-wire zone valves and a gas-fired boiler is shown in Figure 9–38. The operating sequence is similar to that just described with the exception that the high limit/switching relay control is activated by the closing of a contact in the

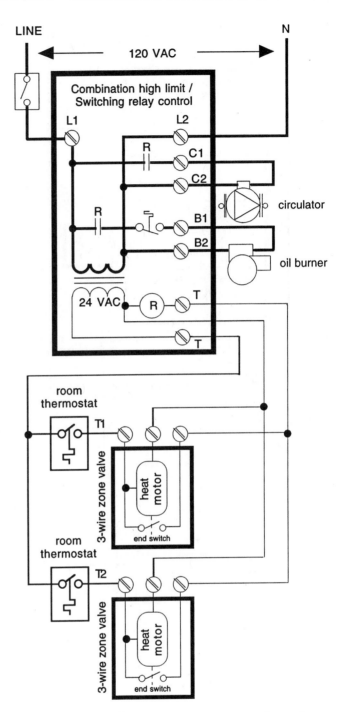

Figure 9–37   Wiring for three-wire zone valves when transformer in boiler control has sufficient external VA capacity to power zone valves

isolation relay, rather than the end switch of a zone valve. This isolation relay is a separate component not supplied with the high limit/switching relay control.

When four-wire zone valves are used, the isolation relay can be eliminated. The isolated end switch of each zone valve is used to complete a circuit across the "T" terminals of the boiler control, thus starting the burner and circulator. This is shown in Figure 9–39.

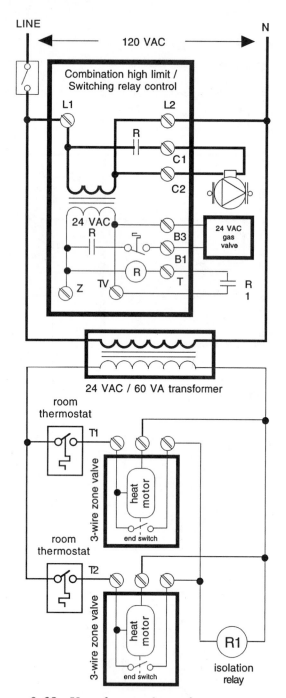

Figure 9–38 Use of external transformer to power zone valves

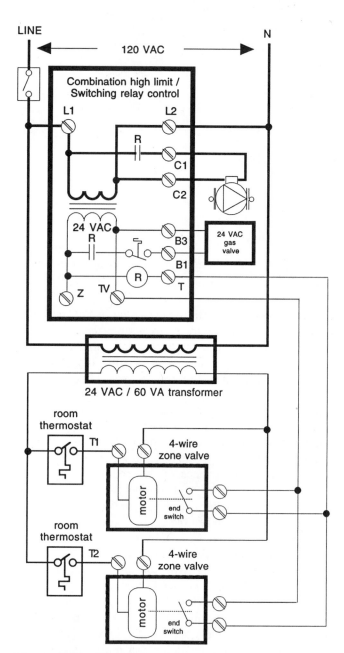

Figure 9–39 Use of four-wire zone valves with isolated end switches

9.6 ELECTRONIC CONTROLS

The rapid advances taking place in electronic controls puts them at the cutting edge of hydronics technology. Electronic controls can provide system regulation well beyond that provided by electro-mechanical controls. Examples include operating up to ten modular boilers, positioning a mixing valve, or varying the speed of a pump in proportion to the instantaneous heating needs of a system. This section discusses several currently available electronic controls as well as new regulating techniques they make possible.

Two major advantages of electronic versus electro-mechanical controls are accuracy and adaptability. The higher accuracy is largely a result of solid-state temperature sensors known as **thermistors**. This small device resembles a glass bead with a diameter of about 0.1 inch, and two attached wire leads. *The electrical resistance of thermistors increases as their temperature decreases,* as shown in Figure 9–40. For certain types of thermistors and temperature ranges, a temperature change of 1 °F can

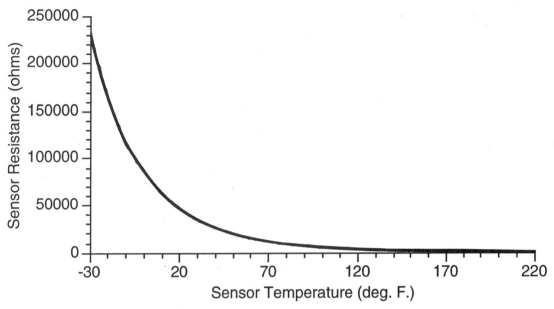

Figure 9–40   Typical relationship between temperature and resistance for a thermistor sensor calibrated for 10,000 ohms resistance at 25 °C

result in a resistance change of well over 100 ohms. This is easily detected by the control's circuits that monitor the sensor's resistance. The combined accuracy of thermistor sensors and internal electronics result in controls that often have rated accuracies of ½ °F. This is significantly better than the accuracy offered by most electro-mechanical controls, and more than sufficient for most heating and cooling control applications.

The small size of thermistor beads allow them to be mounted in a wide variety of housings. The housings protect the delicate bead and its connections to larger lead wires. The thermistor bead and its housing constitute the sensor. Some housings are designed for mounting on flat or curved surfaces. Others are designed to be threaded directly into a pipe or tank fitting. If necessary, thermistor sensors can be located up to several hundred feet from the control itself. A sampling of thermistor sensors is shown in Figure 9–41a. Proper mounting of a thermistor sensor on the outside surface of a pipe is shown in Figure 9–41b.

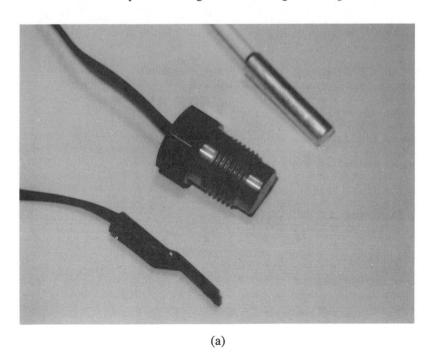

(a)

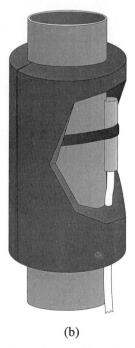

(b)

Figure 9–41   (a) Examples of thermistor sensor housings, (b) a thermistor sensor mounted to the outside of a pipe. Note the insulation covering the sensor. Courtesy of tekmar Controls.

Electronic controls usually have more adjustable settings than electro-mechanical controls. This often allows a single control to be configured for several different applications. The temperature range over which an electronic control can operate is often wider than that of an electro-mechanical control. These two features may allow installers and wholesalers to stock a single model of electronic control rather than several specialized electro-mechanical controls.

Electronic Room Thermostats

One of the first heating controls to enjoy the benefits of electronics was the room thermostat. Currently nearly every manufacturer of room thermostats has several different electronic models available. The most up to date type uses a thermistor sensor combined with a programmable **microprocessor**. They also contain internal clock circuitry and the ability to back up the programmed operating instructions in the event of a power outage. Other common features include built-in time delay(s) to prevent short cycle operation of the heat source, and the ability to disable the programming keys to prevent tampering. An example of an electronic room thermostat is shown in Figure 9–42.

Most electronic thermostats need to be programmed before they can be used. This is usually done by pressing buttons in various combinations while being prompted by a digital display on the thermostat. The program can be as simple as maintaining a constant temperature 24 hours a day, or as complex as having several temperature changes per day and an entirely different set of conditions for weekends. Once the temperatures and times have been entered, the thermostat provides the necessary on/off operation to *attempt* to control the system as desired. The

word *attempt* is important because the thermostat alone cannot force the system (or building) to operate according to the program. For example, if a 5 °F setback is programmed to occur at 11 PM and end at 5 AM in a building with high thermal mass and low heat loss, the building may not drop in temperature more than 2 °F or 3 °F, even if the thermostat prevents all heat input during the setback period.

Most electronic thermostats draw their minimal operating power from the 24 VAC transformer in the controlled device. This allows for both power supply and on/off control through a single pair of wires. Other models operate on batteries that require periodic replacement.

The output of electronic thermostats can be a contact closure of a small internal relay, or solid state switching action. The latter is created when a component called a triac goes into its conducting mode. As it enters this mode, the triac changes from a state of very high resistance to one of very low resistance. This completes the control circuit through the thermostat.

Electronic thermostats cost more than electro-mechanical thermostats, but offer more features. Their ability to automatically follow a prescribed temperature setback schedule can result in energy savings *when combined with a heating system and building that can respond to the desired schedule.*

If an electronic thermostat is being considered, be sure to select one that has the features needed. Many electronic thermostats are designed around combination heating/cooling systems or heat pumps. They often have features such as fan control, that are generally not needed for hydronic systems, but could still be of use if central cooling is also provided. Special models are available for heating-only applications using one or two stages of control.

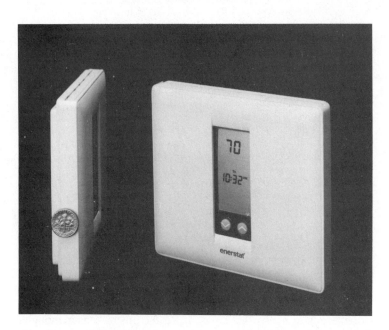

Figure 9–42 Example of a programmable electronic room thermostat. Courtesy of Enerstat Corporation.

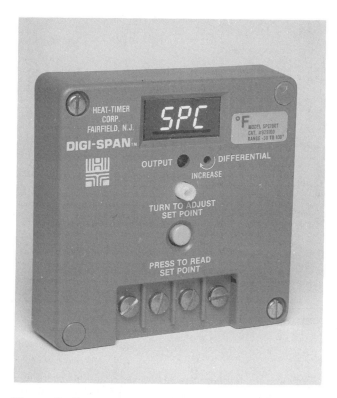

Figure 9–43   Example of an electronic single stage setpoint control. Courtesy of Heat Timer Corp.

Electronic (Single-stage) Setpoint Controls

An electronic setpoint control provides on/off contact closure output based on the temperature of its sensor and its programmed setpoint. Like its electro-mechanical counterpart, output is limited to this simple on/off action. However, several additional features make the control more versatile. These include wider ranges of adjustment for both setpoint and differential, the ability to operate the relay on a rise or drop in temperature, and a digital readout for displaying the sensor temperature. Microprocessor-based setpoint controls offer even more attributes, such as the ability to work in either °F or °C, or to

execute a programmable time delay function. Applications for these controls include operating a heat source, pump, diverting valve, or almost any type of electrically-driven device at a preset fluid temperature. An example of an electronic, single-stage setpoint control is shown in Figure 9–43.

Electronic setpoint controls also have some *disadvantages* relative to electro-mechanical controls. Because of the physical size of the relays used, they cannot usually handle line voltage currents as great as those handled by electro-mechanical controls. In some cases the relay contacts are only rated to operate with control voltages up to 24 VAC. If a line voltage device such as a pump or blower is to be controlled, an additional line voltage rated relay will be required between the control and the device.

Electronic setpoint controls are also more expensive than electro-mechanical setpoint controls. This higher cost can be justified when the application requires features such as accuracy, time delay, adjustable differential, etc. In other applications where accuracy of ± 3 °F to 5 °F is acceptable, and where rapid temperature changes are not likely, the higher cost may not be justified.

The solid state components used in electronic controls, especially microprocessors, are more susceptible to damage from voltage spikes. Although manufacturers have improved the ability of these controls to survive moderate voltage spikes, accidental connection of line voltage to low voltage terminals, or a high voltage surge from a nearby lightning strike could still cause damage.

Electronic (Two-stage) Setpoint Controls

The concept of multi-stage control has already been discussed. The output of an electronic two-stage setpoint control consists of sequential operation of two switches or relay contacts. The first stage contacts will always be the first to close and the last to open. The second stage contacts close when and if necessary to deliver more heating capacity. A two-stage electronic setpoint control allows great flexibility in setting the cut-in temperature of each stage, the differential of each stage, and between stages. A diagram depicting some setting options is shown in Figure 9–44.

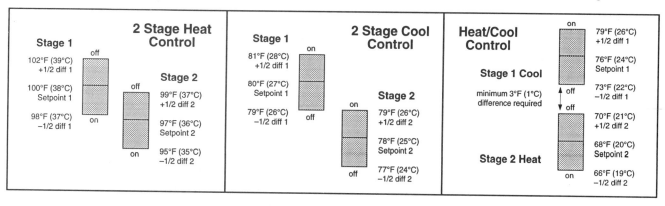

Figure 9–44   Temperature and differential settings of an electronic two-stage setpoint control. Courtesy of tekmar Controls.

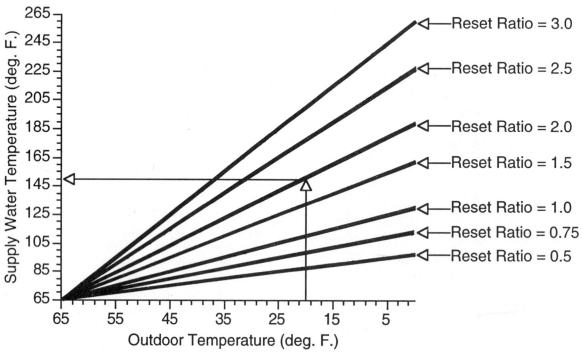

Figure 9–45 Supply temperature determined from combination of outdoor temperature and reset ratio setting of control. In this case a control set to a reset ratio of 2.0 would calculate a supply temperature of 150 °F when the outdoor temperature is 20 °F.

The primary application for two-stage controls is to activate two stages of heating (or cooling) to meet changing load conditions. The heat sources activated by each stage do not have to be the same type or capacity. The flexible set-up allowed by electronic two-stage controls allows the operation of each stage to be matched to the operating characteristics of each controlled device.

Electronic Outdoor Reset Controls

The function of a reset control is to continuously calculate what the supply water temperature to the distribution system should be, and then steer the system toward that temperature. The inputs to the control are the readings from two temperature sensors, one mounted outside, the other on the supply pipe to the distribution system. These temperature readings are combined with the reset ratio setting to calculate the required supply water temperature at any given time. Figure 9–45 shows how the supply temperature is calculated based on outdoor temperature and the value set for the reset ratio.

Keeping the system on or near the calculated supply temperature 24 hours a day can be done in a number of ways. Most of the current methods use electronic controls. These include boiler temperature reset, motorized mixing valves, pulse width modulation of heat input, and variable speed injection pump systems. Each method will be examined separately.

Boiler Temperature Reset

The simplest method of implementing reset control is to turn the heat source on and off as necessary to maintain a given outlet temperature. A reset control with an on/off contact closure is required. An example of such a control and its associated sensors is shown in Figure 9–46a. A schematic showing the location of the sensors is given in Figure 9–46b.

This approach must be combined with constant circulation of system water to minimize temperature variations in the heated space. The boiler limit control should be set at least as high as the reset water temperature under design load conditions. This prevents the limit control from prematurely shutting off the burner if the calculated reset temperature is higher than its setting.

Since the system circulator needs to operate continuously, it cannot be wired to both "c" terminals in the high limit/switching relay control. Figure 9–47 shows the circulator wired so it will operate whenever the master switch to the boiler is on. It should be noted that some reset controls provide a separate contact closure for automatically turning the system's circulator on and off as required. In such cases neither of the "c" terminals in the high limit/switching relay control need to be used.

Most currently available reset controls require an input voltage of 24 VAC to operate their internal electronics and relays. This requires the installation of a separate

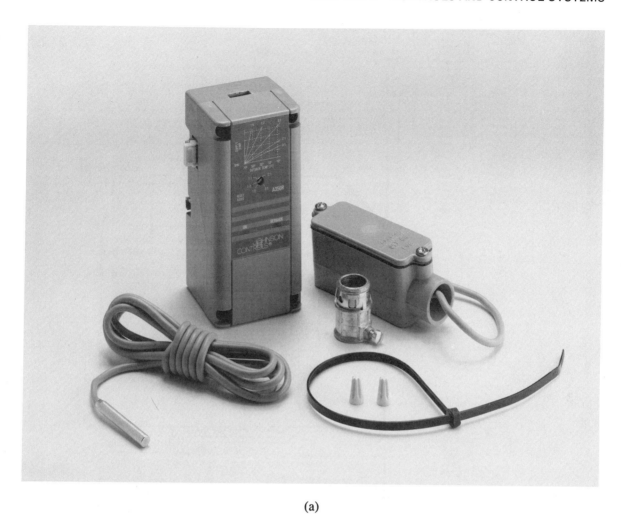

(a)

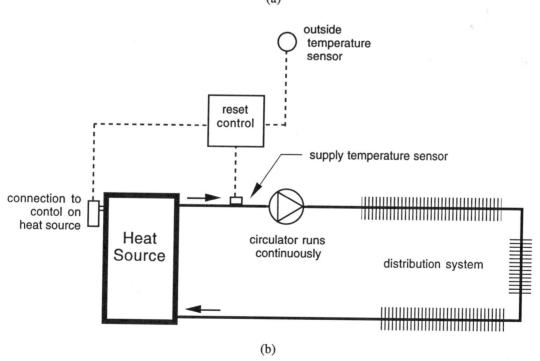

(b)

Figure 9–46   (a) Example of an electronic reset control with its associated outdoor and supply water temperature sensors. Note the reset lines shown on the front of the control. Courtesy of Johnson Controls. (b) Schematic showing installation of this control.

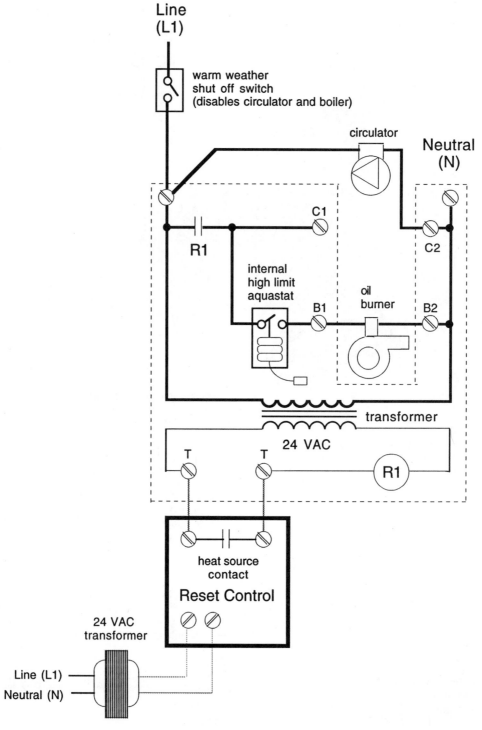

Figure 9–47   One possible method of connecting a reset control to boiler having high limit/switching relay control

24 VAC class 2 transformer. *Power for the reset control should not be drawn from the transformer of the boiler limit control.* Voltage spikes or other types of electrical disturbances could cause operational errors or physical damage.

The output relays of many currently available reset controls are also rated to operate at a maximum of 24

VAC. If line voltage loads are to be controlled, external relays with 24 VAC coils and suitable line voltage current ratings must be used between the reset control and line voltage device.

The type of heat source will often dictate the range over which the system temperature can be reset. *Conventional boilers should not be reset to temperatures*

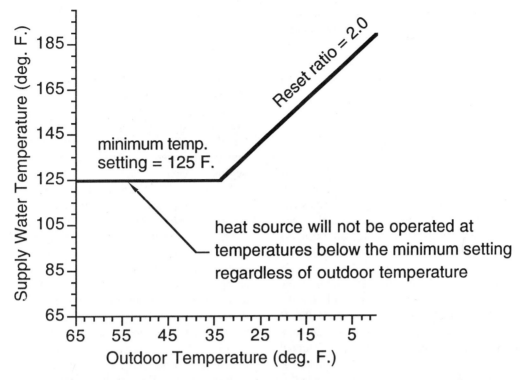

Figure 9–48   Control operation if a minimum boiler temperature is set

lower than the dewpoint temperature of their exhaust gases. For gas-fired boilers this is typically around 130 °F. For oil-fired boilers it is slightly lower. Heat sources such as electric boilers, heat pumps, or other non-combustion systems do not have this limitation.

Electronic reset controls usually have an adjustable setting for minimum supply water temperature. This allows them to be used with all types of heat sources. If the calculated supply temperature is lower than the setting for minimum supply water temperature, the control will attempt to constantly maintain the minimum setting. If the outside temperature drops and the calculated value becomes larger than the minimum setting, the control will resume normal reset action. This is depicted in Figure 9–48.

If the boiler operates at minimum water temperature in a system with constant circulation, overheating of the building will likely occur during mild weather. Overheating can also result from internal heat gains for sun, people, fireplaces, etc. *A provision must be made for overriding the calculated (reset) water supply temperature when the room temperature is at or above comfort setpoint.* Some reset controls can be equipped with an optional room temperature sensor that overrides the reset function and prevents the boiler from firing if the room is above a preset temperature. This is a very desirable feature in most residential and light commercial applications. It should be used whenever it is available. Another way to prevent overheating is to wire a conventional room thermostat in series with the on/off output of the reset control as shown in Figure 9–49.

If this method is used, the thermostat should be set slightly above the normal comfort temperature during cold weather. This keeps the contacts of the thermostat closed, allowing the reset control to maintain control except during an occasional period of overheating. During milder weather, when the reset control is attempting to

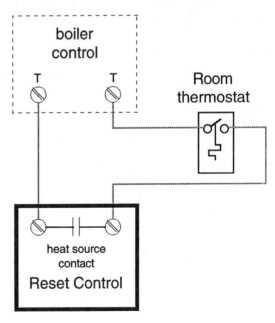

Figure 9–49   Use of a room thermostat to override reset control during mild weather or periods of high internal heat gain

maintain a minimum supply water temperature, the room thermostat will open the circuit between the reset control and boiler, shutting off the heat source. Since the contacts of the reset control remain closed during these periods, the heat source functions as if it was controlled only by a room thermostat.

Motorized Mixing Valves

Chapter 5 described both three-way and four-way mixing valves and their use in blending water temperatures to achieve a specific outlet temperature. These valves can be operated manually, but attempting to accurately track a reset line would require constant minor adjustments 24 hours a day, throughout the heating season. Obviously this is not practical.

By adding a slow-turning reversible gear motor to the valve, its position and thus its outlet temperature can be precisely adjusted by an electronic reset controller. If either the outdoor temperature or boiler outlet temperature change, the control can compensate by making frequent but small adjustments to the valve shaft. This allows the reset control to keep the supply water temperature on or very near the calculated value at all times. It is almost like an autopilot system for hydronic heating. An example of a mixing valve fitted with a motor actuator is shown in Figure 9–50a. A typical piping schematic for this type of application is shown in Figure 9–50b.

Typical mixing valve motors turn very slowly, taking from two to three-and-a-half minutes to rotate a valve across its full travel of about 90 degrees. Since hydronic systems do not require wide temperature changes over short periods of time, it is not necessary to have faster valve rotation. In fact the slow rotation adds stability to the system by allowing adequate time for feedback from the supply sensor located downstream of the mixing valve. The valve motors are equipped with end switches that prevent them from stalling at either end of their travel. Some also have cam-operated auxiliary switches that can be used to activate a heat source as the valve begins to open.

Some manufacturers use heat motors rather than conventional gear motors to rotate the mixing valve. The reset controller sends pulses of electrical energy to the heat motor, causing it to warm up and begin rotating. The longer the duration of the pulses, the further the heat motor rotates. When the pulse length is decreased or stopped, the heat built up in the heat motor dissipates, and a spring rotates the valve stem in the opposite direction. The housing of the heat motor can get quite warm when the valve is near its fully open position. Full rotation of the valve can take five minutes or more, depending on ambient temperature.

Reset controllers that operate valve motors have several adjustments and options. Besides allowing a wide range of adjustment for the reset ratio, they can be set up to automatically assume a nighttime temperature setback, operate the heat source, turn the circulator off during warm weather, and even periodically exercise the circulator and mixing valve during non-operational periods to prevent seizing. All these actions take place under the

(a)

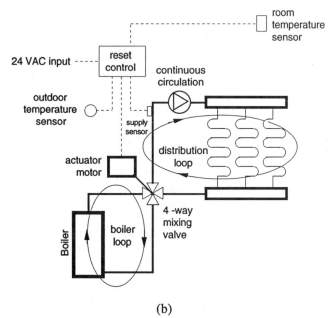

(b)

Figure 9–50   (a) Example of a four-way valve fitted with a motor actuator (b) schematic of typical piping system using a four-way mixing valve.

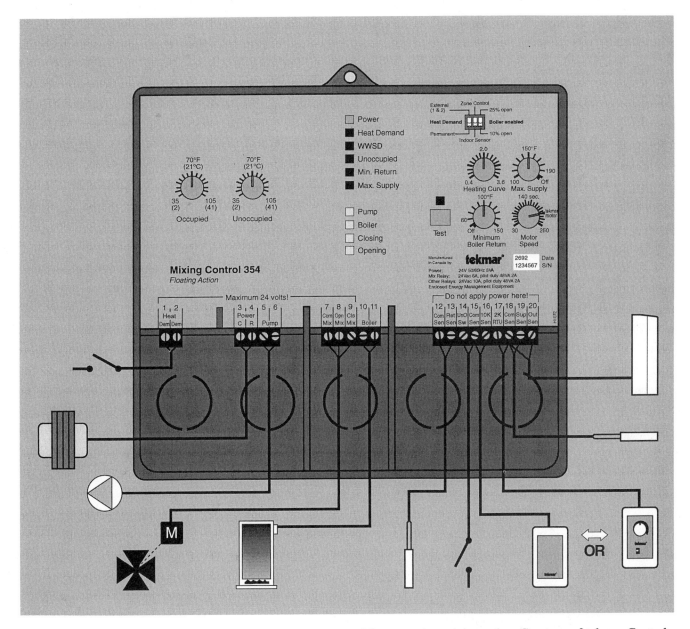

Figure 9–51 Example of a microprocessor-based reset control for operating mixing valve. Courtesy of tekmar Controls.

direction of an internal microprocessor. An example of such a control is shown in Figure 9–51.

Injection Mixing Control

Still another method of implementing reset control of the supply water temperature is to inject hot water into a continuously circulating distribution loop to maintain its temperature at the proper value. As the outdoor temperature decreases, the amount of hot water injected must also increase. This is done by lengthening the duration, or on-cycle, of the device that injects the heat. A common electric zone valve can be used to control hot water injection into the distribution system as shown in Figure 9–52.

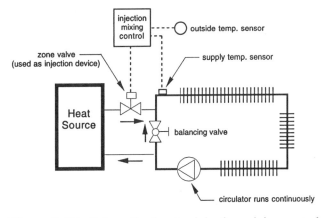

Figure 9–52 Schematic showing injection mixing control

The **injection mixing control** adds heat to the continuously circulating distribution loop by opening the zone valve. Flow between the heat source and distribution loop is induced by the pressure differential created by the partially closed balancing valve. The more this valve is closed, the greater the flow rate of hot water through the zone valve (when it is open). By controlling the length of the open cycle of the zone valve, the controller can increase the rate of heat injection to balance the rate of heat dissipation by the distribution system. This in turn maintains the building at a steady temperature.

As hot water enters the tee downstream of the zone valve, it is immediately blended with the cooler return water returning from the heat emitters. This prevents a slug of hot water from traveling out to the distribution system. The sensor measuring supply water temperature is located downstream of the mixing tee. It detects when the water temperature is slightly above the calculated temperature, and closes the zone valve to prevent further heat injection. As heat is released from the distribution system, the control reopens the valve as necessary to maintain the correct water temperature. The control also closes the zone valve if the supply temperature exceeds a high limit set by the installer. An example of an injection mixing reset control is shown in Figure 9–53.

This type of injection mixing system requires careful adjustment of the balancing valve. *The final adjustments to the balancing valve must be made when the distribution system is up to its normal operating temperature.*

This is particularly important when a radiant floor heating is being controlled. The usual procedure starts with the balancing valve and the zone valve fully open. This will result in the maximum ratio of return water to hot water entering the mixing tee. The valve should be very slowly closed while monitoring the mixed temperature exiting the mixing tee. When the exiting temperature is up to the maximum allowable distribution temperature, the valve is at its proper setting. If the return temperature from the distribution system is typical of normal operating conditions, the valve should not require any further adjustment. After making the final adjustment, the handle of the balancing valve should be removed to prevent tampering.

Variable Speed Injection Mixing

An alternate method of injection mixing control uses a variable speed circulator in place of the zone valve shown in Figure 9–52. The speed of the circulator is continually changed by the controller as necessary to increase or decrease the water temperature in the distribution system. This results in continuous heat injection at the same rate the system dissipates heat to its load. Fluctuations in water temperature in the distribution system are virtually eliminated by this technique. A piping schematic for this concept is shown in Figure 9–54. An example of a **variable speed injection mixing** control is shown in Figure 9–55.

Figure 9–53 Example of an injection mixing control. Courtesy of tekmar Controls.

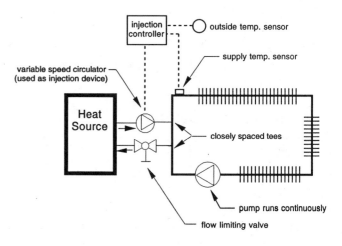

Figure 9–54 Use of a variable speed injection pump for reset control of distribution water temperature

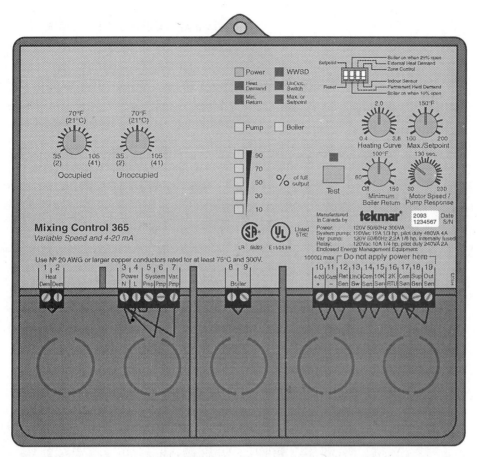

Figure 9-55 Example of a variable speed injection mixing control. Courtesy of tekmar Controls.

Currently available variable speed injection mixing controls modify the shape of the AC voltage signal sent to the circulator in order to control its speed. Only circulators with permanent split capacitor/impedance-protected motors (typical of most small wet rotor circulators) can be used as the variable speed circulator.

Because there is no positive flow shut-off between the heat source and the distribution loop, the tees that connect into the distribution loop should be spaced as close as possible. This makes the pressure at the two tees about equal, and therefore does not induce any significant flow between the heat source and distribution loop during low load conditions.

Both types of injection mixing systems shown will return the cooler return water from the distribution loop directly back to the heat source. This could present a problem if a low temperature distribution system such as radiant floor heating is used with a conventional gas or oil-fired boiler. Case study #2 presented later in this chapter shows a means of combining these subsystems without creating boiler condensation.

Other Features of Reset Controls

Any building with a heating load that depended only on the temperature difference between the inside and outside air could (theoretically) be totally regulated by a reset control once the correct reset ratio was set. Most building heating loads, however, are affected by internal heat gains from sunlight, lights, people, fireplaces, etc., that reduce the heating load below what the reset control thinks it should be at the prevailing outdoor temperature. Consider a passive solar house with auxiliary hydronic heating on a cold but *clear* winter day. The resistance of the cold outdoor temperature sensor is telling the reset control the building needs heat. The sunlight entering the large south-facing windows, however, is already providing plenty of heat. Since the reset controller has no way of detecting this, it continues to supply heat as if there were no solar heating taking place. Overheating, discomfort, and wasted fuel are the result.

This problem can be corrected using an indoor temperature sensor called a **room temperature unit (RTU)**, that communicates with the reset controller. The simplest type of RTU is a specially housed thermistor sensor located in the heated space. Other RTUs have adjustments and setback capability that rival dedicated electronic thermostats. An example of an RTU is shown in Figure 9–56.

If the indoor temperature climbs above the RTU setting, the normal reset control action is overridden, and heat input is reduced or stopped until room temperature returns to normal. The use of this indoor sensor, although

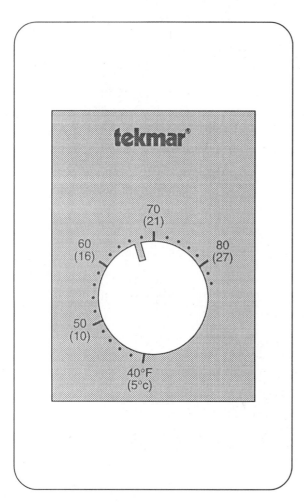

Figure 9–56 Example of a room temperature unit. Courtesy of tekmar controls.

deemed optional by control manufacturers, is very necessary in buildings with passive solar heating or other significant sources of heat gain. Most reset controllers, including those for boilers, mixing valves, and injection mixing systems, can accept input from an RTU.

Another feature provided on many reset controllers is called **warm weather shutdown**. This allows the controller to stop the circulator in the distribution system whenever the outdoor temperature rises to a value where no heating is required. The pump is automatically turned back on when the outdoor temperature drops below the setting on the controller. In the absence of this feature, an outdoor thermostat can perform a similar function. Its action should, however, be coordinated with that of the reset controller to prevent the latter from attempting to add heat to the system when the circulator is off.

Several microprocessor-based reset controls also have the ability to control domestic water heating when an indirectly-fired storage water heater is used. When domestic water heating is required, the reset operation is overridden to provide suitably high boiler temperatures for fast heating of hot water.

Reset Control of Modular Boiler Systems

Modular boiler systems were discussed in Chapter 3. They consist of two or more smaller boilers with a combined output equal to the design heating load of the building. Because these systems can be controlled in stages, they are able to maintain high operating efficiency even at low partial load conditions. Controls are the key to operating modular boiler systems at peak efficiency.

The control must perform two basic functions. First, it needs to continuously calculate the required supply water temperature based on the outside temperature and the reset ratio setting. Second, it needs to determine which boiler(s) should be operating at any given time to maintain this temperature. In making decisions about which boiler(s) should be operating, the control must also attempt to balance the run time of each boiler over an extended period of time. This allows all the boilers to be serviced at the same time. It also improves the likelihood of equal service life for each boiler.

Microprocessor-based controls are well-suited to this task because they can be precisely programmed to match the specific requirements of the modular boiler system. Controls are currently available to regulate systems ranging from two to ten boilers. This is more than adequate for systems in the scope of this book.

An example of a unit control that can handle up to a four-boiler system is shown in Figure 9–57. If fewer than four boilers are used, the additional stages can be disabled. This control performs both the temperature reset and firing priority functions described earlier. When any one boiler accumulates 48 more operating hours than the others, it becomes the last boiler to be fired. The boiler with the least accumulated running time becomes the next boiler to be fired.

This control can also be configured for a fixed lead boiler while rotating the other three. This mode is ideal when a condensing boiler is used for low load conditions, during which low supply temperatures enhance the formation of condensate in the active boiler. Other features include the ability to turn the system pump(s) on and off, correct system output for internal heat gains, and put the system into a reduced temperature setback mode.

9.7 DESIGNING CONTROL SYSTEMS

Entire books have been written describing methods for creating the control logic required for various heating system applications. Obviously this book cannot go into such detail. Fortunately, however, the level of sophistication required to control most hydronic heating systems is relatively low. By consistently applying a few basic

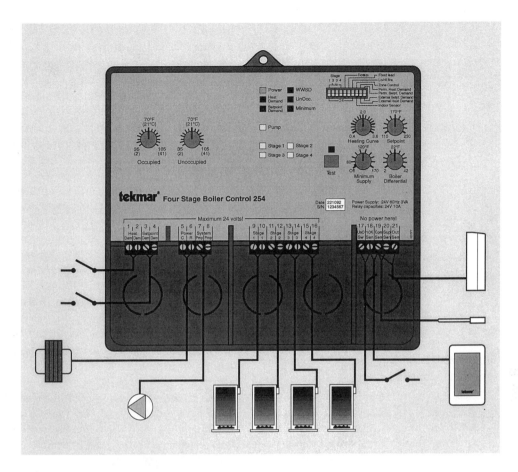

Figure 9–57   Microprocessor-based four-stage modular boiler control. Courtesy of tekmar Controls.

principles, the necessary control logic can usually be created. The principles are as follows:

Principle #1. When a device is to operate only when two or more conditions are met (as evidenced by closed switch or relay contacts), these contacts should be wired in *series*.

For example, the relay coil (R1) shown in Figure 9–58 will be energized, and remain energized, only if all three series-connected contacts are closed at the same time.

Principle #2. When a device is to operate when any one or more of several possible conditions are met (as evidenced by closed switch or relay contacts), these contacts should be wired in *parallel*.

For example, the relay coil (R1) shown in Figure 9–59 would be energized, and stay energized, whenever any one or more of the three contacts are closed.

Principle #3. The majority of any hard-wired control logic that must be created for a particular application should be done in the low voltage part of the system (e.g., the lower part of the ladder diagram). This approach is

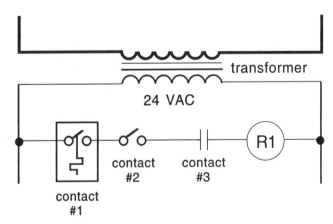

Figure 9–58   All three series-connected contacts must be closed at the same time to energize relay coil R1.

less costly because the low voltage and currents allow smaller components with lower amperage ratings to be used. It is also inherently safer if a short-circuit should occur.

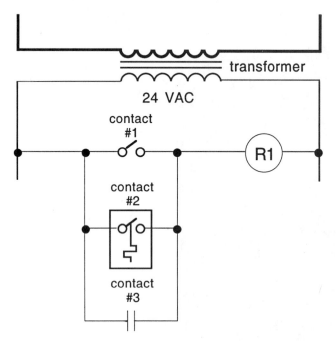

Figure 9-59 Relay coil R1 will operate when any of the three parallel-connected contacts are closed.

Principle #4. Whenever possible, attempt to build custom control systems, or portions thereof, around readily available unit controls. This saves both labor and material costs. In some cases two or more unit controls may have to be combined, or some separate general purpose relays added. Use the same type of control components (same brand) from one system to the next whenever possible.

Do not design around components that may be difficult to replace in the event they fail.

Principle #5. Do not assume the transformer supplied as part of a heat source control is capable of powering several other controls in the system. This transformer may not be designed to output significantly more VAs than required by the heat source's own control components. Adding several relays or powering a mixing valve from this transformer could overload it and burn it out. The best approach is to install a separate transformer for the auxiliary control components. Control transformers with 24 VAC secondary voltage are available in a wide range of VA ratings from about 20 VA through 100 VA. Some are designed for simple mounting on conventional 4 by 4 electrical junction boxes as shown in Figure 9–60. Others have brackets allowing them to mount inside control enclosures. As a general rule, avoid paralleling the secondary sides of two or more transformers to avoid phase mismatching problems.

Principle #6. *Always document the final control system design in the form of a ladder diagram, with a legend to identify all schematic symbols.* Provide a brief written description of the operation of the control system and instructions for any periodic adjustments. Include a table listing part numbers and the sources of supply for all components. A copy of this information should be left with the owner of the system, and a copy filed by the designer. Although this documentation sounds like a lot of work, it can become quite simple by using a computer to store drawing details, parts lists, and descriptions that can often be repeated in part from one job to the next.

Figure 9-60 Example of a control transformer designed to mount to a standard 4 by 4 electrical junction box. Courtesy of White Rogers.

Lack of good documentation for the control system is a sure recipe for costly problems on the job.

9.8 EXAMPLES OF CONTROL SYSTEMS

This section describes two control systems for hydronic heating in detail. These systems incorporate many of the design principles and control components previously discussed. Keep in mind there may be other ways of obtaining equivalent system control, and the approaches shown are only representative.

Example 9-1: A residential hydronic system will consist of two heating zones of finned-tube baseboard, and a third zone for domestic water heating. Each zone will be controlled by a three-wire, 24 VAC zone valve. The system's circulator must run when any of the zones calls for heat. A gas-fired boiler with a 24 VAC gas valve will be used. Each zone valve draws 17 VA when operating. This will require the use of a separate control transformer to provide sufficient power at secondary voltage. The piping schematic for the system is shown in Figure 9–61.

A ladder diagram showing the layout of the control system is shown in Figure 9–62.

Discussion: When either of the two room thermostats or the aquastat of the domestic water tank call for heat, they energize the heat motor of their respective zone valves. After two to three minutes opening time, the end switch in the zone valve closes, completing a voltage signal to the coil of the isolation relay R1. The R1 contact closes across the "T" terminals of the boiler's high limit/ switching relay control. The internal relay of this control enables two simultaneous actions. One contact of the internal relay starts the circulator. Another completes a 24 VAC circuit through the motor of the flue damper driving it open. When the flue damper blade reaches its fully open position, it closes an internal limit switch that places 24 VAC across the gas valve of the boiler allowing it to fire. If the boiler reaches its high limit setting before the call for heat is satisfied, the aquastat contacts open, shutting down the gas valve and allowing the flue damper to close. If the boiler temperature subsequently drops through the 8 °F differential of the high limit control, the flue damper will reopen, and the boiler resume firing. When all calls for heat from either of the room thermostats or the aquastat of the hot water tank are satisfied, all zone valves close, their end switches open, and the gas valve, damper, and circulator are shut off.

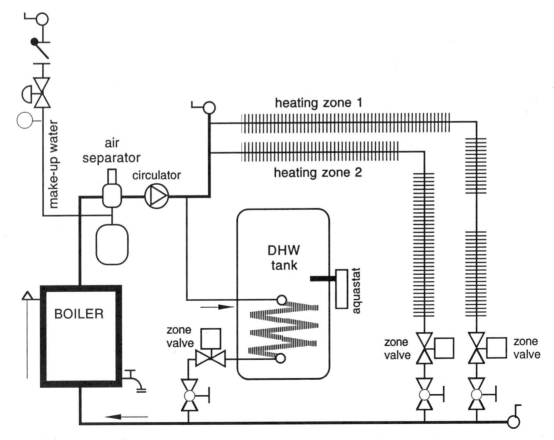

Figure 9–61 Piping schematic for system of Example 9.1

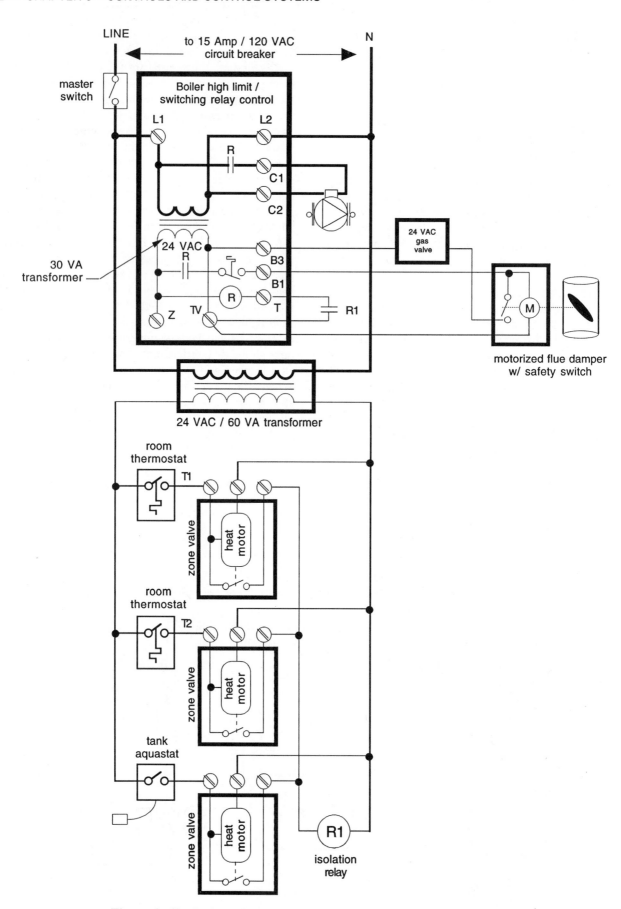

Figure 9–62 Ladder diagram showing control wiring for Example 9.1

Notice that a separate 60 VA rated transformer was used to drive the zone valves. This was necessary because the total VA rating of the three zone valves (51 VA) exceeded the 30 VA external load limit of the boiler's high limit control. The isolation relay R1 effectively isolates the transformer powering the zone valves, from the transformer in the limit control. Use of an even larger VA transformer would be advisable if the system could potentially be expanded in the future.

Example 9.2: A hydronic system will provide radiant floor heating for both the basement and first floor of a house. The basement is to be kept at a slightly lower temperature than the first floor most of the time. A planned future expansion of the building will use finned-tube baseboard convectors controlled as a separate zone. The system will also provide year round domestic water heating using an indirectly-fired storage tank. The heat source is a packaged oil-fired boiler with factory supplied limit control. A piping schematic for the system is shown in Figure 9–63.

The control schematic is shown in the form of a ladder diagram in Figure 9–64.

Discussion: Burner firing is enabled by any device that completes a circuit across the "T" terminals in the boiler's high limit/switching relay control. The primary circulator P1 will also operate whenever there is a completed circuit across the "T" terminals. The burner will stop firing if the boiler reaches its upper temperature limit, or if there is no further call for heat from any of the connected loads.

Upon a call for heat by the DHW tank, boiler firing is enabled by one set of contacts in the DPST tank aquastat. The other set of contacts starts secondary circulator P4 to draw water through the coil of the tank. When the tank aquastat reaches its upper limit setting, circulator P4 is shut off. Burner firing ceases unless another load is calling for heat.

Upon a call for heat from the zone of finned-tube baseboard, the contacts in thermostat T1 energize the coil of general purpose relay R1. One set of R1 contacts enable burner firing. The other set of R1 contacts start secondary pump P2 to circulate hot water to the zone.

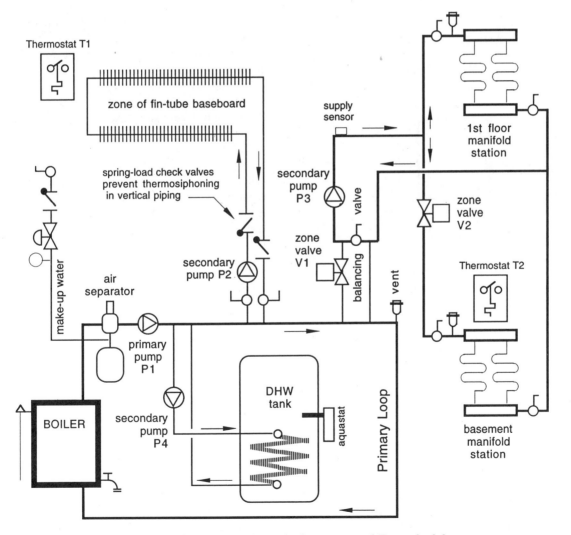

Figure 9–63 Piping schematic for system of Example 9.2

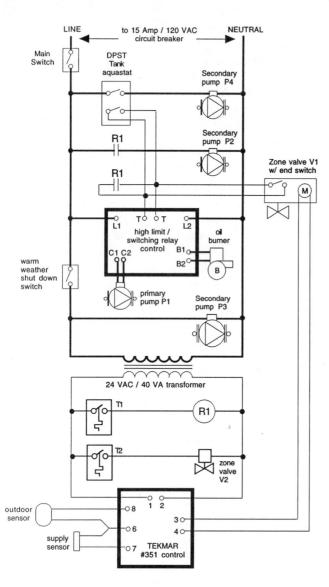

Figure 9-64 Ladder diagram showing control wiring for Example 9.2

When room thermostat T1 is satisfied, circulator P2 is shut off. Burner firing ceases unless another load is calling for heat.

Upon a call for heat from the tekmar 351 injection mixing control, zone valve V1 opens allowing hot water to flow into the radiant floor distribution circuit. The end switch of valve V1 enables boiler firing as long as the valve remains open. When the injection mixing control detects the supply water temperature is at the proper temperature, at its upper limit setting, or the room is warm enough, zone valve V1 is closed. Burner firing ceases unless another load is calling for heat.

Room thermostat T2 is mounted in the basement. In this system it serves as a temperature limiting device only, and cannot enable burner operation. If the basement air temperature drops, it completes a circuit through zone valve V2 allowing warm water to flow through the radiant floor circuits in the basement.

During warm weather, the secondary circulator P3 and the transformer supplying power to the injection mixing control are manually shut off by the warm weather shut down switch. The main switch shuts down power to the entire system when necessary for servicing.

SUMMARY

This chapter has examined several of the basic electro-mechanical and electronic controls used in hydronic heating systems. The controls shown and described are but a small sampling of the hundreds of control devices available for potential use in hydronic systems. New devices enter the market constantly.

Hydronic system designers should always investigate the details of any control device being considered for possible use. They must understand exactly how it works and how it should be set up for the application at hand. Be aware of any limitations such as voltage or current ratings, fixed versus adjustable differentials, accuracy, ability to get repair parts if needed, etc.

Always design control systems with installation and service personnel in mind. Make them as simple as possible while still providing the necessary control actions. Install the various components in a neat and well organized layout. Label all components. Keep circuits carrying line voltage separated from those operating at lower control voltage. Make use of multiple wire colors to designate different parts of the control system.

The importance of proper documentation of the control system cannot be overemphasized. Very minor problems can create hours of frustration and unnecessary component replacement when the service person does not have an understandable diagram of the entire system and a description of how it is supposed to work.

KEY TERMS

Ampacity
Aquastat
Bimetal element
Contactor
Differential
DIN rail
Electro-mechanical controls
End switch
Hard wired control logic
Heat anticipator
Heating curve
Injection mixing control
Ladder diagram
Line voltage
Low (control) voltage

Microprocessor
Modulating control
Multi-zone relay center
Normally closed contacts
Normally open contacts
On/off control
Operating mode
Outdoor reset control
Poles (of a switch)
Primary side (voltage)
Reset line
Reset ratio
Relay
Relay socket
Room temperature unit (RTU)
Room thermostat
Secondary side (voltage)
Setpoint control
Setpoint temperature
Stages
Tankless coil
Thermistor
Throws (of a switch)
Time delay relay
Transformer
Triple action control
VA rating
Variable speed injection mixing
Warm weather shutdown

CHAPTER 9 QUESTIONS AND EXERCISES

1. A heating system is designed to provide 180 °F water to the distribution system when the outside temperature is −15 °F. The building requires no heat input when the outside temperature is 65 °F. What would be the proper reset ratio setting for this system?
2. Explain the difference between the reset line for a hydronic system designed around finned-tube baseboard versus one designed around a heated floor slab.
3. Explain why a typical on/off control system will slightly overshoot the desired temperature setting of the room thermostat. How can this be minimized?
4. Explain the difference between the number of poles and number of throws on a switch. Sketch a schematic symbol for:
   a. A DPST switch
   b. A SPDT switch
   c. A SP8T (rotary) switch
5. Why is low voltage (24 VAC) used for control wiring rather than line voltage?
6. Explain the effect the thermal mass of a hydronic system has on the operation of the control system and the temperature swings inside the building. Describe a situation where a system with a high thermal mass is desirable. Describe another where it is undesirable.
7. Explain the advantages of modulating control versus on/off control when attempting to match the heat output of a hydronic system to the heat loss of a building.
8. Why do some reset controls provide a low limit setting for boiler temperature?
9. Explain the operating principle of electro-mechanical setpoint controls using a fluid-filled bulb sensor.
10. Explain how constant operation of the system circulator improves control system operation and comfort.
11. Complete the ladder diagram shown in Figure 9–65 such that the following control action is achieved. When switch S1 is set in the off position, a red 24 VAC indicator light is on. When the switch is set to the on position, a green 24 VAC indicator light is on, *and* a line voltage circulator is running. Label all components you sketch in the diagram.
12. An existing residential hydronic system consists of a single heating zone and a tankless coil for producing domestic hot water. Its current wiring schematic is shown in Figure 9–66. The system is to be converted to use an indirectly fired storage water heater. Two 24 VAC zone valves (with isolated end switches) will be installed to control water flow to space heating and through the coil of the new water heater. Show how you would modify the control system to

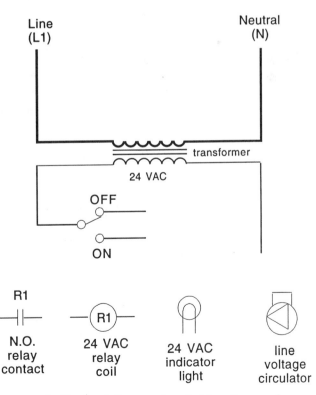

Figure 9–65 Partially complete ladder diagram for Exercise 11

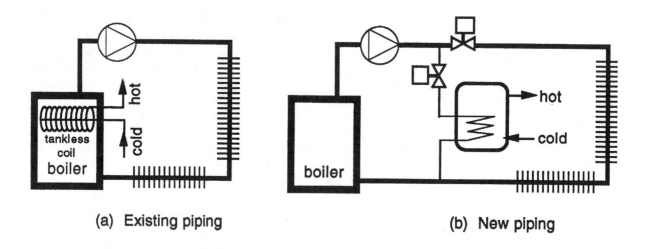

(a) Existing piping   (b) New piping

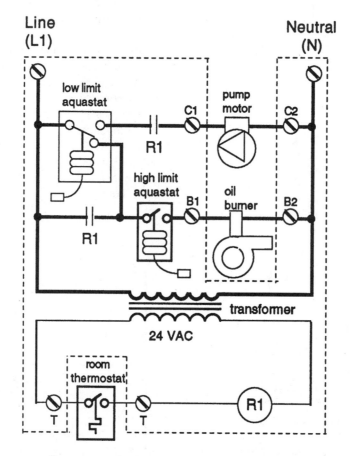

(c) Existing wiring using a triple action control

Figure 9–66 Diagrams for Exercise 12. (a) Existing piping, (b) new piping, and (c) existing electrical wiring.

discontinue firing the boiler for a low limit condition, and instead fire it on a call for heat from either the room thermostat or the aquastat of the new storage tank. You may make modifications within the triple action control if required. Sketch out a new overall ladder diagram after you have made the necessary modifications and additions.

13. A control system having the ladder diagram shown in Figure 9–67 is proposed to operate a three-zone hydronic system using a separate circulator for each zone. Identify any electrical errors in this diagram. Describe what would happen if the identified error were present when the system was turned on, or what is unsafe about the error. Sketch out how you would

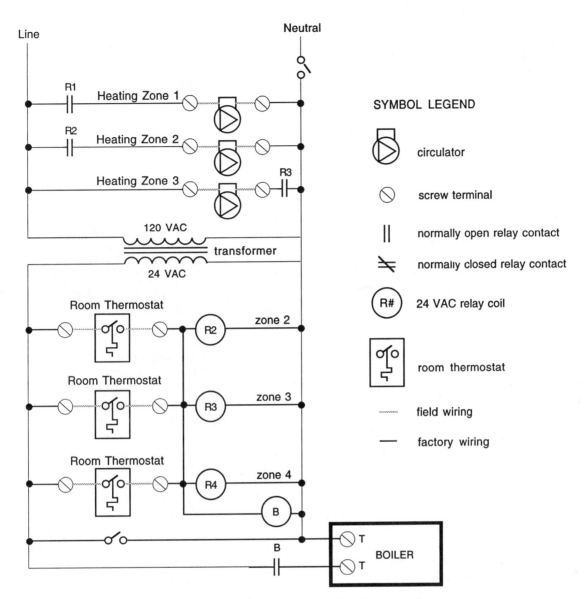

Figure 9–67 Ladder diagram for Exercise 13

modify the schematic for proper and safe operation of the three-zone system.

14. Sketch out a ladder diagram for the following control application.

On a call for space heating from a low voltage room thermostat, a boiler with a high limit/switching relay control as shown in Figure 9–29b is fired, and a pump (P1) starts to circulate water through a fan-coil unit. Because the water takes a few minutes to heat up, the blower of the fan-coil is not to operate until the water entering it has reached 140 °F. When the room thermostat is satisfied, the burner must stop firing, but the circulator and blower of the fan-coil should continue to operate until the water entering the fan-coil has dropped to 120 °F.

You may use any combination of low voltage or line voltage controls you desire. Compare your final design to those of other students, and decide who has the best design based on cost, simplicity, serviceability, and reliability.

10 RADIANT FLOOR HEATING

OBJECTIVES

After studying this chapter you should be able to:

- Explain how radiant heating affects comfort
- Compare radiant floor heating to other types of heating methods
- Identify good opportunities for radiant floor heating
- Describe the components used in a hydronic radiant floor heating system
- Discuss the installation procedure for slab-on-grade systems
- Discuss the installation procedure for thin-slab systems
- Describe the various types of dry system installations
- Compare the various methods of tubing placement
- Calculate the performance of a radiant floor heating system
- Compare the various system piping options used with floor heating
- Assemble control systems for radiant floor heating
- Avoid common errors in design and installation of floor heating systems
- Use the RADFLOOR program in the Hydronics Design Toolkit

10.1 INTRODUCTION

Heating a room by warming its floor is a concept that goes as far back as the Roman Empire. In those ancient buildings, floor heating was accomplished by directing the exhaust gases from wood fires to flow beneath the raised stone floors. Although it seems unlikely such systems could have achieved the even temperature distribution or accurate control possible today, the intricacy of their construction testifies to the fact they offered comfort unattainable by simpler means.

The arrival of forced circulation hydronic heating in the U.S. during the 1930s provided the needed link between producing heat in a boiler and efficiently transporting this heat to a floor slab. During this time steel pipe was threaded together to form distribution circuits that were then cast into concrete floor slabs. Some of those systems are still in use today. The rapid

growth in the copper water tube market in the 1940s further expanded the possibilities for hydronic floor heating. Thousands of systems using copper tube were constructed during the post-World War II housing boom. Many are still in use today.

During the 1960s and 1970s, hydronic floor heating steadily fell victim to American market forces that began substituting easier and less expensive heating options in place of proven comfort. As labor costs increased and central cooling using forced-air distribution systems became available, new hydronic floor heating systems became fewer and fewer. Interestingly, during this same time, new thermoplastic materials were under development that would eventually bring about a resurgence of interest in hydronic floor heating. After years of development, material such as cross-linked polyethylene (PEX) tubing arrived on the European market during the early 1970s. This tubing first appeared on the U.S. market during the early 1980s. PEX and other polymer-based tubing materials have revolutionized the installation of hydronic floor heating, providing fast installation and long service life. Over the last two decades, *billions* of feet of polymer tubing have been used for hydronic floor heating applications worldwide.

Today, hydronic floor heating is being used in all types of buildings including houses, schools, offices, aircraft hangers, fire stations, and more. When combined with modern control technology, hydronic floor heating represents what many, including the author, consider to be the ultimate form of heating. People are quickly rediscovering the comfort the Romans experienced nearly 2,000 years ago.

The many benefits offered by floor heating include:

- A system that creates unsurpassed thermal comfort
- A totally out-of-sight heating system that does not restrict furniture placement
- A durable system, as rugged as the floor itself
- A system with no operating noise
- A system that creates no adverse drafts or dust movement
- A system that can efficiently operate with low temperature heat sources
- A system that minimizes air temperature stratification from floor to ceiling
- A system that reduces energy usage compared to other heating options

This chapter discusses both the physiological and technological aspects of hydronic radiant floor heating. It shows how to lay out tubing circuits and analyze their thermal and hydraulic performance. Methods for installing the tubing in concrete slabs and on wood-frame floors are covered. The chapter concludes with a case study that shows the overall design, documentation, and controls of a floor heating system.

10.2 THE PHYSIOLOGY OF RADIANT FLOOR HEATING

Our bodies are constantly producing heat through a process called metabolism. In order to remain comfortable, this heat must be dissipated at approximately the same rate it is generated. *The feeling of thermal comfort could be described as being unaware of how or where the body is losing heat.*

The priority of the body is to maintain the temperature of its critical organs in the central torso. When placed in an environment that allows heat to be released from its surface faster than it is being produced internally, the body responds by reducing blood flow to its extremities (e.g., hands and feet). The feet in particular are most affected because they are furthest removed from the central body. To make matters worse, the feet are usually in contact with cooler surfaces that tend to draw heat away by conduction. They also have a greater surface area per unit of mass than do other parts of the body. To put it in mechanical terms, *the feet are like large radiators fastened to cold walls at the far end of a hydronic distribution system.* What heat they receive from the bloodstream can be easily dissipated to cool surroundings without having to "operate" at a very high surface temperature. When the surface temperature of our feet drops below normal we experience some degree of discomfort.

The head, on the other hand, usually has a good supply of heat-carrying blood. It is also insulated, to varying degrees, by hair. This combination allows the head to be comfortable at surrounding air temperatures several degrees cooler than near the feet. Most people feel comfortable and more alert when air temperatures at head level are in the low to mid-60 °F range.

The air temperature in the unoccupied zone above head level has a very minor impact on comfort but a significant impact on heat loss from the ceiling. It therefore makes sense to keep the air temperature near the ceiling as low as possible without affecting the occupied zone of the room.

Figure 10–1 shows a graph of how air temperature should vary from floor to ceiling in order to achieve *ideal* comfort at a relaxed activity level. Such a graph is call a **room air temperature profile**. Note the temperature

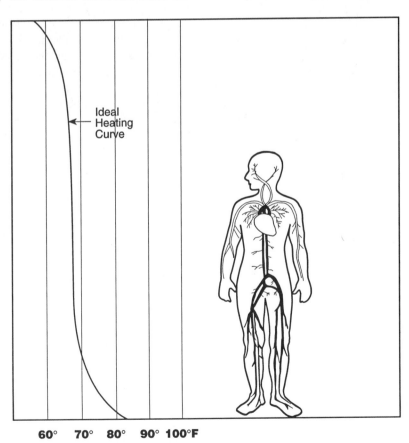

Ideal
Heating
Curve

60°   70°   80°   90° 100°F

Figure 10–1 Ideal air temperature distribution in a heated room. Courtesy of Wirsbo Co.

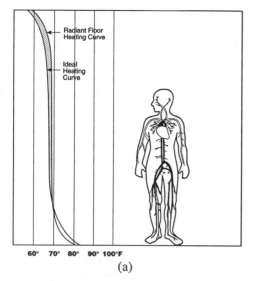

(a)

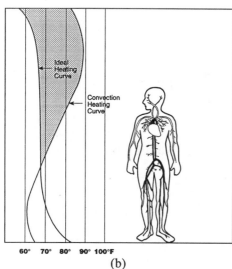

(b)

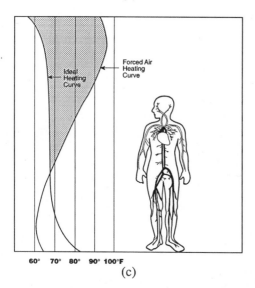

(c)

Figure 10-2 Comparison of temperature profiles of three types of heating delivery systems. (a) Radiant floor heating profile, (b) baseboard convector profile, (c) forced-air profile. Courtesy of Wirsbo Co.

near the floor needs to be higher than average. This, in effect, makes it harder for the lower extremities of the body to release heat. It forces them to maintain a surface temperature high enough to avoid discomfort.

The degree to which a heating system provides thermal comfort is in large part determined by the room air temperature profile it establishes and maintains in a given space. Profiles similar to that shown in Figure 10-1 are desirable. Unfortunately, many of the common methods of adding heat to a room result in significant deviations from this ideal.

Figure 10-2 compares typical room air temperature profiles created by three common methods of heat delivery: floor heating, baseboard convectors, and forced-air. Notice the profiles shown for baseboard convectors and forced air systems indicate lower than ideal temperatures near the floor and higher than ideal temperatures near the ceiling. This is a result of directly heating the air. The lower density of warm air causes it to rise toward the ceiling. The coolest air in the room sinks to floor level, exactly where it is least desired.

Floor heating, on the other hand, creates a temperature profile that is very close to ideal. This is because 80% to 90% of the heat output from the warm floor surface is in the form of **infrared radiation**. Although not visible to the eye, infrared radiation is in all other aspects identical to what we call light. It travels in straight lines from the point where it is emitted toward all other surfaces in the room that are at lower temperatures. As it travels between surfaces, only a small portion of this radiation is absorbed by the air molecules in the room. The rest is either absorbed when the radiation contacts an object or surface, or is reflected along another path. Considering the radiation travels at 186,000 miles per second, it does not take long for reflected radiation to be totally absorbed by objects and surfaces in a room. *When absorbed, infrared radiation ceases to exist as radiation and is instantly converted into thermal energy (heat)*, raising the temperature of the surface that absorbed it. As the objects and surfaces are warmed, they in turn reradiate part of their heat to any surrounding objects or surfaces at lower temperatures. They also release some heat to the room's air through natural convection. However, the convective air currents in a room heated by a warm floor are relatively weak. This is desirable because it reduces temperature stratification, especially in rooms with high ceilings.

The reader should note the room air temperature profiles shown in Figure 10-2 are for comparison only. In any given room the air temperature profile will be influenced by air motion, ceiling height, and insulation levels. Rooms with high rates of air flow and well-designed ducting systems will have less temperature stratification due to vigorous recirculation of air. Rooms with high ceilings or large window areas will tend to experience greater temperature stratification.

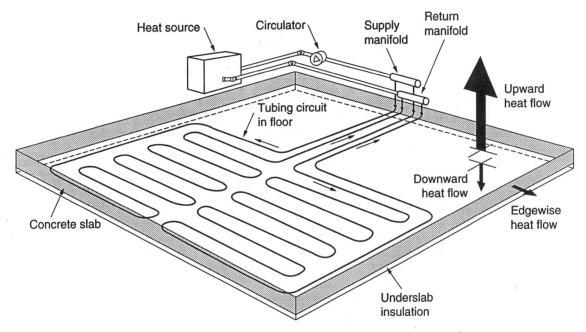

Figure 10–3   Basic layout of a hydronic radiant floor heating system

10.3   METHODS OF HYDRONIC RADIANT FLOORING HEATING

All hydronic floor heating systems consist of tubing circuits embedded in, or fastened to, some type of floor system. When warm water from a heat source is circulated through the tubing, heat is conducted vertically and horizontally away from the tubing by its embedment material. In a properly designed system, the majority of this heat travels to the top of the floor and is released into the room, primarily as infrared radiation. A smaller portion of the heat output is lost through the bottom and edges of the floor. A simplified concept of a modern hydronic floor heating system is shown in Figure 10–3.

Several specialized methods have been developed that allow hydronic radiant floor heating to be incorporated in many types of new and existing buildings. In this book the different approaches will be classified as follows:

- Slab-on-grade systems
- Thin-slab systems
- Dry systems

Each method holds certain technical or economic advantages in certain situations. Each will be detailed in separate sections of this chapter.

10.4   SLAB-ON-GRADE SYSTEMS

The most economical hydronic floor heating systems are those where a concrete slab is already planned as the floor structure. In such applications, the extra cost to make the floor a heated slab is limited to the tubing, underslab insulation, and the labor to place these materials. A cross section of a hydronically heated 4-inch concrete slab-on-grade floor is shown in Figure 10–4. A cut-away view is shown in Figure 10–5.

The tubing near the left side of Figure 10–4, and in Figure 10–5, is secured directly to the welded wire reinforcing using wire ties. Nylon pull ties can also be used when specified by the tubing manufacturer. Typical spacing for these ties is 24 to 30 inches on straight runs, and about 6 to 12 inches apart on curves. A close up of an installed wire tie is shown in Figure 10–6.

The tubing on the right side of Figure 10–4, and in Figure 10–7, is secured to the welded wire reinforcing by a nylon pipe clip that snaps onto the welded wire reinforcing. These clips are located at approximately the

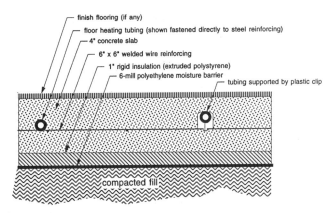

finish flooring (if any)
floor heating tubing (shown fastened directly to steel reinforcing)
4" concrete slab
6" x 6" welded wire reinforcing
1" rigid insulation (extruded polystyrene)
6-mill polyethylene moisture barrier
tubing supported by plastic clip
compacted fill

Figure 10–4   Cross section of a typical hydronic floor heating installation in a 4-inch concrete slab

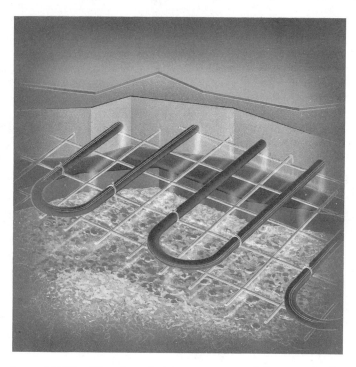

Figure 10–5   Cut-away view of PEX tubing installed in a concrete slab-on-grade floor. Courtesy of Wirsbo Co.

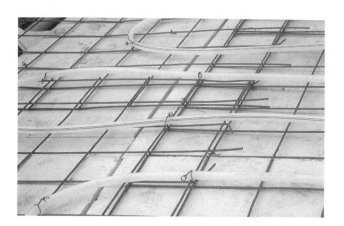

Figure 10–6   Securing PEX tubing to welded wire reinforcing using wire ties

Figure 10–7   Securing PEX tubing to welded wire reinforcing using plastic clips

same spacing as wire ties. Specialized tube clips are also available that screw directly into the under-slab insulation in cases where no wire reinforcing is needed.

Still another method of securing the tubing to the floor uses plastic tracks that are secured to the under-slab insulation with barbed staples as shown in Figure 10–8. These rails allow the tubing to be easily placed at different spacings when required. The tubing is uncoiled and snapped into the rails using light foot pressure as shown in Figure 10–9.

The preferred method(s) of attachment varies among tubing manufacturers. *In some cases the tubing's warranty is voided unless a specific method of attachment is used. Always use a fastening system endorsed by the tubing manufacturer.*

The tubing shown in Figure 10–4 is located at 12 inches on center horizontal spacing. This is a typical spacing that allows tubing such as PEX and polybutylene to be easily bent without kinking. It also coincides with the 6 in by 6 in grid of the welded wire reinforcing. In

Figure 10-8 Example of a plastic rail for holding floor heating tubing in place. Note the plastic staple securing the track to the foam insulation. Courtesy of REHAU, Inc.

some cases closer tube spacing is used to boost heat output. In others, the tubing can be placed 18 or even 24 inches on center when lower heat outputs are sufficient. The effect of tube spacing on thermal performance will be covered in Section 10.8.

The depth at which the tubing is cast into the slab inevitably varies depending on how the concrete is placed. The tendency is for the tubing to end up near the bottom of the slab because the welded wire reinforcing is not supported or properly lifted as the concrete is placed. Experience shows the system will still function well with the tubing near the bottom of the slab. A slightly higher water temperature is required as compared to installations where the tubing is approximately centered within the depth of the slab. The time required to warm the surface of the slab back to normal operating temperature after a setback period will also increase when the tubing is placed deeper in the slab.

The rigid foam insulation under the slab greatly reduces the downward heat flow from the slab into the soil. Many older radiant floor systems were installed with no such insulation. In such cases, the soil under the slab will absorb much of the heat transported to the slab until the soil finally achieves a near-equilibrium condition with

Figure 10-9   Tubing being pressed into the plastic rail using light foot pressure. Courtesy of REHAU, Inc.

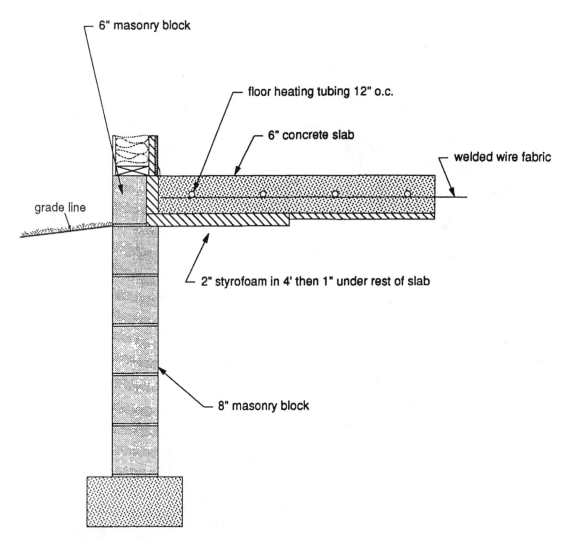

Figure 10–10   One method of incorporating edge and under-slab insulation into a building's foundation design

the slab. This greatly increases the sluggishness of the system during warm-up periods. It also results in greater downward heat losses. *All presently installed heated slabs should have a minimum of 1-inch extruded polystyrene or equivalent (moisture-proof) insulation beneath all interior areas.* The exception is at bearing pads that support columns, or footings that support bearing walls where the insulation should be omitted for structural reasons.

The edge of the slab is an area especially prone to high heat loss because of its greater exposure to lower ambient temperatures. Ideally, the under-slab insulation should be increased to 2 inches of extruded polystyrene or equivalent R-10 insulation within 4 feet of the edge of the slab. Two inches should also be used as a **thermal break** at the edge of the slab whenever possible. A detail showing one method of incorporating edge insulation is shown in Figure 10–10. Notice how the thickness of the exterior frame wall effectively covers the exposed edge of the 2-inch edge insulation. Other insulation details are possible depending on the nature of the construction. *The heating system designer should advise the architect of the need*

to incorporate suitable edge insulation as early as possible in the planning stages of the building.

Installation Procedure for Slab-on-Grade Floors

The installation of floor heating tubing in a slab-on-grade floor is an integral part of the construction sequence of a building. It must be preplanned and closely coordinated with the other trades involved in the construction. The importance of preplanning cannot be overemphasized. *Without an accurate plan showing the location and length of all piping circuits, as well as the location of the manifold stations, even an experienced installer can spend hours attempting installation by trial and error.* The results are still likely to be less than ideal. The installation procedure to be described assumes a layout plan has been prepared, and thus the installer's main function is to place the tubing according to this plan. An example of a layout plan for a small house is shown in Figure 10–11. Section

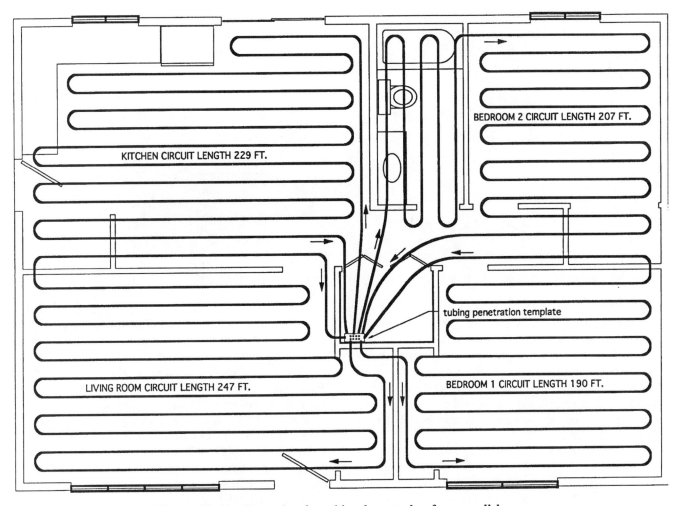

Figure 10–11   Example of a tubing layout plan for a small house

10.7 will describe how to prepare a layout plan accounting for several technical and architectural considerations.

Preparation

Before the radiant floor tubing is installed, all under-slab plumbing or electrical conduit should be in place. All under-slab utility trenches should be filled and compacted. The final grade under the slab should be accurately leveled allowing for the thickness of the slab and the under-slab insulation. This is especially critical near the edges of the slab where thicker insulation is used. Any loose rocks should be raked off, providing a smooth stable surface. A minimum 6-mil polyethylene vapor barrier should be laid on the soil, with all edges over-lapping at least one foot. The under-slab and edge insulation should be neatly placed with tongue and grooved joints engaged. The welded wire reinforcing should be placed and lapped 6 inches on all joints. Adjoining sheets of the wire reinforcing should be wired together. On windy days, the wire reinforcing should be laid on top of the foam insulation as soon as possible to prevent it from being blown away by a strong gust. The installer should double check the eventual location of all control joints, bearing pads, fastener penetrations, plumbing penetrations, or other features that could potentially interfere with tubing placement.

Locating the Manifold Stations

Begin the tubing installation by locating the **manifold station(s)**. These are the areas where the tubing will penetrate up through the slab surface and eventually connect to either supply or return piping. They should be clearly marked on the layout plan. In many cases manifold stations are located in studded wall cavities. Because these walls are probably not in place at this point, *it is crucial that piping penetrations are precisely located from the building's floor plan.* Since the eventual location of most walls cannot be changed, inaccurate placement of tubing penetrations through the slab could cause some of the tubes to fall outside the wall's thickness, requiring an aesthetically unpleasant cover-up detail.

It is helpful to construct a **template block** that will guide the tubing neatly through the surface of the slab. An example of a template block is shown in Figure 10–12.

Figure 10–12   Example of a template block used to guide tubing up through surface of the floor slab

They can be made out of a 2 × 4 or 2 × 6 to match the width of the wall in which the manifold station will be located. They are made by drilling two rows of holes through the 2 × 4 or 2 × 6 block, one for the supply ends of the floor circuits, the other for the return ends. The spacing of the holes should correspond to the spacing of the connections on the manifolds. Two inches center-to-center is common.

The template block is mounted on two 2 × 4 wooden stakes driven into the soil beneath the slab. Using a transit, these stakes are marked and cut so the top of the template block will be flush with the top of the floor slab when fastened to the top of the stakes. This allows the concrete to be neatly troweled around the template. It also provides for neat and properly located piping penetrations. Notice the template block shown in Figure 10–12 has holes large enough to accept plastic **bend supports** for the tubing. The bend supports allow the tubing to make a tight right angle turn in the slab without kinking.

Laying Out the Tubing

Using the previously prepared layout plan, locate all corners and return bends of the first tubing circuit. Use the 6 in by 6 in grid formed by the welded wire reinforcing as a reference. Mark these locations using a bright spray paint on the foam insulation. You are now ready to begin placing the tubing.

Select a continuous coil of tubing that is at least 5 to 10 feet longer than the horizontal length of the circuit. This allows extra pipe to rise up to the manifolds. Many tubing suppliers print sequential length markings on the tube in increments of 3 feet. By subtracting the length marking at one end of the coil from the marking at the other end, it is easy to estimate the coil length. If you have several coil lengths to choose from, pick the one that minimizes waste. Although some tubing manufacturers do

offer joint systems rated for permanent embedment in a concrete slab, it is generally best to avoid the need for such joints whenever possible.

The tubing must be unrolled off the coil to prevent it from twisting. It should never be pulled off the side of the coil. A convenient tool is a tubing uncoiler, as shown in Figure 10–13. This device is simply a rotating cradle for the tubing coil. Often times uncoilers can be borrowed from tubing suppliers. The uncoiler should be placed several feet away from where the tubing is being fastened down. Several feet of tubing should be pulled off the coil to allow slack while positioning the tubing.

Push one end of the tubing up through a bend support and template block, and pull it up at least one foot above the expected height of the manifold. This allows for slight height adjustment of the manifolds if necessary.

Begin fastening the tubing down along the preplanned circuit path. Typical fastener spacing should be 24 to 30 inches on straight runs, with three fasteners on return bends. Use additional fasteners anyplace the tubing tends to bow upward. *If the tubing is not adequately secured it will be pushed around as the concrete is placed.* Continue to work your way along the circuit fastening as you go. Carefully bend the tubing around corners and return bends. If the tubing is very cold, bend it slowly to avoid kinking. Keep an eye out for any sharp ends on the welded wire mesh that could gouge the tubing. Bend them out of the way. As you approach the end of the circuit, you will be near the manifold station where you began. Cut the coil allowing sufficient length to go through another bend support and back up to at least one foot above the expected manifold height. Be sure the return pipe goes up through the hole either in front of or behind the hole for the other end of the circuit. This keeps the piping straight and neat as it rises toward the manifolds. It is also a good idea to label each end of each circuit with a number and arrow indicating the planned

Figure 10–13   An uncoiler allows tubing to be pulled off the coil without twisting.

flow direction. The numbers should correspond to those on the layout plan. Repeat this procedure for each circuit the manifold station will supply.

After all circuits are in place they should be pressure tested to at least 50 psi for 24 hours. There are several ways of doing this. One method is to temporarily connect all the circuits into a series string using soft temper copper tubing bent into U-shapes and pushed into the ends of the tubing circuits as shown in Figure 10–14. Use ¹/₂-inch copper tube for ⁵/₈-inch I.D. tubing, or ³/₈-inch

copper tube for ¹/₂-inch I.D. tubing. These tubing sizes result in a tight press fit to PEX tubing. Use a small hose-clamp to secure all joints. The last circuit in the string should be adapted to a pressure gauge. The beginning of the string can be adapted to either a Schrader valve for compressed air, or a boiler drain if water will be used for the pressure test. Do not use water if it will remain in the tubing during subfreezing weather.

When the tubing is first pressurized it begins to slowly expand. This will sometimes cause a drop in the reading on the pressure gauge, and is not a cause for concern. The temporary joints set up for the pressure test should be checked for air leaks by brushing on a solution of dishwashing detergent and looking for bubbles. Usually any pressure drops are due to a leak in the temporary connections set up for the pressure test itself, or a loose connection at a manifold. With the tubing under pressure, walk around the area looking and listening for any leaks. Given the high quality standards of today's tubing materials, and with reasonable care during installation, leaks should be extremely rare.

Often the tubing will have to pass beneath the eventual location of a sawn **control joint** in the slab. Typically these joints are sawn to a depth of about 20% of the slab thickness soon after the concrete has set. Their function is to create a weak point in the slab to allow it to crack along a predictable line (the sawn control joint). Obviously, if the saw blade contacts a tube it could create a leak that would be very difficult to repair.

One way to prevent this from happening it to *minimize the number of locations where the tubing will pass beneath control joints*. This will be discussed further in Section 10.7. It is usually not possible to avoid routing tubing under all control joints. When it does become necessary to cross a control joint, the tubing and welded wire reinforcing should be held down close to the insulation board under the joint location. Since the saw only penetrates about 20% of the slab thickness, this

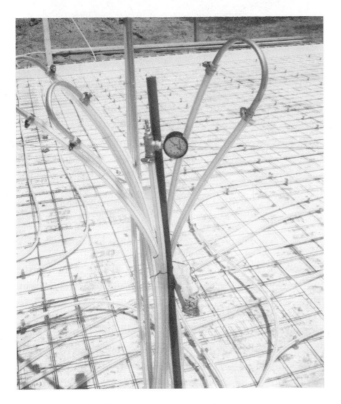

Figure 10–14   Temporary connections for pressure testing tubing circuits

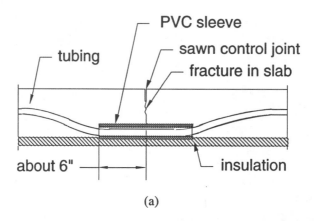

(a)

(b)

Figure 10–15 (a) Cross section showing tubing placed beneath a sawn control joint, (b) sleeves of plastic piping over tubing at location of a control joint.

should provide a generous distance between the bottom of the cut and the top of the tube. Another important detail is to sleeve the tubes for approximately 6 inches on either side of the control joint as shown in Figures 10–15a and b.

The sleeves can be cut from PVC or polyethylene tubing two pipe sizes larger than the heating tubing. They can either be slipped over the pipe as it is placed, or sawn in half and then reassembled around the tubing using duct tape. Their function is to provide a hollow cavity around the tubing under the joint. This relieves any bending

stresses on the tube caused by potential movement of the slab at the control joint. This same detail can be used at expansion joints or joints between successive pours.

When placing the concrete, try to have it flow parallel with the tubes as much as possible. This minimizes the tendency for the concrete to push the tubes sideways. One person should be given the sole task of pulling the welded wire mesh up to about the half depth of the slab while the concrete placement advances as shown in Figure 10–16. A blunt steel hook about 3 feet long with a loop-type handle on the end works well and minimizes back strain.

Figure 10–16   Use of a steel hook to lift the welded wire mesh with attached tubing during the pour

Once the concrete is placed, it can be screeded and troweled as normal. Care should be taken not to allow a power trowel to slide into the tubes where they penetrate the slab. If the tubes bend enough, the blades of the power trowel could hit and sever the tubes, causing a difficult and costly repair.

After the slab is poured, construction scheduling may result in a waiting period of several weeks before the manifolds themselves can be installed. During this time the ends of the tubing should be capped or taped off to prevent dirt or dust from entering. One possible way to do this is simply leave the temporary pressure testing connections in place.

An experienced installer can usually describe other tricks to the trade they have developed for efficient placement of the tubing. With a little practice the overall procedure is quite simple. With proper planning and materials, a crew of two can often place and pressure test 2,000 or more linear feet of tubing in a day.

10.5   THIN-SLAB SYSTEMS

There are many buildings where radiant floor heating would be desirable, but the existing or planned floor construction will not support the weight of a 4-inch or thicker concrete slab. Most typical are the wood-framed floor decks of residential or light commercial structures. Often the solution is to use a **thin-slab system** to cover the floor heating tubing that has been fastened directly to the subflooring. Thin slabs range from about 1.25 to 1.5 inches in thickness, and can be poured using several different materials. A typical cross section of a thin-slab floor heating system is shown in Figure 10–17. A cutaway view is shown in Figure 10–18.

Thin-slab systems use the same types of tubing that are used for slab-on-grade applications. The overall distribution system usually consists of several tubing circuits that begin and end at one or more manifold station(s). These stations are usually located in the

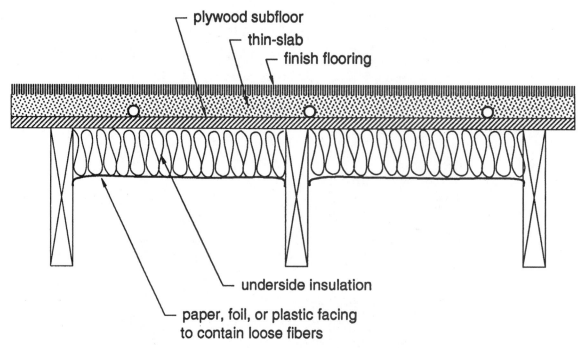

Figure 10–17   Cross section of a typical thin-slab floor heating system

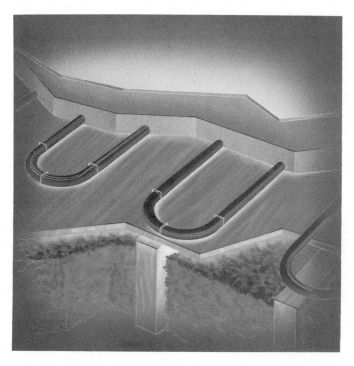

Figure 10–18   Cut-away view of PEX tubing installed in a thin-slab system over a wood-framed floor. Courtesy of Wirsbo Co.

framing cavity of a studded interior wall as shown in Figure 10–19.

A panel or hinged door is usually installed directly in front of the manifold station to allow full access for flow balancing or maintenance. Whenever possible, the manifold stations should be located so their access panels do not interfere aesthetically with the interior design of the building. *Making the door open to the inside of a closet is often a good choice.* Wherever the manifold station is located, be sure there is good access and room to work with tools if necessary.

Some manufacturers offer manifold stations complete with a metal cabinet and hinged door as shown in Figure 10–20. Such assemblies slide between, and are fastened

Figure 10-19 A manifold station located within a 2 × 4 wall partition. Note bend supports where tubing enters slab. Courtesy of Harvey Youker.

to, the wall studs. Be sure to allow space on one side of the manifold station for the supply and return piping connecting it to the remainder of the system.

The tubing is fastened directly to the plywood subflooring according to the layout plan. The location of straight tubing runs can be quickly laid out with a chalk line. The more intricate curves or bends can be marked with a lumber crayon. Fastening is usually done with pneumatically driven staples or nailed-down plastic tube clips. In either case it is critical that the tubing is not damaged by the installation of the fasteners. When pneumatic stapling is used, the tubing manufacturer or supplier often provides a specially set up stapler that straddles over the tubing and controls the depth of the staple to prevent it from pinching the tubing. Typical spacing for fasteners is 24 to 30 inches on straight runs, and as necessary on return bends and corners. *Because of the limited thickness of a thin slab, it is critical that all tubing lay tightly against the subflooring.* Figure 10–21 shows tubing circuits secured to the subfloor using pneumatically driven staples.

Planning for Thin Slab Installations

Because a thin slab is an integral part of the building, it must be preplanned to avoid problems during installation. Ideally the decision to use a thin-slab floor heating system should be made during the early planning stages of the building. Two important aspects must be addressed at this stage: The extra weight the system will impose on the floor structure and the extra height it will require.

The extra weight or **floor loading** created by a thin-slab system depends on the material used and its thickness.

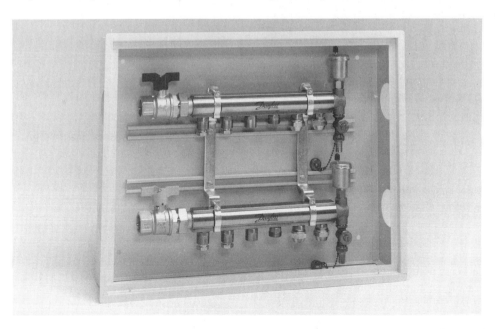

Figure 10-20 Example of a preassembled manifold station complete with metal cabinet and hinged access door. Courtesy of Danfoss, Inc.

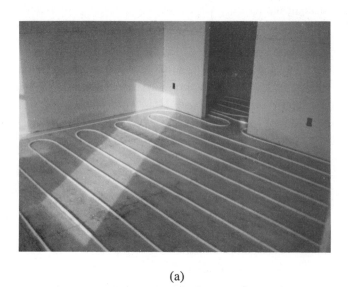

(a)

(b)

Figure 10–21 (a) A tubing circuit stapled down to a plywood floor deck awaiting the pour, (b) close-up of a pneumatically driven staple. Notice the staple does not pinch the tubing.

Most of the currently used materials, including gypsum-based underlayments and lightweight concrete, add from 12 and 15 pounds per square foot to the dead load of the floor. This extra loading may require stronger floor framing than needed for a conventional floor. In some cases joist size might have to be increased, or joist spacing decreased. If in doubt, the floor system should be evaluated for stress and deflection by someone competent with structures. *To simply assume the floor will handle the extra loading could easily create a situation that violates building codes.*

The extra height associated with the thin slab will affect several other aspects of the construction. The following details must take the added thickness into account:

- Overall height and headroom requirements of stairs
- Rough opening height of windows and doors
- Raised thresholds for exterior doors
- Height of kitchen and bathroom cabinets
- Height of plumbing fixtures such as bathtubs and showers
- Height of floor flanges for toilets

Ideally all these items will be adjusted to accommodate the extra thickness of the thin slab during the planning stages of the building. When the decision to incorporate thin-slab floor heating is made *after* construction has started, these details can create major difficulties, and add significant cost to the project.

The common method of adjusting kitchen and bathroom cabinets for thin-slab systems is to install wooden sleepers on the floor under the front and side edges of the cabinets, and omit the slab from these areas. Floor flanges for toilets can also be raised the thickness of the thin slab using plywood shims. These details are shown in Figure 10–22.

Thin-slab Materials

The materials used for embedding tubing for thin-slab systems fall into two categories:

- Self-leveling **gypsum-based underlayments**
- Portland cement-based **lightweight concrete**

Each material has advantages over the other. The decision on which material to use must reflect cost, availability, ease of installation, possible exposure to moisture, and choice of finish floor materials.

Gypsum-based Underlayments

Gypsum-based underlayments have been used for many years to level old floors during remodeling. They are now being widely marketed for thin-slab floor heating applications. They consist of a mixture of a proprietary gypsum-based cement, sand, water, and additives to

(a)

(b)

Figure 10–22   (a) Sleepers used to raise kitchen cabinets to allow for thickness of thin slab. Courtesy of Dick Davis. (b) Plywood shims used to raise the height of a toilet flange to allow for thickness of thin slab.

improve flexibility and reduce shrinkage. The resulting material has a dry density of about 115 lb/ft³. When poured to a finished thickness of 1.5 ins, this adds about 14.4 lb/ft² to the dead loading of the floor. Most gypsum-based underlayments must be installed by a certified crew trained by the manufacturer.

The following installation sequence is typical for gypsum-based thin slabs.

After all tubing has been fastened to the floor and pressure tested, the installer sprays the floor with a combination sealer and bonding agent as shown in Figure 10–23. This material limits water absorption by the subflooring. It also strengthens the bond between the plywood and the thin slab.

The gypsum-based underlayment is prepared in an outside mixer, and pumped into the building through a

Figure 10–23   Combination sealer/bonding agent sprayed on subflooring prior to pouring thin slab.

Figure 10–25   Installer places first lift of gypsum-based underlayment. Note its thickness is approximately the same as the tube's diameter.

large hose as shown in Figure 10–24. The material has the consistency of pancake batter as it is spread on the floor.

The installer moves back and forth across the room as the material flows from the hose. Its consistency allows it to flow around the tubing, and into any small gaps such as under the edge of drywall. This is very desirable because it improves conduction of heat away from the tubing and provides a good seal against air infiltration along the lower edge of exterior walls. A wooden float is used to ensure the material is evenly placed. Its thickness at this stage is equal to the diameter of the tubing. This layer is called the first "lift" of the material. See Figure 10–25.

The first layer of material will be dry enough to walk on in about two hours. It undergoes slight vertical shrinkage as it dries. This shrinkage becomes noticeable adjacent to the tubing, which does not shrink. The second lift is now poured to a depth sufficient to cover the tubing to a minimum depth of ³/₄ inch as shown in Figure 10–26. A wooden float suspended on two small pins is

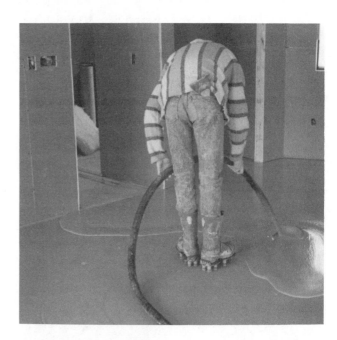

Figure 10–26   Installer places second lift of gypsum-based underlayment over partially hardened first lift.

Figure 10–24   Gypsum-based underlayment is mixed outside and pumped in through a hose.

used to ensure uniform thickness. As the second lift dries, its vertical shrinkage is uniform because the first lift removed differential shrinkage caused by the tubes.

Most gypsum-based underlayments can be walked on a few hours after being applied. However, depending on site conditions, they can take five days or more to dry to the point where finish floor materials can be applied. The drying is often accelerated by operating the heating system after the slab has been poured. The dryness of the material can be checked by taping down a piece of transparent polyethylene sheeting and watching for the formation of water droplets on its underside. *Be absolutely sure the material is thoroughly dry before installing finish flooring materials.*

The primary advantages of gypsum-based underlayments are the speed and ease of their installation. Because they are pumped in through a hose, and essentially self-leveling, the labor required to place gypsum-based underlayments is considerably less than that required for concrete. A trained crew of three can easily install 2,500 ft² of gypsum-based thin slab in a day. Some interior work can usually resume the next day. Gypsum-based underlayments are also less prone to shrinkage cracking than is a thin layer of concrete. Their ability to decrease air infiltration by sealing the small cracks along the base of walls and partitions is also an advantage.

The final compressive strength of gypsum-based underlayments is usually in the range of 1,500 to 2,500 psi depending on the proportions of the mix. This is adequate to support foot and light equipment traffic during construction. Care must be taken, however, not to gouge the completed floor with heavy or sharp objects. *Gypsum-based underlayments are not designed to function as the permanent wearing surface of the floor.* After drying, gypsum-based underlayments can still be damaged by prolonged contact with water. A minor leak from a plumbing fixture or manifold connection will cause the material to soften similar to how gypsum wallboard reacts to persistent contact with water. The makers of gypsum-based underlayments often specify the thin slab be sealed with specific products to protect them against moisture from spills or mopping that may penetrate the finish floor. These sealants also prevent fine dust from working up through carpeting over time.

Gypsum-based underlayments do not have thermal conductivities as high as concrete. This creates a more noticeable temperature distribution between the tubes and theoretically requires a slightly higher water temperature for equivalent heat output.

Lightweight Concrete Thin Slabs

The main alternative to gypsum-based underlayments is lightweight concrete. A typical mix will have a dry density of about 100 to 110 lb/ft³. This adds 12.5 to 14 lb/ft² to the dead loading of the floor when applied at a thickness of 1.5 ins. Because it is made with Portland cement, lightweight concrete is very resistant to moisture damage. It can also serve as a wearing surface, although thin slabs are usually covered with a finish floor in most residential and light commercial applications. Depending on the mix formulation, lightweight concrete will usually have a higher thermal conductivity than gypsum-based underlayments, although still not as high as concrete made with crushed stone aggregate.

The main ingredients in lightweight concrete floor toppings are Portland cement, sand, a lightweight coarse (3/8-in) aggregate, chopped nylon fibers, and several

Portland cement (94 lb. bags)	5.5 bags
Sand	1160 pounds
Water	37 gallons
Norlite lightweight aggregate (3/8")	900 pounds
Hycol (water reducing agent)	15.5 ounces
Daravair (air entraining agent)	4.0 ounces
WRDA-19 (superplasticizer)	51.7 ounces
Fiber mesh	1.5 pounds

Figure 10–27 Suggested mix design for lightweight concrete. Courtesy of the Stadler Corporation. *Note:* Care must be taken not to add too much water to the mix. This will always cause more shrinkage cracks and in many cases will result in a very dusty surface.

additives to improve flexibility and reduce shrinkage. The mix design listed by one tubing manufacturer is given in Figure 10–27. The quantities given are for 1 yd³ of lightweight concrete with a (28 day) compressive stress of 3,000 psi, and a dry density of 105 lb/ft³.

The cost of obtaining and placing lightweight concrete will depend on local availability and labor rates. The heating system designer should compare its installed cost against that of gypsum-based underlayments in the locality of the project.

Lightweight concrete can be placed from wheelbarrows or pumped through a hose using a grout pump. It is not self-leveling and therefore must be accurately screeded to the proper height. One cubic yard will cover about 210 square feet of floor when applied at a thickness of 1.5 inches. *It is crucial the installation crew is sufficient in number and experience to be able to place and level the concrete at about the same rate it is delivered.* When delivery is from a batch plant, the crew should be able to handle at least 4 cubic yards in no more than an hour.

Control joints should be planned to limit eventual cracks to a prescribed path. The joints can be sawn after the concrete is placed if the tubing does not pass beneath the cut, and the saw can be properly positioned. An alternative is to use thin plastic angle strips or lightly oiled wooden strips that are fastened to the subfloor to divide the overall slab into several small areas. The height of these strips should be about 3/8 inch less than the thickness of the slab. This sufficiently weakens the slab so any cracking will occur along the path of the strip. Doorways are often good locations for control joints. This breaks each room into a separate small slab area as shown in Figure 10–28, which can absorb more building movement without random cracking. Some type of a bond-breaker such as a concrete form release agent or polyethylene strip should also be used wherever the edge of the thin slab adjoins walls or other fixed objects. If a form release agent is used, be sure it is chemically compatible with the tubing material.

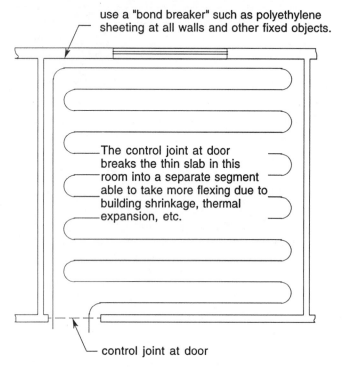

use a "bond breaker" such as polyethylene sheeting at all walls and other fixed objects.

The control joint at door breaks the thin slab in this room into a separate segment able to take more flexing due to building shrinkage, thermal expansion, etc.

control joint at door

Figure 10–28 Break individual rooms into independent thin slabs using control joints and bond breakers at walls. This allows slabs to absorb building movement without random cracking.

In general, the smaller the areas defined by the partitions and control joints, and the lower the water content of the concrete, the less likely random cracking is to occur. The effect of small hairline cracks on thermal performance is insignificant. Cracks can, however, be a cause of concern for finish floor materials such as ceramic tile. If such materials are going to be used, it is wise to consult the flooring goods manufacturer for recommendations.

When a thin-slab distribution system is operating, part of the heat output goes up into the room, and part goes down into the space below the floor. The proportions of upward versus downward heat flow depend on the R-value of the material(s) above and below the slab, and on the temperatures of the spaces above and below the floor. If, for example, a thin-slab system is covered by a thick carpet and pad on the top, while the underside of the subflooring is exposed to a cool basement, a major part of the heat output from the floor will be in the downward direction. This can cause the basement temperature to be higher than desired, while the room(s) above the floor remain uncomfortably cool. To prevent this, the underside of the floor deck must be insulated to minimize downward heat flow as shown in Figure 10–29.

The amount of insulation installed beneath the floor depends on the temperature of the space beneath the floor, and the R-value of any finish floor materials above the thin slab. A rule of thumb is to *provide a minimum of five times more R-value beneath the slab than above it.* The R-value of some common finish floor materials can be found in Figure 10–49.

The following suggested R-values have worked well in the author's experience:

1. Floors over heated space: R-11
2. Floors over partially heated basements: R-19
3. Floors over vented crawlspaces or other exposure to ambient conditions: R-30

The insulation can be any suitable product that is relatively easy to install, will stay in place, and retain its

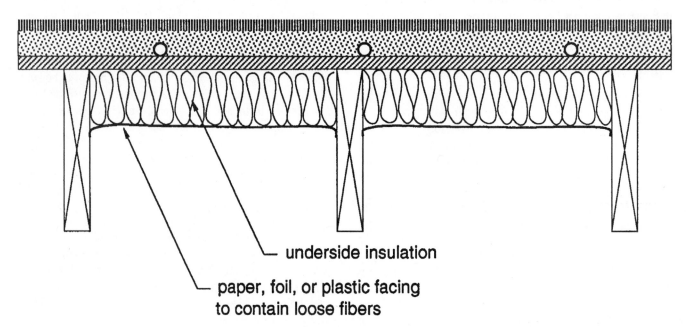

underside insulation

paper, foil, or plastic facing to contain loose fibers

Figure 10–29 Placement of fiberglass batts to limit downward heat flow

R-value. Fiberglass batts are a good choice. If the batts are left exposed to the basement below they should have either a plastic, paper, or foil facing to prevent loose fibers from drifting through the air. It is usually better to install the batts after the wiring and plumbing systems are in place. The cost of the insulation and its installation should be included in the estimated cost of the system unless it is otherwise accounted for in the construction budget.

The Youker System

The placement of concrete thin slabs in buildings where interior partitions are already erected can be slow due to constant maneuvering around walls, doorways, etc. An innovative approach to this problem was suggested by Harvey Youker, a long-time colleague who has installed many hydronic floor heating systems. I will refer to it as the "Youker system" to give credit where it is due.

The Youker system requires an early commitment to the use of the thin-slab approach, and a finalized floor plan showing the exact position of walls, fixtures, etc. It is implemented as soon as the plywood floor deck of the building is in place. The location of all tubing circuits and partitions are first marked on the floor deck. The deck is then covered with a transparent 6-mil polyethylene moisture barrier, or sprayed with a suitable sealer/bond breaker to limit water penetration into the plywood and prevent bonding between the plywood and concrete. 2 × 4 or 2 × 6 bottom plates are fastened to the floor at all locations where walls or partitions will eventually be placed. Penetrations for plumbing stacks, ventilation ducts, etc., that will not be routed through partitions are located and blocked-off with 2 × framing lumber. These blocks can later be lifted out or sawn out as necessary for the passage of mechanical systems. All tubing circuits are placed on the floor, routed to the location of the manifold stations, and pressure tested as usual. The edges of all wall plates and blocks are then coated with a suitable bond-breaker material. The floor is now ready for the placement of the concrete. Because there are very few, if any, objects sticking up above the bottom wall plates, the concrete can be rapidly placed and screeded. The 2 × wall plates serve as well-positioned screed guides for a 1.5-inch thick slab. The concrete can be floated and troweled to the extent required by the finish floor materials. The next day framing can continue on the now flat floor deck. Illustrations showing the concept of the Youker system are given in Figures 10–30a through 10–30d.

10.6   DRY SYSTEMS

Some installation methods for floor heating tubing do not use any poured materials, and are often called **dry systems**. Such systems add very little weight to the floor structure. This allows them to be used in both new and retrofit applications that otherwise would be eliminated based on the load carrying ability of the floor.

Most dry systems involve the use of aluminum **heat transfer plates** that conduct heat laterally away from the tubes. In effect, these plates replace the lateral heat conduction provided by slabs. Some methods locate the plates above the plywood subfloor, others below it. Cross sections of two such systems are shown in Figure 10–31.

In an **above deck dry system**, the fins of the heat transfer plates are supported by sleepers that have been previously glued and nailed to the floor deck. The gap between the sleepers provides a cavity for the tube and the U-shaped channel of the heat transfer plate. The plates are installed only on straight tubing runs. They are placed into the groove between the sleepers and stapled down on *one side only*. This allows the plates to move slightly as they change temperature. It also allows the plates to lay down without wrinkling as the upper cover sheet is applied. After the heat transfer plates are in place, the tubing is uncoiled and pushed down into the U-shaped channels. No adhesive is used. After pressure testing, the floor assembly is covered with a 3/8-inch plywood cover sheet, and eventually with the finish flooring. Corners and return bends can be formed using a plunge router to mill out the sleepers for the tubing, or by fastening down pre-cut curved blocks.

Above deck dry systems are very labor intensive to install. They also require a significant amount of material for sleepers and the cover sheet layer. Preplanning is essential. Tube layout should favor long straight runs with a minimum of corners and return bends. Accurate measurements of tubing position are important, especially after the tubing is hidden by the cover sheet. If hardwood floors are nailed down over such a system it is crucial not to drive nails into the tubing. *The sleepers and cover sheets should be glued as well as nailed down to avoid the possibility of squeaky floors.* The sleepers and cover sheet raise the floor height by at least one extra inch. This in turn requires the same adjustment to door heights, stairs, etc., as mentioned for thin-slab applications.

Below deck dry systems tend to be less expensive and less complicated than above deck dry systems. They also lend themselves to retrofit applications without disturbing the existing flooring materials. In most cases the tubing is cradled by the heat transfer plates that are stapled to the bottom of the floor decking between the floor joists, as shown in Figure 10–32.

The installation of a below deck dry system begins by drilling two holes side by side through the floor joists to allow the tubing to be pulled to each of the joist cavities. See Figure 10–33. These holes should be at least $1/2$ inch larger in diameter than the outside diameter of the tubing. They should be kept in a straight line by snapping a chalk line along the joists. Ideally the holes should be located at half the depth of the joists to minimize weakening the

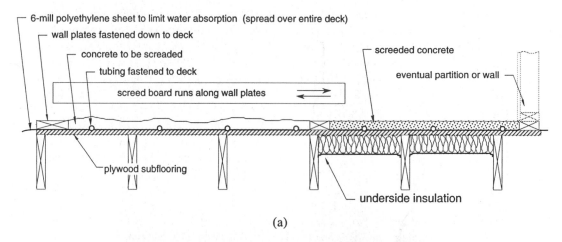

(a)

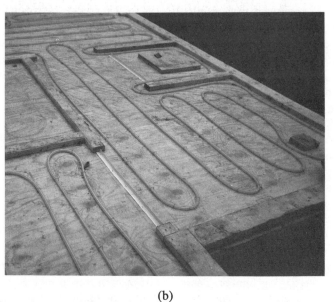

(b)

(c)

(d)

Figure 10-30   (a) Cross section of the Youker system. Note that no walls have been erected prior to pour. (b) Tubing fastened in place over plywood floor deck that has been previously covered with polyethylene film. Bottom plates of walls fastened in place. Courtesy of Walt DeLong. (c) "Pea stone" mix concrete poured over tubing. Note plywood scrap beneath nose of wheelbarrow to protect tubing. (d) Concrete is screeded flat using bottom wall plates as screed guides. Note previously poured floor area in background.

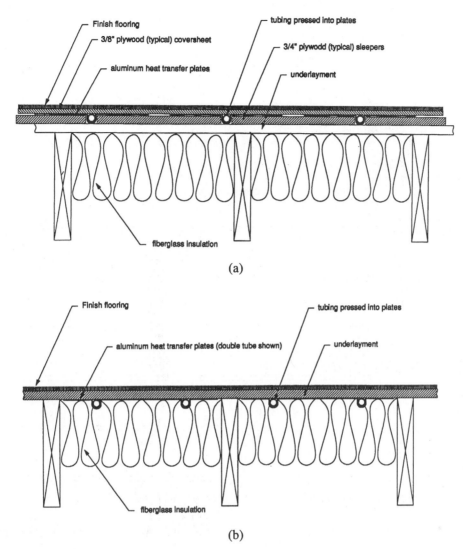

(a)

(b)

Figure 10–31   Cross sections of dry systems for radiant floor heating (a) using heat transfer plates above the subfloor, (b) using heat transfer plates below the subfloor.

Figure 10–32   Aluminum heat transfer plates supporting tubing to the underside of a floor deck. Courtesy of Radiant Technology, Inc.

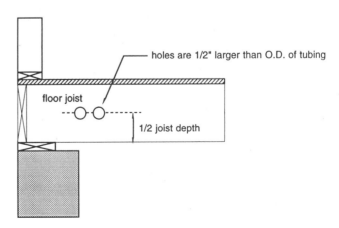

Figure 10–33   Oversized holes through floor joists allow tubing to pass to each joist cavity.

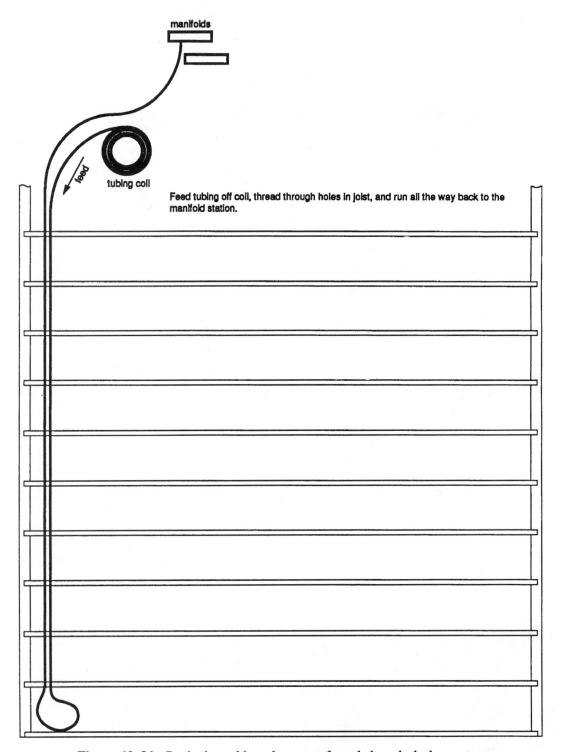

Figure 10–34 Beginning tubing placement for a below deck dry system

structure. *In no cases should the bottom edge of the joists be drilled out or notched.*

The tubing is first pulled to the furthest joist cavity through the oversized holes in the joists as shown in Figure 10–34. The tubing required to run up and back along a joist cavity is then pulled into the cavity where it will be placed. It is pushed up to the underside of the floor deck and a plate is stapled in place to hold it. Work progresses along the joist cavity to the opposite end where a return bend is formed. This procedure is repeated as the return side of the tubing loop is secured to the deck. A minimum 1/4-inch gap should be left between all heat transfer plates to allow for thermal expansion. The installation continues from one joist cavity to the next as shown in Figure 10–35 until the entire floor area is completed.

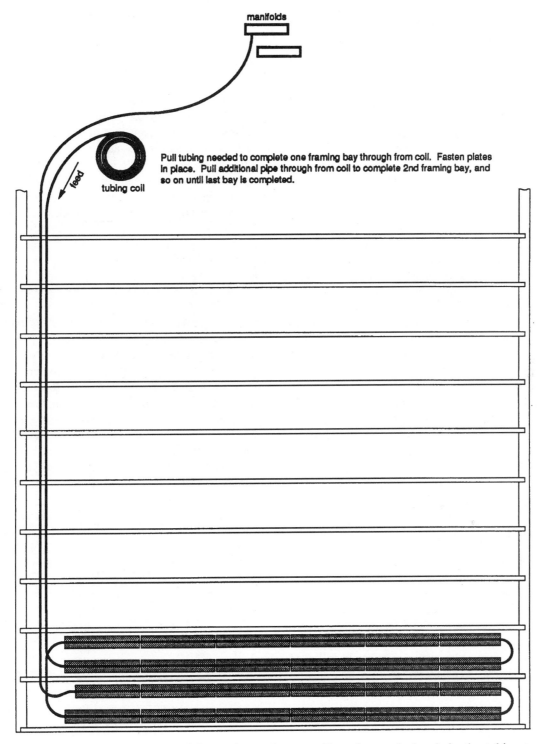

Figure 10–35 Work progresses from one joist cavity to the next. Note the gentle bends in the tubing to prevent kinking.

All curved sections of tubing should make gentle bends to prevent the possibility of kinking. Any pinching or binding of the tubing as it passes through the joists should be avoided because it can lead to expansion noises as the tubing heats and cools.

A number of factors can complicate this installation process. These include interference from plumbing, wiring,

or ducting in the joist cavities; structural blocking or bridging; and nails projecting down through the decking. It is imperative to coordinate with the other trades working on the floor so interference to tube placement can be minimized. Retrofit jobs should also be carefully inspected for such interference before committing to this approach. One final point to ponder: *Be sure all trades working on the project are made aware of the presence of the tubing beneath the floor. Warn them not to drill or saw holes in the floor before verifying the exact position of the tubing.*

Plateless Dry Systems

Some trade information shows tubing directly stapled to the underside of the floor without the use of heat transfer plates. In the author's opinion this is a very questionable practice. Such installations will inevitably result in portions of the tubing not in contact with the underside of the floor deck. This problem is worsened when the tubing is filled with hot water. Heat transfer becomes highly dependent on natural convection, which is a relatively weak process considering the small surface area of the tubes. Even tubes touching the floor deck have very little actual contact area. The wooden deck is much less efficient than an aluminum plate at conducting the heat laterally away from the tubes. The only way to compensate for these inherent heat transfer limitations is to significantly raise the water temperature in order to drive the heat out of the tubing at a sufficient rate. Temperatures in the range of 150 °F to 160 °F may be required. This eliminates the use of most low temperature heat sources. Such high temperatures might also lead to stability problems with certain subflooring materials.

10.7 TUBE PLACEMENT CONSIDERATIONS

A critical aspect of planning a hydronic radiant floor heating system is deciding how to divide the floor area of the building into smaller areas that are each heated by individual piping circuits. Once these areas are defined, the actual path of the piping circuits serving each area must be planned and analyzed. The end result of this process will be a layout plan that can be easily followed on the job. Experienced installers can attest that such a plan can saves hours of frustration and prevents costly errors on the job.

The piping circuit layout of a given system must account for a number of factors including:

- The dimensions of the spaces
- The design heating loads of the spaces
- Ease of tubing installation
- Any objects that interfere with tubing placement
- The position of the manifold stations
- The position of any control joints
- How the building will be zoned for two or more temperatures
- The types(s) of floor coverings used

- The length of the piping circuits
- The horizontal spacing of the tubing
- The position of the cold walls
- The possibility of fasteners being driven into the floor

With all these considerations, it is safe to say any given designer will probably produce a layout plan that differs somewhat from that of another designer. Several different layouts for the same building could in fact work, although there would likely be some variations in comfort, hydraulic efficiency, and the ease or cost of installation.

Tubing layouts are developed by applying several general principles to the extent allowed by a specific building. Often these principles involve thermal and hydraulic performance considerations that are detailed in later sections of this chapter. In some cases the principles compete against each other for inclusion in the final design. Inevitably tradeoffs must be made to accommodate the needs of each building.

Principle #1. *Install the tubing so the hottest fluid is first routed along the exterior wall(s) of the room.* This allows the highest heat output to occur where the losses from the room are greatest. It is similar to the idea of locating baseboard convectors at the base of exterior walls. The higher floor surface temperatures partially compensate for the lower surface temperatures of exterior walls and windows. Figure 10–36 shows three different routing paths for the tubing in a room having one, two, or three exposed walls. Notice in all three cases the warmest water is first routed along the exposed walls. The

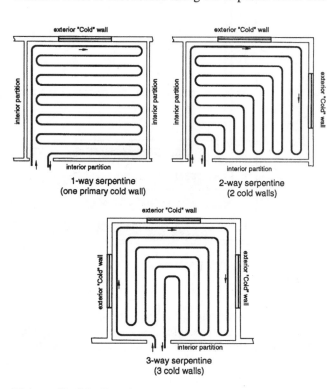

Figure 10–36 Routing paths that direct warmest water near exposed walls

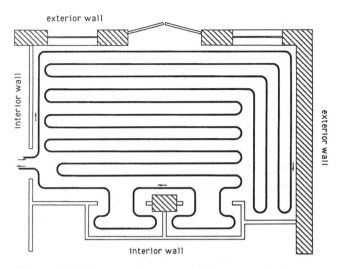

Figure 10–37 Two-way serpentine creates perimeter zone along exposed walls while interior of room uses one-way serpentine to maximize straight runs and minimize bends.

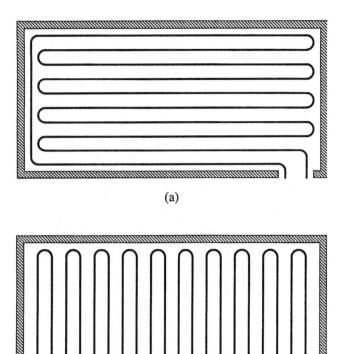

(a)

(b)

Figure 10–38 (a) Preferred tubing layout maximizes straight runs and minimizes return bends and corners, (b) Poor layout creates many more return bends.

circuit then works its way back toward the interior area of the room.

These paths are suggested for relatively small rooms with this type of exposure. In larger rooms it may prove more feasible to provide a **perimeter zone** by routing tubing parallel to the cold exterior walls until the path has reached 4 to 6 ft into the room, then switch to a one-way **serpentine** to minimize short segments and return bends. An example of such a layout is shown in Figure 10–37.

Principle #2. *Whenever possible, arrange the tubing to maximize straight runs and minimize return bends and corners.* This principle speeds installation and reduces the possibility of damage due to kinking the tubing. The narrower a room is relative to its length, the more important this principle becomes. Figure 10–38 illustrates it for a room twice as long as it is wide.

Principle #3. *Limit individual circuit lengths of ⁵/₈-in or ³/₄-in nominal I.D. tubing to 400 ft. For ¹/₂-in I.D. or smaller tubing, a suggested maximum circuit length is 300 ft.* These numbers limit the head losses as fluid flows through the tubing circuits to reasonable values that can be handled by small wet rotor circulators.

One can approximate the circuit length required to cover a certain floor area by dividing the floor area (ft²) by the spacing of the tubing (ft). The results of these calculations are shown for several tube spacings in the table of Figure 10–39.

Principle #4. *Design circuits so room-by-room zoning is possible even if it is not immediately used.* This is especially important for rooms that will be adjusted for different temperatures. For example, it usually makes sense to use separate piping circuits for a bathroom and an adjacent bedroom. Most people like the bathroom to be somewhat warmer for showers and baths, and prefer

Center-to-center spacing of tubing	Linear feet of tubing required per square foot of floor area
4 inches	3 feet
6 inches	2 feet
8 inches	1.5 feet
9 inches	1.33 feet
12 inches	1.0 feet
15 inches	0.8 feet
18 inches	0.67 feet

Figure 10–39 Linear footage of tubing required for various tube spacings

the bedroom to be slightly cooler for sleeping, or to reduce energy usage when the room is unoccupied.

One feature of modern radiant floor heating systems is the ability to add zone control by screwing small 24-volt heat motors called **telestats** onto the manifold valves that control flow to each circuit served by the

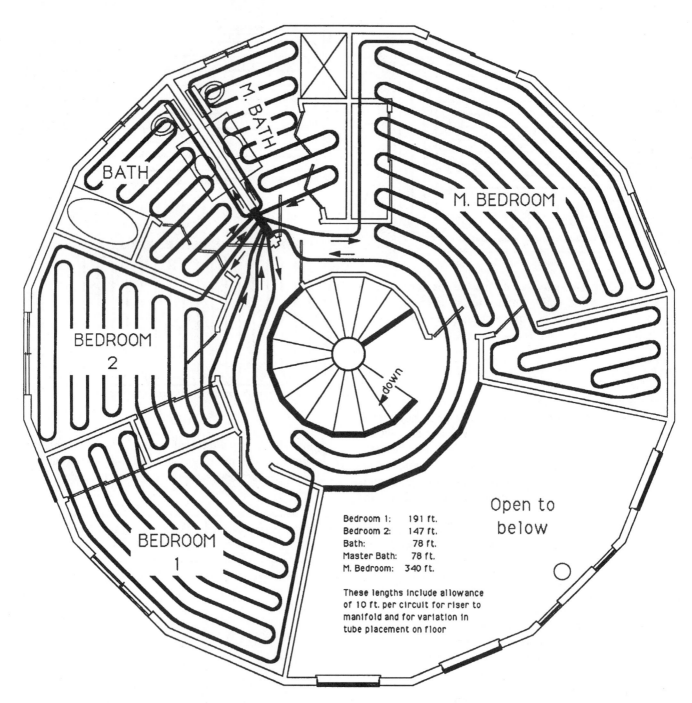

Bedroom 1:   191 ft.
Bedroom 2:   147 ft.
Bath: 78 ft.
Master Bath: 78 ft.
M. Bedroom: 340 ft.

These lengths include allowance
of 10 ft. per circuit for riser to
manifold and for variation in
tube placement on floor

Figure 10–40   Example of room-by-room circuit layout

manifold. A system that has tubing circuits arranged on a room-by-room basis can easily be modified for independent zone control by adding a telestat and room thermostat for each room. In anticipation of future zone control, some installers run thermostat cable from the location of the manifold station to a small wall-mounted junction box in each room served by the manifold station. The junction boxes can be covered with blank face plates until a thermostat is added. The small cost of installing the thermostat cables while the building framing cavities are still exposed saves considerable time over installing the cable in a finish building.

This principle is often the dominant concept that determines how a building such as a house is divided into individual circuits. Figure 10–40 shows the concept of room-by-room circuit design for an unusually-shaped house.

Principle #5. *Minimize situations where the tubing has to pass through control joints in the slab.* In most cases some tubing will have to cross through control joints. When it does it should be held low in the slab to prevent

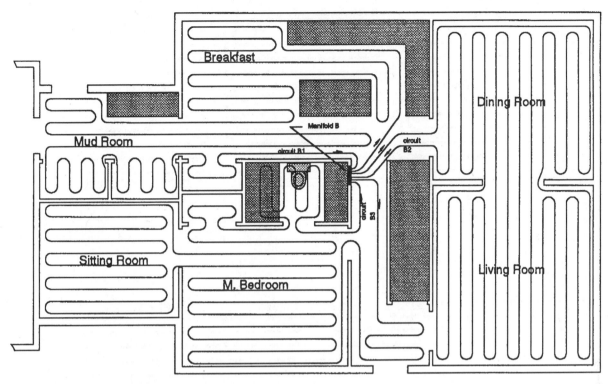

Figure 10–41   Example of tubing layout from a centrally located manifold station

any possible contact with the saw. It should also be protected by a plastic pipe sleeve to relieve stresses that could result if the slab does flex slightly at the control joint. A detail of tubing passing through a control joint is shown in Figure 10–15.

Principle #6. *For slab-on-grade systems, tubing placement should correspond with the grid formed by welded wire reinforcing whenever possible.* This allows for efficient tubing placement and easy installation of fasteners. Standard welded wire reinforcing uses a 6 in by 6 in grid pattern. It is also available, although less commonly stocked, in a 4 in by 4 in pattern.

Principle #7. *Install manifold stations in central locations to minimize tubing runs between the manifold station and the room(s) to be heated.* Think of the manifold station as the hub of a wheel, with the supply and return pipes of individual circuits as the spokes. By keeping the manifold station centrally located to a group of rooms, long, closely spaced tubing runs along hallways can be avoided. This saves tubing and thus reduces cost. Quite often a group of bedrooms with one or two bathrooms can be efficiently supplied from a single manifold station located in a partition near the central hallway. Another common grouping is the typical living room, kitchen, dining room arrangement. Of course there will always be exceptions. Figure 10–41 shows a tubing layout with a centrally located manifold station.

Principle #8. *Whenever possible avoid layouts that require tubes to cross over each other.* This is an absolute must for thin-slab systems. If unavoidable, crossovers can

be made when the tubing is installed in a 4-in or thicker concrete slab.

Principle #9. *Avoid routing the tubing where the likelihood of fasteners penetrating the floor is high.* This reduces the chance of the tubing being punctured after it is installed. Areas where fasteners are likely include beneath interior partitions, under doors with thresholds, or where equipment will be secured to the floor. There is no way to guarantee a fastener will not penetrate a tube at some point in the life of the building. An accurate layout plan showing the location of the tubing, along with avoiding high probability areas is the best defense.

One solution that has worked on numerous slab-on-grade projects is to coordinate with the building designer so that interior partitions over slab areas are bonded to the floor with construction adhesive rather than mechanical fasteners. This usually allows much simpler layouts and faster installation because the tubing can pass repeatedly beneath interior partitions. *Again it is vitally important all trades involved in the project are informed about not penetrating the slab at certain locations.*

Figure 10–42 is a comparison of two different tubing arrangements for a group of rooms or offices with a single exposed wall. In Figure 10–42a the tubing has been arranged so only one passage beneath the interior partition separating rooms is required. Assuming the location of the crossing is avoided, fasteners could be driven as required to anchor the partition to the slab. Because the circuit passes from one room to the next in series, however, the heat output of the floor in the first room will

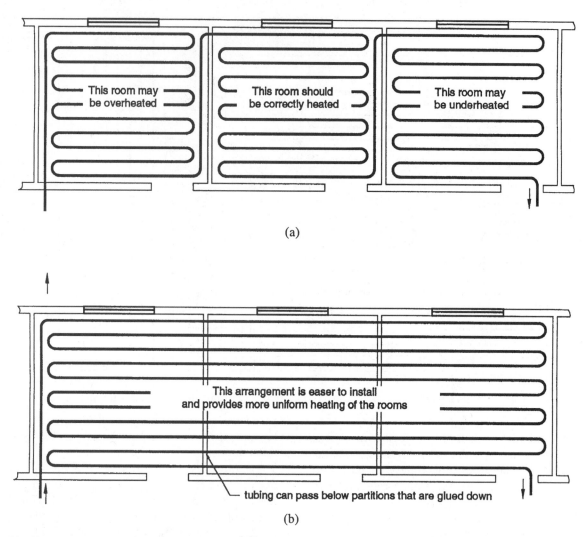

(a)

(b)

Figure 10–42 Comparison of tubing layouts. (a) Pattern that minimizes routing under partitions to minimize potential puncture from a fastener, (b) preferred pattern allowed by securing partitions to slab with construction adhesive.

be the greatest, while that in the last room will be the least. Since the rooms may have nearly identical loads, this could overheat the first room and underheat the last room. This piping arrangement also requires more bends and corners and thus will take more time to install.

Figure 10–42b shows the tubing circuits repeatedly crossing under the partitions that have been secured to the slab with construction adhesive. This arrangement provides more even heating of the rooms. It also can be installed faster because of longer straight runs and fewer bends and corners.

Drawing the Layout Plan

A critical aspect of planning a floor heating system is producing the overall tubing layout drawing. Often the final layout is determined by sketching a proposed layout, analyzing the thermal and hydraulic performance of the piping circuits, and then, if necessary, modifying parts of the layout.

Start with an accurate scaled floor plan of the building. Plans are usually available for new or proposed construction from either the owner or designer. A drafting supply store or print shop can make copies of the floor plan so the originals can be returned unmarked. When plans do not exist, the building must be measured and a scaled floor plan drawn. To some (those who have never tried to install a properly designed floor heating system), this is viewed as a waste of time. The truth is this time will ultimately be returned in the form of a faster/error-free installation, with less wasted material.

A common scale to work with is ¼ in = 1 ft. Graph paper with ¼-in squares can be used as a convenience. The plan should be drawn accurately. At a minimum it should indicate the location off all walls, doorways, cabinets, bathroom fixtures, and closets. The type of finish floor used in different areas of the buildings should also be noted. The approximate location of the heat source should be shown to help lay out the piping to the manifold station(s). This plan, along with a listing of window and door sizes, can also be used to determine the design heating loads of the rooms.

The plan should be drawn dark. When finished it can be overlaid with a translucent sheet of drafting vellum on which the circuit layout can be sketched. If changes are made, the vellum can be erased or replaced without affecting the floor plan drawing.

Start the layout by locating the manifold stations (see Principle #6). These locations will serve to anchor the floor circuits. Their location should minimize the distance to the various room circuits.

Assess the general areas that will be serviced by individual circuits. Principle #4 will influence circuit placement on a room-by-room basis. Plan on having two or more piping circuits in larger rooms. Decide what type of piping layout is appropriate for the type of exterior exposure the room has.

Begin the drawing by locating the piping line nearest the exposed wall(s) of the room. To avoid carpet tack strips or future holes drilled for stereo cables, etc., keep the pipe a minimum of 6 inches away from the wall. Using a ruler, and a proposed tube spacing of 12 inches, make tick marks where the straight runs will fall on the floor plan. If, for example, one of the straight runs is right along the edge of the opposite wall of the room, the starting line near the exterior wall may have to be shifted slightly to fit the overall serpentine pattern in the room. Pay particular attention to getting the piping with the warmest fluid directly to the exterior wall as discussed in Principle #1. Also give consideration to where the tubing will enter and leave the room. Avoid routing it where fasteners might be driven into the slab.

Some tubing manufacturers offer templates with rows of prearranged return bends for a serpentine layout pattern as shown in Figure 10–43. These drawing aides can significantly speed up the layout process. With experience these templates can be used to make quick checks on how the tubing will fall relative to walls or other obstacles.

Avoid routing tubing under cabinets, *or where refrigerators or freezers will be located.* For slab-on-grade floors it is desirable to route tubing under bathtubs or shower enclosures if it does not interfere with drainage piping. This may not be possible for thin slab or above deck dry system installations because the bathroom fixtures may be installed prior to placement of the floor heating tubing.

After a circuit has been sketched, it should be measured for length. The fastest way to do this is to break it up in straight segments, semicircles, and miscellaneous curved segments. In many cases, there will be several straight segments of the same length, especially when a one-way serpentine is used. When this occurs, one need only measure one straight length and multiply by the number of times it is repeated. The return bends are also a fixed length quantity. *Each 180 degree return bend will have a length approximately equal to the spacing of the tubing times 1.6.* The number of return bends can be added

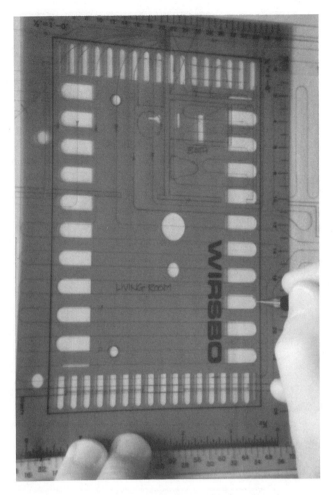

Figure 10–43   Use of a special template to draw serpentine patterns for tubing layout

up and multiplied by this length. The length of the miscellaneous curves is the most difficult to estimate. One tool that is helpful is a small measuring wheel. This tool is available from drafting supply stores. It is rolled along the path to be measured, and the distance read off the appropriate scale on the dial. Figure 10–44 shows how a piping circuit should be thought of for measuring purposes.

After the lengths of all straight sections, return bends, and miscellaneous curves have been added together, *it is prudent to add 10 feet to the total plan length to account for the length of tubing required to reach the risers.* This extra length also allows for small measuring errors, or variations in field measurements. The calculated length of each circuit should be double checked. Remember, it is much less costly to make a mistake on paper, and correct it than it is to end up several feet short of the return manifold as the tubing is being fastened down.

The use of a computer-aided drafting (CAD) program can be a tremendous time saver for producing piping layouts for radiant floor heating projects. Since each CAD system has its own particular commands and techniques, it is impossible to specify the best approach. A common

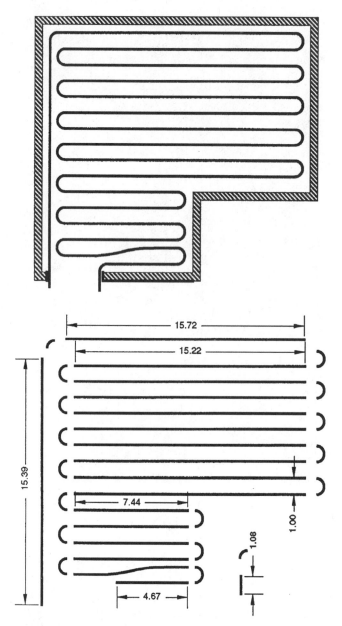

Figure 10-44 (a) Tubing circuit layout in a room, (b) how the circuit should be thought of for measuring purposes.

method is to set up a separate drawing layer over the floor plan with a grid of equally spaced lines that would represent the centerlines of the straight runs of tubing. Other lines can be set up at right angles to these lines to represent the end of the return bends. Using the CAD program's fillet tool, it is possible to add the quarter or semicircular bends and corners. The piping layout can be moved or adjusted as a whole or in sections using a command that groups the line and circular objects together. Perhaps the ultimate feature offered on many CAD systems is the ability of the program to automatically and precisely measure the lines designated as parts of a piping circuit. The circuit is thus measured as soon as it is drawn or modified.

10.8 THERMAL PERFORMANCE OF FLOOR HEATING CIRCUITS

Because hydronic floor heating systems are an integral part of a building rather than a manufactured heat emitter assembly, their thermal performance is affected by many factors including:

- The type of floor system (e.g., slab-on-grade, thin-slab, dry system)
- The type of floor covering(s) used
- The effectiveness of the underside insulation
- The size, spacing, and depth of the tubing
- The bond between the tubing and the material that conducts heat away from it
- How the piping circuit is routed in the room

An accurate theoretical analysis of the performance of a radiant floor heating system requires a level of detail and computer modeling well beyond what can be used as a practical design method. What will be presented is a method that combines data and procedures from several industry sources, into a reasonably accurate design tool.

Heat Flux

The heat output of a floor heating system is usually expressed as the rate of heat flow/ft^2 of heated floor area. This is often called the **heat flux** from the floor. In the U.S., the common units for heat flux are Btu/hr/ft^2 of floor area. Most of the heat output from a floor heating system should flow upward from the floor. This is called upward heat flux, and is represented by q_u. The heat that flows downward per unit of floor area is called downward heat flux and is represented as q_d.

The *average* upward heat flux *required* in a given room can be found by dividing its design heating load by the square footage of floor area *available* for heat output. The **available floor space** is often less than the total floor area due to the presence of cabinets, kitchen islands, shower stalls, etc.

(Equation 10.1)

$$q_u = \frac{\text{average upward}}{\text{heat flux}} = \frac{\text{design heat load of space}}{\text{available floor area of space}}$$

Example 10.1: A room has a calculated design heating load of 5,000 Btu/hr. The room measures 12 ft by 18 ft and contains a 2-ft wide floor cabinet along the entire length of the 12-ft wall under which no heating tubing will be installed. Calculate the required average upward heat flux for the room.

Solution: The available floor area is the floor area of the room minus the floor area occupied by the cabinet:

Available floor area = (12)(18) − (2)(12) = 192 ft²

The required average upward heat flux is:

$$q_u = \frac{5,000 \text{ Btu/hr}}{192 \text{ ft}^2} = 26 \text{ Btu/hr/ft}^2$$

Keep in mind the upward heat flux calculated from Equation 10.1 is the average value required in the space. The heat output of the floor near the beginning of the piping circuit will be higher than this value because of the higher water temperature in the tubing. The output near the end of the circuit will be lower because the water cools down as it passes along the circuit.

In addition to upward heat flow, it is important to account for the downward heat flow during the design process. The better the underside of the floor is insulated, the lower the downward heat flow will be. For slab-on-grade applications, the average downward heat flux can be estimated using the following equation:

(Equation 10.2)

$$q_d = \frac{4.17(P_e)(T_{slab} - T_{outside})}{(R_{edge} + 5)(A)}$$

where:

q_d = estimated average downward heat flux (Btu/hr/ft²)
P_e = exposed perimeter of the slab (e.g., exposed to outside) (ft)
T_{slab} = estimated operating temperature of floor slab at design conditions (°F)
$T_{outside}$ = outdoor design air temperature (°F)
R_{edge} = R-value of any edge insulation used
A = available floor area of the space (ft²)

Example 10.2: A 12 ft by 18 ft room is exposed along one 12-ft wall and one 18-ft wall. The slab edge insulation is 2-in extruded polystyrene (R-11). The slab temperature at design conditions will be about 105 °F. The outside design temperature is 0 °F. Estimate the downward heat flux using Equation 10.2:

Solution: Substituting the data into Equation 10.2 yields:

$$q_d = \frac{4.17(12 + 18)(105 - 0)}{(11 + 5)(12 \cdot 18)} = 3.8 \text{ Btu/hr/ft}^2$$

In cases where the floor looses heat to an air space below, the following equation can be used to estimate the downward heat flux:

(Equation 10.3)

$$q_d = \frac{q_u(R_{ff} + 0.61) + T_{air\,top} - T_{air\,down}}{R_{downward}}$$

where:

q_d = average downward heat flux (Btu/hr/ft²)
q_u = average upward heat flux (Btu/hr/ft²)
R_{ff} = R-value of the finish floor material on the top of the slab
$T_{air\,top}$ = air temperature in space above the floor (°F)
$T_{air\,down}$ = air temperature in space beneath the floor (°F)
$R_{downward}$ = total R-value of material(s) beneath the floor slab

Example 10.3: Estimate the downward heat flux for the floor assembly shown in Figure 10–45. Assume the upward heat flux is 26 Btu/hr/ft².

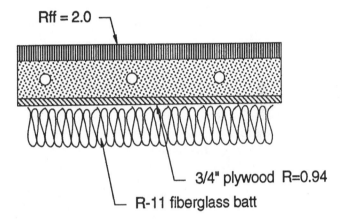

air temperature in room = 68 deg. F.

Rff = 2.0

3/4" plywood R=0.94

R-11 fiberglass batt

air temperature below floor = 60 deg. F.

Figure 10–45   Floor construction for Example 10.3

Solution: The total downward resistance is the sum of the resistance of the fiberglass batt and the 3/4-in plywood subfloor: $R_{downward}$ = 0.94 + 11 = 11.94. Substituting the values into Equation 10.3:

$$q_d = \frac{26(2.0 + 0.61) + 68 - 60}{11.94} = 6.35 \text{ Btu/hr/ft}^2$$

The Thermal Resistance of a Heated Floor Slab

The heat that passes from the water in the tubing, into the room can be thought of as passing through several thermal resistances. These include the convective boundary layer along the inside wall of the tube, the wall of the tube, the slab material, the finish flooring material(s), and finally the air film at the top of the floor. For simplicity these resistances will be lumped into three values:

1. The **slab resistance** (includes the effective resistance of the slab and tubing)

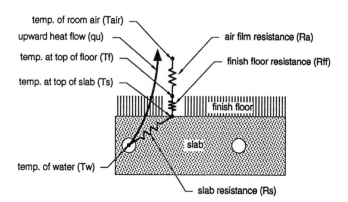

temp. of room air (Tair)
upward heat flow (qu)
temp. at top of floor (Tf)
temp. at top of slab (Ts)

air film resistance (Ra)
finish floor resistance (Rff)

finish floor

slab

temp. of water (Tw)

slab resistance (Rs)

Figure 10–46 Representation of thermal resistances and temperatures in the path of heat flow from the water in the tubes to the room air

2. The **finish floor resistance** (includes all layers of the finish flooring system)
3. The **air film resistance** (includes convective and radiative heat transfer)

These resistances, along with the temperatures between each resistance, are represented in Figure 10–46.

Of these three resistances, the slab resistance is the most difficult to estimate. Its value depends on tube spacing, depth, size, conductivity of pipe wall, slab material, and also the rate of upward and downward heat flow. Some empirical values for the slab resistance (R_s)

have been established based on detailed thermal models and computer simulations. These values are not available for all combinations of tube sizes, materials, spacings, slab materials, etc.

The following procedure can be used to estimate the slab resistance (R_s) for some of the more common floor heating applications using polymer tubing:

Step 1. Estimate the upward and downward heat fluxes from the floor under design load conditions.

The required average upward heat flux can be found using Equation 10.1.

The downward heat flux can be estimated using Equation 10.2 or 10.3.

Step 2. Calculate the ratio of the upward to downward heat fluxes.

(Equation 10.4)

$$r_{u/d} = \frac{q_u}{q_d}$$

If the ratio is greater than 10.0, assume it is equal to 10. Good design practice requires this ratio to be a minimum of 5. If it is less than 5, additional underside insulation should be used.

Step 3. Use $r_{u/d}$ to find the slab resistance from the graph in Figure 10–47 or 10-48 depending on the nominal I.D. tube size used.

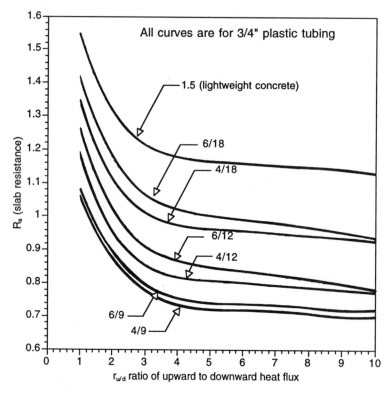

Note: The number before the slash indicates the slab thickness in inches.
The number after the slash indicates the tube spacing in inches.

Figure 10–47 Graph for determining slab resistance (R_s) for ³/₄-inch plastic tubing in concrete slabs

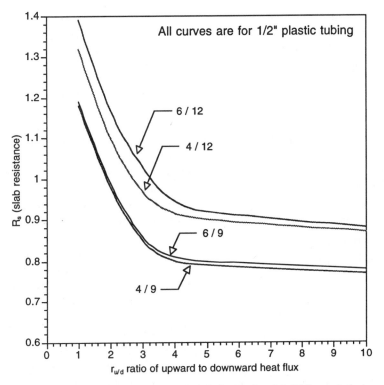

Note: The number before the slash indicates the slab thickness in inches.
The number after the slash indicates the tube spacing in inches.

Figure 10–48   Graph for determining slab resistance (R_s) for ¹/₂-inch plastic tubing in concrete slabs

The thermal resistance of finish flooring materials varies widely. A representative sample of several finish flooring materials and their thermal resistances is given in Figure 10–49. Thermal resistances of other floor products can be obtained from suppliers of floor heating equipment, or from the manufacturers of the flooring products.

The thermal resistance of the finish flooring material(s) will greatly affect the required supply water temperature of the heating system. The lower the thermal resistance of the finish floor, the lower the required supply water temperature.

A bare or painted concrete slab has a finish floor resistance of zero. Such a floor will operate at the lowest possible water temperature for a given rate of heat output. Finish flooring such as ceramic tile, or other thin masonry materials, have relatively low thermal resistances *if they are fully bonded to the slab with mastic or thin-set mortar.* These materials will also allow the system to operate at relatively low supply water temperatures. Wood flooring and especially thick carpet and pad creates relatively high thermal resistance. Such materials act like an insulation blanket laid over the floor, restricting upward heat flow to a fraction of what would be obtained from a bare slab. *When high resistance finish floor materials are used, the output of the floor heating system may have to be supplemented by other heat emitters.* Attempting to force heat up through such high resistance flooring materials at a rate equal to the design heat loss of a room will require very high water temperatures in all but extremely well-insulated rooms. This in turn will significantly increase downward and edgewise heat losses from the slab. The slab could even be cracked from the high thermal stresses that can develop. As a rule, such situations are best handled by using the floor to provide only a portion of the room's design heating load, while the remainder is supplied from another type of heat emitter.

Finish floor	Thermal resistance
bare (or painted) slab	$R_{ff} = 0.0$
¹/₈" vinyl tile or sheet	$R_{ff} = 0.21$
³/₈" ceramic tile	$R_{ff} = 0.22$
¹/₄" nylon level loop carpet (glued to slab)	$R_{ff} = 1.36$
¹/₂" nylon carpet over ¹/₂" urethane pad	$R_{ff} = 3.14$
³/₈" laminated oak flooring over ¹/₈" foam pad	$R_{ff} = 0.94$

Figure 10–49   Representative thermal resistances of several finish floor materials

The air film resistance at the top of the floor can be assumed to be 0.61 (°F/hr/ft²/Btu). This value is suggested by ASHRAE for still air films on horizontal surfaces. When air motion across the floor is present, this value can be as low as 0.5.

The upward heat flux from one square foot of heated floor can be calculated using the sum of all three resistances.

(Equation 10.5)

$$q_u = \frac{(T_w - T_{room\,air})}{(R_s + R_{ff} + R_{air\,film})}$$

where:

q_u = average upward heat flux (Btu/hr/ft²)
T_w = water temperature in the tubing (°F)
$T_{room\,air}$ = air temperature in the space being heated (°F)
R_s = slab resistance
R_{ff} = total R-value of the finish floor material(s)
$R_{air\,film}$ = R-value of the air film at the top of the finish floor

This equation can be rearranged to find the water temperature necessary to produce a given upward heat flux:

(Equation 10.6)

$$T_w = T_{room\,air} + q_u(R_s + R_{ff} + R_{air\,film})$$

where:

T_w = water temperature in the tube
$T_{room\,air}$ = air temperature in the space being heated (°F)
q_u = average upward heat flux (Btu/hr/ft²)
R_s = slab resistance
R_{ff} = total R-value of the finish floor material(s)
$R_{air\,film}$ = R-value of the air film at the top of the finish floor

Example 10.4: Assume a heated slab-on-grade floor is required to produce an upward heat flux of 25 Btu/hr/ft². The downward heat flux is estimated at 3.5 Btu/hr/ft². The slab is 4-ins thick and has ³/₄-in plastic tubing spaced 12 ins on center. The finish floor is ¹/₈-in vinyl tile. The room is to be maintained at 68 °F. Determine the average water temperature required in the piping circuit.

Solution: In order to find the slab resistance for Equation 10.6, the ratio of the upward to downward heat flow must first be calculated. Using the data supplied:

$$r_{u/d} = \frac{25}{3.5} = 7.14$$

This value, along with the curve for a 4-in slab with 12-in tube spacing in Figure 10–47 can be used to find the slab resistance. The estimated value from the graph

is: $R_s = 0.795$. The finish floor resistance can be found in Figure 10–49. For ¹/₈-in vinyl tile $R_{ff} = 0.21$. The air film resistance will be assumed at 0.61.

Substituting this data into Equation 10.6:

$$T_w = 68 + 25(0.795 + 0.21 + 0.61) = 108.4 \text{ °F}$$

Discussion: Keep in mind this is the *average* water temperature required in the circuit. The supply temperature would be this value plus half of the temperature drop of the circuit. The return temperature would be this value minus half of this temperature drop. A method for assessing the temperature drop along the circuit will be presented shortly.

Example 10.5: Repeat the previous example using ¹/₄-in nylon carpet as the finish flooring material. What average water temperature is now required for the same 25 Btu/hr/ft² upward heat flux?

Solution: The only value that will change is the thermal resistance of the finish flooring. This value can be found in Figure 10–49 as $R_{ff} = 1.36$. Substituting into Equation 10.6:

$$T_w = 68 + 25(0.795 + 1.36 + 0.61) = 137.1 \text{ °F}$$

Discussion: The revised average water temperature shows how *the choice of finish floors significantly affects the operating temperature of the system.*

Equation 10.6 is a powerful tool for evaluating the consequences of various combinations of required upward heat flux, finish floor materials, and water temperatures. It can quickly point out unreasonable operating requirements before one proceeds with a detailed system design.

Temperature Drop Along a Floor Heating Circuit

It has already been stated that the heat output from floor areas near the beginning of the piping circuit will be greater than those near the end of the circuit. This is a result of decreasing water temperature along the circuit. It is necessary to account for this temperature drop in determining the *total* heat output from the floor served by a particular piping circuit. The following equations can be used to find the outlet temperature of the piping circuit based on its construction, length, inlet temperature, and flow rate. *They assume water is used as the heat transfer fluid, and that the piping is arranged in a serpentine pattern.*

(Equation 10.7a)

$$T_{out} = T_{room\,air} + (T_{in} - T_{room\,air})e^{-b}$$

(Equation 10.7b)

$$b = \frac{(a)(L)}{500(f)} \quad \text{for } f > 0$$

(Equation 10.7c)

$$a = \left(\frac{1}{R_s + R_{ff} + R_{air\ film}} \right)\left(1 + \frac{q_d}{q_u} \right)$$

where:

T_{out} = outlet water temperature from the floor circuit (°F)

T_{in} = inlet water temperature to the floor circuit (°F)

$T_{room\ air}$ = air temperature in the room being heated (°F)

L = length of the piping circuit (ft)

f = flow rate through the circuit (gpm)

q_u = required upward heat flux at design load conditions (using Equation 10.1)

q_d = downward heat flux at design load conditions (using either Equation 10.2 or 10.3)

R_s = slab resistance

R_{ff} = total thermal resistance of the finish floor material(s)

$R_{air\ film}$ = thermal resistance of the air film at the top of the floor

e = 2.71828 (mathematical constant)

Note that it is necessary to first find the value of a (using Equation 10.7c), then use it to find the value of b (using Equation 10.7b). The value of b can then be substituted into Equation 10.7a to find the outlet temperature of the circuit.

These equations were derived assuming that a serpentine pattern of a tubing circuit could be (conceptually) stretched out into a long straight line. The floor area associated with the tubing would also be stretched out into a long straight ribbon centered on the tubing. The length of the ribbon would equal the length of the tubing circuit. The width of the ribbon would equal the spacing between the tubes in the floor slab. Another assumption that is reasonable for serpentine piping patterns, is that the heat flow from the sides of this ribbon would be very small and could be ignored for estimating purposes. Figure 10–50 illustrates this concept.

Once the outlet temperature of the circuit is determined, the total heat output can be found with the familiar sensible heat rate equation from Chapter 4:

(Equation 10.8)

$$Q_{total} = 500(f)(T_{in} - T_{out})$$

The total heat output includes both the upward and downward heat flows. The upward heat flow into the room can be found using the following equation:

(Equation 10.9)

$$Q_{upward} = \frac{Q_{total}}{\left(1 + \frac{q_d}{q_u} \right)}$$

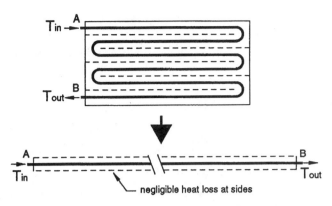

Figure 10–50   Stretching a serpentine piping path into a long straight line

where:

Q_{upward} = total upward heat flow from the circuit (Btu/hr)

Q_{total} = total (upward + downward) heat flow from the circuit (Btu/hr)

q_u = required upward heat flux (using Equation 10.1) (Btu/hr/ft²)

q_d = downward heat flux (using either Equation 10.2 or 10.3) (Btu/hr/ft²)

Example 10.5: A floor heating circuit consists of 300 ft of ³/₄-in polymer tubing embedded in a 4-in concrete slab at a spacing of 12 ins. The required upward heat flux at design conditions is (q_u = 25 Btu/hr/ft²). The estimated downward heat flux at these same conditions is (q_d = 3.5 Btu/hr/ft²). The finish floor is ¹/₈-in vinyl tile (R_{ff} = 0.21). Water enters the beginning of the circuit at 110 °F and 1.5 gpm. The desired room air temperature is 68 °F. Find the outlet temperature from the circuit and the total and upward heat flows.

Solution: Using the ratio of q_u/q_d, and Figure 10–47, the slab resistance is estimated at 0.795. The value of a can now be found using Equation 10.7c:

$$a = \left(\frac{1}{0.795 + 0.21 + 0.61} \right)\left(1 + \frac{3.5}{25} \right) = 0.706$$

The value of b can be found can be found using Equation 10.7b:

$$b = \frac{(0.706)(300)}{500(1.5)} = 0.282$$

Finally the outlet temperature can be found using Equation 10.7a:

$$T_{out} = 68 + (110 - 68)e^{-0.282} = 99.7 \ °F$$

The *total* heat output of the circuit is found using Equation 10.8:

$$Q_{total} = 500(1.5)(110 - 99.7) = 7,725 \text{ Btu/hr}$$

The upward portion of the total heat flow is found using Equation 10.9:

$$Q_{upward} = \frac{7,725}{\left(1 + \dfrac{3.5}{25}\right)} = 6,776 \text{ or approximately } 6,800 \text{ Btu/hr}$$

The downward heat flow would be the difference between the total heat output of the circuit and the upward heat flow:

$$Q_{down} = 7,725 - 6,776 = 949 \text{ Btu/hr}$$

Discussion: The upward heat flow of approximately 6,800 Btu/hr is what can be reasonably expected from this piping circuit under these operating conditions. *This output may not exactly match the design heating load of the room.* It may be greater than or less than the design heating load. If smaller, the designer must make some modification(s) to boost heat output. These might include raising the supply water temperature, increasing the circuit's flow rate, decreasing tubing spacing, increasing underside insulation, or a combination of these changes. If the heat output was greater than the design load of the room, the designer could try lowering the flow rate to the circuit using a balancing valve, or possibly increasing the spacing between the tubing.

Equations 10.7a, b, c, 10.8, and 10.9 provide a way to predict the actual heat output of a particular piping circuit accounting for most of the construction and operating factors. If the heat output is not adequate, or if it is significantly higher than required, these equations can be used to evaluate different options until the heat output is reasonably close to the room's design heating load.

The RADFLOOR Program

The Hydronics Design Toolkit contains a program called RADFLOOR that can evaluate the equations presented in this section. It allows several different core floor constructions to be combined with several different finish floor options. It also allows several fluids other than water to be used. This program allows rapid evaluation of various combinations of floor constructions and operating conditions until a suitable circuit configuration is found. A screen shot of the program using the data from Example 10.5 is shown in Figure 10–51.

Design Assistance from Manufacturers

Today several manufacturers of hydronic floor heating equipment offer design assistance either directly or through their sales representatives. This assistance can be in the form of heat output charts, in-house computer analysis, or software that can be purchased by the system designer. Some companies will take the plans of the

```
- RADFLOOR -   F1-HELP     F3-SELECT                    F10-EXIT TO MAIN MENU
  -INPUTS-
   ENTER Temperature of fluid to floor circuit,(80 to 150).  110       deg.
   ENTER Flowrate through the floor circuit,(.5 to Limit)..  1.5       gpm
   ENTER Room air temperature,(50 to 85)...................  68        deg. F
   ENTER Floor area served by floor circuit,(50 to 1000)...  300       ft²
   ENTER Design heating load of space served,(500 to 20000) 7500       btu/hr
   ENTER Downward + edgewise floor heat loss,(100 to 10000) 1050       btu/hr
   ENTER Length of tubing in floor circuit,(25 to 500).....  300       ft
   SELECT Radiant floor construction type,( F3 to select).. CORE #2
   SELECT Finish flooring material(s),( F3 to select)...... FINISH #2
   SELECT Fluid type,( F3 to select)....................... water

  -RESULTS-
   Floor circuit inlet temperature........  110.0  deg. F
   Floor circuit outlet temperature........  99.5  deg. F
   Temperature drop through floor circuit...  10.5  deg. F
   UPWARD heat output from floor circuit...  6,794  BTU/hr
   DOWNWARD heat output from floor circuit.    951  BTU/hr

    MESSAGE
   Use the UP, DOWN, LEFT, RIGHT ARROW keys to move the highlighting bar
  over an input. Type desired value. F10 to return to MAIN MENU. F3 to SELECT
```

Figure 10–51 Screen shot of the RADFLOOR program from the Hydronics Design Toolkit building and design a complete radiant floor heating system (around their products).

A comparison of these design methods finds they all differ somewhat in their prediction of heat output for a given floor system. This can be attributed to different theoretical models, assumptions about material properties, and differences in products from one manufacturer to the next. Undoubtedly the design assistance offered by manufacturers will continue to improve as more research is incorporated into practical design tools. An example of a heat output table supplied by a manufacturer is shown in Figure 10–52.

SLAB ON GRADE APPLICATION

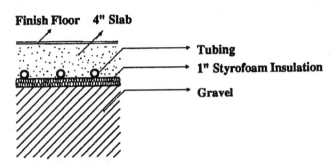

Pextron Tubing 5/8" OD at 110°F Water Supply Temperature and 68°F Room Temperature					
Finish Floor	Sp. Dim. (inches)	Max. Circuit Length(Ft.)	Max. Floor Area Coverage (Sq. Ft.)	Floor Surface Temperature	Output BTU/Hr/Sq.Ft.
Ceramic Tile 3/8"	3	400	100	86	37
	6	400	200	84	32
	9	400	300	81	27
	12	400	400	79	22
Natural Stone/ Marble 1/2"	3	400	100	88	39
	6	400	200	86	36
	9	400	300	84	33
	12	400	400	83	29
PVC/Linoleum Tile 1/8"	3	400	100	84	32
	6	400	200	79	22
	9	400	300	76	16
	12	400	400	74	12
Solid Hardwood 3/4"	3	400	100	79	22
	6	400	200	76	17
	9	400	300	74	12
	12	400	400	73	9
Laminated/Floating Wood 5/8"	3	400	100	78	20
	6	400	200	75	15
	9	400	300	73	11
	12	400	400	72	8
1/2" Carpeting with 3/8" Jute Pad	3	400	100	74	12.5
	6	400	200	72	9
	9	400	300	71	6
	12	400	400	70	5
3/4" Carpeting with 3/8" Jute Pad	3	400	100	73	11
	6	400	200	72	8
	9	400	300	71	6
	12	400	400	70	4

Figure 10–52   Example of a heat output table for various type of radiant floor heating systems. Courtesy of Stadler Corporation.

10.9 SYSTEM PIPING AND CONTROL OPTIONS

There are several piping options for connecting hydronic heat sources to radiant floor heating systems. All must address the fact that floor heating systems usually require water temperatures significantly lower than those of other heat emitters. All must also have an associated control strategy that will stabilize the system and guide it through occasional transient conditions. This section will show several methods for system piping and control. These include:

- Direct piping
- Three-way motorized mixing valves
- Four-way motorized mixing valves
- Tempering tank
- Injection mixing systems
- Thermostatic mixing valve and telestats
- Multi-temperature systems

Each method will be illustrated with a basic piping schematic and control diagram. *These diagrams will show the general concept, but not every individual component required by the system.* The reader is encouraged to cross reference previous chapters on control and piping design to fill in the details for these basic design strategies.

Manifold Stations

A common part of modern floor heating systems is the manifold station. It consists of one supply **manifold** and one return manifold. All piping circuits in the floor begin at a supply manifold and end at the return manifold. Each circuit is therefore in parallel with the other circuits attached to the manifold. The two manifolds are supported by a common set of brackets, usually in the stud cavity of an interior wall. Manifold stations should always be accessible. A common approach is to provide an access panel or door through which the entire manifold station can be inspected, and any needed adjustments made. Figure 10–53 shows a typical manifold station being installed in a wall cavity.

The number of piping circuits supplied by a manifold station can range from one to ten or more depending on total flow rate. Some manifolds must be specifically ordered with a given number of circuit connections. Others are assembled on the job site using modular components. Figure 10–54 shows an example of a manifold system consisting of brass sections serving either two or three circuits each. These sections can be screwed together to form a larger manifold if necessary. Figure 10–55 is an example of a modular manifold system assembled from individual high temperature thermoplastic components to suit the needs of the system.

Figure 10–53 Example of a two-circuit manifold station inside a wall cavity. Note the access door and trim have not yet been installed.

Most manifold systems are equipped with balancing valves for each circuit they serve. These valves allow individual flow adjustment of each circuit. As part of their design procedure, some manufacturers even specify the exact setting of each balancing valve so the flow rate in each circuit will match design specifications. Some manifold systems even include a flowmeter for each circuit. Other manifold trim includes an air venting device, thermometers, and drain/purging valves.

Most manifold systems also allow any one or more of their circuits to be adapted for control by a room thermostat, using a device called a telestat shown in Figure 10–56. This small 24 VAC heat motor simply screws onto the manifold valve for the circuit to be controlled by the room thermostat. When the thermostat completes the circuit through the telestat, it slowly opens the manifold valve over a period of three to five minutes. Telestats are available with and without isolated end switches, the same as with zone valves. When piping circuits are laid out on a room-by-room basis, the use of telestats on the manifold(s) makes it possible to control every room as a separate zone. This does, however, add to the cost of the system and may not be justified in all cases.

Sometimes the number of circuits used in a system, or their placement in a building, will require two or more manifold stations. When the inlet temperature of all areas of the floor is about the same, the individual manifold stations should be piped in parallel with each other as shown in Figure 10–57.

When inlet temperatures to various parts of the floor are different due to different floor coverings, one manifold station may have to operate at a significantly different temperature than another. There are a number of methods for doing this that will be presented later in this section.

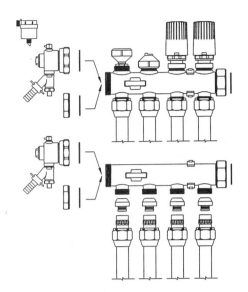
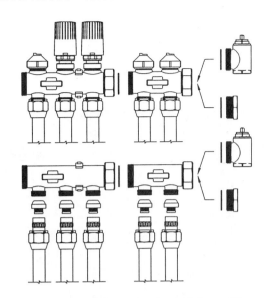

Figure 10–54   Example of a brass manifold system assembled from two, three, and four circuit modules with associated end caps, angle valves, and telestats. Courtesy of Wirsbo Co.

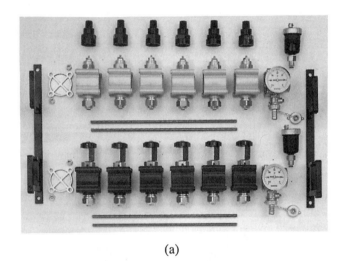

(a)

(b)

(c)

Figure 10–55   Example of a modular thermoplastic manifold system. (a) Individual parts ready for assembly, (b) modules are sealed together using O-rings and held together with threaded rods, (c) the completed manifold station ready to mount. Courtesy of Euro-tech, Inc.

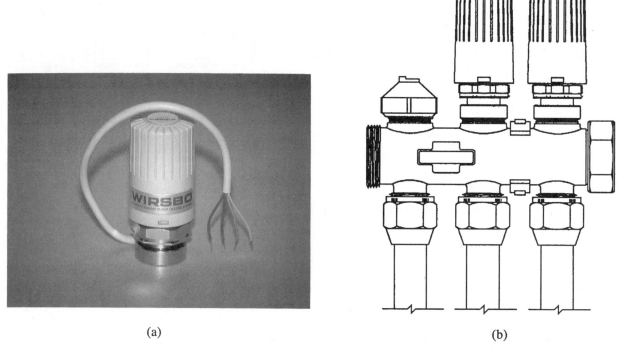

(a)   (b)

Figure 10–56   (a) Example of a 24 VAC telestat, (b) telestats screwed onto a manifold. Courtesy of Wirsbo Co.

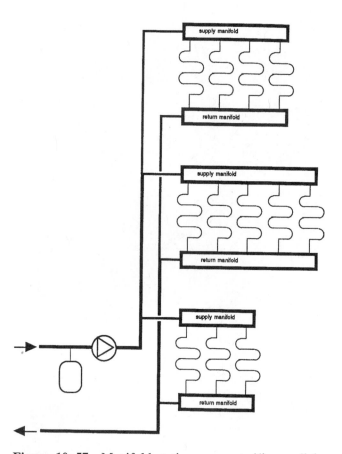

Figure 10–57   Manifold stations connected in parallel. Each stations receives approximately the same supply water temperature.

Direct Piping Systems

The simplest piping design for floor heating is when the manifold station(s) can be directly connected to the heat source as shown in Figure 10–58.

This type of piping design can only be used with heat sources that can continuously operate at the low return water temperatures typical of floor heating systems, without damage from flue gas condensation. Such heat sources include condensing boilers, hydronic heat pumps, electric boilers, hot water tanks, and solar collectors. *Conventional gas or oil-fired boilers should not be directly piped to floor heating systems that have return water temperatures below the dewpoint of the boiler's exhaust gases.* To do so will cause continuous condensation of the water vapor on the boiler sections and flue pipes. The resulting corrosion can cause dangerous leaks in the flue piping in a matter of weeks! Several of the

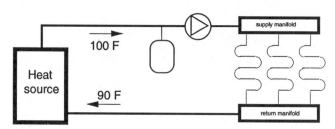

Figure 10–58   Direct connection of heat source and floor heating distribution system

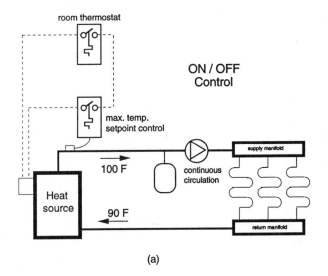

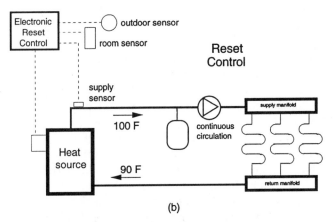

Figure 10–59 Control concepts for direct piping systems. (a) On/off control using a room thermostat and high limit setpoint control in series, (b) reset control.

piping options presented later in this section are designed to prevent flue gas condensation.

The system circulator in this type of system should run continuously during the heating season. The heat source can be controlled by either of the methods shown in Figure 10–59.

The on/off method uses a standard room thermostat wired in series with a temperature-limiting setpoint control. Upon a call for heat from the room thermostat, the heat source is turned on. It continues to operate until either the room thermostat is satisfied or the setpoint control detects supply water temperature has risen to a preset maximum value. If the latter condition occurs, the setpoint control breaks the control circuit to the heat source, shutting it off. When the temperature of the continuously circulating water drops by the differential setting of the setpoint control, the heat source is restarted

(assuming the room thermostat is still calling for heat).

The setpoint control protects the slab from excessively high water temperatures that could occur in systems with oversized heat sources during startup conditions. During these times, the large thermal mass of a slab-type system will significantly delay the increase in room air temperature. Since the room thermostat cannot sense the high water temperature in the piping circuits, it cannot turn the heat source off. In extreme situations, unprotected floor slabs have been known to crack due to high thermal stresses from excessively high water temperatures in the piping circuits.

The other option is to use an electronic reset control to operate the heat source as necessary to maintain the calculated supply temperature. This option provides heat input to the system in direct response to the prevailing heating load. This minimizes temperature fluctuations in the heated space. The strategy of reset control was discussed in detail in Chapter 9.

Three-way Motorized Mixing Valves

A three-way mixing valve can be used to lower the water temperature to the floor heating system by blending a portion of the return water from the distribution system with the hot water from the heat source. The rest of the low temperature return water flows directly back to the heat source. *Because of the likelihood of flue gas condensation, three-way mixing valves should not be used to control low temperature floor heating systems using conventional gas or oil-fired boilers.* Three-way mixing valves are a good choice for reducing the temperature of water supplied from a storage tank such as those used for electric thermal storage or solar energy systems. The most common control method uses an electronic reset control to operate the motorized actuator, which adjusts the position of the valve's shaft, as discussed in Chapter 9. A schematic for a three-way mixing valve, along with representative system temperatures, is shown in Figure 10–60.

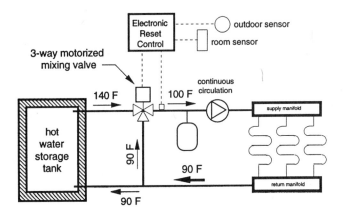

Figure 10–60 Use of a motorized three-way mixing valve to control water temperature to a floor heating distribution system

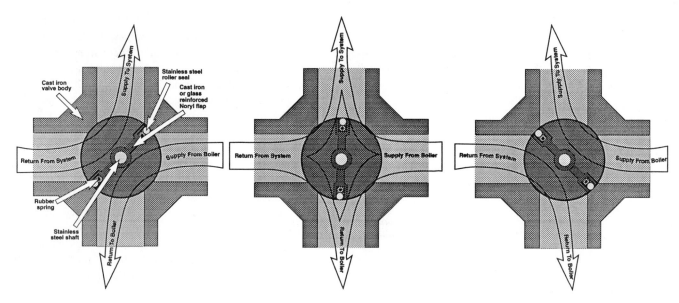

Figure 10–61 Cut-away view of a four-way mixing valve at various shaft positions. Courtesy of tekmar Controls.

Four-Way Mixing Valves

Four-way mixing valves can provide low temperature water to the distribution system while also maintaining a conventional boiler above its dewpoint temperature. This is done by mixing a certain amount of hot water from the boiler with the cool return stream from the distribution system before it reenters the boiler as illustrated in Figure 10–61.

The ability to maintain a relatively high return water temperature makes four-way valves the preferred mixing valve for interfacing a conventional gas or oil-fired boiler with a low temperature floor heating system. As with a three-way valve, heat output to the distribution system is regulated by a reset control driving a motorized actuator that adjusts the position of the valve's shaft in response to outdoor temperatures.

A piping schematic for a four-way mixing valve is shown in Figure 10–62a. Note the location of the circulator in the system. This circulator runs continuously during the heating season. In systems using cast-iron boilers, this circulator can induce adequate flow in the boiler loop using the pressure differential of the partially closed mixing valve. Experience has shown, however, *when low mass boilers are used, a second circulator should be placed between the boiler and the four-way valve* as shown in Figure 10–62b. This circulator should operate whenever the low mass boiler is firing to prevent high thermal stresses and associated expansion noises in the boiler.

Tempering Tank Systems

The piping schematic shown in Figure 10–63 uses an insulated tank as a mixing chamber between the heat source and the distribution system.

To prevent a conventional gas or oil-fired boiler from operating in a condensing mode, a bypass pipe with a ball valve is used to shunt part of the hot supply water to a tee where it mixes with the cool return water from the tank. This raises the temperature of the water entering the boiler above dewpoint. The flow balancing valve is used to create a pressure drop suitable to limit the amount of cool water exiting the tank toward the boiler. These two valves must be adjusted by a trial and error process to obtain the desired return temperature to the boiler.

The volume of the tempering tank affects the length of the boiler on-cycle. The water in the tank must absorb any portion of the heat output from the heat source that is not going to the distribution system at the time. The larger the storage tank, the longer the boiler on-cycle will be. A suggested minimum firing time under partial load conditions is ten minutes. Longer on-cycles result in greater seasonal efficiency because the boiler is closer to steady state conditions more of its operating time.

Equation 10.10 can be used to estimate the length of the boiler on-cycle for a given tank size, heat source capacity, and heating load condition:

(Equation 10.10)

$$\text{On time} = \frac{\Delta T (8.33)(V)}{Q_{\text{boiler}} - Q_{\text{load}}}$$

where:

On time = approximate run time of the boiler, per operating cycle (hrs)

ΔT = temperature differential through which the tank rises during the on-cycle (°F)

V = volume of water in the tempering tank (gals)

Q_{boiler} = gross output of the boiler (Btu/hr)

Q_{load} = heating load required by the distribution system (Btu/hr)

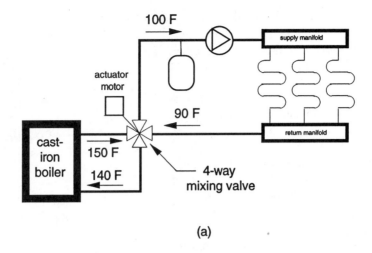

(a)

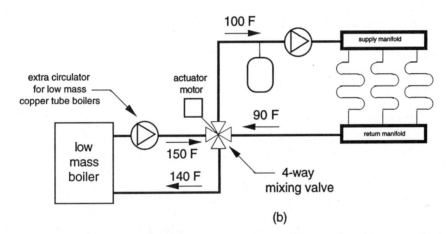

(b)

Figure 10–62   Use of a four-way mixing valve to control water temperature to floor heating system. (a) Single circulator used with a high mass boiler, (b) two circulators used with a low mass boiler. Notice the relatively high return water temperature to the boilers.

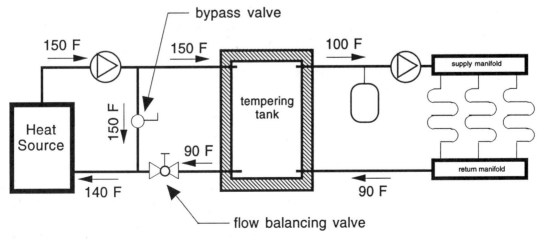

Figure 10–63   Tempering tank concept along with representative operating temperatures. Note boiler bypass to limit flue gas condensation.

Example 10.6: A flooring heating system uses a 120-gal tempering tank as shown in Figure 10–63. The boiler's gross output is 60,000 Btu/hr. The heating load required by the building at a particular time is 35,000 Btu/hr. The setpoint control on the tank is set to stop boiler firing when the tank reaches 110 °F, and restart if the temperature drops to 100 °F. Estimate the approximate on-time of the boiler under these conditions.

Solution: The differential the tank passes through between burner shutdown and start-up is 110 – 100 = 10 °F. Substituting this into Equation 10.10:

$$\text{On time} = \frac{10(8.33)(120)}{60,000-35,000} = 0.4 \text{ hours or 24 minutes}$$

Discussion: Equation 10.10 assumes the boiler reaches full gross output as soon as the burner is fired. This of course is not true because of the warm-up time required by the thermal mass of the boiler and its water. The actual on-cycle time will therefore be slightly longer than that predicted by Equation 10.10.

There are several variations of this basic piping design depending on the type of controls used and how the system is to be operated. Figures 10–64 through 10–66 illustrate three different strategies. One should carefully weigh the cost of an insulated tempering tank and other required components against other piping/control options before committing to this approach.

In the system shown in Figure 10–64, the tank setpoint control fires the boiler as necessary to maintain the tempering tank within its operating temperature range. Upon a call for heat from the room thermostat, the distribution circulator is operated. *Note:* If a line voltage thermostat was used to operate the distribution pump, the transformer and relay could be eliminated.

The arrangement shown in Figure 10–65 allows continuous circulation in the distribution system through the bypass port of a three-way zone valve. When the room thermostat calls for heat, the boiler is fired, and the zone valve reroutes all distribution water through the tank. Because this arrangement does not maintain tank temperature at a fixed lower limit, there will be a longer delay in heat delivery because of the thermal mass of the tank water.

The system shown in Figure 10–66 uses a reset control to maintain the tank temperature as necessary for the prevailing outdoor temperature. Because the tank is always at the correct temperature for the current heating load, constant circulation between the tank and distribution system can be maintained without overheating. If the room temperature sensor detects overheating from internal heat gains, it can suspend boiler operation. Some reset controllers also can automatically shut off the distribution pump during warm weather. Of the three methods pre-

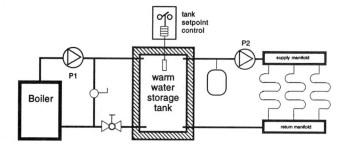

(a)

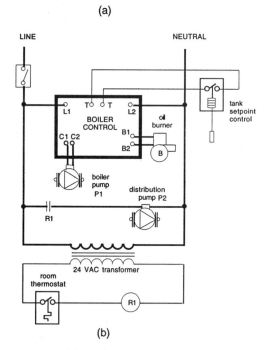

(b)

Figure 10–64 On/off control of distribution pump, with tank maintained at minimum temperature by boiler. (a) Piping schematic, (b) control schematic.

sented, the reset control will provide the most stable interior temperatures.

Injection Mixing Systems

Another method for piping a heat source to a floor heating distribution system is called **injecting mixing**. The basic concept is to add hot water to the continuously circulating floor heating circuits by opening and closing a zone valve that separates the distribution loop from the heat source. Three piping arrangements for injection mixing are shown in Figure 10–67.

Notice the hot water that passes through the zone valve is blended in the tee downstream of the valve with cooler return water from the floor circuits. This prevents a hot slug of water from being sent out to the floor circuits while the zone valve is open. Flow between the heat source and distribution loop is induced by the partially closed balancing valve. The more this valve is closed, the

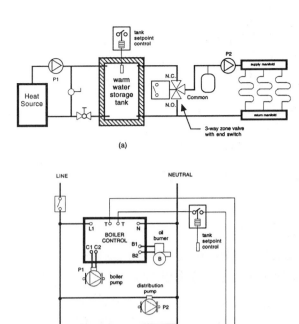

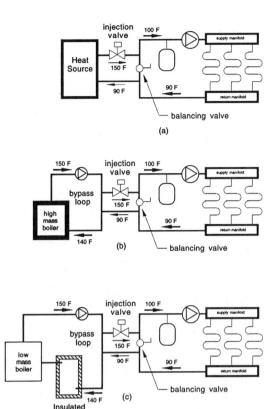

Figure 10–65   Use of three-way diverting valve to maintain constant circulation in the distribution system. Boiler is fired only upon a call for heating. (a) Piping schematic, (b) control schematic.

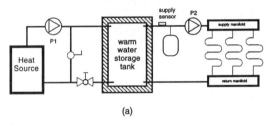

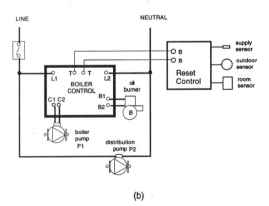

Figure 10–66   Use of an outdoor reset control to maintain the tempering tank at the proper temperature for the constantly circulating distribution system. (a) Piping schematic, (b) control schematic.

Figure 10–67   Three possible piping arrangements for injection mixing control. (a) Piping suitable for heat sources compatible with low return water temperature, (b) piping for use with non-condensing boiler with high thermal mass, (c) added thermal mass (storage tank) to prevent short cycles of a low mass boiler.

greater the flow rate of hot water through the zone valve when it is open. The longer the zone valve is held open, the warmer the distribution loop gets. The opening and closing of the zone valve can be controlled by either an on/off setpoint control or an outdoor reset control. A wiring schematic for the on/off control option is shown in Figure 10–68.

The distribution circulator runs continuously during the heating season. When the room thermostat calls for heat, and the distribution loop is below its setpoint temperature, the zone valve is opened. An end switch in the zone valve signals the boiler and primary pump P1 to operate. When the room thermostat is satisfied or the distribution loop reaches its operating setpoint temperature, the zone valve closes, and the boiler and primary pump P1 are turned off. The system thus maintains the supply temperature to the floor circuits within a preset operating range whenever the room thermostat calls for heat.

When outdoor reset control is used, the distribution loop temperature is maintained at a temperature based on the current outdoor temperature. The reset control opens

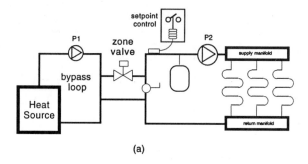

(a)

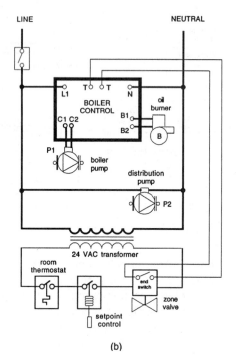

(b)

Figure 10–68 (a) Piping schematic and, (b) control wiring for injection mixing using on/off control

the zone valve as necessary to maintain the supply water temperature within a narrow differential range of this continuously calculated optimum temperature. The boiler and primary pump P1 are turned on by an end switch in the zone valve. Reset control will produce less temperature swing in the heated space because the floor is always at, or close to, its ideal operating temperature for the current heating load.

Either type of injection mixing system requires careful adjustment of the balancing valve. *The final adjustments to the balancing valve must be made when the distribution system is up to its normal operating temperature.* Because of their large thermal mass, this is particularly important with slab-type floor heating systems. The balancing procedure starts with the balancing valve and the zone valve fully open. This allows the maximum amount of cool return water to enter the tee downstream of the zone valve. The balancing valve is then slowly closed while monitoring the mixed temperature exiting the mixing tee.

When this temperature is up to the setpoint temperature, the valve is at its proper setting. If the return temperature from the distribution system is typical of normal operating conditions, the valve should not require any further adjustment. The handle of the balancing valve should be removed to prevent tampering.

Another variation of injection mixing is to use a variable speed injection pump in place of the zone valve. The pump continuously injects heat into the distribution loop. Its speed is controlled by an electronic reset control in response to outdoor temperature. This concept was discussed in Chapter 9.

Thermostatic Mixing Valve with Individual Zone Control

Some buildings using radiant floor heating are best controlled by individual thermostats in each room or space. An example would be a passive solar house where solar gains affect different rooms at different times of the day. Another example is a building where the temperature in different rooms will be adjusted on a daily basis based on occupancy and use. If all zones in such buildings continually circulate water at a single supply temperature, some are likely to overheat while others remain too cool. One method of controlling this type of system is to use a thermostatic mixing valve, as discussed in Chapter 5, as the central water temperature control device, while also equipping each zone circuit with a telestat controlled by a room thermostat. The thermostatic mixing valve delivers a constant supply water temperature to the floor circuits. *In this type of application, the thermostatic mixing valve should be set to deliver the water temperature required under design load conditions.* This allows for fast pickup following a setback period. Any given telestat will close its manifold valve as its associated room thermostat approaches setpoint temperature. A piping schematic of this concept is shown in Figure 10–69a.

This piping schematic uses a concept known as primary/secondary piping to connect the radiant floor heating distribution system with the boiler. This configuration allows a relatively large amount of hot water to bypass the first tee leading to the tempering valve, and thus mix with the cooler water returning from the floor circuits, before the cool water can enter the boiler and create condensation of flue gases. As was the case with injection mixing, the use of a low thermal mass boiler in this type of application may require the addition of an insulated thermal storage tank in the boiler (primary) circuit. The purpose of this tank is to add thermal mass to the primary circuit to prevent short but frequent cycling of the boiler. A modified schematic using such a tank is shown in Figure 10–69b.

The controls for the system shown in Figure 10–69a must operate circulators P1 and P2, and also enable boiler

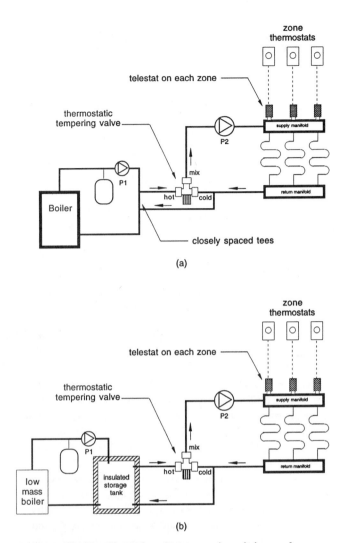

Figure 10–69 Use of a thermostatic mixing valve and thermostats/telestats for individual room temperature control. (a) Piping for use with a high mass boiler, (b) adding a thermal storage tank for use with a low mass boiler.

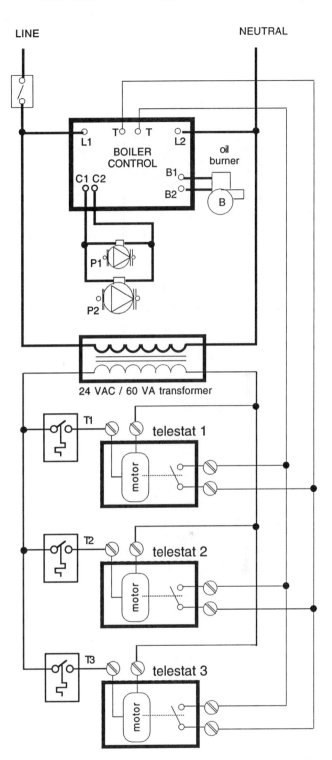

Figure 10–70 Control schematic for system shown in Figure 10–69a

firing when any one or more of the zone thermostats calls for heat. Telestats are available with an isolated end switch and operate with a 24 VAC input signal. This makes the control circuit almost identical to that used with four-wire zone valves as discussed in Chapter 9. A wiring schematic for this type of system is shown in Figure 10–70. For this type of application, some manufacturers also have specialized zone control panels with indicator lights that show which zones are on at any given time.

The controls for the system shown in Figure 10–69b would use a setpoint control to fire the boiler and operate circulator P1 as necessary to maintain an acceptable temperature range in the thermal storage tank. Circulator P2 would operate whenever one or more of the zones is calling for heat. A control schematic for this system is shown in Figure 10–71.

Multi-temperature Floor Heating Systems

When two or more manifold stations require significantly different operating temperatures, a thermostatic mixing valve may be required to maintain one or more of the

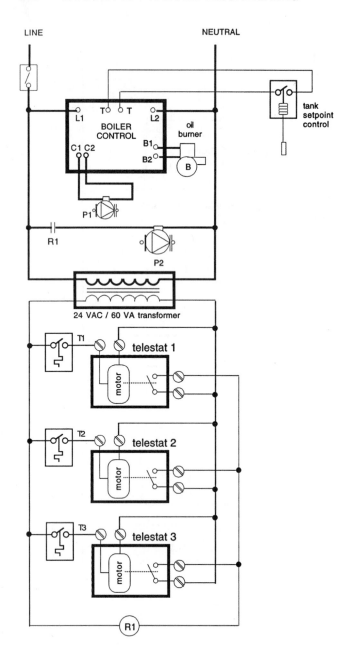

Figure 10–71 Control schematic for system shown in Figure 10–69b

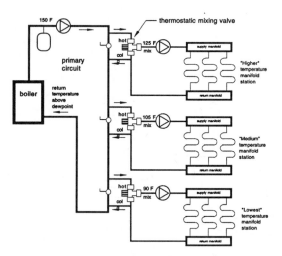

Figure 10–72 Using thermostatic mixing valves to operate manifold stations at different temperatures. Temperatures shown are representative only.

manifold stations at a lower temperature. One method of piping such a system is shown in Figure 10–72.

This arrangement uses primary/secondary piping to bypass a portion of hot boiler water directly into the return side of the boiler. If properly designed, this allows the return temperature to the boiler to remain above the dewpoint temperature of the exhaust gases.

Notice the higher temperature manifold station is the first to tap into the primary loop. The farther downstream a manifold station is connected to the primary loop, the cooler the primary water will likely be. However, if the upstream portions of the system are not operating (e.g., if their associated controls are not calling for heat), the primary temperature will not be lowered. Therefore it is

critical that each secondary circuit leading to a manifold station be equipped with a thermostatic valve that can automatically compensate for such fluctuations in the primary loop temperature. The ball valves located between the supply and return tees to each secondary loop provide a means of adding a slight pressure differential across the thermostatic valve ports. Once the system is operating, they are partially closed only as necessary for proper operation of the thermostatic valve.

The lower a manifold station's supply temperature requirement, the lower the ratio of hot water to return water entering its thermostatic mixing valve. Thermostatic valves always attempt to keep their outlet temperature at the valve's temperature setting, even if the temperature from the heat source is varied by a reset control or other means. As the boiler's outlet temperature is reduced, the thermostatic mixing valves eventually become fully open to supply water and fully closed to cooler return water. Their return water port is totally closed. At this point the distribution loops with mixing valves will behave as if they were directly connected to the primary circuit. If nonthermostatic mixing valves were used, the temperature of the entire system will be reduced as the heat source temperature is reduced. The rate of temperature change in each distribution loop may, however, not follow the ideal reset line for each particular zone. It is possible, although somewhat expensive, to add a motorized actuator driven by a separate reset control, to each three-way valve. This provides optimum control of each manifold station.

This concept can also be applied to systems that use a combination of floor heating and other (higher temperature) heat emitters. Figure 10–73 illustrates this with a combination of floor heating and finned-tube baseboard. Note: There are several other ways of integrating low temperature floor heating distribution systems with high temperature loads. Some will be discussed in more detail in Chapter 11.

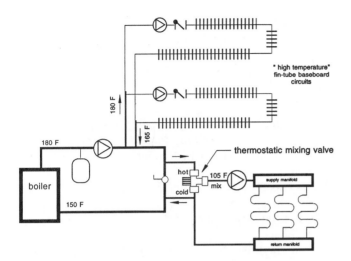

Figure 10–73   One method of combining low temperature floor heating and high temperature heat emitters in the same system

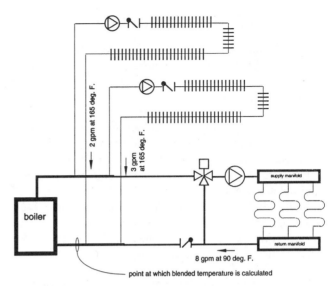

Figure 10–74   Piping system for Example 10.7

Return Water Temperature

In multi-temperature distribution systems, the return temperature to the heat source will depend on the return temperatures from each of the loads, as well as their return flow rates. *When a conventional boiler is used as the heat source, it is critical the return temperature be high enough to prevent continuous condensation in the boiler.*

The return temperature can be estimated using the following equation:

(Equation 10.11)

$$T_{mix} = \frac{(f_1)(T_{R1}) + (f_2)(T_{R2}) + (f_3)(T_{R3}) + \ldots + (f_n)(T_{Rn})}{f_1 + f_2 + f_3 + \ldots + f_n}$$

where:

T_{mix} = return temperature from the multi-temperature distribution system (°F)

$f_1, f_2, f_3, \ldots f_n$ = flow rates from each of the entering fluid streams (gpm)

$T_{R1}, T_{R2}, T_{R3}, \ldots T_{Rn}$ = temperatures of each of the entering fluid streams (°F)

The mixed temperature can also be found using the MIX **program** in the Hydronics Design Toolkit.

Example 10.7: A multi-temperature distribution system incorporates a floor heating manifold station and two zones of fin-tube baseboard as shown in Figure 10–74. The floor heating system returns water at 8 gpm and 90 °F. The two baseboard circuits return water at 165 °F and

operate at flow rates of 3 gpm and 2 gpm respectively. Determine the return water temperature into the heat source under these conditions. Is condensation in the boiler likely to occur?

Solution: Substituting the data in Equation 10.11:

$$T_{mix} = \frac{(8)(90) + (3)(165) + (2)(165)}{8 + 3 + 2} = 118.8 \; °F$$

Discussion: This return temperature is likely to cause condensation formation in either a gas-fired or oil-fired boiler. One way of correcting this problem is to use a four-way rather than three-way mixing valve on the floor heating system, as shown in Figure 10–75. Another way would be to use a primary/secondary piping arrangement as shown in Figure 10–73.

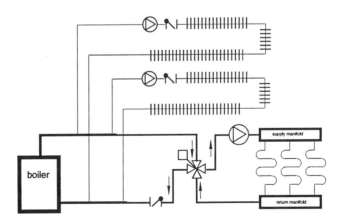

Figure 10–75   Piping system showing use of four-way rather than three-way mixing valve

10.10 FLOOR HEATING CASE STUDY

To conclude the chapter, a hydronic floor heating system will be designed for a small slab-on-grade house. The following steps will be undertaken as part of the design process.

1. Determine the room-by-room design heating loads
2. Prepare a floor circuit layout drawing
3. Measure/calculate the length of each circuit
4. Estimate the flow rate in each circuit based on the circulator specified
5. Calculate the temperature drop and heat output of each piping circuit
6. Draw the system's piping schematic
7. Draw the system's control schematic

The following information and data has been obtained for the house for which the system is being designed:

- 26 ft by 38 ft single-level ranch
- 4-in concrete slab-on-grade floor insulated with 1-in extruded polystyrene (bottom and edges)
- Overall R-value of walls = 15.0
- Overall R-value of ceiling = 25.0
- Unit R-value of all windows = 2.0
- Unit R-value of exterior door = 5
- Air infiltration estimated at one-half air change per hour

- Inside design temperature = 68 °F
- Outdoor design temperature = –20 °F

The following information describes the intended heating system:

- Five-circuit hydronic floor heating system
- Manifold station centrally located as shown on plan
- $1/2$-in nominal (I.D.) PEX tubing used for all floor circuits
- 1-in copper tube used to pipe manifold station back to mechanical room
- $1/4$-in level loop carpet glued to slab in living room and bedrooms
- $1/8$-in vinyl tile glued to slab in bathroom and kitchen/hall area
- Injection mixing via reset control
- 30,000 Btu/hr gas-fired (non-condensing) boiler with automatic flue damper
- Circulator is a Grundfos model UP26-64
- 20-gal indirectly-fired storage water heat piped as a separate load

The floor plan for the house is shown in Figure 10–76. The plan has been broken up (with dashed lines) into zones that correspond to the areas serviced by each floor heating circuit.

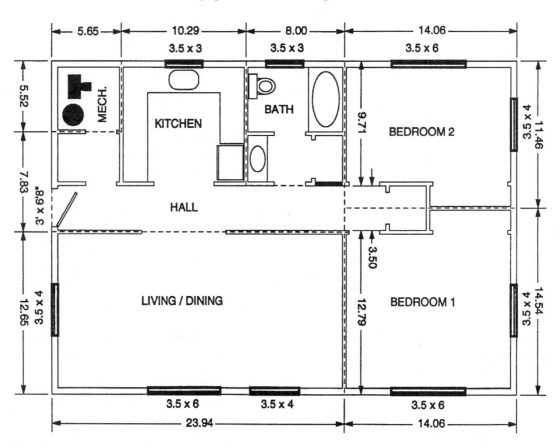

Figure 10–76 Floor plan of house for case study. Dashed lines indicate the zones assigned to each floor heating circuit.

Step 1. The following design heating loads were obtained for each room using the ROOMLOAD program in the Hydronics Design Toolkit and the stated data for the house:

Living/dining	7,450 Btu/hr
Bedroom 1	5,840 Btu/hr
Bedroom 2	5,170 Btu/hr
Bath	1,870 Btu/hr
Kitchen/hall	4,310 Btu/hr
Design Load =	24,640 Btu/hr

This is a very small design heating load, reflective of the small building size.

Step 2. The floor circuit layout is given in Figure 10–77. The circuits were based on keeping rooms such as the bathroom and bedroom on separate circuits so future zone control is an option. The tubing layout uses 12-in spacing. Notice the tubing does cross under some interior partitions. These crossings could have been avoided, but the complications involving tubing layout would be significant. A decision was made to route the

piping as shown and indicate the crossing on the completed slab to prevent any fasteners from being driven near the tubing.

The tubing layout directs the warmest water near the exterior walls first. A two-way serpentine pattern is used adjacent to the exterior walls. The circuit then works its way back toward the interior of the rooms as a one-way serpentine. The placement of the manifold station allows short tubing runs to each of the room circuits. The access panel could open to the inside of the closet in the bathroom. Notice the piping is not routed beneath the kitchen or bathroom cabinets. It is also omitted from the small mechanical room where the heat losses from the boiler jacket will sufficiently heat the space.

Step 3. The measured circuit lengths shown on the layout plan are as follows:

Living/Dining	297 ft
Bedroom 1	198 ft
Bedroom 2	147 ft
Bath	69 ft
Kitchen/Hall	157 ft
Total =	868 ft

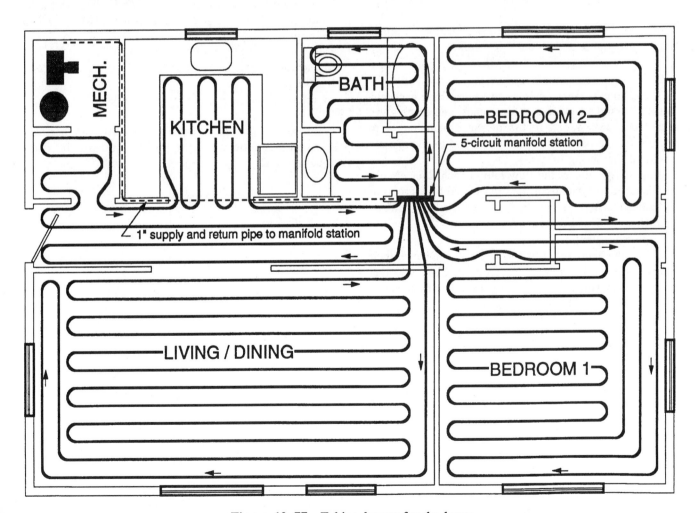

Figure 10–77   Tubing layout for the house

The valve in each manifold, for each circuit will be treated as equivalent to a 1/2" globe valve

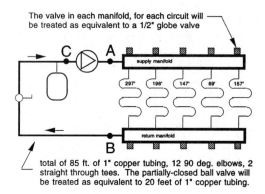

total of 85 ft. of 1" copper tubing, 12 90 deg. elbows, 2 straight through tees. The partially-closed ball valve will be treated as equivalent to 20 feet of 1" copper tubing.

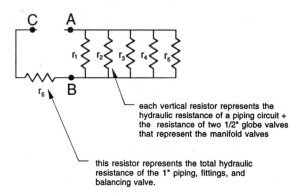

each vertical resistor represents the hydraulic resistance of a piping circuit + the resistance of two 1/2" globe valves that represent the manifold valves

this resistor represents the total hydraulic resistance of the 1" piping, fittings, and balancing valve.

Figure 10–78 Representing the floor heating circuits and supply piping as hydraulic resistors

These lengths were determined by breaking each circuit into straight segments, return bends, and corners. An extra 10 ft was added to the measure plan length to account for the risers to the manifolds and to compensate for slight measuring errors.

Step 4. To estimate the flow rate in each floor circuit, the distribution system will be represented by a group of hydraulic resistors as shown in Figure 10–78. Methods from Chapter 6 will then be used to reduce the circuit to a single equivalent resistance.

Before the hydraulic resistances can be combined, their values must be determined. Recall from Chapter 6 that the hydraulic resistance is defined by the following equation:

(Equation 6.11)

$$r = (\alpha c L)$$

where:

α = fluid properties factor (dependent on fluid type and temperature)
c = pipe size constant
L = equivalent length of a piping string including fitting and valves (ft)

Assuming 100 °F water temperature as typical of a slab-type flooring heating system, the value of α is found from figure 6-17: $\alpha = 0.052$.

The c-value for 1/2-in tubing is found from Figure 6-18: $c = 0.33352$

The equivalent length of each of the piping branches requires a count of the piping length plus all the fittings and valves in each branch. The table in Figure 10–79 shows this data.

The hydraulic resistance of each piping string is found by multiplying the total equivalent lengths by α and c.

Living room circuit $r_1 = 327\ (0.052)(0.33352) = 5.67$
Bedroom 1 $\qquad r_2 = 228\ (0.052)(0.33352) = 3.95$
Bedroom 2 $\qquad r_3 = 177\ (0.052)(0.33352) = 3.07$
Bath $\qquad\qquad r_4 = 99\ (0.052)(0.33352) = 1.72$
Kitchen/hall $\qquad r_5 = 187\ (0.052)(0.33352) = 3.24$
1-in supply
+ return piping $\quad r_6 = 135.9\ (0.052)(0.01776) = 0.126$

Resistors r_1 through r_5 are in parallel, and can therefore be combined into a single equivalent resistor using Equation 6.22:

$$R_{\text{equivalent}_{\text{parallel}}} = \left[\left(\frac{1}{r_1}\right)^{0.5714} + \left(\frac{1}{r_2}\right)^{0.5714} + \left(\frac{1}{r_3}\right)^{0.5714} + \dots + \left(\frac{1}{r_n}\right)^{0.5714} \right]^{-1.75}$$

piping string	pipe + equiv. lgth. of fittings	total equivalent length
living room circuit	297 + 2(15)	327 ft. of 1/2" tube
bedroom 1	198 + 2(15)	228 ft. of 1/2" tube
bedroom 2	147 + 2(15)	177 ft. of 1/2" tube
bath	69 + 2(15)	99 ft. of 1/2" tube
kitchen/hall	157 + 2(15)	187 ft. of 1/2" tube
1" supply/return piping	85 + 12(2.5) +2(.45) + 20	135.9 ft. of 1" tube

Figure 10–79 Summary of equivalent length of piping branches

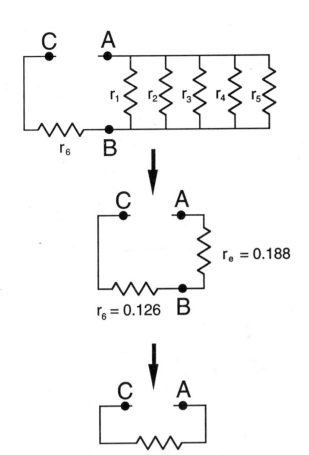

$$R_t = 0.188 + 0.126 = 0.314$$

Figure 10-80   Reducing the hydraulic resistor diagram to a single equivalent resistance (R_t)

$$R_e = \left[\left(\frac{1}{5.67}\right)^{0.5714} + \left(\frac{1}{3.95}\right)^{0.5714} + \left(\frac{1}{3.07}\right)^{0.5714}\right.$$

$$\left. + \left(\frac{1}{1.72}\right)^{0.5714} + \left(\frac{1}{3.24}\right)^{0.5714}\right]^{-1.75} = 0.188$$

The PARALLEL program in the Hydronics Design Toolkit could also be used to find this equivalent resistance.

This equivalent resistance can now be added to the resistance (r_6) that is in series with it to obtain the total equivalent resistance of the distribution circuit. The process of reducing the hydraulic resistor diagram to a single equivalent resistance is shown in Figure 10–80.

The system curve for the distribution system can now be sketched by plotting the following equation:

$$H_L = 0.314(f)^{1.75}$$

The system curve and pump curve have been plotted in Figure 10–81. Their intersection falls at a flow rate of 8.5 gpm and head of 13.4 ft.

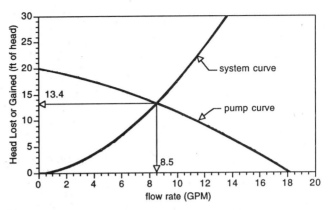

Figure 10–81   Graph of system curve and pump curve showing the flow rate and head across the circulator

The alternative to sketching these curves is to use the PUMPCURV and PUMP/SYS programs in the Hydronics Design Toolkit to find the operating point.

Using Equation 6.23, the system flow rate of 8.5 gpm can now be broken into the individual flow rates in each of the floor circuits:

(Equation 6.23)

$$f_i = f_{total}\left(\frac{R_e}{r_i}\right)^{0.5714}$$

where:

f_i = flow rate in a given parallel piping string (gpm)
f_{total} = system flow rate (gpm)
R_e = equivalent resistance of all the parallel hydraulic resistors
r_i = hydraulic resistance of a given parallel piping path

The flow rates in each of the floor circuits are as follows:

$$f_1 = 8.5\left(\frac{0.188}{5.67}\right)^{0.5714} = 1.21 \text{ gpm} \quad \text{(Living/dining)}$$

$$f_2 = 8.5\left(\frac{0.188}{3.95}\right)^{0.5714} = 1.49 \text{ gpm} \quad \text{(Bedroom 1)}$$

$$f_3 = 8.5\left(\frac{0.188}{3.07}\right)^{0.5714} = 1.72 \text{ gpm} \quad \text{(Bedroom 2)}$$

$$f_4 = 8.5\left(\frac{0.188}{1.72}\right)^{0.5714} = 2.4 \text{ gpm} \quad \text{(Bath)}$$

$$f_5 = 8.5\left(\frac{0.188}{3.24}\right)^{0.5714} = 1.67 \text{ gpm} \quad \text{(Kitchen/hall)}$$

Room name	Design load	active floor area	upward heat flux[1]	slab edge length	downward heat flux[2]	qu/qd[3]	Rs[4]	Rff[5]	length	Flow rate	a[6]	b[7]
Liv./Din	7451	303	24.6	36.6	5.8	4.23	0.91	0.21	287	1.21	0.715	0.34
Bdrm 1	5841	204	28.6	28.6	6.75	4.23	0.91	0.21	188	1.49	0.715	0.18
Bdrm 2	5168	161	32.1	25.5	7.62	4.21	0.91	0.21	137	1.72	0.715	0.114
Bath	1868	56.7	32.9	8	6.8	4.84	0.90	0.21	59	2.4	0.702	0.035
Kitch/Hall	4314	160	27.0	18.1	5.44	4.96	0.90	0.21	147	1.67	0.70	0.123

TOTAL   24642 Btu/hr

Room name	Tin = 110F $T_{out}/Qt/Qu$[8]	Tin = 120F $T_{out}/Qt/Qu$[8]	Tin = 125F $T_{out}/Qt/Qu$[8]
Liv./Din	97.9/7324/ 5924	105/9068/ 7334	108.4/9940/8040
Bdrm 1	103.1/5154/4168	111.4/6382/5162	115.6/6995/5657
Bdrm 2	105.5/3892/3145	114.4/4818/3893	118.9/5282/4268
Bath	108.6/1733/1436	118.2/2146/1779	123/2353/ 1950
Kitch/Hall	105.1/4059/3378	114/5025/ 4182	118.4/5509/4585
TOTAL UPWARD HEAT	18051 Btu/hr	22350 Btu/hr	24500 Btu/hr

Footnotes:
1. Equation 10.1
2. Equation 10.2
3. Column 3 divided by column 5
4. From figure 10.48
5. From figure 10.49
6. From equation 10.7c
7. From equation 10.7b
8. From equation 10.7a / equation 10.8 / equation 10.9

Figure 10–82 Table summarizing calculations for the floor heating circuits

As a check, these flow rates should add up to the total system flow rate. They add to 8.49 gpm, which is slightly off from 8.5 gpm because the calculated values were rounded off.

An alternative to these calculations is to use the PARALLEL program in the Hydronics Design Toolkit.

Step 5. The heat output of the circuits can now be estimated for various inlet temperatures using the methods from section 10.8. These calculations can be lengthy if done manually. The RADFLOOR program in the Hydronics Design Toolkit can save a considerable amount of time. Proceed by gathering the information needed to evaluate Equations 10.7, 10.8, and 10.9 for estimated values of supply water temperature. Once the upward heat flows are determined for each room, they can be compared to the design heating loads of the rooms. If most of the upward heat outputs are lower than the room heating loads, the supply temperature should be increased slightly and the heat outputs reevaluated. If the heat outputs are mostly too high, the supply water temperature can be lowered and the heat outputs recalculated. Using this process, an appropriate supply temperature can usually be found within two or three tries.

The table in Figure 10–82 tabulates the calculations for each room. The numbers above the columns reference the footnotes that describe how the value was obtained.

Discussion: The first inlet temperature assumed was 110 °F, which proved to be too low. The heat outputs of the floor circuits were significantly lower than the design heating loads of the rooms. The next inlet temperature tried was 120 °F. Again the outputs were slightly lower than the room loads, but by a much smaller margin. The last input temperature tried was 125 °F. This yielded a total system output about equal to the calculated design load of the building. Notice, however, that at 125 °F inlet temperature some of the rooms have a heat output slightly greater than their design load, while others have an output slightly lower than their load. The differences are relatively small. Experience shows these minor differences tend to compensate for each other from one room to the next. Remember the heat loss calculations could also have enough error to make these small differences irrelevant.

Of all the rooms, bedroom 2 has the greatest difference between its design heating load and the heat output from its floor. One could compensate for this in a number of ways:

1. Add a supplementary heat emitter (perhaps a length of radiant baseboard along one wall)
2. Use closer tube spacing (perhaps 9 inches on center)
3. Use two tubing circuits (the higher flow rate and average water temperature yields greater heat output)
4. Use larger tubing (resulting in greater flow rate and greater heat output)
5. Decrease the heat loss of the room by adding insulation
6. Try a higher supply water temperature to the manifold and reduce the flow rate through the other circuits using the manifold valves

It is left an exercise for the reader to investigate these options.

The return temperature from the manifold station is calculated using Equation 10.11:

$$T_{mix} = \frac{\begin{array}{c}(1.21)(108.4) + (1.49)(115.6) + (1.72)(118.9)\\ + (2.4)(123) + (1.67)(118.4)\end{array}}{1.21 + 1.49 + 1.72 + 2.4 + 1.67}$$

$$= 117.9\ ^\circ F$$

This will be the approximate temperature of the water returning from the floor heating system under design load conditions.

Step 6. The piping schematic for the system is shown in Figure 10–83. It incorporates a separate high temperature zone for domestic water heating using an indirectly-fired storage water heater. Notice a bypass loop is used to

supply the floor heating circuits. This bypass water mixes with, and raises the temperature of, the water returning from the floor heating system preventing boiler condensation.

The flow rate across the zone valve can now be calculated by using Equation 10.11 at the junction marked A in Figure 10–84.

When the equation at the bottom of Figure 10–84 is solved, the flow rate through the zone valve is determined to be 1.43 gpm, assuming the primary loop flow rate was previously calculated at 12.8 gpm. The return temperature to the boiler could be calculated by applying Equation 10.11 at the junction marked B in Figure 10–84.

$$x = \frac{160(12.8 - 1.43) + 117.9(1.43)}{12.8} = 155.3\ ^\circ F$$

This temperature is well above the dewpoint temperature of a gas-fired boiler. It will decrease slightly when the distribution loop is operated at a lower temperature during warmer weather. If the return water temperature from the floor circuits drops to 80 °F, the return water temperature to the boiler is still 151 °F.

Step 7. The control schematic for this system is shown in Figure 10–85. The floor heating system is controlled

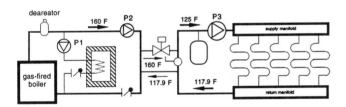

Figure 10–83   Piping schematic for overall system

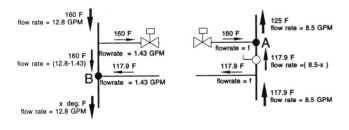

$$125 = \frac{117.9\,(8.5 - f) + 160\,(f)}{8.5}$$

Figure 10–84   Flow rates and temperatures at the junction of the floor heating distribution system and bypass piping

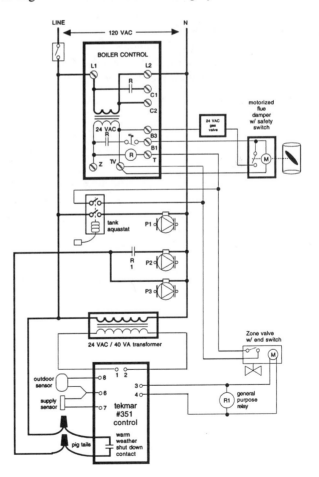

Figure 10–85   Control schematic for the system shown in Figure 10–83

by a reset control that operates the zone valve allowing hot water to be injected into the distribution loop. This control also senses room temperature and can stop heat input if internal heat gains cause overheating. It also contains a line voltage rated contact that can be used to turn off circulators P2 and P3 during warm weather. When this control determines heat is needed, the zone valve is opened. An end switch in the zone valve enables boiler firing. The relay R1 operates the bypass pump P2 only when floor heating is needed.

The reset ratio of the injection mixing control is estimated by dividing the range of water temperature change (from design load to no heating load) by the range of outdoor temperature change.

$$\text{Reset ratio} = \frac{125 - 65}{65 - (-20)} = 0.7$$

The control will be initially set to this value. Slight adjustments above or below this value can be made once the system is in operation.

Domestic water heating is controlled by a double pole aquastat mounted on the tank. One set of contacts in the aquastat completes a line voltage circuit to the tank's circulator (P1). The other contact completes a 24-volt circuit to enable boiler firing.

Because an automatic flue damper is used, the boiler's gas valve cannot operate until the limit switch in the damper closes, indicating the damper is fully open.

SUMMARY

Because hydronic radiant floor heating systems are an integral part of a building, they require more planning compared to other types of hydronic systems. The time committed to this planning, however, is time well spent. It will easily return itself in the following ways:

- An accurate layout plan speeds installation and reduces chance of errors
- Identification of problem areas (low heat output due to high resistance flooring, etc.)
- Identification of building features that interfere with installation (control joints, etc.)
- An accurate list of circuit lengths from which to order tubing coils with *minimal waste*
- Plan for piping installation and setup of controls
- Accurate sizing of the heat source
- Proper sizing of circulators and piping

A point worth making is that hydronic floor heating systems tend to be very "forgiving" systems. Minor differences between calculated loads and estimated heat

outputs from the floor tend to be insignificant in most projects. This statement is not meant to endorse careless design or installation, but rather to calm concerns over obtaining a precise match between calculated heat output and heating load in every room. Remember the methods used to calculate both the heating load and floor heat output are subject to inaccuracies. Neither can totally account for all the peculiarities of every building in which hydronic floor heating can be used. These calculations do, however, serve as a guide to judge the reasonableness of a proposed design, and check for large or unanticipated performance problems.

The technology of hydronic floor heating is constantly evolving. New products and design methods appear each year. The reader is encouraged to contact several manufacturers of flooring heating equipment to stay abreast of the latest materials and design methods.

KEY TERMS

Above deck dry system
Air film resistance
Available floor space
Below deck dry system
Bend supports
Control joint
Dry system
Finish floor resistance
Floor loading
Gypsum-based underlayments
Infrared radiation
Injection mixing
Lightweight concrete
Heat flux
Heat transfer plate
Lightweight concrete
Manifold
Manifold station
MIX program
Perimeter zone
RADFLOOR program
Room air temperature profile
Serpentine
Slab resistance
Telestat
Template block
Thermal break
Thin-slab system

CHAPTER 10 QUESTIONS AND EXERCISES

Note: Questions and exercises requiring the Hydronics Design Toolkit are indicated with marginal symbol.

1. Why is it desirable to have warmer surfaces and warmer air near the lower extremities of the body?

2. What is an advantage of radiant floor heating compared to finned-tube baseboard convectors in rooms with high ceilings?

3. Describe some considerations when locating a manifold station in a building.

4. Why is insulation needed beneath a thin-slab floor heating application even when the space beneath the floor is maintained at normal comfort temperatures?

5. Describe the function of control joints in thin-slab installations.

6. What are some of the architectural considerations (things the building designer needs to know about) associated with using a thin-slab floor heating system?

7. Why are gypsum-based underlayments installed in two lifts rather than a single pour?

8. What is the minimum thickness of gypsum-based underlayments over the top of the tubing?

9. Discuss the advantages and disadvantages of using lightweight concrete rather than gypsum-based underlayments for thin-slab applications.

10. Why should floor piping circuits route the warmest water near the exterior wall?

11. A kitchen measures 10 ft by 12 ft with an island measuring 3 ft by 5 ft. The kitchen has a design heating load of 4,000 Btu/hr when it is 68 °F inside and 0 °F outside. What is the required average upward heat flux from a floor heating system at design load conditions?

12. Describe what happens to the supply water temperature requirement of a floor heating system when a bare concrete slab floor is covered with 3/8-in ceramic tile. How does this compare to the supply water temperature requirement if the floor is covered with 1/4-in carpet?

13. The kitchen described in Exercise 11 has a slab-on-grade floor with one 10-ft wall and one 12-ft wall exposed to the outside. The edges of the slab are insulated with 2 ins of extruded polystyrene insulation. The slab is expected to operate at 105 °F. Estimate the downward heat flux.

14. Redraw the floor plan shown in Figure 10–86 to a scale of 1/4 in = 1 ft on a sheet of graph paper. Estimate dimensions if necessary. Choose a location for the manifold station. Lay out a floor heating circuit for the living/dining room, assuming a tube spacing of 12 ins on center. Estimate the length of the circuit you have drawn.

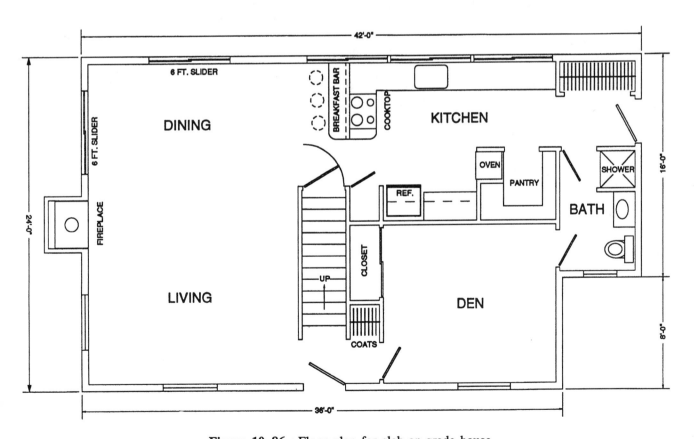

Figure 10–86   Floor plan for slab-on-grade house

15. Calculate the design heating load of the living/dining room shown in Figure 10–86. Use the same R-values and other assumptions from the case study of Section 10.10. Calculate the following:
    a. The required average upward heat flux at design load
    b. The average downward heat flux at design load
    c. The ratio of upward to downward heat flux
    d. The value of slab resistance R_s assuming $1/2$-in plastic tubing, 12 ins on center in a 4-in concrete slab.
    e. The value of a using Equation 10.7c, assuming no finish flooring
    f. The value of b using Equation 10.7b. Use your calculated length and assume a flow rate of 1.5 gpm
    g. The outlet temperature and upward heat output assuming an inlet temperature of 100 °F

16. A room that will use floor heating has an average upward heat flux requirement of 28 Btu/hr/ft². The upward to downward heat flow ratio is 8.0. It uses $3/4$-in tubing, 12 ins on center in a 6-in concrete slab. Estimate the required water temperature in the floor for the following finish floors assuming the room air temperature is 68 °F:
    a. $3/8$ ceramic tile
    b. $1/8$ sheet vinyl
    c. $1/2$-in carpet over $1/2$-in urethane padding

17. Water from four individual tubing circuits enters the return manifold at the following conditions:

Flow rate (gpm)	Temperature (°F)
1.8	102
2.3	97
1.05	94
1.3	99

Estimate the outlet temperature of the mixed stream from the manifold station.

18. Lay out the remaining floor piping circuits for the plan shown in Figure 10–86.

19. Use the lengths of the circuits obtained from Exercise 18 in place of the circuit lengths shown in the case study section of this chapter. Complete the revision of the calculations using the same assumptions and equipment of the case study.

20. Repeat Exercise 17 using the MIX program from the Hydronics Design Toolkit.

DISTRIBUTION PIPING SYSTEMS 11

OBJECTIVES

After studying this chapter you should be able to:

- Define seven unique distribution piping systems for hydronic heating
- Recognize the type of distribution piping used in existing systems
- Explain the iterative processes involved in designing a distribution system
- Discuss the pros and cons of zoning with building owners
- Assess the suitability of a particular distribution piping system for a given system application
- Select distribution piping systems well-matched to the intended heat emitters
- Use previously discussed analytical tools to design a distribution system
- Draw system piping schematics incorporating a specific piping design
- Recognize potential problems associated with improper piping design
- Describe the advantages of reverse-return versus direct-return systems

11.1 INTRODUCTION

The purpose of a hydronic distribution system is to transport heated water to each heat emitter so that it can properly heat its assigned area. The decision of what type of distribution system to use must consider factors such as:

- The required heating zones in the building
- How zoned heat output will be controlled
- The temperature requirements of the heat emitters
- The head loss characteristics of the heat emitters
- The cost of the piping system
- The size and operating cost of the system's circulator(s)
- The ability to expand the system in the future

This chapter discusses all the standardized piping approaches applicable to residential and light commercial systems. The strengths and weaknesses of each approach

are examined. Design procedures are discussed and examples are given.

This chapter also ties together much of the material from earlier chapters into true system design. It is important the reader be familiar with concepts such as series and parallel piping arrangements, head loss, hydraulic resistance diagrams, pumps curves, thermal performance of heat emitters, controls, and other material presented in earlier chapters.

11.2 SYSTEM EQUILIBRIUM AND ITERATIVE DESIGN

All hydronic systems, regardless of how they are designed, always attempt to stabilize themselves to a unique combination of operating temperatures and flow rate(s). At these conditions, the system is said to be in equilibrium. The system may or may not provide the proper heating of the building even when operating at equilibrium. In animated terms, the system "does not care" if proper heating of the building is occurring, it only "cares" about adjusting itself to and maintaining itself at equilibrium conditions.

The task of the designer is to assemble components into a system so equilibrium occurs at values that also allow the building to be properly heated. This needs to be accomplished within an allowable budget, with the highest possible fuel efficiency, and so the resulting system will last as long as possible with minimal maintenance.

Thermal Equilibrium

Thermal equilibrium occurs when the heat emitters release heat at exactly the same rate the heat source adds heat to the system. While it occurs, the fluid temperature at any given point in the system will remain constant.

An analogy to thermal equilibrium between the heat source and heat emitters is that of pouring water into a bucket containing several holes along a vertical line up its side, as shown in Figure 11–1. The holes represent the ability of the heat emitters to release heat into the building. The water being poured into the bucket represents

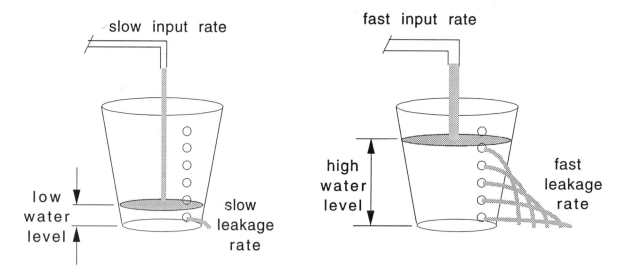

Figure 11–1 Analogy between fluid leaking from a bucket with holes, and thermal equilibrium between a hydronic heat source and the heat emitters of a distribution system

the heat input from the heat source. The height of the water in the bucket, at any time, represents the temperature of the fluid in the distribution system.

When water is slowly poured into the bucket, it can run out the lower few holes such that the water level rises only a short distance above the bottom of the bucket. As the rate at which water is added increases, the water level in the bucket rises until the rate of leakage again balances with the rate of water input. Further increases to the water input rate correspond to further rises in the water level. If the input rate is reduced, the water level will drop until equilibrium conditions between input and leakage are again established.

In a similar manner, the fluid temperature in a hydronic system will increase or decrease as necessary to balance the rate of heat input with the rate of heat release. Most control components merely interrupt this process as necessary to prevent wide differences between what the designer thinks the system should be doing and what it naturally wants to do, or to prevent unsafe operating conditions.

Perfect thermal equilibrium is seldom achieved in most hydronic systems. This is due to a number of factors such as on/off operation of the heat source, partial load conditions, and so-called transient operation while the thermal mass of the system components and the water they contain is being heated or is cooling down. System design traditionally attempts to achieve thermal equilibrium when the system operates under continuous maximum heating load. This is why the maximum load is often called the "design" load.

Hydraulic Equilibrium

Hydraulic equilibrium is achieved when the head added by the circulator equals the head dissipated by the flow resistance of the piping system. This occurs almost

instantly when the circulator is turned on. It occurs regardless of how the piping system is designed. Chapters 6 and 7 showed how to predict where hydraulic equilibrium will occur for any given piping system and circulator combination.

The objective is to design the piping system, and select the circulator, so the operating flow rate is reasonably close to flow rate assumed when the heat emitters were selected. This presents a paradox because the heat emitters will have a significant effect on the flow resistance of the system, and hence its flow rate. To further complicate matters, the heat output of the heat emitters also depends on flow rate.

This type of design situation, where the selected equipment depends on the overall performance of the system, yet the system performance depends on the selected equipment, occurs frequently in many areas of engineering. The solution to this situation lies in a concept known as **iteration**.

Iterative Design

Iterative design involves making successively refined estimates of equipment and system performance, each time reducing the differences between the two. Eventually the operating conditions that provide equilibrium of both the equipment and the overall system are found, or very closely estimated.

For hydronic systems the process of iterative design typically goes as follows:

Step 1. Begin by making a tentative selection of heat emitters for the rooms to be heated based on their design heat loads, and on an estimated average water temperature. A common practice is to assume the average water temperature in the system will be about 5 °F to 10 °F lower than the supply temperature from the heat source.

Step 2. The system flow rate for water can be estimated using Equation 11.1:

(Equation 11.1)

$$f_{system} = \frac{Q_T}{500(\Delta T)}$$

where:

f_{system} = intended flow rate of the system (gpm)
Q_T = total design heating load served by the system (Btu/hr)
ΔT = intended temperature drop of the system (°F)

If a fluid other than water is used, the flow rate can be estimated using Equation 11.2:

(Equation 11.2)

$$f_{system} = \frac{Q_T}{8.01(D)(c)(\Delta T)}$$

where:

f_{system} = intended flow rate of the system (gpm)
Q_T = total design heating load served by the system (Btu/hr)
ΔT = estimated temperature drop of the system (°F)
D = fluid's density at the average system temperature (lb/ft³)
c = fluid's specific heat at the average system temp (Btu/lb/°F)

If the heat emitters are connected in series, f_{system}, will be the flow rate through each heat emitter. If the heat emitters are connected in parallel, the intended flow rate through each heat emitter can be estimated using Equation 11.3:

(Equation 11.3)

$$f_i = f_{system}\left(\frac{Q_i}{Q_T}\right)$$

where:

f_i = intended flow rate through a particular heat emitter (gpm)
f_{system} = intended (total) flow rate of the system (gpm)
Q_i = design heat output of the particular heat emitter (Btu/hr)
Q_T = total design heating load served by the system (Btu/hr)

Equation 11.3 simply divides the system flow rate in proportion to the heat output rate assigned to each heat emitter.

Step 3. A piping distribution system is now selected to connect the heat emitters with the heat source. Several options will be shown in later sections of this chapter.

Some are better suited than others, depending on the type of heat emitters used.

Step 4. Pipe sizes should be selected based on keeping the flow velocity under 4 ft/sec to avoid flow noise. The PIPESIZE program in the Hydronics Design Toolkit can be used for making these selections.

Step 5. Using the methods of Chapter 6, the piping system can now be described as a hydraulic resistance diagram. After reducing this diagram to a single equivalent hydraulic resistance, the system curve can be sketched. The PIPEPATH program in the Hydronics Design Toolkit can also be used to determine hydraulic resistances, and to produce a system curve.

Step 6. The pump curves of one or more circulators can now be matched against the system curve. Recall from Chapter 7 that the intersection of the system curve and the pump curve yields the operating flow rate for the combination. *Chances are the system flow rate for any one pump will not exactly match the intended flow rate estimated in step 2. This is normal and to be expected as part of the design process.* The PUMPCURV and PUMP/SYS programs in the Hydronics Design Toolkit can be used for this step.

Step 7. If all heat emitters are in a series circuit with the circulator, this step can be skipped. If some or all of the heat emitters are located in parallel branch circuits, the methods of Chapter 6, or the PARALLEL program in the Hydronics Design Toolkit, can be used to find the flow rate in each branch.

Step 8. The thermal performance of each heat emitter can now be verified at the actual flow rate and entering water temperature present at their location in the system. If the heat output of a given heat emitter is too low, a number of options are possible. These include using a higher capacity heat emitter, increasing system operating temperature, or modifying the piping to obtain a higher flow rate. If the output of a heat emitter is more than 10% above design load, a lower capacity unit can be used, or, in the case of a parallel system, a balancing valve can be used to reduce its flow rate.

Step 9. Depending on what, if any, changes are made to the system, the designer may need to go back to step 3 and repeat the process of determining the new flow rates in the modified system. *This is iteration.* In most cases only one or two iterations need to be made to find a design that can provide the necessary heat output with the selected heat emitters, piping system, and circulator. The stated programs in the Hydronics Design Toolkit can be a tremendous time saver in making the necessary calculations.

11.3 ZONING

The idea of dividing a building into two or more areas that have independent control of heating input has been mentioned in earlier chapters. These so-called zoned systems offer the potential to achieve two benefits not available in single-zone systems. These benefits are:

1. To conserve energy by reducing the temperature of unoccupied zones
2. To meet individual comfort requirements through independent control of heat input

Hydronic systems can be easily zoned. There are several ways of constructing multi-zone systems depending on the specific method of control and number of zones required.

People sometimes get the idea the more zones a system has, the better it is designed. This is *not* necessarily true. The *potential benefits* of **zoning** can be overestimated in terms of how the system will actually work and in how the building occupants will operate the system.

The Effect of Thermal Mass

Many people, including heating professionals, often have the belief that zoned systems are somehow able to lower the temperature of a room (or group of rooms) down to 55 °F when the occupants leave for work, school, etc., and then bring it right back up to 70 °F when they return. They somehow think the temperature of the zone can be turned up or down just like turning lights on or off. If this could be accomplished, the energy conservation potential of zoned systems would be enormous.

Unfortunately, such wide and rapid changes in temperature are seldom possible because of the thermal mass of the heating system, as well as the building itself. The greater the thermal mass of the distribution system and building components, and the lower the rate of heat loss from the building, the slower the temperature will drop when the room thermostat is turned down. Studies has shown that well-insulated buildings with high thermal mass might only experience a temperature drop of 2 °F to 3 °F over a mid-winter nighttime period lasting eight to ten hours, *even with the heating system completely off.* If, for example, one were to reduce the thermostat setting in such a building from 70 °F to 65 °F at bedtime, the inside air temperature would not decrease to the reduced setpoint before morning. The energy savings potential of this situation would be very limited. If the thermostat were turned down 10 °F or even 20 °F, the results would be the same. All the thermostat does is disable operation of the heating system until the building has cooled to its lowered setting. If the building will not drop in temperature fast enough, the amount the thermostat is turned down is irrelevant.

The thermal mass of the zone and its heating system further complicates the situation when morning arrives and the occupants want the building quickly restored to normal comfort temperature. High thermal mass systems, like hydronic radiant floor heating in a 4-inch or thicker concrete slab, can take several hours to bring the building back to normal comfort temperature following a prolonged setback period.

The bottom line is the energy conservation potential of a high thermal mass-zoned system is very dependent on the duration of the reduced thermostat setting. When a zone is consistently maintained at a reduced temperature day after day, the energy conservation potential can be significant. However, attempts at frequent and wide temperature changes are usually futile. This characteristic should be explained to building owners who may not realize how the system responds to changes in the thermostat setting.

Inter-zone Heat Transfer

Another factor effecting zoning performance is **inter-zone heat transfer**. This refers to heat transfer through *interior* partitions as the building attempts to equalize temperature differences from one room to the next. If a particular room is kept at 60 °F, for example, while an adjacent room is kept at 75 °F, heat will flow from the warmer to the cooler room. This partially defeats attempts to maintain temperature differences by zoning. The greater the thermal resistance of the exterior envelope of the building, the harder it is to maintain significant temperature differences between rooms separated by uninsulated interior partitions.

Rooms or otherwise zoned areas that will have significant temperature differences (more than 10 °F) should have insulated interior partitions. The doors that separate a cooler zone from a warmer one will obviously have to remain closed when these temperature differences are to be maintained.

Occupant Management of Zoned Systems

A final consideration that often determines the cost effectiveness of zoning is the willingness of the occupants to regulate the system. It might seem obvious that an owner who is willing to pay more for an extensively zoned hydronic system would be willing to regulate it. Experience shows this is not always true. Today's fast-paced lifestyles mean people often forget to regulate their heating systems, especially when their schedules vary from day to day. This point is made not to discourage zoning, but rather to encourage an honest discussion on how the system will be used, before committing several hundred dollars for extensive zoning.

What Areas Should Be Zoned?

The zoning of a building should reflect not only the activities and schedules of the occupants, but factors such as internal heat gain from the sun, equipment, or supplemental heat sources like a fireplace.

Rooms with large south-facing windows can overheat from solar heat gains, while the northern rooms remain too cool for comfort. Good zoning design allows heat input to the cooler rooms even when the south-facing rooms are satisfied. In some cases, the rooms heated by solar gains will change over the course of the day as the sun moves across the sky. Zoning design that allows rooms to adapt to these changes will provide stable comfort.

In commercial buildings, rooms that contain heat producing equipment such as computers, vending machines, or cooking facilities may be good candidates for separate zoning control. Rooms with minimal exterior exposure will have small heating loads relative to their floor area. When such rooms are controlled as separate zones, overheating can be minimized.

When zoning is properly applied it can significantly enhance operation of the system and improve owner satisfaction. Good design requires careful thought about how the building will be used at present and in the future. The designer should carefully discuss the options with the building owner before committing to a particular distribution system.

11.4 SINGLE SERIES CIRCUITS

The simplest distribution system is a single piping loop that progresses from the heat source, through each heat emitter, and back to the heat source. System operation is often controlled by a single regulating device such as a room thermostat. This type of distribution system is called a **single series circuit**. *Single series circuits are appropriate for buildings in which all rooms experience similar changes in their heating loads as outdoor conditions vary.* An example of a single series circuit of finned-tube baseboard convectors is shown in Figure 11–2.

Single circuit systems can also be designed around a combination of heat emitters including panel radiators and fan-coils.

Because heat input to the entire building is regulated based on the temperature at one location in the building, overheating or underheating of rooms other than where the air temperature is sensed is very possible. *It is crucial*

that all heat emitters be sized for the load of their respective rooms, and to the water temperature at their location within the piping circuit. The latter condition is often overlooked when heat emitters are sized based on a single average water temperature.

Chapter 8 demonstrated the differences in sizing finned-tube baseboard units based on a single average water temperature, versus the temperature of the water at the location of each baseboard within the circuit. Using the single average water temperature method, the baseboards near the beginning (hottest) part of the circuit tend to be oversized and thus put out too much heat. Those near the end of the circuit tend to be undersized and thus put out too little heat. These tendencies will also be true for other types of heat emitters if sized at a single average water temperature.

To prevent this from happening, *the designer must keep track of the water temperature as it cools down from one heat emitter to the next.* Begin at the location of the first heat emitter and progress around the circuit treating the outlet temperature of each heat emitter as the inlet temperature to the next. If a piping segment between heat emitters is longer than 20 feet, or is otherwise expected to have a significant heat loss, the designer should calculate the temperature drop along the pipe and correct the inlet temperature to the next heat emitter accordingly. Recall that methods for estimating pipe heat loss were presented in Chapter 8. The PIPELOSS program in the Hydronics Design Toolkit can also be used for such estimates.

The SERIESBB program in the Hydronics Design Toolkit can be used to quickly determine the required lengths of finned-tube baseboard in a series circuit, or portion thereof. A screen shot of the program is shown in Figures 8–16 and 8–17.

The limiting factors in designing single circuit systems are temperature drop and flow resistance. In the U.S., series circuits have traditionally been designed around a temperature drop of about 20 °F. *However, there is nothing magical about this number.* Systems can be designed to operate with much smaller or larger temperature drops.

In Europe, systems with temperature drops of 30 °F to 40 °F are used successfully. The advantage of having a high temperature drop is the flow rate can be relatively low. This in turn allows small diameter piping and a smaller, less power-consuming circulator to be used. The disadvantage of a high temperature drop system is that heat emitters near the end of the circuit will have to be significantly larger in order to meet design load conditions with the lower temperature water. There is also the concern of maintaining the return temperature to a conventional boiler high enough to prevent condensation.

Systems with low temperatures drops usually have relatively high flow rates that may require increased pipe sizes to keep the flow velocity below 4 ft/sec. They also require larger circulators. The heat emitters will, however, have higher average temperatures, and thus can be smaller in size.

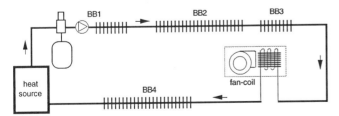

Figure 11–2 Example of a single series circuit distribution system

It is impossible to say what the optimum temperature drop is for a single series circuit. The answer depends on the cost of piping, circulators, heat emitters, electricity, and several other factors. The reader is encouraged to use a design tool like the SERIESBB program to experiment with different combinations of flow rate and temperature drop.

Series circuits that contain one or more heat emitters with high flow resistance characteristics (such as fan-coils) must be carefully evaluated for potential problems. A heat emitter with high flow resistance can greatly restrict flow rate through the rest of the system, limiting its total heat output.

Another potential problem is excessive flow velocity through the small tubes or valves in the restrictive heat emitter. This can create flow noise and eventual erosion corrosion. A high head circulator will probably be required. High head differentials across a circulator encourage cavitation, especially in high temperature systems. Pumping heads in excess of 30 feet should be carefully scrutinized for cavitation potential. In many cases a different type of piping system will prove more appropriate for connecting heat emitters with high flow resistance characteristics.

A definite drawback of a single zone series circuit is that zone control is limited to the control features present on the heat emitters. In the case of finned-tube baseboard, heat output can be reduced up to about 40% by closing the dampers on the enclosure. When fan-coils are used, the blower speed can be reduced to limit heat output. Both these methods, however, require manual adjustments to the heat emitters in response to changing load conditions. Many occupants either do not realize these adjustments can be made or soon get tired of making them. The resulting overheating and underheating is begrudgingly tolerated to the detriment of future referrals for the installer.

11.5 SINGLE CIRCUIT/MULTI-ZONE (1-PIPE) SYSTEMS

A variation on the basic series circuit design allows for one or more of the heat emitters connected to a single piping circuit to be independently controlled. It involves the use of diverter tees as discussed in Chapter 5. These tees are used to divert a portion of the water flowing in the main piping circuit through a branch circuit that includes the heat emitter. The concept is illustrated in Figure 11-3.

The diverter tees can be used individually or in pairs. When two diverter tees are used, they set up a greater pressure difference across the branch circuit. This in turn creates a higher flow rate in the branch. Single diverter tee arrangements are often sufficient for low resistance heat emitters, such as a few feet of finned-tube baseboard

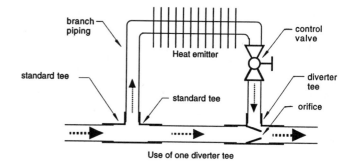

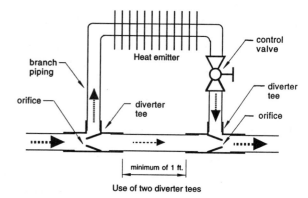

Figure 11-3 Use of diverter tee(s) and a control valve to regulate flow through a heat emitter

or a panel radiator. Double diverter tees may be needed on heat emitters with higher flow resistances such as fan-coils. When double diverter tees are used, there should be a minimum of one foot of pipe between the tees to allow turbulence created by the first tee to partially dissipate before the flow enters the second tee.

The control valve is used to control the flow rate through the branch circuit containing the heat emitter. This in turn limits or totally stops heat output from the heat emitter. The valve can be a simple manually-adjusted globe or angle valve adjacent to or even built into the heat emitter. Manual valves will require frequent adjustment if the room is to maintain a constant temperature during changing load conditions. If automatic control of the heat output from the heat emitter is desired, a non-electric thermostatic radiator valve can be used These valves were discussed in Chapter 5.

Systems employing diverter tees are often called **1-pipe systems**. They offer tremendous zoning flexibility at relatively low cost. Any one or more of the heat emitters on the piping circuit can be selected to have independent temperature control simply by piping it into the circuit using diverter tee(s), and some type of control valve.

Figure 11-4 shows a single circuit system where one of the heat emitters is connected using a diverter tee arrangement while the other baseboards are piped in the standard series configuration. This is a convenient way to limit the heat output of one heat emitter without significantly affecting the remainder of the heat emitters.

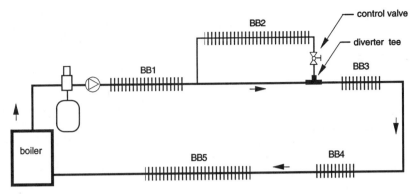

Figure 11–4   Single heat emitter connected for independent zone control using diverter tee and control valve

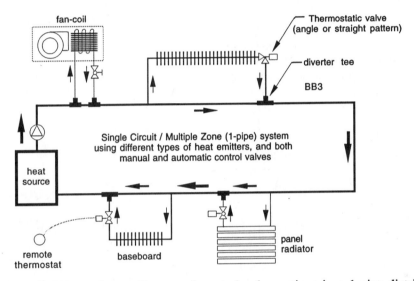

Figure 11–5   Use of different heat emitters and control valve options in a 1-pipe distribution system

This arrangement could be used, for example, in a guest bedroom that only needs to be heated to normal comfort temperatures a few days each year.

When properly designed, 1-pipe systems can include all types of heat emitters and control valve options, as shown in Figure 11–5.

It is important to realize that nonelectric thermostatic radiator valves do not control the operation of the heat source or the system circulator. They cannot fire the boiler and turn on the circulator when a given room needs heat. For hot water to flow through a heat emitter regulated by a nonelectric control valve, there must be hot water flowing through the main piping circuit. In this sense the control valves serve only as temperature limiting devices for the rooms where they are located.

When all heat emitters are piped into the main circuit using diverter tees, constant circulation of heated water in the main piping circuit is often necessary to ensure heat is available at any heat emitter when its control valve opens. *The main piping circuit should be well insulated to limit heat loss.*

To further reduce piping heat loss, and reduce temperature swings in rooms, a reset control can be used to operate the heat source. This control will maintain the circulating water only as hot as necessary for the prevailing outdoor conditions.

When constant circulation is used along with a gas or oil-fired boiler, the boiler should be piped into the system as shown in Figure 11–6. This so-called primary/secondary arrangement prevents heated system water from constantly circulating through the boiler during its off-cycle, reducing heat loss up the chimney. Primary/secondary systems will be discussed in detail in Section 11.10.

In systems that have a mixture of series and diverter tee-connected heat emitters (such as shown in Figure 11–4), constant circulation will not work because the series-connected heat emitters will produce a constant heat output even when not needed. These systems must rely on a room thermostat to stop the circulator, preventing overheating of the spaces served by the inline-connected units.

One technique that provides the ability to heat a space above its normal comfort temperature when desired is to

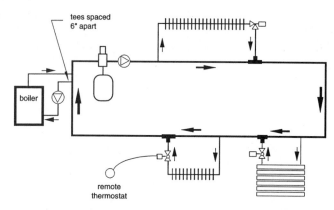

Figure 11-6 Connecting a boiler to a constantly circulating 1-pipe distribution system using primary/secondary piping

intentionally oversize the heat emitter and connect it to the main piping circuit, using diverter tees and a control valve. Depending on the amount of oversizing, the heat emitter can raise the temperature of the room several degrees above normal comfort temperature when the control valve is set fully open. This extra capacity also provides for fast temperature pick-up following a set-back period. If a thermostatic radiator valve is used, the room temperature can still be automatically limited when the high temperatures are not desired. This concept is well applied in bathrooms where higher than normal air temperatures can quickly be established before taking a shower or bath, and then reset to normal. It can also be used in spaces that need to be quickly heated, but may otherwise be kept at a reduced temperature. Examples include workshops, part-time offices, and recreation rooms.

Whenever diverter tees and control valves are used to reduce or totally stop water flow to a heat emitter, it crucial that the heat emitter and the piping leading to it cannot freeze during cold weather. Some thermostatic radiator valves have a freeze-proof minimum setting that will attempt to maintain the air temperature in the space above freezing, even when the setting knob appears fully closed. If a manual control valve is used, it may have to be left slightly open during very cold weather. If there is a chance the valve may be unintentionally closed, it may be a good idea to remove the handle.

Design of 1-pipe Systems

The design of a 1-pipe system using diverter tees is necessarily more complex than that of a single series circuit. The major difference is that the flow rate through each heat emitter can be different depending on its flow resistance and the type and number of diverter tees used to connect it to the main circuit.

It is convenient to think of each use of the diverter-tee arrangement as a small segment of two parallel piping strings. Using the methods of Chapter 6, the hydraulic

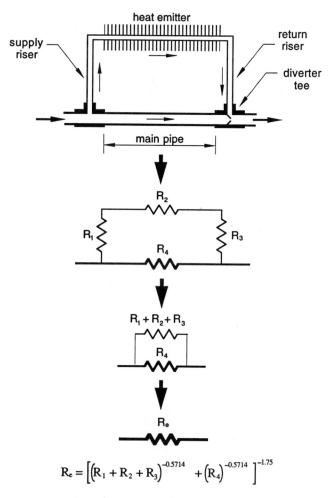

$$R_e = \left[\left(R_1 + R_2 + R_3 \right)^{-0.5714} + \left(R_4 \right)^{-0.5714} \right]^{-1.75}$$

R_1 = hydraulic resistance of supply riser and fittings
R_2 = hydraulic resistance of heat emitter
R_3 = hydraulic resistance of return riser and fittings
R_4 = hydraulic resistance of main pipe and diverter tee orifices

Figure 11-7 Systematically reducing a diverter tee arrangement down to a single equivalent hydraulic resistance

resistance of each parallel string can be found and then reduced to a single equivalent resistance. Since the equivalent resistances of each diverter tee arrangement will be in series with each other, they can be added to get the total equivalent resistance of the circuit. This resistance can then be combined, either analytically or graphically, with the pump curve to obtain the overall system flow rate. The designer then works backwards, reexpanding the equivalent resistances into their original parallel resistances to find the flow rate through each heat emitter. This concept is illustrated in Figures 11-7 and 11-8.

When the flow rate through each heat emitter is determined, it can be combined with the inlet temperature and performance data of the heat emitter to estimate its heat output and the temperature drop.

The outlet temperature of the heat emitter will not be the same as the outlet temperature from the tee connecting

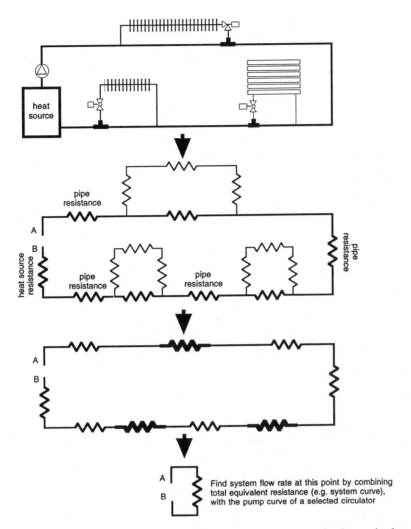

Figure 11–8   Systematically reducing a 1-pipe distribution system down to a single equivalent hydraulic resistance

its return riser to the main pipe. This is because a portion of the water will have bypassed the heat emitter and mixed with the water returning from the heat emitter. The blended temperature leaving the downstream tee can be found using the equation shown in Figure 11–9.

flow rate = f_1
Temp = T_{R1}

flow rate = f_2 flow rate = $f_1 + f_2$
Temp = T_{R2} Temp = T_{mix}

$$T_{mix} = \frac{(f_1)(T_{R1}) + (f_2)(T_{R2})}{f_1 + f_2}$$ (Equation 11.4)

Figure 11–9   Equation used to determine the blended outlet temperature from a tee

The outlet temperature from the tee can now be used as the inlet temperature to the next heat emitter. The blended temperature leaving the tee can also be found using the MIX program in the Hydronics Design Toolkit.

The MONOFLO Program

The Hydronics Design Toolkit contains a program called MONOFLO that is specifically designed to determine the equivalent hydraulic resistance of a diverter tee arrangement with any type of heat emitter and riser piping. The hydraulic resistance of several combinations of Bell & Gossett MONOFLO® tees is automatically determined by selecting the size and number of the tees from a pop-up list. The user enters the size and length of the risers and the hydraulic resistance of the heat emitter. The latter can be obtained using either the PIPEPATH or HYRES programs. When the diverter tee arrangement is fully specified, the program will also determine the proportions of the entering flow that go through the heat emitter and main pipe. A screen shot of the MONOFLO program is shown in Figure 11–10.

```
 - MONOFLO  -  F1-HELP      F3-SELECT                    F10-EXIT TO MAIN MENU
┌──-INPUTS-═══════════════════════════════════════════════════════════════════┐
│ ENTER Hydraulic resistance of branch circuit,(.01 to 10).. .1                │
│ ENTER Temperature of system fluid,(80 to 250)............. 150     deg. F    │
│ ENTER length of pipe between tees,(1 to 100).............. 10      ft.       │
│ ENTER System flow rate into tee,(.5 to 100).............. 10       gpm       │
│ SELECT Monoflow tee configuration,(press F3 )............. single 3/4"       │
│ SELECT System fluid,(press F3 )........................... water            │
└──────────────────────────────────────────────────────────────────────────────┘
┌──-RESULTS-══════════════════════════════════════════════════════════════════┐
│ Equivalent hydraulic resistance of monoflow block........ 0.044             │
│ Flow rate in branch circuit.............................. 5.73 gpm          │
│ Flow rate between tees................................... 4.27 gpm          │
└──────────────────────────────────────────────────────────────────────────────┘



┌── MESSAGE ═══════════════════════════════════════════════════════════════════┐
│ Use the UP or DOWN ARROW keys to move the highlighting bar over an input,     │
│ then type in a value. Press F10 to exit to MAIN MENU. Press F3 to SELECT.     │
└────────────────────────────────────────────────────────────────────────────────┘
```

Figure 11–10   Screen shot of the MONOFLO program from the Hydronics Design Toolkit

Like single series circuits, 1-pipe systems are best applied with medium to high distribution water temperatures. This is because the temperature drop around the circuit tends to be higher than when parallel zone circuits are used. If a low supply temperature is used, the heat emitters will have to be significantly larger to be able to meet design load conditions, especially near the end of the circuit. This may eliminate the possibility of using finned-tube baseboard. It may also create problems with fan-coils because the discharge air temperature could be low enough to cause discomfort if directed toward occupants. Also, when a conventional boiler is used, the return water temperature must be high enough to prevent flue gas condensation.

11.6 MULTI-ZONE/MULTI-CIRCULATOR SYSTEMS

Multiple-zone systems using **multiple circulators** have been used in both residential and light commercial buildings for many years. They consist of two or more parallel piping loops, each with its own circulator and check valve, connected to a common heat source. Each zone circuit can have several heat emitters connected in series or with diverter tees. A schematic representation of a four-zone system is shown in Figure 11–11. In this case, three of the zones are used for space heating, while the remaining zone is used for domestic water heating.

Notice the zone supplying the domestic water heater is connected closest to the boiler. While this is not absolutely necessary, it does reduce heat loss from the outer portions of the common piping when only domestic water heating is required, during the summer for example.

The Importance of Check Valves

It is crucial that a self-closing check valve is used on every zone circuit in this type of system. The check valves prevent water from flowing backward through inactive circuits when other circuits are operating. Without these valves, warm return water flowing backward through inactive zone circuits will cause heat to flow from the heat emitters into areas where it is not needed. Reverse flow through an inactive zone circuit is shown in Figure 11–12.

The *type* of check valve used is also important. It should either be a spring-loaded inline check valve as shown in Figure 5–42, or a flow check valve as shown in Figure 5–46. Either of these valves can prevent both reverse flow and undesirable thermosiphoning of hot water through inactive zone circuits. The latter occurs because the higher temperature water in the heat source has a lower density than the cooler water in the distribution piping. Without a valve that imposes a small resistance to forward flow, hot water will migrate upward into inactive zone circuits, causing heating when it is not needed. This effect is especially undesirable during the summer when the boiler operates only for domestic water

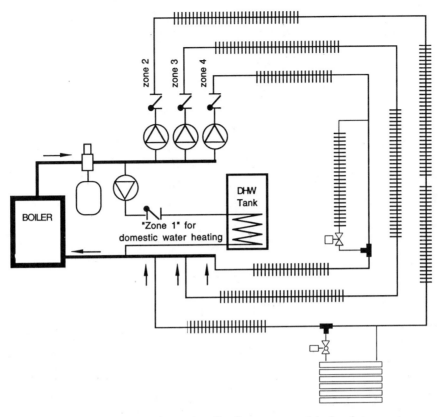

Figure 11-11   Piping schematic of a four-zone multi-circulator system

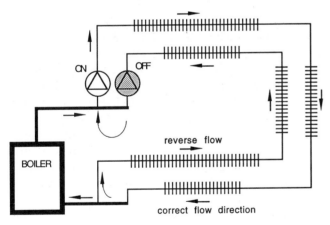

Figure 11-12   Reverse flow through inactive zone when check valves are not used

heating, and heat leaking from the heat emitters adds to the building's cooling load.

There are several advantages to using multiple pumps in zoned systems:

• The head loss and flow rate requirements of a given zone circuit tend to be less than for single circuit systems. This usually allows the use of smaller piping and circulators. Small wet rotor circulators that draw between 50 and 100 watts are often adequate. Since each circulator operates only when its associated room

thermostat is calling for heat, a savings in seasonal electric consumption will likely be realized compared to a single larger circulator.

• The possibility of a pump failure presents fewer potential problems. In a multiple-circulator system this only shuts down heat flow to one of the zone circuits. In a single circulator system, the entire building could be without heat.

• Multiple-circuit zoned systems have the advantage that each circuit receives approximately the same supply water temperature. This tends to keep the average water temperature of each circuit slightly higher than if all heat emitters were served by a single circuit. As a result, the size of the heat emitters may be slightly smaller.

• When a heat source with a relatively low flow resistance is used, flow rates in each zone of a multiple pump system are fairly stable regardless of what zone(s) are operating. This allows for better balancing and consistent heat output from each zone circuit. It also eliminates the possibility of flow noise caused by a large pump pushing water through a single active zone circuit while the other zones are closed off by valves.

• Recently-introduced control products such as the multizone relay center described in Chapter 9 make the electrical connections for multi-circulator systems very simple.

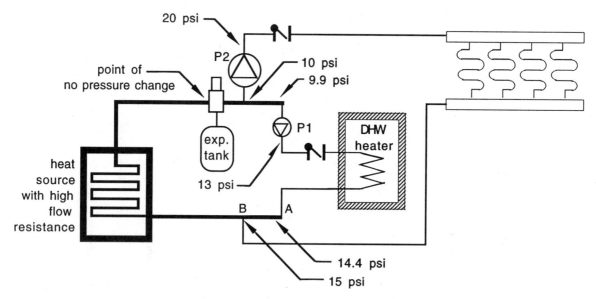

Assume circulator P1 has a maximum head (at zero flow) of 4.5 psi

Figure 11–13   Hypothetical system in which flow through low head circulator (P1) is blocked by high head differential across the heat source produced by circulator (P2)

A Situation to Avoid

One precaution that must be observed in multiple circulator systems is not to use high head and low head circulators in the same system with a heat source having high flow resistance characteristics. If the flow resistance of the heat source is high enough, and the maximum head of the smaller circulator is low enough, a pressure distribution similar to that shown in Figure 11–13 could occur.

Notice the pressure at point B on the return manifold is greater than at point A, *even with the smaller circulator operating at maximum head (e.g., zero flow rate)*. Under these conditions, flow cannot occur from A to B. The reverse pressure differential would in fact close the check valve on the domestic water heater zone. In this hypothetical system, flow through the domestic water heater would totally stop whenever the larger circulator (P2) was operating. A primary/secondary piping arrangement (to be discussed later) could be used to solve this problem.

If a conventional boiler with a low flow resistance was used, most of the head produced by the larger circulator would be dissipated by the distribution system. The pressure differential across the boiler would be quite low, and the smaller circulator (P1) would likely be able to produce a flow through the domestic water heater coil.

Design of Multi-circulator/Multi-zone Systems

The design of a multi-zone system using individual circulators is similar to the procedure used for a single circuit system. The difference is in how the flow resistance of the heat source and common piping is treated. Because the pressure drop across the heat source and common piping is a function of flow rate, it will change depending on how many zones are active. For heat sources with low flow resistances such as cast-iron sectional boilers, the change in pressure drop across the boiler will be minimal. It will not significantly affect the flow rate in an active circuit as other circuits turn on and off. However, when a heat source with higher flow resistance, such as a copper tube boiler or heat exchanger is used, the increased pressure drop that occurs when all zones are on can significantly affect the flow rate in each circuit.

The conservative design approach is to include the flow resistance of the heat source and common piping, *assuming all zones are operating*, into the overall resistance of each zone circuit. This approximates the situation that is likely to occur under design load conditions. It allows the heat emitters to be sized at their minimum operating flow rate within the system. This concept is illustrated in Figure 11–14.

Equation 11.5 can be used to modify the hydraulic resistance of the heat source and common piping to represent full flow conditions:

(Equation 11.5)

$$r_{corrected} = r_{calculated} \left(\frac{f_{total}}{f_{zone}} \right)^2$$

where:

$r_{corrected}$ = hydraulic resistance of the heat source and common piping at full system flow

$r_{calculated}$ = hydraulic resistance of the heat source and common piping as normally calculated

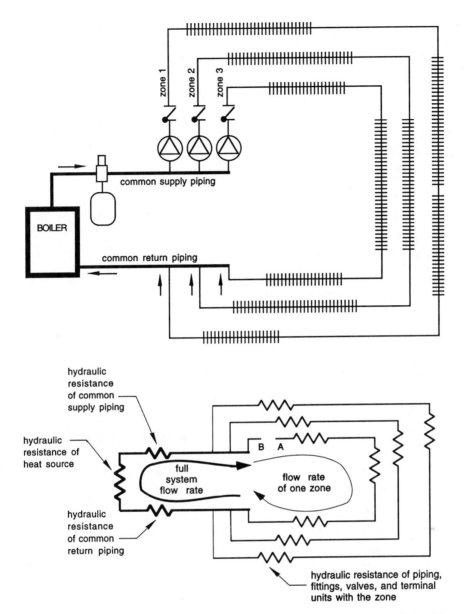

Figure 11–14   Assume full system flow rate in the heat source and common piping. Include this resistance in the overall resistance of each zone circuit.

f_{total} = estimated system flow rate through the heat source when all zones are on (gpm)

f_{zone} = estimated flow rate in the zone circuit being designed (gpm)

Example 11.1: Estimate the flow rate in the zone circuit shown in Figure 11–15 *when all zones are operating*. The heat source is a copper water tube boiler with a hydraulic resistance of 0.3 as determined using the ℍ𝕐ℝ𝔼𝕊 program in the Hydronics Design Toolkit. The tentative circulator selection is a Grundfos UP15-42. The zones are intended to have a temperature drop of 20 °F. The design heating load supplied by the boiler is 70,000 Btu/hr. The design load supplied by the zone is 25,000 Btu/hr. The water temperature supplied from the boiler is 160 °F.

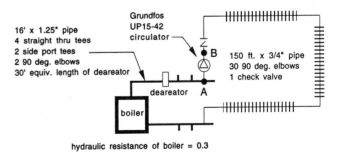

Figure 11–15   Piping schematic for Example 11.1

Solution: It is necessary to assemble information from Chapter 6 to develop the hydraulic resistance diagram of the zone circuit.

The average water temperature in the zone will be about:

$$T_{ave} = 160 - \frac{20}{2} = 150 \ °F$$

The α value of water at this temperature is estimated from Figure 6–17 as: $\alpha = 0.047$

The c-value for $^3/_4$-in copper tube is: $c = 0.061957$

The c-value for the 1.25-in copper tube is: $c = 0.0068082$

The total equivalent length of the zone piping and fittings is: $150 + 30(2) + 3 = 213$ ft

The total equivalent length of the common piping is: $16 + 4(.6) + 2(3) + 2(5.5) + 30 = 65.4$ ft

The hydraulic resistances of the zone circuit and common piping can each be calculated using Equation 6.11: $r = (\alpha cL)$

For the zone circuit:
$r_1 = \alpha cL = (0.047)(0.061957)(213) = 0.62$

For the common piping:
$r_2 = \alpha cL = (0.047)(0.0068082)(65.4) = 0.0209$

A hydraulic resistor diagram showing these resistances is given in Figure 11–16.

The total system flow rate can be estimated using Equation 11.1

$$f_{system} = \frac{70,000}{500(20)} = 7.0 \ gpm$$

The portion of the system flow rate needed by the individual zone is found using Equation 11.3:

$$f_{zone} = 7.0 \left(\frac{25,000}{70,000} \right) = 2.5 \ gpm$$

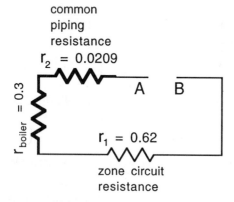

Figure 11–16 Hydraulic resistance diagram for Example 11.1

The hydraulic resistance of the boiler and common piping need to be added together and then corrected using Equation 11.5:

$$r_{boiler+common \ piping} = r_{boiler} + r_{common \ piping} = 0.3 + 0.0209$$
$$= 0.3209$$

$$r_{corrected} = 0.3209 \left(\frac{7.0}{2.5} \right)^2 = 2.52$$

Notice the corrected hydraulic resistance of the circuit is considerably greater than the original calculated value. This accounts for the additional head loss of the boiler and common piping when all zone circuits are operating. The total hydraulic resistance of the zone circuit is found by adding the corrected resistance of the boiler plus common piping, to that of the zone piping:

$$r_{zone \ circuit} = (r_{boiler} + r_2)_{corrected} + r_1 = 2.52 + 0.62 = 3.14$$

The system curve can now be represented as:

$$H_L = 3.14(f)^{1.75}$$

This system curve is plotted in Figure 11–17 along with the pump curve for the Grundfos UP15-42 circulator to obtain the approximate flow rate in the zone circuit when all zones are operating.

Discussion: Notice the actual flow rate expected in the zone (2.4 gpm) is only slightly less than the flow rate based on obtaining a 20 °F drop (2.5 gpm). This indicates the tentative circulator selection is well suited for this zone.

Example 11.2: Repeat Example 11.1 assuming a cast-iron sectional boiler with a hydraulic resistance of 0.002 is used instead of the high flow resistance boiler.

Solution: The total hydraulic resistance of the boiler and common piping is now:

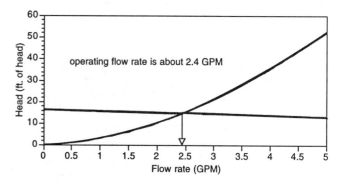

Figure 11–17 Plot of system curve and pump curve for Example 11.1

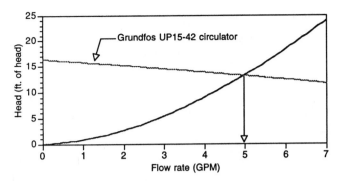

Figure 11–18   Operating point for zone circuit of Example 11.1 using a sectional cast-iron boiler and Grundfos UP15-42 circulator

$$r_{\text{boiler+common piping}} = r_{\text{boiler}} + r_{\text{common piping}} = 0.002 + 0.0209$$
$$= 0.0229$$

This resistance is corrected for full flow conditions using Equation 11.5:

$$r_{\text{corrected}} = 0.0229\left(\frac{7.0}{2.5}\right)^2 = 0.179$$

The total hydraulic resistance of the zone circuit is found by adding the corrected resistance of the boiler plus common piping, to that of the zone piping:

$$r_{\text{zone circuit}} = (r_{\text{boiler}} + r_2)_{\text{corrected}} + r_1 = 0.179 + 0.62$$
$$= 0.799$$

The revised system curve and pump curve for the Grundfos UP15-42 pump are shown in Figure 11–18.

Discussion: The same circulator now yields a flow rate of about 5 gpm compared to only 2.4 gpm when used with the high flow resistance boiler of Example 11.1. This significant increase is due solely to the lower flow resistance of the cast-iron sectional boiler. *It demonstrates the need to carefully scrutinize multi-zone/multi-circulator systems using high flow resistance heat sources.* The 5 gpm flow rate will result in a higher than expected average circuit temperature with a temperature drop around the circuit of only about 10 °F. This will result in greater heat output from the circuit.

11.7 MULTI-ZONE SYSTEMS USING ZONE VALVES

Another option for multiple-circuit zoned systems is the use of electric zone valves and a single circulator. A piping schematic for this approach is shown in Figure 11–19.

When the room thermostat in a given zone calls for heat, its associated zone valve begins to open. When the valve reaches its fully open position, an internal end switch closes to signal the circulator and boiler to operate. When the thermostat in the space served by the zone is satisfied, power to the zone valve is shut off, and an internal spring closes the valve. The wiring of various types of zone valves was discussed in Chapter 9.

Since each zone valve remains closed until a call for heat from a room thermostat opens it, *there is no need for check valves as with multiple-circulator systems.*

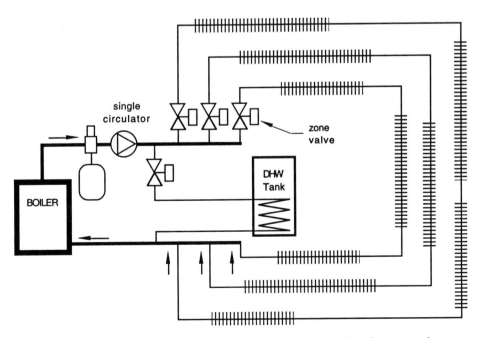

Figure 11–19   Example of a four-zone system using electric zone valves

There is a major difference in the hydraulic characteristics of single circulator/zone valve systems compared to multi-circulator systems. This can be seen by considering what happens as the system operates with one, then two, then three zone valves open at the same time.

With one zone valve open the system operates as if it were a single circuit. Since a larger pump is likely to be used on this type of system, there may be a substantial flow rate in the single active circuit. The flow rate may be high enough that it causes the zone valve to make noticeable flow noises unless the system is properly designed.

When a second zone valve opens, the two open circuits are in parallel with each other, and the flow will divide according to the hydraulic resistance of each branch circuit. Equation 6.23 could be used to predict these branch flow rates. The flow rate through the first zone circuit will be lower than when it was the only active circuit. The *system* flow rate, however, will increase.

As additional zones open up, each circuit becomes a parallel circuit path with the other active circuits. The flow rate in each circuit will drop somewhat, while the overall system flow rate will continue to increase.

To be conservative, this type of system should be designed based on the flow rates that will exist in each zone circuit when all zones are on. Example 6.11 presents a complete analysis of a three-zone system based on constructing a hydraulic resistance diagram. This method allows one to compare the performance of the system with several different circulators to find a reasonable design solution for a given application.

Use a Circulator with a Flat Pump Curve

For zone valve type systems, it is important to use a circulator with a relatively "flat" pump curve. The flatter the curve, the more stable the flow rates in each zone circuit as the other circuits cycle on and off. This is illustrated in Figure 11–20 for a hypothetical system with three identical zone circuits and two different circulators.

Notice if the pump with the flat curve is used, the head across the pump changes very little as additional zone circuits open up. This allows the flow rate in each active zone circuit to stay relatively stable because the pressure difference between its starting point and ending point stays about the same. If a pump with a "steep" curve is used, there will be a considerable drop in the head across the pump as additional zones open up. This will lead to significant changes in flow rate in a given zone circuit as other zones open up. When a high head pump operates against a single open zone circuit, the flow rate can become excessive and create flow noise. The zone valve manufacturer should be able to specify the maximum flow rate to prevent excessive noise or other problems associated with a high flow velocity.

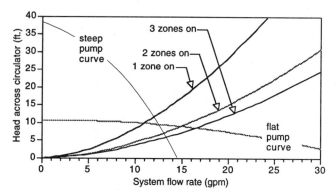

Figure 11–20 System curve for a system with three identical zone circuits and a common pump. "Steep" and "flat" pump curves also shown.

One disadvantage of this type of system is a circulator failure will prevent heat flow to the entire system. Another disadvantage is the single larger circulator must operate even when only one zone circuit is active. This will likely result in greater power consumption relative to using several smaller zone circulators.

11.8 PARALLEL DIRECT-RETURN SYSTEMS

Parallel distribution systems are constructed so that every heat emitter receives water from a common supply pipe and returns it to a common return pipe. A schematic layout for a **parallel direct-return system** is shown in Figure 11–21.

In a direct return system, the closer a given heat emitter is to the heat source, the shorter its overall piping path leading back through the heat source and circulator. The length of this path has a major effect on the flow rate through the heat emitter. The longer the equivalent length of the path, the greater its hydraulic resistance, and hence the lower its flow rate.

If we assume a situation where all heat emitters have identical hydraulic resistances, the flow rate through each unit would get progressively lower as we move farther away from the heat source. Its thermal performance would also decrease due to the lowered flow rate. This situation must be corrected to avoid serious heat distribution problems.

The balancing valves present on each heat emitter can be used to correct this effect. By partially closing the balancing valve associated with a particular heat emitter, the hydraulic resistance of that unit's parallel piping path can be increased to compensate for its distance away from the heat source. Typically the balancing valves closer to the heat source are closed more than those farther out.

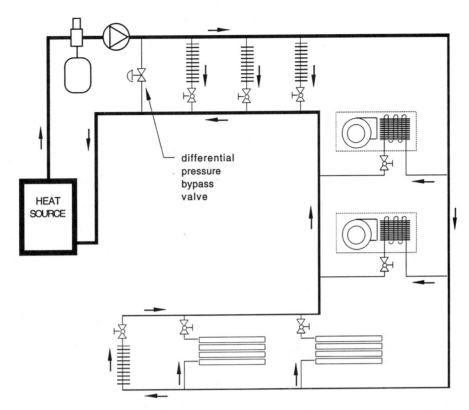

Figure 11–21   Piping schematic of a parallel direct-return system

The exact amount a given valve is closed depends on the hydraulic resistance of the heat emitter, as well as its position in the system. For example, the valve on a high flow resistance fan-coil near the heat source may not have to be closed as much as that associated with a short length of finned-tube baseboard farther out in the system.

The pipe size of the supply and return mains at any point in the system should be determined by the flow rate present at that point. As the supply main passes by each heat emitter, a portion of its flow is routed through the heat emitter. Thus the flow rate that continues is reduced. This same situation occurs as the return main collects additional flow from each heat emitter it passes on its way back to the heat source. *Good design practice is to keep the flow velocity in all locations under 4 ft/sec.*

One advantage of a parallel distribution system is each heat emitter receives supply water at close to the same temperature. This will result in a smaller temperature drop across the overall distribution system. *Because the temperature drop in parallel distribution systems can be kept relatively low, they are especially well suited for low temperature heat emitters.* In some cases the small temperature drop may also allow the use of smaller heat emitters.

There will be some heat loss from the piping mains that will lower the supply temperature to the heat emitters farther out in the system. How much heat is lost depends on the water temperature, piping material, length, insulation,

and ambient air temperature. The PIPELOSS program in the Hydronics Design Toolkit can be used to evaluate the loss of any given pipe segment. Obviously, losses from piping outside of heat space should be minimized by installing adequate insulation.

Another advantage of parallel systems is each heat emitter can be separately controlled by either a manual valve, nonelectric thermostatic valve, or electric valve and associated room thermostat. This allows for excellent zone control possibilities.

As the flow through individual heat emitters is reduced or stopped, the system flow rate will decrease and the head across the circulator will increase. This in turn will increase the flow rate in the heat emitters that remain open. If only one or two parallel piping paths are open, and a constant speed circulator is used, flow noise could develop. For this reason a **differential pressure control valve** is often used to prevent the pump from operating at pressure differentials above a certain setting. The valve, discussed in Chapter 5, allows water to bypass the distribution system entirely by flowing from near the pump discharge port, through the valve, and directly to the return main near the heat source. Bypass flow begins whenever the pressure differential across the valve reaches its set opening pressure. This arrangement also protects the pump from operating in a no-flow mode if all the heat emitters are closed at the same time. This protection can also be accomplished using electrical

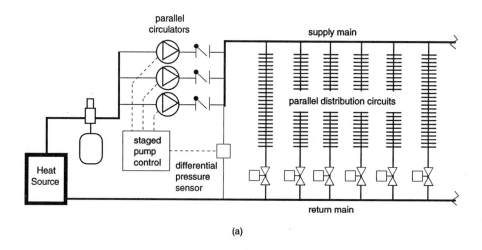

(a)

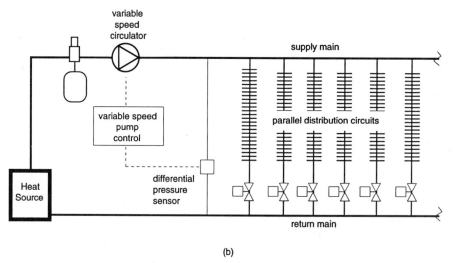

(b)

Figure 11–22 Use of (a) staged parallel pumps, or (b) single variable speed pump to keep head across distribution system relatively constant as flow through branch circuits varies.

controls that only allow the circulator to operate when one or more of the heat emitters is calling for heat.

It is possible to use two or more circulators in parallel and turn them on and off as necessary to meet the flow demands of the system. This is called staged parallel pumping, and is shown in Figure 11–22a. Notice each pump must be equipped with a check valve to prevent reverse flow when it is not operating. Off-the-shelf controls for this type of arrangement in smaller residential and light commercial systems are not widely available.

The ultimate extension of the concept of staged pumping is to use a single pump that is run at the precise speed necessary to maintain a given head differential across the system. This is shown in Figure 11–22b. This approach is now routinely used in larger commercial systems in the U.S. In Europe, small wet rotor circulators are available with built-in electronics that allow them

to vary their speed to maintain a constant pressure differential. These circulators are not commonly used in the U.S. at this time. In the near future, however, such variable speed pumping systems should be available for residential and light commercial systems in the U.S.

One popular type of parallel piping design uses pre-assembled manifold stations as the beginning and ending of individual piping circuits to each heat emitter. Such hardware is extensively used for radiant floor heating, and is shown in several figures in Chapter 10. It is also used in conjunction with radiant baseboard systems, as shown in Figure 8–42. The manifolds are mounted in a central location in the building, with the supply and return runs to each heat emitter made using flexible PEX or PB tubing. The manifolds provide a convenient location for flow balancing of all parallel circuits they serve. They also provide a mounting location for low voltage heat motor actuators called telestats, used for automatic zone control. This approach to parallel system design can be used with any type of heat emitter, from low temperature radiant floor circuits to higher temperature baseboard systems. The PEX or PB tubing is flexible enough to fit through many spaces that cannot be accessed to install rigid piping. This type of system will undoubtedly gain further acceptance in the U.S., especially in retrofit applications.

Design Procedure for Parallel Direct-return Systems

Parallel direct-return systems can be analyzed using the concepts of series and parallel hydraulic resistances as discussed in Chapter 6. The following procedure outlines a typical approach.

Step 1. Heat emitters should be selected based on providing design heat output at a nominal operating flow rate.

Step 2. A tentative piping layout should be sketched for the particular heat emitters used and their placement in the building. The lengths of the piping segments and the number/type of fittings required should be estimated.

Step 3. Select a tentative temperature drop for the system under design load conditions. A practical value for parallel systems is 10 °F. The *system* flow rate (for water) can then be estimated using Equation 11.1:

(Equation 11.1)

$$f_{initial} = \frac{Q_T}{500(\Delta T)}$$

where:

$f_{initial}$ = first estimate of system flow rate (gpm)
   Q_T = total design heating load served by the system (Btu/hr)
   ΔT = selected temperature drop of the system (°F)

The tentative flow rate through each heat emitter can be estimated using Equation 11.3:

(Equation 11.3)

$$f_i = f_{initial}\left(\frac{Q_i}{Q_T}\right)$$

where:

f_i = estimated flow rate through a particular heat emitter (gpm)
$f_{initial}$ = first estimate of *system* flow rate (gpm)
   Q_i = design heat output of the particular heat emitter

Q_T = total design heating load served by the system (Btu/hr)

Step 4. Pipe sizes can now be selected based on keeping the flow velocity in all pipes less than or equal to 4 ft/sec. The PIPESIZE program in the Hydronics Design Toolkit can assist.

Step 5. The pipe sizes, and their estimated lengths, can now be combined with the sketch of the piping layout to construct a hydraulic resistance diagram. The PIPEPATH program in the Hydronics Design Toolkit can be used to find the hydraulic resistance of each piping path. The HYRES program can be used to estimate the hydraulic resistance of heat emitters, or other piping components not listed in the PIPEPATH program, based on flow resistance versus flow rate data supplied by their manufacturer.

Step 6. The resistor diagram can be systematically reduced to a single equivalent resistance using the methods for series and parallel hydraulic resistors from Chapter 6.

Step 7. The single equivalent resistance can be used to sketch a system resistance curve. This can be overlaid with the pump curve of a proposed circulator to find the actual system flow rate. Again this step can be quickly accomplished with the PUMPCURV and PUMP/SYS program in the Hydronics Design Toolkit.

Step 8. When the system flow rate is found, Equation 11.1 can be used to verify the temperature difference present under design load conditions. If the temperature difference is significantly *greater* than the initially selected value (e.g., more than 20%), a circulator capable of increasing the flow rate is worth considering. Likewise, if the system temperature drop is significantly *lower* than the initial estimate, a smaller circulator should be considered. If the temperature drop is within 20% of the initially selected value, the circulator being checked is probably suitable.

Step 9. Knowing the system flow rate, the flow rate through each heat emitter can be found by repeated use of Equation 6.23. When all these flow rates are determined, they should be added up as a check. Their total should equal or be very close to the system flow rate found in step 6. If this does not check out, there is probably a mistake in the calculations.

Step 10. The flow rates through the individual parallel piping paths and heat emitters should be checked for excessive flow velocity (over 4 ft/sec). These flow rates should also be checked against the tentatively estimated values. If a given flow rate is higher than its initial estimate, it can probably be reduced using a balancing valve. If it is too low, a number of corrective options are available. These include using a large branch piping size or a different heat emitter with a lower flow resistance.

Step 11. After checking the flow rate and expected thermal performance of each heat emitter, and making any changes to the pipe sizes or heat emitters, the hydraulic resistor diagram can be modified to reflect these changes. The modified resistor diagram can then be reanalyzed beginning with step 6. This process can be repeated until the resulting system performs reasonably close to expectations. With practice this can usually be accomplished within two iterations of the calculations.

Example 11.3: A parallel distribution system is proposed to serve four heat emitters. A diagram showing the proposed piping layout, and giving information about pipe sizes, lengths, and fittings is shown in Figure 11–23. Estimate the flow rate required in each heat emitter to allow it to release its design heating output with a 10 °F drop between the supply and return main. Construct a hydraulic resistance diagram of the system and find the flow rate that will exist in each branch. Determine which branches will require balancing.

Other data selected for the system are listed below:

Supply water temperature = 160 °F
Design temperature drop across all heat emitters = 10 °F
Average water temperature = 160 – 10/2 = 155 °F
Piping materials: Type M copper tube in sizes indicated

Panel radiator: Design output = 9,000 Btu/hr
          Hydraulic resistance = 0.05
Fan-coil: Design output = 10,000 Btu/hr
          Hydraulic resistance = 0.10
Baseboard 1:   Design output = 8,000 Btu/hr
          Length = 24 ft
Baseboard 2:   Design output = 13,000 Btu/hr
          Length = 39 ft

Solution: The entire solution to this problem will be shown in detail. In practice, many of these calculations can be expedited using the Hydronics Design Toolkit.

The total design heat output of the system is obtained by summing the design outputs of the four heat emitters: 9,000 + 10,000 + 8,000 + 13,000 = 40,000 Btu/hr.

The system flow rate required to deliver this amount of heat with a 10 °F temperature drop can be found using Equation 11.1:

$$f_{initial} = \frac{Q_T}{500(\Delta T)} = \frac{40,000}{500(10)} = 8.0 \text{ gpm}$$

The flow rates through the individual heat emitters can be found using Equation 11.3. The calculations are shown below:

$$f_{panel\ radiator} = 8.0\left(\frac{9,000}{40,000}\right) = 1.8 \text{ gpm}$$

$$f_{fan\text{-}coil} = 8.0\left(\frac{10,000}{40,000}\right) = 2.0 \text{ gpm}$$

$$f_{baseboard\ 1} = 8.0\left(\frac{8,000}{40,000}\right) = 1.6 \text{ gpm}$$

$$f_{baseboard\ 2} = 8.0\left(\frac{13,000}{40,000}\right) = 2.6 \text{ gpm}$$

Before determining hydraulic resistances, the following data was determined from Chapter 6 (or with the PIPEPATH program):

c-value for 1-in copper pipe = 0.01776 (from Figure 6–18)
c-value for 3/4-in copper pipe = 0.061957 (from Figure 6–18)
c-value for 1/2-in copper pipe = 0.33352 (from Figure 6–18)
α-value for water at 155 °F = 0.046 (from Figure 6–17)

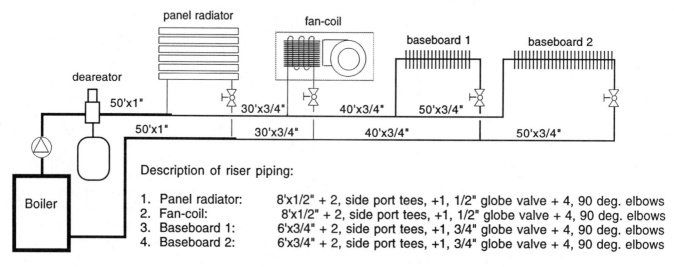

Figure 11-23 Piping layout and data for Example 11.3

The equivalent lengths of the various piping components were determined from Figure 6–21. The total equivalent lengths of the piping segments are as follows:

1. 50 ft × 1 in supply and return main:
   50 + 4(2.5) = 60 ft
2. 30 ft × 3/4 in supply and return main:
   30 + 4(2.0) = 38 ft
3. 40 ft × 3/4 in supply and return main:
   40 + 4(2.0) = 48 ft
4. 50 ft × 3/4 in supply and return main:
   50 + 4(2.0) = 58 ft
5. Deaerator is equivalent to 11 ft of 1-in pipe
6. 1/2-in panel radiator risers:
   8 + 2(2) + 15 + 4(1) = 31 ft
7. 1/2-in fan-coil risers:
   8 + 2(2) + 15 + 4(1) = 31 ft
8. 3/4-in baseboard 1 risers (not including baseboard itself):
   6 + 2(3) + 20 + 4(2) = 40 ft
9. 3/4-in baseboard 2 risers (not including baseboard itself):
   6 + 2(3) + 20 + 4(2) = 40 ft

The hydraulic resistances of the main piping segments and the parallel branch piping are determined using the above data and Equation 6.11:

$$r_1 = (\alpha cL) = (0.046)(0.01776)(60) \qquad = 0.049$$

$$r_2 = (\alpha cL) = (0.046)(0.061957)(38) \qquad = 0.108$$

$$r_3 = (\alpha cL) = (0.046)(0.061957)(48) \qquad = 0.137$$

$$r_4 = (\alpha cL) = (0.046)(0.061957)(58) \qquad = 0.165$$

$$r_5 = (\alpha cL) = (0.046)(0.01776)(11) \qquad = 0.009$$

$$r_6 = (\alpha cL) = (0.046)(0.33352)(31) + 0.05 = 0.53$$

$$r_7 = (\alpha cL) = (0.046)(0.33352)(31) + 0.10 = 0.58$$

$$r_8 = (\alpha cL) = (0.046)(0.061957)(40 + 24) = 0.182$$

$$r_9 = (\alpha cL) = (0.046)(0.061957)(40 + 39) = 0.225$$

The hydraulic resistance diagram and its step-by-step reduction to a single equivalent resistance is shown in Figure 11–24.

The single equivalent resistance describes the system curve. It can be combined with the pump curve of a selected circulator to find the operating point of the system. The system curve and the pump curve for a Grundfos UP15-42 circulator is shown in Figure 11–25.

To find the flow rate in the branch piping paths, the resistor diagram must be carefully reexpanded. Work from the simplified resistor diagram back toward the more complex diagram. Each time a known flow enters a group of two parallel resistors, use Equation 6.23 to divide it. Figure 11–26 shows the total system flow divided among the first two parallel hydraulic resistors.

The flow rate of 3.71 gpm is the flow rate through the panel radiator. The flow rate of 4.7 gpm is the total flow rate to the remaining three heat emitters. To further divide the 4.7 gpm flow rate, resketch the next pair of parallel resistances that previously were combined using the parallel resistance formula. This is shown in Figure 11–27.

The flow rate of 2.03 gpm is the flow rate through the fan-coil. The flow rate of 2.67 gpm continues along to supply the two remaining heat emitters. To further divide the 2.67 gpm flow, resketch the next pair of parallel resistances that previously were combined using the parallel resistance formula. This is shown in Figure 11–28.

The table in Figure 11–29 summarizes the differences between the intended design flow rates (those that would result in a temperature drop of 10 °F across each heat emitter) and those that can be expected to occur before balancing.

Discussion: Notice the flow rate through the panel radiator will be considerably higher than its design flow rate for a 10 °F temperature drop. This will increase its heat output slightly. The balancing valve on the panel radiator will have to be closed somewhat to reduce this flow rate. The flow rate in baseboard 2 is considerably lower than its design value. The reason for this low flow rate is the high flow resistance imposed by the lengths of supply and return main piping leading to baseboard 2. At this flow rate, the baseboard will not produce the heat output expected of it. By closing down the balancing valve on the panel radiator, the flow rate through baseboard 2 can be increased. The unbalanced flow rates in the fan-coil and baseboard 1 are reasonably close to their respective design values. These heat emitters can be expected to produce a drop close to 10 °F.

This has been a long and detailed example. It shows the capability of the methods presented in Chapters 6 and 7 to handle a relatively complex system. It also shows that obtaining a system flow rate close to the design expectation does not guarantee the flow rate in each heat emitter will be as expected. This is why balancing valves should always be used for each heat emitter in any type of parallel piping system.

The programs in the Hydronics Design Toolkit along with simple hand sketches of the resistor diagrams, will allow this problem to be completed in about half an hour. This time investment returns several insights into how the system will operate and confirms the adequacy of the selected circulator.

11.9 PARALLEL REVERSE-RETURN SYSTEMS

The second method for laying out a parallel system is known as reverse-return piping. Figure 11–30 shows a

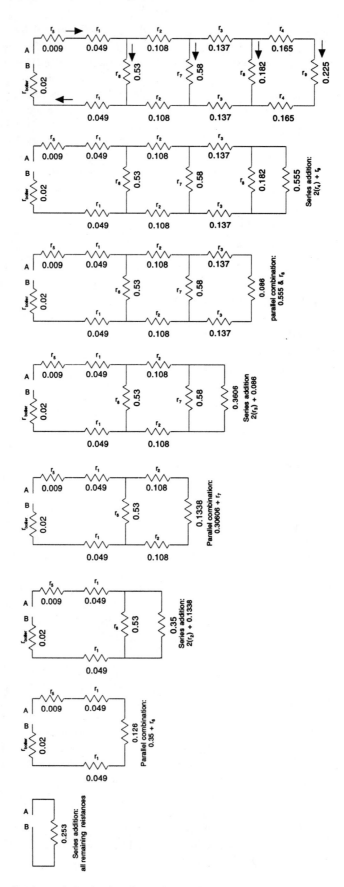

Figure 11-24 Step-by-step reduction of the hydraulic resistor diagram into a single equivalent hydraulic resistance for Example 11.3

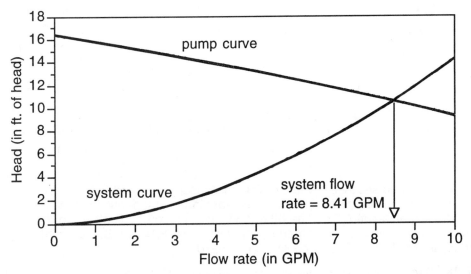

Figure 11–25   Finding the system flow rate for Example 11.3 using the system curve and pump curve

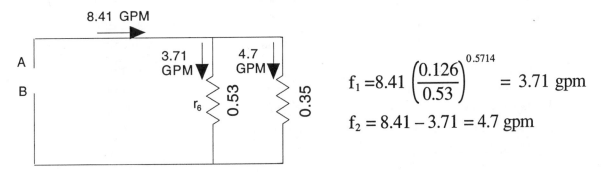

$$f_1 = 8.41 \left(\frac{0.126}{0.53} \right)^{0.5714} = 3.71 \text{ gpm}$$

$$f_2 = 8.41 - 3.71 = 4.7 \text{ gpm}$$

The equivalent resistance of the parallel resistances 0.53 and 0.35 was previously calculated as 0.126. The flow rate in the first parallel branch is found using equation 6.23. The other flow rate is found by subtracting the first flow rate from the total entering flow rate. The PARALLEL program in the Hydronics Design Toolkit could also be used.

Figure 11–26   Beginning to reexpand the resistor diagram of Example 11.3 to obtain branch flow rates

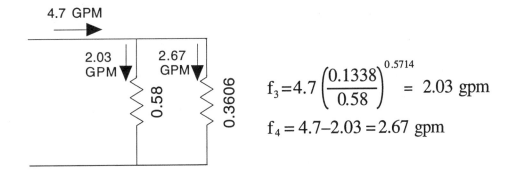

$$f_3 = 4.7 \left(\frac{0.1338}{0.58} \right)^{0.5714} = 2.03 \text{ gpm}$$

$$f_4 = 4.7 - 2.03 = 2.67 \text{ gpm}$$

The equivalent resistance of the parallel resistances 0.58 and 0.3606 was previously calculated as 0.1338. The flow rate in the first parallel branch is found using equation 6.23. The other flow rate is found by subtracting the first flow rate from the total entering flow rate. The PARALLEL program in the Hydronics Design Toolkit could also be used.

Figure 11–27   Continuing to reexpand the resistor diagram for Example 11.3 to obtain the next branch flow rates

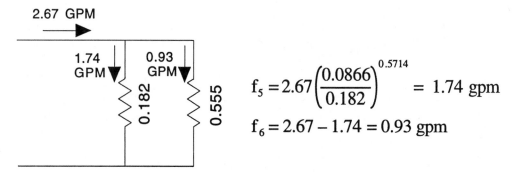

$$f_5 = 2.67 \left(\frac{0.0866}{0.182} \right)^{0.5714} = 1.74 \text{ gpm}$$

$$f_6 = 2.67 - 1.74 = 0.93 \text{ gpm}$$

The equivalent resistance of the parallel resistances 0.182 and 0.555 was previously calculated as 0.0866. The flow rate in the first parallel branch is found using equation 6.23. The other flow rate is found by subtracting the first flow rate from the total entering flow rate. The PARALLEL program in the Hydronics Design Toolkit could also be used.

Figure 11–28 Reexpanding the resistor diagram of Example 11.3 to find the final branch flow rates

	Intended Design Flow Rate	Calculated Flow Rate (before balancing)
System Flow rate	8.0 gpm	8.41 gpm
panel radiator flow rate	1.8 gpm	3.71 gpm
fan-coil flow rate	2.0 gpm	2.03 gpm
baseboard 1 flow rate	1.6 gpm	1.74 gpm
baseboard 2 flow rate	2.6 gpm	0.93 gpm

Figure 11–29 A comparison of intended design flow rates and those calculated in Example 11.3

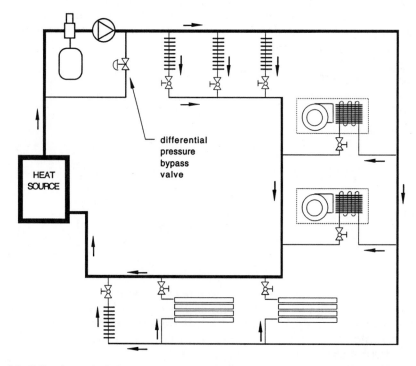

Figure 11–30 Modification of piping schematic of Figure 11–22 into a parallel reverse-return system

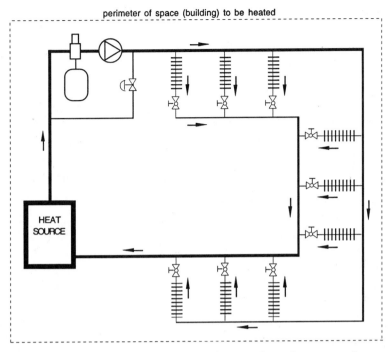

Figure 11–31   Preferred layout of a reverse-return system, with supply and return mains arranged around the perimeter of the heated space

modification of the direct return system of Figure 11–21, making it into a reverse-return system.

Notice the heat emitter closest the heat source along the supply main is also the farthest from the heat source along the return main. Likewise the farthest heat emitter on the supply is the closest on the return. This arrangement largely corrects the inherent flow balancing problems of the parallel direct return systems, especially when the heat emitters all have similar hydraulic resistances. *Reverse-return is usually the preferred way of arranging a parallel distribution system.*

Reverse-return systems require different pipe sizes in various parts of the system. The supply main gets progressively smaller as one moves away from the heat source. The return main gets progressively larger as one moves toward the heat source. See Figure 11–30. The

concept is to keep the flow velocity at or below 4 ft/sec to prevent flow noise in all areas of the system.

The ideal arrangement for a **parallel reverse-return system** is to route both the supply and return mains around the perimeter of the area to be heated, as shown in Figure 11–31. If this is not done, there may be a need for considerable extra piping to achieve the reverse return layout as shown in Figure 11–32.

Theoretically, if all branches of a reverse-return system had identical hydraulic resistances, and all pipe size changes in the supply and return mains were symmetrical, the system would be entirely self-balancing. Systems like this are uncommon to say the least. *Because of this, a balancing valve should still be installed in each branch of the system for adjusting the flow rate through individual heat emitters.*

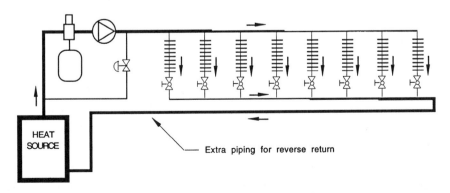

Figure 11–32   Non-looping layout of reverse return system requiring additional piping

Most of the concepts discussed under parallel direct-return systems are also applicable to reverse-return systems. This design is a good choice for low temperature distribution systems. Staged or variable speed pumping is an option. A differential pressure bypass valve should be used to avoid no-flow operation of the pump or flow noise when only one or two zones are active.

Interestingly, the standard approach of setting up a hydraulic resistance diagram of the system is not very useful for reverse return systems. If the number of parallel branches exceeds two, the resulting hydraulic resistor diagram cannot be reduced to a single equivalent resistance using the methods of Chapter 6. The flow rates can still be solved using sophisticated mathematical iteration, but the work is well beyond the scope of this book. The exception to this problem is when the hydraulic resistances of the supply and return piping mains between the parallel branches are very small in comparison to the resistance of the branches themselves, and can therefore be considered insignificant. Unfortunately, this approach is only applicable to manifold type systems where branch connections are very close together.

Design Procedure for Parallel Reverse-return Systems

Since the equivalent resistance methods of Chapter 6 are usually not applicable to reverse-return systems, a somewhat simplified method will be presented. This procedure will result in a conservative (e.g., slightly over-sized) circulator being selected. The first four steps of this method are identical to those used for direct-return systems.

Step 1. Heat emitters should be selected based on providing design heat output at a nominal operating flow rate.

Step 2. A tentative piping layout should be sketched for the heat emitters selected and their placement in the building. The lengths of the piping segments and the number and type of fittings required should be estimated.

Step 3. Select a tentative temperature drop for the system under design load conditions. A practical value for parallel systems is 10 °F. The system flow rate (for water) can then be estimated using Equation 11.1:

(Equation 11.1)

$$f_{initial} = \frac{Q_T}{500(\Delta T)}$$

where:

$f_{initial}$ = first estimate of system flow rate (gpm)
Q_T = total design heating load served by the system (Btu/hr)
ΔT = selected temperature drop of the system (°F)

The tentative flow rate through each heat emitter can be estimated using Equation 11.3:

(Equation 11.3)

$$f_i = f_{initial}\left(\frac{Q_i}{Q_T}\right)$$

where:

f_i = estimated flow rate through a particular heat emitter (gpm)
$f_{initial}$ = first estimate of *system* flow rate (gpm)
Q_i = design heat output of the particular heat emitter
Q_T = total design heating load served by the system (Btu/hr)

Step 4. Pipe sizes can now be selected based on keeping the flow velocity in all pipes less than or equal to 4 ft/sec. The PIPESIZE program in the Hydronics Design Toolkit can assist in this process.

Step 5. *The branch with the highest expected flow resistance should be identified.* This will often be the branch with the most restrictive heat emitter. If all heat emitters are identical, any branch can be used.

Step 6. Sketch a system diagram showing the expected flow rates (as calculated in step 3) present in all piping segments. An example is illustrated in Figure 11–33.

Step 7. Determine the head loss across the pump by totaling the head loss of each piping segment that is in an imaginary circuit formed by the circulator and the *branch with the highest flow resistance,* as shown in Figure 11–34. The PIPEPATH program in the Hydronics Design Toolkit can expedite this process.

Step 8. The head loss obtained in step 7, along with the *full system flow rate,* sets a conservatively high operating point for the circulator. This operating point should fall on or slightly below the pump curve of circulators suitable for use in the system. Since the flow resistance of the other branches will be lower than the branch used for design purposes, the system will likely operate at a flow rate slightly higher than determined by this method.

11.10 PRIMARY/SECONDARY SYSTEMS

Primary/secondary piping design has been used for years in larger commercial hydronic systems. It has recently been applied in smaller commercial and residential systems. An example of a primary/secondary distribution system is shown in Figure 11–35.

This system uses a single piping path known as the **primary circuit** to transport heated water around the perimeter of the building. In many cases, especially commercial systems, primary circuits are set up for continuous circulation during the heating season. Individual **secondary circuits** are connected into the primary circuit using closely spaced tees. The distance between the side

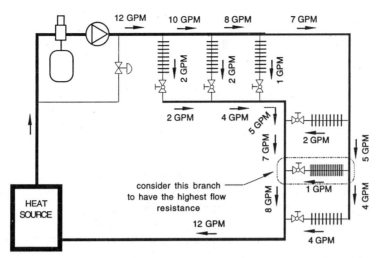

Figure 11–33 Sketch of a piping system showing flow rates found in step 3

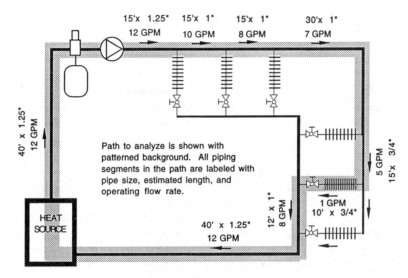

Figure 11–34 Piping path through heat emitter with greatest flow resistance. The total head loss along this path must be determined in step 7.

port centerlines of the tees should not exceed 6 inches. The short length of pipe between the tees produces almost no head loss. This results in almost zero flow from the primary circuit into the secondary circuit, *until the secondary circulator is turned on.* When operating, the secondary circulator provides the head differential necessary for flow in the secondary circuit.

In some respects, a **primary/secondary system** is similar to the 1-pipe diverter tee systems discussed in Section 11.5. Both use a single circuit to transport heated water around the building. Both can control each branch circuit as a separate zone if desired. The difference is that a secondary circulator, rather than a diverter tee, is used to induce flow in a branch circuit. One advantage of using a secondary circulator, rather than diverter tees, is that it can produce a greater flow rate in the secondary circuit. It also consumes power only when flow is needed in a

given secondary circuit. A diverter tee, on the other hand, dissipates pumping energy through its orifice even when its branch circuit is closed off with a valve.

In some systems, the size of both piping and circulators can be reduced by designing the primary loop for a relatively high overall temperature drop of 25 °F or more. This is especially feasible if the loads served by the primary circuit cover a wide range of operating temperatures. Such loads would be connected to the primary circuit in order of descending operating temperature. For example, a storage water heater may be the first load connected to the primary circuit so the hottest water in the system is pumped through its heat exchanger for a fast recovery rate. Finned-tube baseboard circuit(s) may be connected next due to their relatively high water temperature requirement. A circuit used for low temperature radiant floor heating would likely be the last load connected

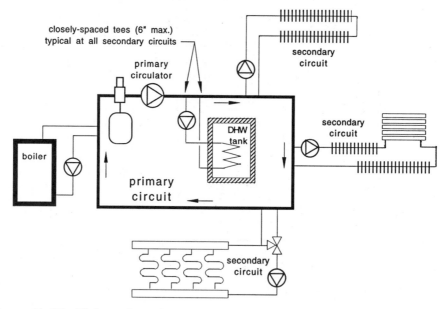

Figure 11–35 Piping schematic example of a primary/secondary distribution system

to the primary circuit. This type of arrangement is depicted in Figure 11–35.

The length of piping used in a well-planned primary/secondary design is usually less than what would be required with separate zone circuits using either zone valves or zone circulators. This is because the primary piping is serving as both the supply and return pipe. The primary circuit should be routed in close proximity to the heat emitters to minimize secondary piping length. It should also be insulated to minimize heat loss, especially when constant circulation is used.

In a typical commercial application of primary/secondary piping, the primary circuit is routed overhead around the perimeter of the building. Often the piping is located in

the mechanical space above a drop ceiling. Secondary circuits are routed down from the primary circuit to serve individual heat emitters or series circuits containing two or more heat emitters. Depending on the total load, either a single boiler or modular boiler configuration can be used to provide heat input to the primary circuit as shown in Figure 11–36.

Because the hotter water in the primary circuit is above the secondary circuits and their associated heat emitters, there is no concern for thermosiphoning of hot water downward into the secondary circuits when they are not operating. However, in systems where the primary circuit is *below* the secondary circuit, as often occurs in smaller residential installations, thermosiphoning can be

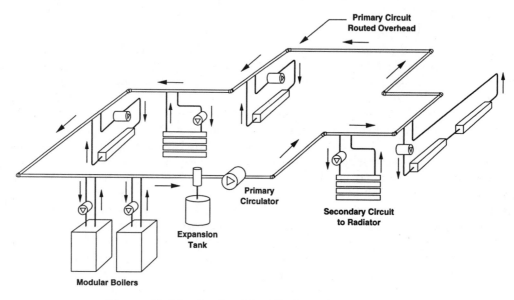

Figure 11–36 Overhead installation of primary circuit

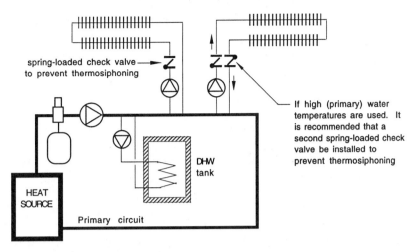

Figure 11–37   Use of spring-loaded check valve(s) to prevent thermosiphoning in upgoing riser of secondary circuit

a problem. It is corrected by installing either a flowcheck valve or spring-loaded inline check valve in the upgoing supply pipe of the secondary circuit as shown in Figure 11–37. In some situations, where primary water temperatures are relatively high, thermosiphoning can still occur in the return pipe of an upgoing secondary loop. In such cases it may be necessary to install a second check valve near the return tee. The head loss of these check valves must be accounted for in sizing the secondary circulator.

The method of connecting the boiler to the primary loop depends on how the system is operated. *When continuous circulation is used, the boiler should be connected as a secondary circuit as shown in Figure 11–35.* This reduces off-cycle heat loss from the boiler. This arrangement should also be used if a modular boiler system is used as shown in Figure 11–36. If the primary circulator is only operated when one or more loads are calling for heat, it is acceptable to pipe the boiler directly into the primary circuit as shown in Figure 11–37.

Design of Primary/Secondary Systems

Because the length of tubing between the tees of each secondary loop is very short, there is no significant head loss and associated pressure difference between them. This allows the primary and secondary loops to act as if they were independent of each other. Each circuit can be designed as a separate series circuit with its assortment of pipes, fittings, valves, and a circulator.

The primary loop flow rate can be established by setting a design temperature drop for the system under maximum loading conditions, using Equation 11.1. If a non-condensing boiler is used, be sure the return temperature from the primary circuit is at least 140 °F.

The design flow rate in each of the secondary circuits can be established by applying Equation 11.1 *to the design load associated with the secondary circuit* rather than the entire system. Once this flow rate is established,

pipe sizes can be selected, and an appropriate circulator chosen. The $\mathbb{PIPEPATH}$ and $\mathbb{PUMP/SYS}$ programs can be used for this step.

The return flow in each secondary circuit blends with the bypass flow in the primary circuit. The blended temperature that continues along the primary circuit can be established in the same manner used for diverter tee systems using the information in Figure 11–9. The $\mathbb{MIX}$ program in the Hydronics Design Toolkit can also be used to find the blended outlet temperature.

If the length of the primary piping from this outlet tee to the next inlet tee is relatively short (20 feet or less is suggested), the blended outlet temperature from one secondary circuit can be used as the inlet temperature to the next secondary circuit. If the length of the primary piping is longer than 20 feet, or if the pipe is otherwise expected to have a significant heat loss, the temperature drop along the primary pipe can be calculated using Equation 8.12 or the $\mathbb{PIPELOSS}$ program. This temperature drop would be subtracted from the blended outlet temperature of the upstream secondary circuit to obtain the inlet temperature to the next secondary circuit.

SUMMARY

This chapter has shown several methods for distributing heated water from a heat source to the heat emitters. Each method has advantages and disadvantages. The following information summarizes these strengths and weaknesses.

1. Single Series Circuits

Advantages:

- Simple design
- Relatively inexpensive

- Best suited to small (single-zone) applications
- Simple control system

Disadvantages:

- Cannot be zoned
- Long circuits require higher pumping heads
- Heat emitters with high flow resistance can significantly increase pumping head
- If a wide system temperature drop is used, heat emitters may need to be larger due to lower average system water temperature

2. Single Circuit/multi-zone (1-pipe) Systems

Advantages:

- Each heat emitter can be separately regulated
- Usually uses less pipe than other multi-zone methods

Disadvantages:

- More complex to design than single series circuits
- No reduction in pumping energy as individual heat emitters are closed off
- Diverter tees create relatively large head loss when no branch flow occurs
- If a wide system temperature drop is used, heat emitters may need to be larger due to lower average system water temperature

3. Multi-zone/Multi-circulator Systems

Advantages:

- Each heat emitter (or room) can be separately zoned
- Uses smaller, less expensive circulators relative to a comparable zone valve system
- If one circulator fails, the rest of system can still operate
- Pumping power usage is lowered as zones are shut off
- Flow rates in each zone remain relatively stable regardless of what other zones are on
- Heat emitters may be smaller if a smaller system temperature drop is used

Disadvantages:

- Uses more piping than single circuit options
- Requires a check valve on each circuit to prevent reverse flow through inactive zones
- Requires more sophisticated controls than single circuit systems
- Potential problems when high and low head circulators are combined in parallel

4. Multi-zone Systems Using Zone Valves

Advantages:

- Each heat emitter (or room) can be separately zoned
- Zone valves do not need check valve to prevent reverse flow through inactive zones
- Tends to be less expensive than multi-circulator systems
- Heat emitters may be smaller if a smaller system temperature drop is used

Disadvantages:

- Uses more piping than single circuit options
- Zones valves tend to require more servicing than circulators
- Single larger circulator must run even if only one zone needs heat
- Variation in zone flow rates depending on what other zones are on
- High head pumps might create flow noise in zone valve if only one zone is open

5. Parallel Direct-return Systems

Advantages:

- Each heat emitter (or room) can be separately zoned
- Approximately the same water temperature is supplied to each heat emitter
- Good for systems using low temperature heat sources
- Good arrangement for heat emitters with high flow resistance characteristics
- Heat emitters may be smaller if a lower system temperature drop is used

Disadvantages:

- Uses more piping than single circuit options
- Will require more balancing than reverse-return systems
- Single circulator must run even if only one zone needs heat
- More complicated and lengthy design procedure often required
- May require differential pressure bypass valve to handle low load conditions

6. Parallel Reverse-return Systems

Advantages:

- Each heat emitter (or room) can be separately zoned
- Approximately the same water temperature supplied to each heat emitter
- Good for systems using low temperature heat sources

- Good arrangement for heat emitters with high flow resistance characteristics
- Heat emitters may be smaller if a lower system temperature drop is used
- Will likely require less balancing than direct-return systems

Disadvantages:

- Uses more piping than single circuit options
- Single circulator must run even if only one zone needs heat
- Often cannot be analyzed with a standard resistor diagram approach
- May require differential pressure bypass valve to handle low load conditions

7. Primary/Secondary Systems

Advantages:

- Each heat emitter (or room) can be separately zoned
- Uses less piping than parallel systems
- Tends to use smaller, less power consuming circulators
- Ideal piping for modular boiler systems
- Good arrangement for heat emitters with high flow resistance characteristics
- Higher flow rate in secondary circuits compared to diverter tee systems

Disadvantages:

- Primary circuit needs to be insulated if constant circulation is used
- Primary circulator must run even if only one zone needs heat
- Upward risers to secondary circuits may need check valves to prevent thermosiphoning

KEY TERMS

1-pipe systems
Differential pressure control valve
Hydraulic equilibrium
Inter-zone heat transfer
Iteration
MONOFLO program
Multi-zone/multi-circulator system
Parallel direct-return system
Parallel reverse-return system
Primary circuit
Primary/secondary system
Secondary circuit
Single series circuit
Thermal equilibrium
Zoning

CHAPTER 11 QUESTIONS AND EXERCISES

Note: Questions and exercises requiring the Hydronics Design Toolkit are indicated with marginal symbol.

1. A thermometer attached to the supply pipe of a hydronic distribution system shows the temperature of the water rising steadily even after the system has been on for several minutes. What can you conclude about the heat output of the boiler relative to the heat loss of the heat emitters?
2. When is hydraulic equilibrium established in a hydronic system? Are there certain types of hydronic systems where hydraulic equilibrium will not be established?
3. Will a hydronic heating system, in all cases, properly heat a building when it attains both thermal and hydraulic equilibrium? Explain your answer.
4. Discuss the feasibility of using a setback thermostat, with a daily setback schedule in a well-insulated building equipped with a hydronically heated 6-inch concrete slab. When would this make sense? When would it not make sense?
5. What is the main advantage of a 1-pipe distribution system relative to a single series circuit?
6. A three-zone distribution system is to supply a total load of 90,000 Btu/hr with a design temperature drop of 15 °F. The design heating loads of the three zones are 25,000 Btu/hr, 50,000 Btu/hr, and 15,000 Btu/hr. Determine the system design flow rate and the design flow rate in each of the zone circuits.
7. Describe why check valves are necessary in each zone circuit of a multi-circulator zoned system.
8. What is the principal advantage of a reverse-return rather than direct-return parallel distribution system?
9. Why are parallel distribution systems good for use with low temperature heat emitters?
10. What is the advantage of designing a primary/secondary system for a high temperature drop in the primary circuit? What precaution must be observed if a conventional boiler is used?
11. What is an advantage of connecting a boiler to a *continuously circulating* distribution system using a primary/secondary arrangement? When is this not necessary?
12. Describe what happens to the flow rate through a particular zone circuit when additional zone circuits (containing zone valves) turn on. What happens to the *system* flow rate as additional zone circuits open up? Justify your answer.
13. Describe why a circulator with a relatively "flat" pump curve is best suited for multi-zone systems using zone valves. What can happen if a circulator with a "steep" pump curve is used in such a system?
14. Why is it necessary to correct the hydraulic resistance of the heat source and common piping for a multi-zone system using Equation 11.5?

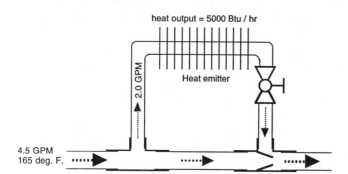

Figure 11–38 Diverter tee piping arrangement for Question 17

15. When the calculated flow rate through a particular parallel branch circuit is significantly lower than the initial design value, what are some things that can be done to increase it?

16. Describe why a multi-circulator zone system is likely to use less electrical energy over a heating season than an equivalent system using a single larger circulator and zone valves.

17. A diverter-tee arrangement operates as shown in Figure 11–38. Calculate the outlet temperature from the diverter tee.

18. Redraw the floor plan shown in Figure 11–39 to a scale of ¼ in = 1 ft. Using the assumptions stated, design a single series loop distribution system using finned-tube baseboard convectors. Use baseboard sizing method #2 from Chapter 8. Neatly draw the layout of the system on the floor plan, showing where all baseboards will be located. As an option, use the programs in the Hydronics Design Toolkit to assist with the calculations. Begin the circuit in the mechanical room and proceed in a clockwise direction around the floor plan.

Assumptions:

Design heating loads:
1. Living/dining/kitchen: 14,800 Btu/hr
2. Bedroom 1: 3,100 Btu/hr
3. Bedrooms 2 and 3: 3,800 Btu/hr
4. Bathroom: 1,400 Btu/hr
5. Mechanical/laundry: 1,600 Btu/hr
   • The baseboard used is described by the lower curve in Figure 8–9.
   • The system is to be piped with ³/₄-in copper tubing. All fittings and valves will also be ³/₄ in.
   • A Grundfos UP15-42 circulator is to be used.
   • The design temperature drop of the system is to be no more than 15 °F.
   • Assume the piping circuit will contain 50 90° elbows, four side port tees, and four gate valves.
   • The boiler is set to supply 160 °F water.

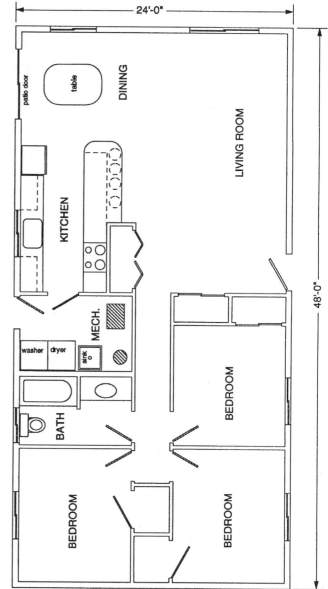

Figure 11–39 Floor plan for Exercises 18, 19, and 20

19. Repeat Exercise 18, only design the circuit to begin in the mechanical room and proceed in a counter-clockwise direction around the building. Compare the new baseboard lengths to those obtained in Question 18.

20. Using the floor plan shown in Figure 11–39, lay out a 1-pipe system that would allow the baseboard in each room to be controlled by its own thermo-static radiator valve. For simplicity, assume the flow rate in each baseboard will be 2 gpm when its valve is open. Sketch the piping layout on the floor plan. Compare the baseboard lengths obtained for this design with those obtained in Exercises 18 and 19.

EXPANSION TANKS 12

OBJECTIVES

After studying this chapter you should be able to:

- Explain the purpose of an expansion tank in a closed-loop hydronic system
- Describe two different types of expansion tanks
- Explain why open-type expansion tanks become water-logged
- Calculate the volume of fluid in a piping system
- Determine the proper location of the expansion tank within the system
- Calculate the required expansion tank volume for a given system
- Determine the required air pressurization of a diaphragm-type tank
- Make use of the EXPTANK program in the Hydronics Design Toolkit
- Avoid incompatibilities between diaphragm materials and system fluids
- Describe how an expansion tank should be mounted

12.1 INTRODUCTION

All liquids used as heat transfer fluids in hydronic heating systems expand when they are heated. This so-called **thermal expansion** is an unavoidable and extremely powerful fact of nature. When it occurs, the molecules that make up the liquid need a larger volume in which to exist.

For all practical purposes, liquids are **incompressible**. A given number of liquid molecules cannot be compacted or squeezed into a smaller volume without tremendous force. Any container *completely* filled with a liquid and sealed from the atmosphere will experience a rapid increase in pressure as the liquid is heated. If this pressure is not somehow relieved, the container will burst, in some cases violently. To prevent this from happening, *all hydronic heating systems must be equipped with a means of accommodating the volume increase of their fluid as it is heated.*

In systems that are open to the atmosphere, such as a non-pressurized thermal storage tank, the volume increase

can be accommodated by leaving the fluid level slightly below the rim of the tank. This allows space for the expanding fluid to "park" its extra volume as shown in Figure 12–1.

In a more typical *closed-loop* hydronic system, the extra space is usually provided by a separate chamber called the expansion tank. This tank contains a volume of air that is compressed somewhat like a spring when the system's fluid expands against it, as shown in Figure 12–2.

This chapter discusses the types of expansion tanks used for closed hydronic systems. It also gives methods for sizing, pressurizing, and locating the tank in the

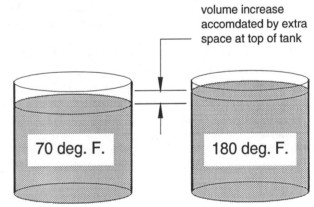

Figure 12–1 Extra space at top of an open tank accommodates expansion of fluid.

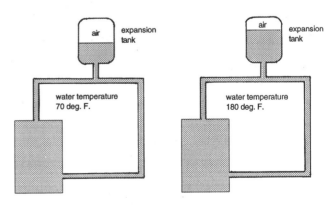

Figure 12–2 Separate expansion tank used to accommodate fluid expansion in a closed-loop hydronic system

system. A program called EXPTANK in the Hydronics Design Toolkit is also introduced as an aid for determining tank size and pressurization.

12.2 STANDARD EXPANSION TANKS

One type of expansion tank used in early hydronic heating systems consisted of a simple open-top drum placed at the high point of the system. It served as a chamber into which water would rise when heated. In some cases this tank was placed in the attic of a building.

This arrangement presented several problems. First, the open tank allowed fluid to escape from the system by evaporation. This water loss would have to be replaced with fresh water containing dissolved oxygen. The oxygen caused corrosion of iron-containing components in the system. Second, the height of the tank limited the pressure and thus the upper operating temperature of the system. Finally, since a tank located in a cold attic was often some distance away from the heated parts of the system, it could freeze in cold weather. This could lead to a ruptured tank and a mess in the spaces below. Needless to say, this type of tank is not used in modern hydronic systems.

The next step in expansion tank technology was the use of a closed tank located above the boiler as shown in Figure 12–3.

The air in this type of **standard expansion tank** is initially at atmospheric pressure. When the system is filled with fluid, the air is trapped in the tank and partially compressed. The higher the system piping rises above the expansion tank, the more the entrapped air is compressed. As the water expands upon heating, additional fluid enters the tank, further compressing the air. *If the tank is properly sized, the air should reach a*

pressure about 5 psi below the setting of the system's pressure relief valve as the water reaches its maximum operating temperature. The 5 psi reduction of the relief valve's rated pressure is to prevent the valve from leaking just below its rated opening condition. This 5 psi reduction also allows the relief valve to be located slightly below the inlet of the expansion tank as is typical in many systems.

Standard expansion tanks were used on thousands of early hydronic systems. Many are still in service today. These tanks were often hung from the underside of floor joists in the basement just above the boiler. A special fitting on the boiler allowed air bubbles, driven out of solution by heating, to rise into the tank. A dip tube extended below the fluid level in the boiler through which hot water left the boiler. The dip tube minimized the amount of air bubbles sent to the distribution system.

One inherent problem with standard expansion tanks is the air and water they contain are in direct contact. As the water in the tank cools off, it has the ability to reabsorb some of the air back into solution. Because the pipe leading from the boiler to the expansion tank is usually short and of small diameter, water from the expansion tank, along with the dissolved air it contains, will be drawn back into the system as it cools during an off-cycle. Upon reheating, the dissolved air will again come out of solution, but now in the system piping. Inevitably part of this air will form larger bubbles and be ejected from the system by air vents. An automatic feed-water valve on the system will then admit small amounts of water to make up for the lost air. The net effect over many heat-up/cool-down cycles is the air in the expansion tank is replaced by water. Eventually the tank becomes **water-logged**, which means the tank has become completely filled with water.

When water-logging occurs there is no longer a cushion of air for the system's water to expand against as it is heated. This causes the system's pressure relief valve to release small amounts of water each time the system heats up. The system's feed-water valve replaces this water as soon as the system cools and its pressure decreases. *This repeated sequence of events can move hundreds of gallons of fresh (oxygen-containing) water through a system during a single heating season.* It greatly increases the chance of serious corrosion damage.

Standard expansion tanks typically need to be drained and refilled with air two times a year to prevent the problems associated with water-logging. Special drain valves that allow air to enter the tank at the same time water is being drained are available for this purpose. These valves also isolate the tank from the rest of the system during this draining operation.

Sizing Standard Expansion Tanks

The size of a standard expansion tank can be calculated using Equations 12.1 and 12.2.

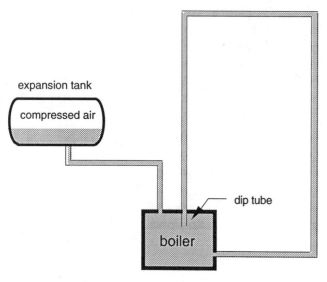

Figure 12–3 Installation of a standard expansion tank

(Equation 12.1)

$$V_t = \frac{V_s\left(\dfrac{D_c}{D_h} - 1\right)}{14.7\left(\dfrac{1}{P_f + 14.7} - \dfrac{1}{P_{RV} + 9.7}\right)}$$

where:

V_t = *minimum* required tank volume (gals)
V_s = fluid volume in the system (gals) (see Section 12.4)
D_c = density of the fluid at its initial (filling) temperature (lb/ft³)
D_h = density of the fluid at the maximum operating temperature of the system (lb/ft³)
P_f = static pressure of the fluid in the tank when the system is filled (psig)*
P_{RV} = rated pressure of the system's pressure relief valve (psig)

(Equation 12.2)

$$P_f = (H_s - H_t)\left(\frac{D_c}{144}\right) + 5$$

where:

P_f = static pressure of the fluid in the tank when the system is filled (psig)
H_s = height of the uppermost system piping above the base of the boiler (ft)
H_t = height of the inlet of the expansion tank above the base of the boiler (ft)
D_c = density of the fluid at its initial (fill) temperature (lb/ft³)

Example 12.1: Calculate the minimum size standard expansion tank for the system shown in Figure 12–4. Water is added to the system at an initial temperature of 60 °F. The maximum operating temperature of the system

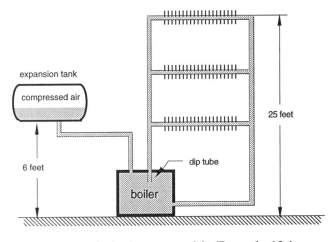

Figure 12–4   System used in Example 12.1

is 200 °F. The boiler is equipped with a 30 psi rated relief valve. The estimated system volume is 35 gals.

Solution: Before using Equation 12.1, the pressure in the tank when the system is filled must be found using Equation 12.2. This equation requires the density of the water at 60 °F, which can be read from Figure 4–5, calculated using Equation 4.6, or found using the FLUIDS program in the Hydronics Design Toolkit. Since the water's density at the maximum operating temperature of 200 °F is eventually required, it is also found at the same time:

$$D_c = 62.36 \text{ lb/ft}^3 \text{ (water's density at 60 °F)}$$

$$D_h = 60.13 \text{ lb/ft}^3 \text{ (water's density at 200 °F)}$$

The pressure at the inlet of the tank is now calculated using Equation 12.2:

$$P_f = (25 - 6)\left(\frac{62.36}{144}\right) + 5 = 13.23 \text{ psig}$$

This value and the remainder of the data is now substituted in Equation 12.1:

$$V_t = \frac{35\left(\dfrac{62.36}{60.13} - 1\right)}{14.7\left(\dfrac{1}{13.23 + 14.7} - \dfrac{1}{30 + 9.7}\right)} = \frac{1.298}{0.156} = 8.3 \text{ gal}$$

Discussion: It is interesting to consider the effect of the height of the system's expansion tank on its volume requirement. If the tank were located closer to the top of the system, the static pressure caused by fluid in the piping above the tank would decrease. This would result in less initial compression of the air in the tank, and thus more room for fluid. In this particular example, if the tank were moved 10 feet higher, its size could be reduced to 5.1 gallons. A lower maximum fluid temperature would also decrease the tank size required.

It should be noted that Equation 12.1 determines the minimum volume of a standard expansion tank. It is certainly possible to use a tank with a larger volume. As the tank's volume increases above its minimum calculated value, the increase in the tank's air pressure as the system's fluid expands will be less. A larger tank, however, provides no tangible benefit to the performance or longevity of the system. The only reason to use a larger tank would be to use a standard production tank size rather than a custom-made tank.

A specialized boiler fitting is often used with standard expansion tanks. This fitting, shown in Figure 12–5, allows air bubbles that accumulate at the top of the boiler to rise up into the expansion tank. This helps replace air drawn out of the tank by reabsorption, and slows the rate

---

* The static pressure at the expansion tank connection when the system is filled is based on the height of the system piping above the expansion tank, the type of fluid used, and a customary allowance for 5 psig pressure at the top of the system for proper operation of the air vents. This pressure can be found using Equation 12.2:

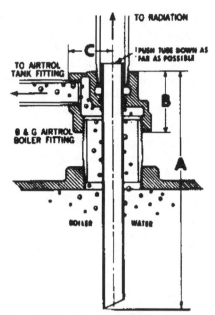

TO RADIATION

PUSH TUBE DOWN AS FAR AS POSSIBLE

TO AIRTROL TANK FITTING

C

B & G AIRTROL BOILER FITTING

B

A

BOILER

WATER

Top Outlet Airtrol boiler fittings should always be installed with a short nipple so that the adjustable dip tube extends well into the boiler.

Figure 12–5 Example of a specialized boiler fitting for use with standard expansion tanks. Courtesy of Whitman and Johnson.

at which the tank fills with water. This boiler fitting contains a short dip tube that extends down below the top of the boiler sections to pick up hot water without air bubbles. It forms the beginning of the distribution piping.

The pipe leading from the boiler fitting to the expansion tank should be sloped at least 1 inch in 5 feet. This allows the bubbles to rise toward the tank. It should also be at least 10 feet long to minimize heat migration toward the tank. Such heating can cause an additional pressure increase in the tank.

A separate tank fitting is sometimes used to connect the riser pipe from the boiler to the tank. This fitting must be selected based on the diameter of the expansion tank. It helps to maintain the proper fluid level in the tank.

In summary: *standard expansion tanks are seldom used on currently installed residential or light commercial hydronic systems.* Their maintenance requirements, size, and associated fittings and valves make them much less desirable than the diaphragm-type tanks discussed in the next section.

12.3 DIAPHRAGM-TYPE EXPANSION TANKS

Many of the shortcomings of standard expansion tanks can be avoided by separating the air and water in an

expansion tank. About 25 years ago, a new type of expansion tank with an internal synthetic diaphragm became available. On one side of the diaphragm is a captive volume of air that has been pre-pressurized by the manufacturer. On the other side is a chamber for accommodating the expanded volume of system fluid. As more fluid enters the tank, the diaphragm flexes, allowing the air volume to be compressed. An example of a small **diaphragm-type expansion tank** and a sequence illustrating the movement of its diaphragm is shown in Figure 12–6.

(a)

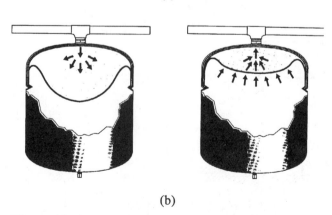

(b)

Figure 12–6 (a) Example of a small diaphragm-type expansion tank. (b) Flexing of the diaphragm as additional fluid enters the tank. Courtesy of Amtrol, Inc.

There are several advantages to using diaphragm-type tanks rather than standard expansion tanks:

- Since the diaphragm separates the fluid from the air, the air volume is protected against reabsorption by the fluid. Diaphragm-type expansion tanks do not have to be periodically drained to prevent water-logging.
- By avoiding water-logging, the possibility of accelerated corrosion caused by addition of fresh water to make up for relief valve losses is no longer a factor.
- The fact that the air volume can be pre-charged to match the static pressure of the system results in a significantly smaller and lighter tank compared to a standard expansion tank.
- Since the air is captive, the tank can be mounted in any position.
- The special boiler and tank fittings required with standard expansion tanks are no longer needed.

Sizing a Diaphragm-type Expansion Tank

A properly sized diaphragm-type expansion tank should reach a pressure about 5 psi lower than the relief valve setting when the system reaches its maximum operating temperature. The 5 psi safety margin prevents the relief valve from leaking just below its rated opening pressure. It also allows for the relief valve to be located slightly below the inlet of the expansion tank.

The first step in sizing a diaphragm-type expansion tank is to determine the proper **air-side pressurization** of the tank, using Equation 12.3.

(Equation 12.3)

$$P_a = (H_s - H_t)\left(\frac{D_c}{144}\right) + 5$$

where:

P_a = air-side pressurization of the tank (psig)
H_s = height of the uppermost system piping above the base of the boiler (ft)
H_t = height of the inlet of the expansion tank above the base of the boiler (ft)
D_c = density of the fluid at its initial (cold) temperature (lb/ft³)

The proper air-side pressurization is equal to the static fluid pressure at the inlet of the tank, plus an additional 5 psi allowance at the top of the system. The pressure on the air-side of the diaphragm should be adjusted to the calculated value *before* fluid is added to the system. This is done by either adding or removing air through the Schrader valve on the shell of the tank. A small air compressor or bicycle tire pump can be used when air is needed. An ac*curate* pressure gauge should be used to

check this pressure as it is being adjusted. Because the required pressure may be relatively low, a standard automotive tire gauge may not be accurate enough.

This air-side pressure adjustment ensures the diaphragm will be fully expanded against the shell of the tank when the system is filled with fluid, but before it is heated. Failure to make this adjustment can result in the diaphragm being partially compressed by the fluid's static pressure before any heating occurs. The full expansion volume of the tank is thus not available as the fluid heats up. An underpressurized tank will act as if undersized, and possibly allow the relief valve to open each time the system heats up. This situation must be avoided.

After the air-side pressurization is determined, Equation 12.4 can be used to find the *minimum* required volume of the expansion tank:

(Equation 12.4)

$$V_t = V_s\left(\frac{D_c}{D_h} - 1\right)\left(\frac{P_{RV} + 9.7}{P_{RV} - P_a - 5}\right)$$

where:

V_t = *minimum* required tank volume (gals)
V_s = fluid volume in the system (gals) (see Section 12.4)
D_c = density of the fluid at its initial (cold) temperature (lb/ft³)
D_h = density of the fluid at the maximum operating temperature of the system (lb/ft³)
P_a = air-side pressurization of the tank found using Equation 12.3 (psig)
P_{RV} = rated pressure of the system's pressure relief valve (psig)

Example 12.2: Determine the minimum size diaphragm-type expansion tank for the system described in Example 12.1. Water is added to the system at an initial temperature of 60 °F. The maximum operating temperature of the system is 200 °F. The boiler is equipped with a 30 psig rated relief valve. The estimated system volume is 35 gals.

Solution: As in Example 12.1, the density of the water at both 60 °F and 200 °F is required. This information can be read from Figure 4–5, calculated using Equation 4.6, or found using the FLUIDS program in the Hydronics Design Toolkit.

$$D_c = 62.36 \text{ lb/ft}^3$$

$$D_h = 60.13 \text{ lb/ft}^3$$

The proper air-side pressurization of the tank is now calculated using Equation 12.3:

$$P_a = (25 - 6)\left(\frac{62.36}{144}\right) + 5 = 13.23 \text{ psig}$$

This information is now substituted into Equation 12.4:

$$V_t = 35\left(\frac{62.36}{60.13} - 1\right)\left(\frac{30 + 9.7}{30 - 13.23 - 5}\right) = 4.38 \text{ gal}$$

Discussion: Notice the required volume of the diaphragm-type tank is about half that required by the standard expansion tank of Example 12.1. This is due to the pre-pressurization of the air side of the tank.

Compatibility of the Expansion Tank and System Fluid

It is very important the diaphragm material used in the tank is chemically compatible with the fluid used in the system. Incompatibilities can result in the diaphragm being slowly dissolved by the fluid. This can create a sludge in the system that gums up system components. When the diaphragm eventually ruptures, the system will quickly water-log, and the vicious cycle between the relief valve and feed-water valve will begin.

The diaphragm materials used in expansion tanks include butyl rubber, **EPDM**, and a material called hydrin. All these materials are compatible with water. Butyl rubber, however, is slowly dissolved by glycol-based antifreezes. EPDM diaphragms are generally compatible with glycol-based fluids. Hydrin is used primarily in closed-loop solar systems when hydrocarbon oils are used as the heat transfer fluid. *The system designer should always verify compatibility between the system fluid and diaphragm material with the tank manufacturer.*

When fresh water is in contact with an expansion tank, as when a domestic hot water tank is also used as a hydronic heat source, the portions of tank that contact the water must be made of a nonferrous material or lined with a noncorrodible material. Standard expansion tanks used in closed-loop hydronic systems have carbon steel shells that will rapidly corrode when continuously exposed to fresh water.

Pressure and Temperature Ratings

Expansion tanks have maximum pressure and temperature ratings. This information is stamped on the label of the tank. Typical ratings are 60 psig and 240 °F. These rating are almost always adequate for residential and light commercial systems. They must be observed, however, in any nontypical applications. For example, if an expansion tank with a 60 psig pressure rating was used on a domestic hot water tank with a 150 psig relief valve, the tank could rupture before the relief valve opened.

(a)

(b)

Figure 12–7 (a) Expansion tank hung from its ¹/₂-inch MPT connection. (b) Horizontally mounted expansion tank. Note hanger strap that supports weight of the tank.

Sizes, Shapes, and Mounting

Diaphragm expansion tanks are available in a variety of sizes and shapes. Tanks with volumes from 1 gallon to about 10 gallons are usually adequate for standard residential and light commercial hydronic systems. These tanks are usually equipped with a ½-inch MPT connection from which they are designed to hang. Because of the water weight in the tank, the piping supporting the tank should be well supported. An alternate method of mounting is to suspend the tank horizontally using metal hanger straps. Again, be sure the weight of the tank and its water is supported by the straps so the inlet connection is not stressed. See Figure 12–7.

When larger expansion tanks are required, floor mounted models are usually the best choice. Because the fluid weight in these larger tanks can be several hundred pounds, it is best to avoid the complexity of overhead mounting.

In all cases it is important the Schrader valve of the expansion tank be accessible in case the air pressure has to be adjusted. *It is also good practice to install a pressure gauge near the inlet of the expansion tank.* This gauge will confirm the static fluid pressure at the tank inlet. The pressure setting of the system's feed-water valve should be adjusted until the pressure at the inlet of the tank matches the air-side pressurization calculated using Equation 12.3. If the tank ever becomes water-logged, this gauge will aid in diagnosing the problem by showing a rapid increase in pressure as the system heats up.

12.4 ESTIMATING SYSTEM VOLUME

The sizing equations for both standard and diaphragm-type expansion tanks require an estimate of system volume. In a typical hydronic system, most of the fluid is contained in the piping and heat source. The fluid volume contained in various sizes of type-M copper and PEX tubing can be calculated by estimating the total length of each tube type/size used in the system, then multiplying these lengths by the appropriate volume factors from Figure 12–8.

For other pipe sizes and materials, Equation 12.5 can be used to estimate the volume factor:

(Equation 12.5)
$$V = 0.04085(d_i)^2$$

where:

V = volume of *one foot* of the tubing (gals/ft)
d_i = *exact inside diameter* of the tubing (ins)

The volume of the heat source can vary over a wide range. Small copper tube boilers may only contain one or two gallons of fluid. Cast-iron sectional boilers may

Tubing size/type	Volume factor (gallons/foot)
1/2" type M copper	0.01319
3/4" type M copper	0.02685
1" type M copper	0.0454
1.25" type M copper	0.06804
1.5" type M copper	0.09505
2" type M copper	0.1647
8mm (nom. 3/8" ID) PEX	0.00491
12mm (nom. 1/2" ID) PEX	0.0092
16mm (nom. 5/8" ID) PEX	0.01387

Figure 12–8   Volume factors for type-M copper tubing

hold 10 to 15 gallons. The only reliable source for this data is the manufacturer's specifications.

The volume of any type of storage tank, such as the tempering tank described in Chapter 10 or the buffer tank described in Chapter 3, must also be included. Thermal storage systems that employ large closed storage tanks will require relatively large expansion tanks.

Most modern heat emitters contain relatively small amounts of water. In the case of either finned-tube or radiant baseboard, the volume can be estimated using the appropriate pipe volume factors in Figure 12–8. Small kick space heaters or wall mounted fan-coils contain very little fluid, probably not more than ¼ gallon each. The fluid content of panel radiators varies with their design and size. Standing cast-iron radiators can contain significant quantities of water. Again, manufacturer's specification sheets are the best source of this data.

Example 12.3: Estimate the volume of the system in Figure 12–9.

Solution: The volume of all piping is estimated using data from Figure 12–8. This is then added to the volume estimates for the heat emitters and boiler.

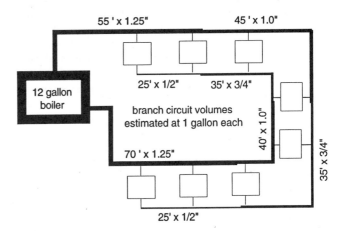

Figure 12–9   Piping diagram of system for Example 12.3

The total piping volume is:

$$v_{piping} = (55 + 70)(0.06804) + (45 + 40)(0.0454)$$

$$+ (35 + 35)(0.02685) + (25 + 25)(0.01319)$$

$$= 14.9 \text{ gals}$$

The total estimated volume of the branch piping and heat emitters is:

$$8(1) = 8 \text{ gals}$$

Assume the boiler volume is listed as 12 gals in the manufacturer's specifications.

The total system volume is the sum of these volumes:

$$14.9 + 8 + 12 = 34.9 \text{ gals}$$

12.5 THE EXPTANK PROGRAM

The Hydronics Design Toolkit contains a program called EXPTANK that can be used to size *diaphragm-type* expansion tanks. A screen shot of the program is shown in Figure 12–10.

The program estimates system volume using inputs of piping length, and a miscellaneous volume entry to account for tanks, heat emitters, etc. It calculates the minimum tank size necessary and its proper air-side

pressurization. Different system fluids can be selected from a pop-up menu. When a fluid other than water is used, the program corrects for the density differences and estimates the volume of antifreeze required based on its concentration.

12.6 PLACEMENT OF THE EXPANSION TANK WITHIN THE SYSTEM

The placement of the expansion tank relative to the circulator will significantly affect the pressure distribution in the system when it operates. *The expansion tank should always be placed near the inlet port of the circulator.* This principle has been applied for years in commercial hydronic systems, yet somehow has been largely ignored in residential installations.

To understand why this rule should be followed, one needs to consider the interaction between the circulator and expansion tank. In a closed-loop piping system the amount of fluid (including that in the expansion tank) is fixed. It does not change regardless of whether the circulator is on or off. The upper portion of the expansion tank contains a captive volume of air at some pressure. The only way to change the pressure of this air is to either push more fluid into the tank compressing the air, or to remove fluid from the tank expanding the air. This fluid would have to come form, or go to, some other location in the system. But since the system's fluid is incompressible, and the amount of fluid in the system is

```
- EXPTANK -    F1-HELP    F3-SELECT                    F10-EXIT TO MAIN MENU
 -INPUTS-
 ENTER Height of uppermost piping above boiler,(.5 to 100).  25        ft.
 ENTER Height of expansion tank above boiler,(.5 to 10)....  7         ft.
 ENTER Highest operating temperature of system,(125 to 250)  200       deg. F
 ENTER Boiler volume + Misc. volumes,(0 to 100)...........   12        gal
 ENTER Pressure relief valve setting,(15 to 60)....         30         psi
 SELECT System fluid,(press F3 to select).................   water
 ENTER Length of 1/2 in. copper pipe in system,(0 to 1000). 0          ft.
 ENTER Length of 3/4 in. copper pipe in system,(0 to 1000). 200        ft.
 ENTER Length of  1  in. copper pipe in system,(0 to 1000). 0          ft.
 ENTER Length of 1¼ in. copper pipe in system,(0 to 1000). 0          ft.
 ENTER Length of 1½ in. copper pipe in system,(0 to 1000). 0          ft.
 ENTER Length of  2  in. copper pipe in system,(0 to 1000). 0          ft.

 -RESULTS-
 Estimated system volume...................................  17.4 gal
 Required expansion tank volume............................  2.2 gal
 Required volume of specified antifreeze...................  0.0 gal
 Pressurization of tank diaphagm...........................  12.8 psi

  MESSAGE
 Use the UP, DOWN, LEFT, RIGHT ARROW keys to move the highlighting bar
 over an input. Type desired value. F10 to return to MAIN MENU. F3 to SELECT
```

Figure 12–10 Screen shot of the EXPTANK program in the Hydronics Design Toolkit fixed, this cannot happen regardless of whether the circulator is on or off. *The expansion tank thus fixes the pressure of the system's fluid at its point of attachment to the piping.* This is called the **point of no pressure change**.

Next consider a horizontal piping circuit filled with fluid and pressured to, say 10 psi, as shown in Figure 12–11. When the circulator is off, the pressure is the same (10 psi) throughout the piping circuit. This is indicated by the solid horizontal pressure line shown above the piping. When the circulator is turned on, it immediately creates a pressure difference between its inlet and discharge ports. However, the expansion tank still maintains the same (10 psi) fluid pressure at its point of attachment. The combination of the pressure difference across the pump, the flow resistance of the piping, and the point of no pressure change gives rise to a new

pressure distribution as shown by the dashed line in Figure 12–11.

Notice the pressure increases in nearly all parts of the circuit when the circulator is operating. This is desirable because it helps eject air from vents. It also minimizes the chances of pump cavitation. The short segment of piping between the expansion tank and the inlet port of the circulator experiences a slight drop in pressure due to flow resistance in the piping. The numbers used for pressure in Figure 12–11 are illustrative only. The actual numbers will of course depend on flow rates, fluid properties, and pipe sizes.

Now consider the same system with the expansion tank (incorrectly) located near the discharge port of the circulator. Figure 12–12 illustrates the new pressure distribution that develops when the circulator is operating.

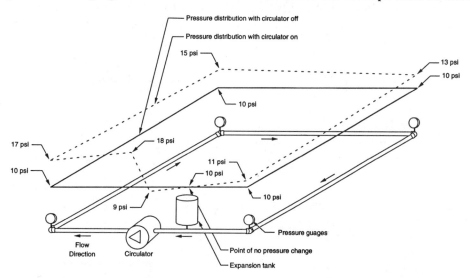

Figure 12–11   Pressure distribution in horizontal piping circuit with circulator on and off. Note the expansion tank is (correctly) located near the inlet port of the circulator.

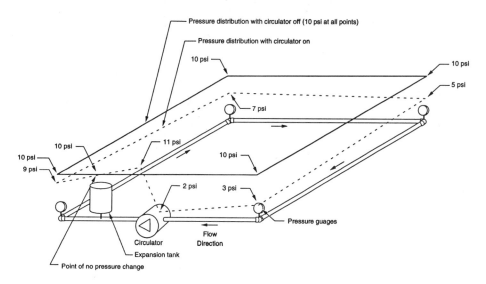

Figure 12–12   Pressure distribution in horizontal piping circuit with circulator on and off. Note the expansion tank is (incorrectly) located near the discharge port of the circulator.

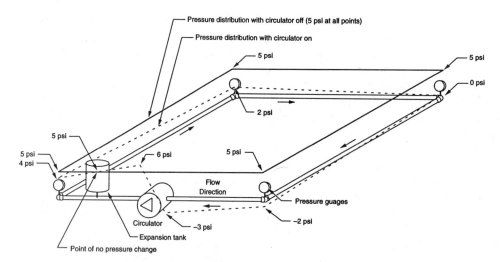

Figure 12–13 Pressure distribution in horizontal piping circuit with low static pressure. Note the expansion tank is (incorrectly) located near the discharge port of the circulator. The pressure in a portion of the piping system becomes subatmospheric when the circulator operates.

Again the point of no pressure change remains at the location where the expansion tank is attached to the system. This causes the pressure in most of the system to decrease when the circulator is turned on. The pressure at the inlet port of the circulator has dropped from 10 psi to 2 psi. This situation is not desirable because it reduces the system's ability to expel air. It can also encourage pump cavitation.

To see how problems can develop, imagine the same system with a static pressurization of only 5 psi. The same 9 psi differential will be established across the circulator when it starts, and the entire pressure profile shown with dashed lines in Figure 12–12 will be shifted downward by (10 – 5) 5 psi as shown in Figure 12–13.

Notice the pressure in the piping between the upper right hand corner of the circuit and the inlet port of the circulator is now below atmospheric pressure. If there were air vents located in this portion of the circuit, this subatmospheric pressure would actually suck air *into* the system every time the circulators operate. The circulator is also much more likely to cavitate in this system.

This situation, unfortunately, has occurred in many residential hydronic systems. Consider the typical residential system piped as shown in Figure 12–14.

In this arrangement the circulator is pumping toward the expansion tank. Although there appears to be a fair distance between the circulator and the tank, the flow resistance created by a typical sectional boiler is very low, and thus, from the standpoint of pressure drop, the circulator's discharge port is very close to the expansion tank. Whenever the circulator operates, this arrangement will cause a drop in system pressure from the expansion tank connection point, around the distribution system to the inlet port of the pump.

Some hydronic systems that are piped as shown in Figure 12–14 have worked fine for years. Others have had

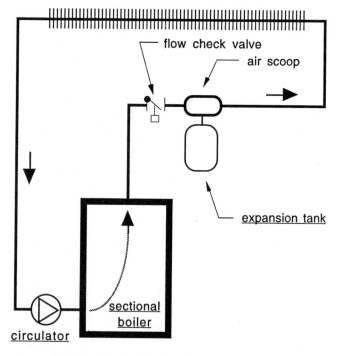

Figure 12–14 Typical piping configuration of a residential boiler. Notice the placement of the expansion tank relative to the circulator.

problems from the first day they were put into service. Why is it that some systems work and others do not? The answer lies in a number of factors that interact to determine the exact pressure distribution in any given system. These include system height, fluid temperature, pressure drop, and system pressurization. The systems most prone to problems from this type of expansion tank placement are those with high fluid temperature, low static pressure, low system height, and high pressure drops around the piping circuit. Rather than gamble on

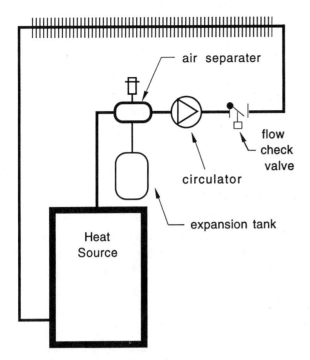

Figure 12-15 Proper arrangement of circulator and flow check valve relative to expansion tank. Notice the inlet port of circulator is now close to the expansion tank connection.

whether these factors will work, it is best to arrange the piping as shown in Figure 12–15.

This arrangement will *increase* the pressure in nearly all parts of the system when the circulator operates. Many systems that have chronic problems with air in the piping can be cured with this type of component rearrangement.

For systems employing conventional sectional boilers with low flow resistance characteristics, the expansion tank can also be placed at the return side of the boiler with virtually identical results. However, this should not be done with heat sources having a high flow resistance characteristic, such as hydronic heat pumps or coiled copper tube boilers.

SUMMARY

The expansion tank is an important part of any closed-loop hydronic heating system. Its sizing, pressurization, and location in the system are easily determined, but unfortunately often misunderstood or ignored. When these factors are determined by unresearched estimates, or the "I've-always-done-it-this-way" approach, serious problems can develop that lead to unsatisfactory system operation and even premature system failure. The reader is encouraged to examine all system piping diagrams, as well as existing installations, for the proper location of the expansion

tank. As previously stated, many systems that have chronic problems with air removal, gurgling sounds in the piping, or low heat output from some heat emitters can often be corrected with a simple relocation of the expansion tank.

KEY TERMS

Air-side pressurization
Diaphragm-type expansion tank
EPDM
EXPTANK program
Incompressible
Point of no pressure change
Standard expansion tank
Thermal expansion
Water-logged

CHAPTER 12 QUESTIONS AND EXERCISES

Note: Questions and exercises requiring the Hydronics Design Toolkit are indicated with marginal symbol.

1. Explain why air slowly disappears from standard expansion tanks.
2. Can a standard expansion tank be placed lower than the boiler outlet? Justify your answer with a sketch.
3. Calculate the minimum size standard expansion tank for a system containing 50 gals of water where the uppermost system piping is 35 ft above the base of the boiler, and the inlet of the expansion tank is 8 ft above the base of the boiler. The water temperature when the system is filled is 50 °F. The maximum operating temperature of the system is 170 °F.
4. Calculate the minimum size diaphragm-type expansion tank for the same system described in Exercise 3. Compare the results with those of Exercise 3.
5. Describe what happens if the diaphragm in an expansion tank ruptures. How could you check the tank to see if it is ruptured?
6. How does heat that migrates from the system piping to the expansion tank affect the pressure of the air in the tank? What could happen if the tank is significantly heated by the system? How can this be avoided?
7. Why is it necessary to adjust the pressure in the air chamber of a diaphragm-type expansion tank on each system? What can happen if this is not done, and the pressure in the diaphragm is less than the static fluid pressure at its mounting location?
8. Why should the expansion tank be located near the inlet port of the circulator?

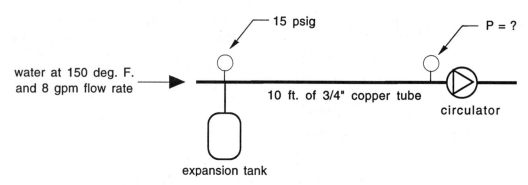

Figure 12-16 Piping arrangement for Exercise 11

9. How does the use of glycol-based antifreeze affect the required size of an expansion tank? Why is this so?

10. The pressure relief valve on a hydronic system with a standard expansion tank opens each time the system heats up. How could the expansion tank be causing this situation? What can be done to correct it?

11. An expansion tank is connected 10 ft upstream from the inlet port of a circulator as shown in Figure 12-16. The pipe between the tank connection and the inlet port of the circulator is 3/4-in copper with 150 °F water flowing through at 8 gpm. Determine the pressure at the inlet port of the circulator assuming the water pressure at the tank connection is 15 psi.

12. In sizing expansion tanks, an allowance for 5 psi at the top of the system is made. Why?

13. What type of diaphragm material should be specified in expansion tanks used with glycol-based antifreeze solutions?

14. Is an expansion tank required in an open-loop hydronic system? How is the expansion of the fluid accommodated in such a system?

15. What is the purpose of the boiler fitting used with standard expansion tanks?

16. Why should expansion tanks be sized so they reach a pressure about 5 psi below the rated pressure of the system's pressure relief valve, as the system reaches maximum temperature?

17. Look up the exact inside diameters of 3/4-in and 1-in schedule 40 steel pipe in a piping handbook. Calculate the volume of fluid they contain per foot.

18. Use the EXPTANK program to determine the size of a diaphragm-type expansion tank required for the system described in Exercise 3. Resize the tank assuming propylene glycol in concentrations of 20%, 30%, 40%, and 50% were used rather than water. What can you conclude from this?

19. Use the EXPTANK program to determine the minimum size of a diaphragm-type expansion tank located 40 ft below the top of a distribution system containing 100 gals of water. Assume the system if filled with 60 °F water and has a maximum operating temperature of 200 °F.

20. Use the FLUIDS program to determine the density of a 30% ethylene-glycol solution at 50 °F and 170 °F and repeat Exercise 4 assuming this fluid is used in the system.

AIR REMOVAL 13

OBJECTIVES

After studying this chapter you should be able to:

- Discuss the different forms in which air can exist in a hydronic system
- Explain the operation of several devices designed to remove air from a system
- Diagnose some of the problems caused by air in hydronic systems
- Describe where air venting devices should be placed in a system
- Lay out hydronic systems that will not experience air problems
- Correct chronic air problems in existing systems
- Describe the construction and operation of a purge cart
- Add antifreeze to a system using a purge cart
- Explain the operation of a microbubble resorber

13.1 INTRODUCTION

The most common complaints about hydronic heating are often related to air in the system. These complaints, expressed by owners and contractors alike, range from occasional "gurgling" sounds in the pipes to a complete loss of heat output.

Air problems are often bewildering. Just when the problem appears to be fixed, it can recur. Many owners and installers eventually give up, thinking the system's air problem simply cannot be corrected. The installer hopes the same mysterious problem will not occur on their next job. A lack of understanding often prevents the true cause of the problem from being diagnosed, corrected, and most importantly, *avoided* in the future!

At some point, every hydronic system contains a mixture of water and air. This is especially true when the system is first filled and put into service. *If properly designed, it should rid itself of most of this air in a few days operation. The system should then maintain itself virtually air-free throughout its service life.* Systems that experience chronic problems with entrapped air usually contain one or more classic design or installation errors.

This chapter provides the facts of how air gets into the system, how it behaves during system operation, and most importantly, how to get rid of it.

13.2 PROBLEMS CREATED BY ENTRAPPED AIR

The full impact of the problems created by air entrapped in a hydronic heating system are often underestimated. Although many people recognize the classic symptom of entrapped air (gurgling sounds in the piping), they are often unaware that entrapped air can lead to several other problems including:

- *Accelerated corrosion of iron or steel components* due to chemical reactions with the oxygen in the air. In systems with chronic air problems, corrosion can proceed at several times its normal rate.
- *Loss of heat output in a heat emitter or an entire system* due to a large pocket of entrapped air. This is often called **air binding**. It occurs in a number of ways. The most common is when an air pocket resides near the circulator's impeller and cannot be cleared by the impeller. Another possibility is when a low head circulator cannot lift the water over the top of a piping path filled with air.
- *Reduced circulator head* because a mixture of water and air bubbles is a compressible fluid. The circulator is unable to efficiently transfer mechanical energy to the fluid. This in turn reduces the system's flow rate and possibly limits its heat output.
- *Improper lubrication of the bushings in wet rotor circulators* because a mixture of water and **microbubbles** creates a foam-like solution near the impeller shaft. This inhibits the formation of a fluid film on which the bushings depend for lubrication.
- *Reduced heat transfer in the boiler* due to the presence of microbubbles along the heated surface. These bubbles interfere with the contact between the water and boiler wall, reducing convective heat transfer.
- *Pump cavitation* can be encouraged by the presence of air bubbles.

13.3 TYPES OF ENTRAPPED AIR

In hydronic systems, air is found in three distinct forms:

- Stationary air pockets
- Entrained air bubbles
- Air dissolved within the fluid

All three forms can exist simultaneously, especially when the system is first started. Each exhibits different symptoms in the system.

Stationary Air Pockets

Since air is lighter than water, it tends to migrate toward the high points of the system. These points are not necessarily at the top of the system. Stationary air pockets can form at the top of heat emitters, even those located near the lower parts of a distribution system. Air pockets also tend to form in horizontal piping runs that turn downward following a horizontal run. A good example is when a pipe is raised to cross over a structural beam, and then dropped to its previous level as shown in Figure 13-1.

When a system is first filled with water, these high points are dead ends for air movement. In some cases the trapped air can displace several gallons of fluid that ultimately must be added to the system. Even after a system is initially purged, stationary air pockets can reform as residual air bubbles merge together and migrate toward high points. This is especially likely in components with low flow velocities where slow moving fluid is unable to push or drag the air along with it. Examples of such components include large heat emitters, large diameter piping, and storage tanks.

Entrained Air Bubbles

When air exists as bubbles, a moving fluid *may* have the ability to carry the bubbles along (entrain them) through the system. **Entrained air** can be both good and bad: Good from the standpoint of transporting air from

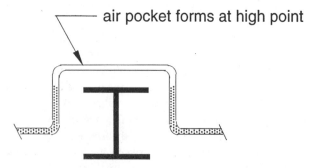

Figure 13-1 A typical location where a stationary air pocket can form

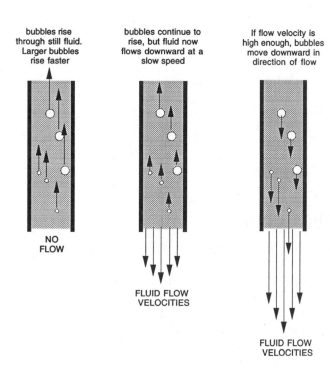

Figure 13-2 The speed of the downward flowing fluid determines if bubbles continue to rise or are pulled along (entrained) in the direction of the flow.

remote parts of the system back to a **central deaerating device,** bad if the air cannot be separated from the fluid in the deareating device, and thus continues to flow through the system.

How well a fluid entrains air is best judged by its ability to move bubbles downward, against their natural tendency to rise. Simply put, *if the fluid moves downward faster than a bubble can rise, it will carry the bubble in its direction of flow.* If entrainment of air through a downward flowing pipe is desired, it is crucial that the fluid's flow velocity is greater than the **bubble's rise velocity** as illustrated in Figure 13-2.

The velocity at which a bubble rises through a fluid depends on several factors. These include the bubble's diameter and density, as well as the density and viscosity of the surrounding fluid. The larger its diameter, the faster the bubble will rise. The higher the viscosity of the surrounding fluid, and the lower its density, the slower the bubble will rise. Of these factors, the diameter of the bubble has the greatest effect on rise velocity. A bubble half the diameter of another will rise at only one-fourth the velocity through the same fluid. The difference in bubble rise velocity can be observed in a glass filled with a carbonated drink.

One type of bubble found in hydronic systems is called a microbubble. They are so small it is often difficult to see a single bubble. Dense groups of microbubbles make otherwise clear water appear cloudy. They can often be observed in a glass of water filled from a faucet that has an aerator device as shown in Figure 13-3.

Figure 13-3   Microbubbles slowly rising through a glass of water

Air Dissolved within the Fluid

Perhaps the least understood form in which air exists in a hydronic system is as **dissolved air**. Molecules of the gases that make up air including oxygen and nitrogen can exist "in solution" with water molecules. These molecules cannot be seen, even under a microscope. *Although water may appear perfectly clear and free of bubbles, it can still contain a significant quantity of air in solution.*

The amount of air that exists in solution with water is strongly dependent on the water's temperature and pressure. The curves shown in Figure 13-4 show the *maximum* dissolved air content of water as a percentage of volume, over a range of temperature and pressure.

Notice that *as the temperature of water increases, its ability to hold air in solution decreases.* This explains why air bubbles appear on the lower surfaces of a pot of water being heated on a stove. It also explains the formation of microbubbles along the inside surfaces of a boiler section. As the water nearest these surfaces is heated, some of the air comes out of solution in the form of microbubbles. The opposite is also true. *When heated water is cooled, it will attempt to absorb air back into solution from any source it can get it from.*

The pressure of the water also has a marked effect on its ability to contain dissolved air. *When the pressure of the water is lowered, its ability to contain dissolved air decreases, and vice versa.* This is the reason carbon dioxide bubbles instantly appear when the cap is removed from a bottle of soda. Removing the cap depressurizes the liquid and reduces the fluid's ability to hold carbon dioxide in solution. This results in the instant formation of visible bubbles. Lowered pressure is also one reason

Microbubbles can also be formed as dissolved air comes out of solution upon heating, or when a component generates significant turbulence. They have very low rise velocities, and are easily entrained by moving fluids. This characteristic makes it difficult to separate microbubbles from the fluid while attempting to vent them out of the system. Separation requires the fluid to move through an area of very low velocity and low turbulence, where the microbubbles have sufficient time to rise and merge into larger bubbles.

Unfortunately many hydronic systems have air separating devices that do not provide sufficiently low flow velocities for efficient microbubble separation. While larger bubbles are captured due to their greater rise velocities, the much smaller microbubbles are swept through before being separated. In some cases this is due to the design of the air separator. In others it is due to the flow velocity through the air separator being too high. Eventually the microbubbles merge into larger bubbles that can be separated and ejected from the system, but this may take several days operation.

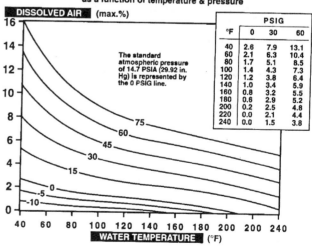

Figure 13-4   Curves showing the maximum solubility of air in water at different temperatures and pressures. Courtesy Spirotherm Corp.

air bubbles are more likely to appear in the upper portions of a multi-story hydronic system. The lower static pressure in the upper part of the system makes it easier for dissolved air to come out of solution. The greater static pressure near the bottom of the system tends to keep the air in solution.

The ability of water to repeatedly absorb and release air can affect the system in a number of ways, some good and some bad. For example, the ability of water to absorb air can be very helpful in removing stationary air pockets from otherwise inaccessible portions of the system (such as a high point in the piping not equipped with an air vent), and transporting the air back to a central deareating device. On the negative side, the ability of water to absorb air is also the primary cause of waterlogging in standard expansion tanks as discussed in Chapter 12.

It is always desirable to minimize the dissolved air content of the system's water. This is accomplished by establishing conditions that "encourage" the air to come out of solution (e.g., high temperatures and low pressures). When conditions are favorable for air to come out of solution, the resulting microbubbles must be gathered together and ejected out of the system through a vent. Devices for doing this are discussed in the next section.

13.4 AIR REMOVAL DEVICES

The air removal devices used in small hydronic systems can be classified as either high point vents or central deareators.

High point vents are intended to release air from one or more high points in the system where it tends to accumulate in stationary pockets. Typical locations for high point vents are at the top of each heat emitter, on supply and return manifolds in floor heating systems, or wherever piping turns downward following an upward or horizontal run. High point vents are particularly useful for ejecting air immediately after the system has been filled with fluid (e.g., at startup or after servicing).

A central deaerator is a device intended to remove entrained air from a flowing fluid and maintain the minimum possible air content in the system. It is usually mounted near the outlet of the heat source and has the entire system flow passing through it.

There are several types of devices used as high point vents and several others used as central deareators. All types relevant to residential and light commercial systems will be discussed separately.

Manual Air Vents

The simplest type of high point venting device is a **manual air vent**. These components are actually small valves with a metal-to-metal seat. They thread into ⅛-inch

Figure 13-5 Example of a manual air vent. Courtesy of ITT Corporation.

or ¼-inch FPT tappings, and are operated with a screwdriver, square head key, or the edge of a dime. They are sometimes referred to as "coin vents" or "bleeders." When their center screw is rotated, air can move up through the valve seat and exit a small side opening. An example of a manual air vent is shown in Figure 13–5.

Manual air vents are commonly located at the top of each heat emitter. They can also be mounted into baseboard tees as shown in Figure 5–28, or any fitting with the appropriate threaded tapping. Because of their small size, care must be taken that joint compound or Teflon tape does not cover their small inlet port.

After a manual vent is opened, and any air below it is released, *a small stream of water will continue to flow through the vent until it is closed*. While it is operating, a person should stay near the vent with a small can ready to catch any ejected fluid before it stains a carpet, etc. This can be difficult (especially for one person), if all manual vents are left open as the system is filled. It is better to move from one vent to the next, operating them in sequence.

Automatic Air Vents

Another type of small high point venting device is called an **automatic air vent**. An example is shown in Figure 13–6. When the upper cap of this device is open, air can pass through and be expelled. When water reaches the vent, an internal washer or disc swells to seal off the opening. This prevents excessive water spillage if the vent is unattended when the system is filled. As more air accumulates beneath the vent, the washer or disc dries out and reopens to expel the air.

The time required for the internal disc or washer to close off the vent is not fast enough to prevent all water loss. Automatic vents may experience a slight water loss from time to time. In some cases installers close off the cap after the system is initially purged of air to prevent any leakage.

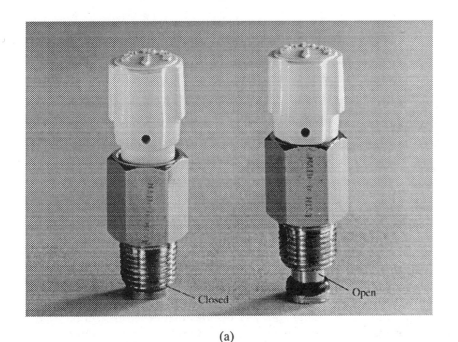

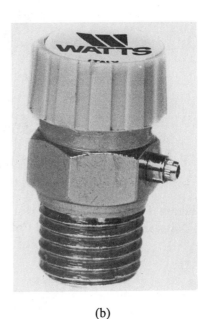

(a)

(b)

Figure 13–6 Examples of (a) push-button air vent, and (b) disk-type automatic air vent. Courtesy of Maid O' Mist and Watts Regulator.

Float-type Air Vents

The need for fully automatic (unattended) venting in locations not easily accessed requires a different type of device. A **float-type air vent** often provides the solution. This device, shown in Figure 13–7, contains an air chamber, a float, and an air valve. When sufficient air accumulates in the chamber, the float drops down and pulls open the valve at the top of the unit. As air is vented, water enters the chamber and lifts the float to close the air valve. This type of vent can operate without leakage.

Most float-type air vents are equipped with a metal cap that protects the stem of the air valve. *It is important this cap remains loose when the vent is put into service.* If the cap is tight, air cannot be ejected.

Float-type vents are designed to thread into baseboard tees or reducer tees the same as manual vents. They are available in different sizes and shapes that allow mounting

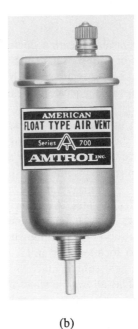

(a)

(b)

Figure 13–7 Examples of float type air vents. Courtesy of Maid O' Mist and Amtrol, Inc.

Figure 13–8 Example of a cast-iron air purger. Courtesy of Watts Regulator, Co.

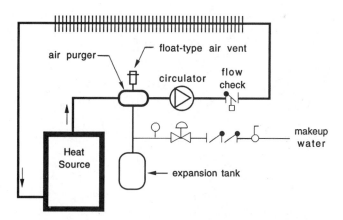

Figure 13–9 Desirable placement of an air purger relative to other components

both horizontal and vertical orientations. Low profile designs are available that allow mounting within the enclosures of heat emitters such as finned-tube convectors or fan-coils.

It is important to realize most automatic and float-type air vents will allow air to flow into the system if the fluid pressure at their location falls below atmospheric pressure. This can result from a number of factors, most notably improper placement of the expansion tank relative to the circulator. It is a deceptive problem because it usually only occurs when the circulator is operating. The best way to prevent this from happening is to *make sure there is at least 5 psi of positive pressure at the top of the system at all times.* This is accomplished by proper setting of the feed water valve as discussed in Chapter 12.

Air Purgers

One type of central deareating device is known as an **air purger** or **air scoop**. This cast-iron component, shown in Figure 13–8, is equipped with an internal baffle that directs bubbles into a separation chamber where they can rise into a float-type air vent screwed into the top of the chamber. The internal baffle also creates a region of lowered pressure that assists in bringing dissolved air out of solution. An arrow on the side of the air purger indicates the direction it must be installed relative to the flow.

To be effective, air purgers should have an inlet flow velocity of 4 ft/sec or less. At higher flow velocities, smaller air bubbles will remain entrained in the flow and be swept through the purger without being extracted.

One of the best locations for an air purger is near the outlet side of the heat source, and close to the inlet port of the circulator. This allows the water entering the purger to be as hot as possible while also being at a relatively low pressure. The air's solubility in water is low at these conditions, and thus bubbles are more likely to form. If make-up water is added to the system through the bottom

of the purger, any bubbles that come in with the water stand a better chance of rising directly into the air purger and being vented before being carried through the system. Figure 13–9 shows the desirable placement of an air purger.

Microbubble Resorbers

A relative newcomer to the American hydronics market, **microbubble resorbers** have received widespread praise for their effectiveness in removing all types of air from hydronic systems. Their internal design provides a region of reduced pressure on the downstream side of several small vertical wires. This encourages dissolved air to come out of solution as microbubbles. The bubbles rise along the wires where they are shielded from higher velocity flow. In the upper part of the resorber the bubbles merge together into a larger air volume. A float-operated air valve opens to eject the air when a sufficient amount has accumulated. An example of a microbubble resorber is shown in Figure 13–10.

A microbubble resorber is capable of maintaining the system fluid in an **unsaturated state of air solubility**. This means the water is always ready to absorb air from any areas of the system where it may be present. Once absorbed, the air is transported by system flow back through the microbubble resorber where it can be separated and ejected from the system. The fluid then returns to its unsaturated state and is ready to absorb more air if necessary. This ability to continually seek out and collect air is very helpful for removing residual air pockets from all areas of the system, especially inaccessible areas that may not be equipped with vents.

Studies have shown microbubble resorbers are able to bring the dissolved air content of the system fluid below 0.5%. The amount of oxygen present in this small air content is of virtually no significance as far as corrosion is concerned.

(a)

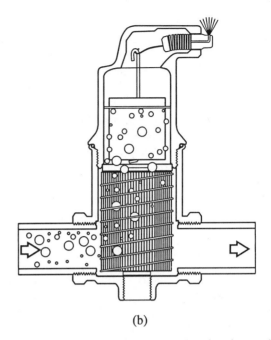

(b)

Figure 13–10   Example of a microbubble resorber. (a) External appearance, (b) cut-away view showing internal design. Courtesy of Spirotherm, Inc.

13.5 CORRECTING CHRONIC AIR PROBLEMS

Properly designed and installed hydronic systems do not require constant manual air venting. A complaint about recurring air noises is symptomatic of one or more underlying design or installation errors. Many of these errors have little intuitive connection with the problems they create and thus are hard to recognize, especially by a novice installer. What follows is a list of the potential causes of recurring air problems, a short description of how the problem develops, and what to do to correct it. Keep in mind any given air problem may be the result of almost any combination of these factors.

Potential Cause #1: *Expansion tank located on discharge side of circulator.*

Description: If the expansion tank is located on the discharge side of the circulator, the pressure in certain locations in the system may drop below atmospheric pressure while the circulator is on. If an air vent is located in this portion of the system, air will be sucked into the system each time the circulator operates. Air can also be sucked in through valve packings or micro-leaks at threaded joints.

Corrective action: This problem is often solved by relocating the expansion tank to the inlet side of the circulator. This fix has been remarkably successful in many systems with chronic air problems. The location of the expansion tank relative to the circulator is definitely one of the first things to check on a problem system.

Potential Cause #2: *Low system pressurization.*

Description: Low system pressurization is usually caused by not setting the make-up water system properly. If, for example, the shut-off valve on the make-up water line is left closed, the system pressure will eventually drop due to minor losses at valve packings, or by air removed at vents. When the pressure at the top of the system drops below atmospheric pressure, air can be drawn in through vents, valve packings, or micro-leaks in threaded joints. This problem can be worsened by having the expansion tank on the wrong side of the circulator as described above.

Corrective Action: Make sure the feed-water valve is adjusted so a minimum pressure of 5 psig is maintained at the top of the system at all times.

Potential Cause #3: *Water-logged expansion tank.*

Description: If there is no air cushion in the expansion tank, or if the tank is too small, the system's pressure relief valve can open each time the system heats up. This allows fresh water (with dissolved air) to enter the system during each heating cycle.

Corrective action: If the system contains a diaphragm-type expansion tank with a ruptured diaphragm, the tank must be replaced. A ruptured diaphragm is indicated by water flowing out of the tank's air valve when its stem is depressed. If the tank is suspected of being undersized, use Equation 12.4 to determine the minimum acceptable tank volume. Be sure the tank is properly pressurized according to Equation 12.3. If the system contains a standard expansion tank, it may require periodic draining to relieve a water-logged condition. Another approach is to replace the standard tank with a diaphragm-type tank.

Potential Cause #4: *Air goes in and out of solution but is not vented from system.*

Description: It is possible for air to come out of solution in the form of microbubbles, and then be absorbed back into solution without being captured and ejected from the system. The microbubbles that form as air comes out of solution are not easily captured by common air purgers.

Corrective action: A microbubble resorber should be installed to efficiently separate and remove dissolved air.

Potential Cause #5: *Lack of high point vents or central deaerator.*

Description: High points in the system that do not have either a manual or automatic air vent can persistently collect air. If the system does not have a micro-bubble resorber, the air pockets can remain in place for weeks.

Corrective action: Install either manual or automatic air vents at all high points and on all heat emitters. Be sure the caps on automatic vents are loose so air can escape.

Potential Cause #6: *Unentrained air bubbles.*

Description: When persistent gurgling sounds are heard in piping, especially piping with downward flow, it is likely the flow velocity is too low to entrain the air and transport it to a central deaerator. The low flow velocity can be caused by a number of factors. These include but are not limited to significantly oversized piping, insufficient circulator size, or excessively long piping circuits.

Corrective action: All down-flowing piping should be sized to maintain a flow velocity of at least 2 ft/sec to effectively entrain air bubbles. If this is not possible, automatic high point vents must be installed at all locations where air can collect.

Potential Cause #7: *Air vents or leaks at the top of open loop systems.*

Description: This problem is often associated with hydronic thermal storage systems that have unpressurized tanks. Consider the schematic shown in Figure 13–11.

Because this is an open loop system, the pressure in all piping above the water level in the tank will drop below atmospheric pressure whenever the circulator stops. Air will attempt to enter the piping at any point it can (above the water level). Air vents located above the water level are the most likely entry points. However, valve packings, pump flange gaskets, and even micro-leaks at threaded joints can all admit air.

Corrective action: This type of system should not have air vents in any piping above the water level in the tank. All piping should be designed for a flow rate of at

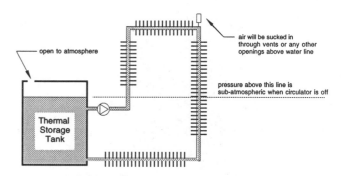

Figure 13–11 Schematic of open loop system showing areas of subatmospheric pressure

least 2 ft/sec to effectively entrain air bubbles and bring them back to the tank where they can be vented.

13.6 FILLING AND PURGING A SYSTEM

Over the years, many installers have developed their own methods of filling and purging hydronic systems as part of their startup procedure. This section will show some common approaches, as well as a new approach using a tool called a purge cart.

Gravity Purging

A good portion of the air inside an empty hydronic system can be expelled by opening the feed-water valve to fill the system, and allowing the air to rise up to and out of the high point vent(s). This approach is known as **gravity purging**. Keep in mind that air pockets will likely form at *all* high points in the system. If all these points are not equipped with a venting device, it may be difficult for the system's circulator to establish flow through the system after filling it, especially if a low head circulator is used. The air pockets may take several minutes to vent through low capacity manual or automatic vents. This method, although simple in concept, can use up a great deal of time, especially if high point vents are not properly located. Figure 13–12 shows some potential problems when gravity purging alone is used.

Forced-water Purging

The amount of air that can be quickly expelled from a hydronic system during filling is substantially increased when the entering fluid has sufficient velocity to entrain air bubbles and carry them to an outlet valve. The greater the water pressure available from a well or water main, the faster **forced-water purging** can push air out of the system.

One method of forced-water purging is to add a boiler drain valve and gate valve to the return line of the boiler

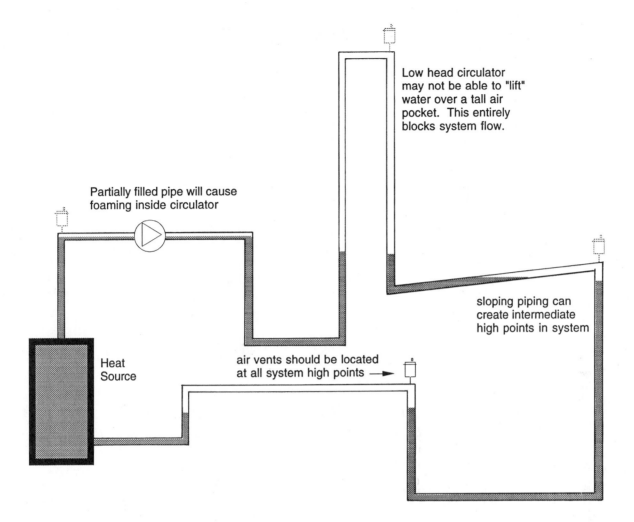

Low head circulator
may not be able to "lift"
water over a tall air
pocket. This entirely
blocks system flow.

Partially filled pipe will cause
foaming inside circulator

Heat
Source

air vents should be located
at all system high points →

sloping piping can
create intermediate
high points in system

Figure 13–12   Diagram showing some locations where air pockets can form and cause difficulty while gravity purging an empty system

as shown in Figure 13–13a. To fill and purge the system, the fast fill level at the top of the feed water valve is lifted to admit water at a high flow rate. Because the gate valve on the return line to the boiler is closed, the entering water begins to fill the boiler, and then, as the boiler is filled, is forced out through the distribution piping. An air vent on the boiler allows the air trapped inside to be released as the boiler fills. The purging flow continues through the system and eventually reaches and exits through the outlet drain valve just downstream of the gate valve. The exiting mixture of air and water is discharged to a drain through a hose. When the discharge stream is free of visible air bubbles, most of the air will have been purged from the system. At this point the fast-fill lever is released and the drain valve is closed.

The placement of the outlet drain valve and gate valve near the *return* connection of the boiler helps ensure debris such as small solder balls or chips of metals will be flushed out of the system, rather than into the boiler where they can cause corrosion. The reader should note the outlet drain valve *must* be open *before* water is admitted

through the feed water valve. If not, the entering water could increase the pressure in the system above the pressure relief valve setting, causing the relief valve to open and defeat the purpose of the high flow rate purging.

A specialty valve that gives the same functionality as the drain valve and gate valve combination is shown in Figure 13–14. This so-called **purge valve** was created to reduce installation time by replacing the drain valve, gate valve, and associated fittings. The vane inside the purge valve can also be used to balance system flow rate when required. The installed position of a purge valve is shown in Figure 13–13b.

Purge valves are usually designed to be soldered into ³/₄-inch copper piping. This size is typical for small and medium size residential systems. Some systems, however, will use larger piping mains. It is generally not a good idea to install a ³/₄-inch purge valve in a larger size pipe because of the potential pressure drop of the internal valve. In these cases it is best to use the combination of a high capacity drain valve and a gate valve the same size as the boiler return line.

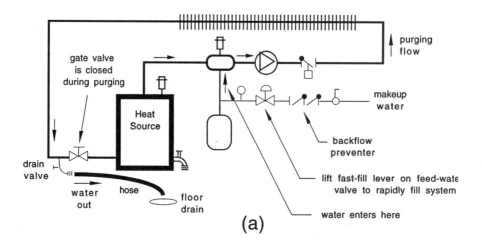

(a)

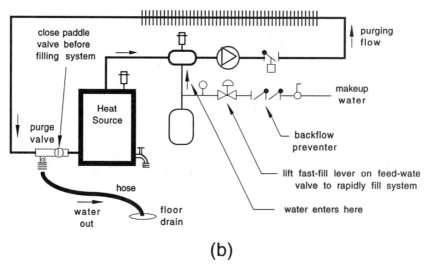

(b)

Figure 13-13 Placement of purging valves (a) using a drain valve and separate gate valve, (b) using a purge valve.

Figure 13-14 Example of a purge valve. Courtesy of Watts Regulator, Co.

To purge a system having parallel piping branches, it is best to close off all but one branch piping path at a time. This forces all purging flow through the open branch at a high flow velocity, improving the chances of displacing entrapped air. Proceed from one branch circuit to the next in sequence. After all branches have been purged individually, open them all up at the same time. The reduced equivalent hydraulic resistance will increase system flow rate and help dislodge air pockets in larger piping and components. This technique is especially helpful on radiant floor heating systems having several parallel branches. The valves on the manifolds can be used to open and close each circuit as needed.

Building and Using a Purge Cart

For about $200 worth of materials, an installer can build a device that makes short work out of filling and

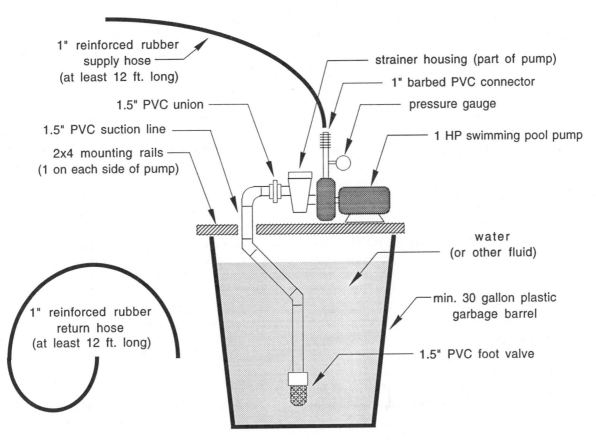

Figure 13–15   The components of a purge cart

purging residential and light commercial hydronic systems. It also serves as a tool for filtering system water, as well as adding antifreeze. This device, referred to as a **purge cart**, consists of a *minimum one horsepower* swimming pool pump mounted on a 30 gallon plastic trash barrel as shown in Figure 13–15.

By installing a union on the pump's suction line, the assembled purge cart can be easily broken down for transportation as shown in Figure 13–16.

The piping arrangement that works best with a purge cart is shown in Figure 13–17. The layout is quite similar to that shown for use with a purge valve. It does, however, allow the piping system to be purged as two parallel loops, one through the heat source, the other around the distribution system. The piping connections for the purge cart hoses are usually made using 1-inch ball or gate valves equipped with 1-inch barbed connectors. The reinforced hoses are secured to the barbed connectors using standard hose clamps.

Procedure for Filling and Purging a System

The following procedure describes how the purge cart is used to fill the system, and remove entrapped air bubbles.

Step 1. Before turning on the purge cart pump, make

Figure 13–16   Purge cart parts broken down for transportation

sure all hoses are secured, the suction line of the pool pump is primed, and the inlet and outlet valves on the system are open. The barrel should be almost full of water and at least one parallel branch of the system should be open. Close one of the isolation valves near the circulator so initial purging is through the heat source. Also make sure the return hose is *held* down inside the barrel, or you could get a short but *intense* shower!

Step 2. Turn on the purge cart pump. The fluid level in the barrel will drop quite rapidly. In 10 to 15 seconds the initially full barrel will be reduced to a minimum pumping level of 4 to 6 inches above the bottom of the barrel. If the heat source volume is more than about 25 gallons, have a hose or bucket ready to add water to the barrel. Initially the air in the heat source will be pushed ahead of the water and exit through the return hose. After a few seconds of pump operation, a solid water stream will appear in the return hose. *Hold the end of the return hose just below the surface of the fluid in the barrel to prevent air from being entrained down into the fluid.* You may notice some occasional air bubbles exiting the return hose as air pockets are displaced.

Step 3. After the heat source has been purged for a minute or so, add water to the barrel until it is almost full. Open the isolation valve on the system's circulator, and close the gate valve near the return connection of the heat source. This allows air and any debris in the distribution system to be purged back to the barrel.

Step 4. Let the purge cart operate for at least ten minutes while flow is through the distribution system. On multiple zone systems it is a good idea to open one parallel piping circuit at a time to achieve a high flow velocity in each circuit. Be aware that zone valves may make a whining sound due to the high flow rate. If individual zone circulators are used, zones can be isolated by closing the isolation flanges or isolation valves flanking these circulators. After each zone circuit has been purged individually, open all circuits to increase the system flow rate. If the system circulator is located as shown in Figure 13–17, it can also be operated if additional flow is desired. The purge cart, however, usually provides plenty of flow.

Step 5. After the distribution piping has been purged for about 15 minutes, switch the purging flow back through the heat source for a minute or so, then back through the distribution system. Finally, open both parallel paths to allow simultaneous flow through both the heat source and distribution system. Allow the purge cart to operate for another two minutes in this mode.

Step 6. The shut-down procedure is simple. Close the *outlet* valve on the system to stop return flow to the barrel. In a few seconds the purge cart pump will have added all the fluid it can to the system as it reaches its maximum (no flow) pressure. This will probably be around 20 psi. Now close the inlet valve, and then shut of the purge cart pump. If the system pressure is higher

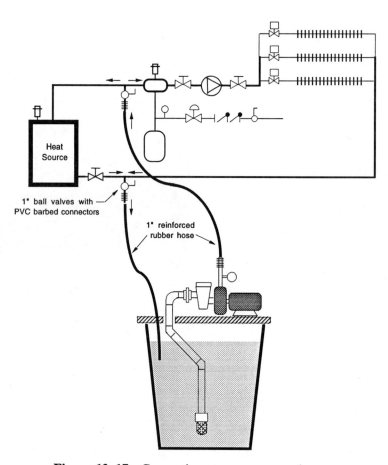

Figure 13–17   Connecting a purge cart to the system

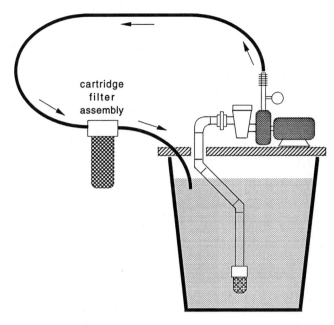

Figure 13-18 Setup for pre-filtering system water using purge cart and cartridge filter

than desired, drain some water from the pressure relief valve or another drain valve. At this point nearly all (non-dissolved) air bubbles should be out of the system.

Pre-filtering Using a Purge Cart

The purge cart can also be used to pre-filter system water to remove some sediments before the water is pumped into the system. This is accomplished by recirculating the fluid through a cartridge filter assembly as depicted in Figure 13–18. A number of different filter cartridges are available with various filtration abilities. It is usually best to start with a coarser filter and work down to smaller particle size limits. Keep in mind some very fine sediments may not be entirely removed.

Adding Antifreeze Using a Purge Cart

The purge cart can also be used to mix and inject antifreeze into a system. Before adding antifreeze, however, it is always advisable to operate the system for a short time with water. This will confirm the presence of any leaks, and the proper functioning of system com-

ponents. If a leak is found, or a system component must be removed, at least antifreeze will not be lost or spilled. Glycol-based antifreezes tend to leave a slippery film on components and tools that eventually turns sticky. It is much easier to correct any piping problems without this added mess.

Step 1. Calculate the system volume using the methods presented in Section 12.4.

Step 2. Temporarily close the system's make-up water valve to prevent drained water from being automatically replaced.

Step 3. Calculate the amount of antifreeze required using Equation 13.1:

(Equation 13.1)
$$V_{antifreeze} = (\%)(V_s + V_{Bmin})$$

where:

$V_{antifreeze}$ = volume of antifreeze required (gals)
$\%$ = desired volume % antifreeze required for a given freeze point (decimal %)*
V_s = calculated volume of the system (gals)
V_{Bmin} = minimum volume of fluid required in the barrel of the purge cart to allow proper pumping (gals)**

Step 4. *Drain an amount of water from the system equal to the required volume of antifreeze calculated in step 3.* This water should be removed from a low point in the system such as the boiler drain valve.

Step 5. Attach the purge cart hoses to the system as shown in Figure 13–17. Keep both inlet and outlet valves closed.

Step 6. *Fill the barrel to its minimum operating level with water,* then pour in the required volume of antifreeze. Turn on the purge cart pump. Open the inlet and outlet valves to the system.

Step 7. Follow the same purging procedure described earlier in this section. Allow the mixture to circulate through the system for at least ten minutes to thoroughly mix the water and antifreeze. Open and close any parallel piping circuits to obtain maximum purging velocity through each circuit.

Step 8. When no visible air is being return from the system to the barrel, close the *outlet* valve on the system. Let the purge cart pressurize the system, then close the

---

* The FLUIDS program in the Hydronics Design Toolkit can be used to find the freeze points of several concentrations of both ethylene glycol and propylene glycol.
** The minimum volume of fluid required in the barrel of the purge cart will depend on how the foot valve is located. It should be determined by experimenting with the cart (using water). Reduce the water level in the barrel until the pump begins to draw air into the foot valve. Make sure the return hose is located to minimize turbulence around the foot valve. Using a permanent marker, put a prominent line on the side of the barrel at least 1 inch above this level. This is the minimum operating level of the barrel. Write "Minimum Operating Level" on the outside of the barrel just above this line. Carefully measure the water in the barrel and note its total volume. *Write this minimum operating volume (gallons) on the side of the barrel* as shown in Figure 13–19.

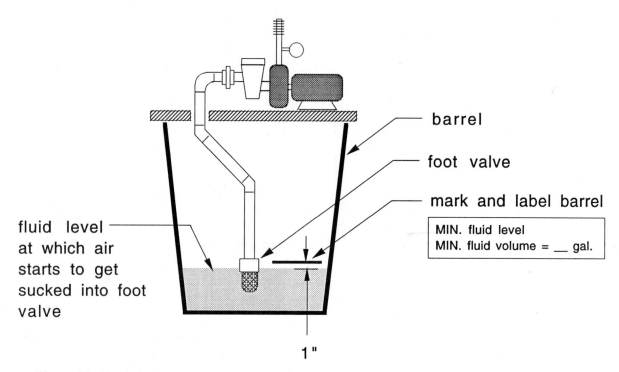

barrel

foot valve

mark and label barrel

MIN. fluid level
MIN. fluid volume = ___ gal.

fluid level
at which air
starts to get
sucked into foot
valve

1"

Figure 13–19 Labeling the purge cart barrel for minimum fluid level and minimum fluid volume

inlet valve. If the system pressure is higher than desired, drain some of the fluid through the relief valve.

Step 9. The system should now contain the desired concentration of antifreeze and be purged of all but dissolved air. Turn off the purge cart and disconnect the hoses. Reopen the shutoff valve on the make-up water line.

Step 10. The volume of mixed fluid in the barrel should be very close to the minimum operating fluid level from which this procedure began. See Figure 13–19. This mixed fluid should be neatly poured into containers and *labeled*. Be sure to include the exact fluid used, the percentage mixture, and the date. This fluid should be saved for the next job *that will use the same fluid and percentage mixture*. It can then be used to make up the minimum operating volume of the barrel. This will reduce the amount of antifreeze used on the next job to just that required by the system. In such a case Equation 13.1 should now be modified to:

(Equation 13.2)
$$V_{anitfreeze} = (\%)(V_s)$$

where:

$V_{antifreeze}$ = volume of antifreeze required (gals)
  % = desired volume % antifreeze required for a
        given freeze point (decimal %)
  V_s = calculated volume of the system (gals)

Keep in mind that a purge cart allows hydronic systems to be filled and operated on construction sites where the building's water supply system may not yet be installed. In buildings with relatively low pressure domestic water system, the purging power of a purge cart will provide faster air removal and less need to manually vent heat emitters or other high points in the piping. Considering its many uses, it can be a valuable investment for hydronic heating contractors.

SUMMARY

This chapter has discussed the various forms in which air can enter and exist in a hydronic heating system. A thorough understanding of how air behaves will allow many of the classic complaints associated with hydronic heating (poor heat output from certain heat emitters and gurgling sounds) to be avoided entirely. This knowledge also gives installers the ability to diagnose and correct previously installed systems exhibiting the classic symptoms of entrapped air. The bottom line is simple: If properly designed and installed, all hydronic systems should automatically rid themselves of almost all internal air after a few days' operation. They should then maintain themselves virtually air-free throughout their service life.

KEY TERMS

Air binding
Air purger (or air scoop)
Automatic air vents

Bubble's rise velocity
Central deaerating device
Dissolved air
Entrained air
Float-type air vents
Forced-water purging
Gravity purging
High point vents
Manual air vents
Microbubble resorber
Microbubbles
Purge cart
Purge valve
Stationary air pockets
Unsaturated state of air solubility

CHAPTER 13 QUESTIONS AND EXERCISES

Note: Questions and exercises requiring the Hydronics Design Toolkit are indicated with marginal symbol.

1. Describe why a water-logged expansion tank can lead to recurring air problems in a hydronic system.
2. What flow velocity is necessary to entrain air bubbles with the flow in a downward flow direction?
3. Will a bubble rise faster through water or a mixture of water and antifreeze? (*Hint:* Consider the viscosity of each fluid).
4. Describe why microbubbles form on the inside surfaces of boiler sections as the boiler heats up. How does this affect heat transfer from the boiler wall to the fluid?
5. Describe the difference between an air purger and a microbubble resorber.

6. Describe how a microbubble resorber is able to remove air pockets from remote areas of a system.
7. Why should a $3/4$-in purge valve not be installed in large diameter main piping?
8. A system is filled with water at 60 °F and 15 psi. The water contains the maximum amount of dissolved air it can hold at these conditions. The water is then heated to 160 °F and maintained at 15 psi. How many gallons of air have been given off in the process?
9. What is the best location for a central deaerator in a typical residential or light commercial system? Why?
10. Describe a situation in which air is drawn into system through a vent each time the circulator operates.
11. How would you test an installed diaphragm expansion tank to see if its diaphragm has ruptured?
12. A customer describes air noises in flow passing through the air purger. What might be preventing the air from being separated and vented?
13. Describe a situation where a purge cart would be desirable over the use of forced-water purging through a feed water valve.
14. Does a purge cart remove dissolved air from water? Why?
15. Why is it necessary to hold the end of the return hose to a purge cart below the fluid level in the barrel?
16. A hydronic system consists of a 15 gal sectional boiler, 250 ft of $3/4$-in copper tubing, 8 heat emitters holding 0.5 gals each, and a 2-gal expansion tank. It is desired to protect the system from freezing to temperatures as low as 0 °F. The minimum usable volume of the purge cart used to fill the system is 4 gals. How much propylene glycol is needed to produce a 20% solution for this system? Use the FLUIDS program in the Hydronics Design Toolkit to determine the freezing point of this mixture.

14 AUXILIARY HEATING LOADS

OBJECTIVES

After studying this chapter you should be able to:

- Calculate the energy required for domestic water heating
- Discuss different methods of obtaining domestic hot water from a hydronic system
- Discuss the advantages of indirectly-fired storage water heaters
- Convert a tankless water heating system for use with a storage tank
- Describe methods for intermittent garage heating using a home's hydronic system
- Modify a hydronic heating system to heat a spa or swimming pool
- Analyze the performance of heat exchangers
- Develop piping and control schematics for hydronic systems with auxiliary loads

14.1 INTRODUCTION

Space heating is the primary load served by residential and light commercial hydronic systems. Through careful design, however, other **auxiliary heating loads** such as **domestic water heating**, garage heating, or heating a pool or spa can often be supplied from the same heat source. This approach reduces installation costs relative to using separate heat sources for each load. It also saves fuel because the single common heat source operates for longer periods of time and therefore at higher efficiency. This chapter describes methods for estimating the heating capacity required by these auxiliary loads. It also describes how to configure system piping and controls to supply such loads.

The feasibility of supplying auxiliary loads depends on their magnitude and when they occur relative to space heating. *The best results are obtained when auxiliary loads do not occur at the same time as design space heating loads.* A good example of such a situation is heating a pool during the summer months using the same boiler that supplies space heating during cooler weather. A poor example would be attempting to meet a large snow-melting load during a cold winter day while space heating demands are high. In this case, a single heat source, unless excessively oversized, would not have the heating capacity to meet both loads at the same time.

If an auxiliary load is relatively small in comparison to the space heating load, it might be met using the surplus capacity of the heat source, especially during milder fall and spring weather. Still another possibility is to prioritize the auxiliary load over the space heating load. This control strategy temporarily suspends space heating until the small auxiliary load is satisfied.

Most auxiliary loads consist of fluids that must be heated, but cannot be mixed with the water used for the space heating portion of the system. In such cases it is necessary to use a heat exchanger to transfer heat from one fluid to another without allowing the fluids to come into contact with each other. This chapter includes a discussion of heat exchanger design and performance.

The detailed analysis of some auxiliary heating loads is too lengthy to be presented in this chapter. In such cases the designer should contact manufacturers of specific types of equipment for further information.

14.2 DOMESTIC WATER HEATING LOADS

Residential and light commercial buildings usually require **domestic hot water** (DHW). Providing domestic hot water is by far the most common auxiliary load to be combined with space heating. This section presents a simple method for estimating daily and peak hourly DHW heating loads. It also gives a rule of thumb for estimating the additional capacity required of the hydronic heat source.

Daily DHW Load Estimates

The *daily* energy requirement for DHW production can be estimated using Equation 14.1

(Equation 14.1)
$$E_{daily} = (G)(8.33)(T_{hot} - T_{cold})$$

where:

E_{daily} = daily energy required for DHW production (Btu/day)

G = volume of hot water required per day (gals)
T_{hot} = hot water temperature supplied to the fixtures (°F)
T_{cold} = cold water temperature supplied to the water heater (°F)

The daily amount of domestic hot water required in a given building is of course heavily dependent on occupancy, living habits, type of water fixtures used, water pressure, and the time of year. The following estimates are suggested as a guideline:

- House: 10 to 20 gallons per day per person
- Office building: 2 gallons per day per person
- Small motel: 35 gallons per day per unit
- Restaurant: 2.4 gallons per average number of meals per day

Example 14.1: Determine the energy used for domestic water heating for a family of four with average water usage habits. The cold water temperature averages 55 °F. The supply temperature is set for 125 °F.

Solution: The daily domestic water heating load is estimated using Equation 14.1:

$$E_{daily} = (4 \times 15)(8.33)(125 - 55) = 34{,}990 \text{ Btu/day}$$

The estimated *annual* usage would be the daily usage times *355* days of occupancy per year (assuming the family is away 10 days per year).

$$\left(34{,}990 \frac{\text{Btu}}{\text{day}}\right)\left(355 \frac{\text{occupied days}}{\text{year}}\right) = 12{,}421{,}000 \frac{\text{Btu}}{\text{year}}$$

$$= 12.4 \frac{\text{MMBtu}}{\text{year}}$$

Discussion: The cost of providing this energy can be estimated by multiplying the annual energy requirement by the cost of delivered energy ($/MMBtu) using the FUELCOST program in the Hydronics Design Toolkit. For example, at $0.10/kwhr and 100% efficiency, electrical energy has a delivered cost of $29.30/MMBtu. The annual energy cost of providing the domestic hot water in this example would be:

$$\left(12.4 \frac{\text{MMBtu}}{\text{year}}\right)\left(29.30 \frac{\$}{\text{MMBtu}}\right) = \$363/\text{year}$$

The DHWLOAD Program

The Hydronics Design Toolkit contains a program called DHWLOAD that calculates the daily and annual energy required for domestic water heating. The program also estimates the standby heat loss of a cylindrical DHW tank based on its size and insulation. The total energy cost associated with producing domestic hot water is the sum of the energy required to heat the water plus the energy required to make up for standby heat losses. This program allows one to quickly compare the energy requirements of DHW water usage and tank insulations. It can be used in conjunction with the FUELCOST program to quickly compare the *costs* of producing domestic hot water with various fuels.

DHW Usage Profiles

The *rate* at which domestic hot water is used varies considerably with the type of occupancy. A typical *residential* domestic hot water **usage profile** is shown in Figure 14–1.

Notice there are two distinct periods of high demand: One during the wake-up period, the other during the early evening. The greatest period of usage in this profile is from 7 PM to 8 PM. However, this still represents only 11.6% of the total daily DHW demand. For a family using 60 gallons of DHW per day, this would represent a peak usage of just under 7 gallons per hour. The average *rate* at which energy is required for DHW production during this peak hour (assuming cold water and hot water temperatures of 55 °F and 125 °F, respectively), can be found using Equation 14.1:

$$E_{daily} = [(0.116)(60)](8.33)(125 - 55)$$

$$= 4{,}058 \simeq 4{,}100 \text{ Btu/hr}$$

The first factor in this equation, 0.116, represents the peak demand of 11.6% of the total daily water volume.

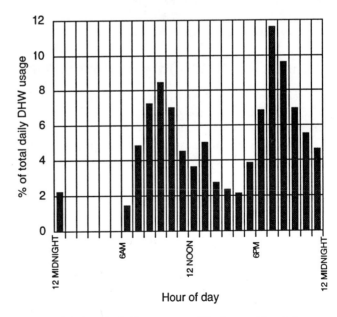

Figure 14–1 Typical usage profile for residential domestic hot water. Source: ASHRAE data.

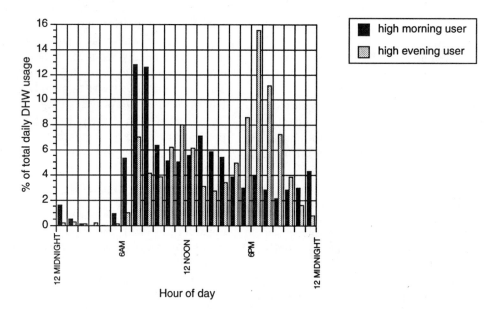

Figure 14–2 Comparison of hourly usage of domestic hot water for a high morning user versus a high evening user. Source: ASHRAE data.

If the heat source was able to supply an extra 4,100 Btu/hr *above the space heating load*, this peak DHW requirement could be met, even on continuous basis if required.

The occurrence of two peak demand periods, one during the wake-up period, the other following the evening meal period, is common in residential buildings. Some households tend to be "high morning users" while others are "high evening users." Figure 14–2 compares these two categories. Notice peak demand for the high evening user is only about 15.5% of the total daily demand.

Data published by the American Society of Heating, Refrigerating, and Air Conditioning Engineers (ASHRAE) indicate probable peak household hot water demands of 6 to 10 gallons *per hour*.

Rule of Thumb for DHW Heating Capacity

The extra boiler capacity required to provide DHW in a combined space heating/DHW system should not exceed the DHW energy usage during the hour of peak demand.

Two considerations suggest this rule of thumb is relatively conservative:

1. The DHW stored in the tank of an indirectly-fired water heater provides a reserve for periods of exceptionally high demand. In such systems, it is not uncommon that 60% to 80% of the tank's volume can be delivered *before* the delivery temperature is significantly affected by the incoming cold water.
2. Modern controls for combined space heating/DHW systems can be set to give a priority to DHW production. This is accomplished by turning off the circulator(s) supplying space heating loads until the

DHW subsystem has recovered to its setpoint temperature. This recovery period usually lasts only a few minutes. The fact that no space heating is supplied for this short time is unnoticable in most buildings.

Hourly DHW load profiles can be very different in buildings such as motels or restaurants, where very high peak demands occur during wake-up and meal periods. DHW systems for such buildings often have dedicated heat sources and require careful analysis using methods that are beyond the scope of this book.

14.3 TANKLESS WATER HEATERS

The traditional method of producing DHW with a space heating boiler uses a heat exchanger coil mounted inside the boiler. These so-called **tankless coils** are made of finned copper tubing rolled into a compact helical shape as shown in Figure 14–3a. They are inserted into a special chamber in the boiler block where they are totally surrounded by hot boiler water as shown in Figure 14–3b. A gasketed bulkhead forms a watertight seal at the side of the boiler block. Cold water is drawn through the coil whenever a hot water fixture valve is open. The entering cold water makes only one pass through the tankless coil on its way to the hot water fixtures.

Since hot water might be required at any time of day, the boiler water surrounding the tankless coil must be kept hot *at all times*. A triple aquastat control as described in Chapter 9 is used to fire the boiler whenever its water temperature drops below a preset lower limit. During warm weather, when there is no demand for space heating, the boiler water must be kept at a relatively high

(a)   (b)

Figure 14–3   (a) Examples of tankless coil water heaters. Courtesy of Amtrol, Inc. (b) A tankless coil being inserted into a boiler. Courtesy of Weil McLain Co.

temperature solely for the purpose of supplying DHW. The standby heat losses associated with this requirement substantially lower fuel efficiency. Values as low as 35% have been observed during warm weather operation of tankless coil systems.

Tankless coils can also cause noticeable variations in the temperature of hot water delivered to fixtures, especially if the fixture is operated for several minutes. This is caused by fluctuations in boiler temperature between firing cycles. Since the amount of water contained in the coil is very small, it has very little thermal mass to help absorb energy without a significant temperature change. This problem can be partially compensated for by installing a thermostatic mixing valve on the outlet of the coil.

The inherent inefficiency of maintaining hot water in the boiler at all times, along with the ever-increasing cost of fuel, is making the traditional use of tankless coil water heaters economically obsolete. Some state energy codes are in fact proposing to ban their use in new installations, except in systems where a storage tank will also be used.

Combining a Tankless Coil with a Storage Tank

Many of the disadvantages of tankless coil water heaters can be offset by incorporating a storage tank into the

system. A schematic of this concept is shown in Figure 14–4. When used to retrofit a standard tankless coil installation, this modification will markedly improve annual fuel efficiency, as well as reduce temperature fluctuations at fixtures.

For this arrangement to function properly, *it is crucial the triple aquastat control on the boiler be either internally reconfigured to function as a high limit control only, or be replaced with a conventional high limit control.* The feasibility of converting a triple aquastat control will depend on its make and model. After this modification is made, the boiler should fire only upon a call for space heating *or* a call for water heating from the tank's setpoint control. It no longer maintains a constant minimum operating temperature.

Upon a call for domestic water heating, a small circulator moves cooler water from near the bottom of the storage tank through the tankless coil. The heated water from the coil should be returned to a *separate connection* near the top of the storage tank. This ensures it will mix with water in the tank, and thus provide a stable supply temperature to the fixtures. The circulator continues to run until the tank thermostat is satisfied. The larger the volume of the storage tank, and the wider the differential of the tank's setpoint control, the longer the operating cycle of the boiler. Longer firing cycles mean higher efficiency. When the storage tank reaches its setpoint

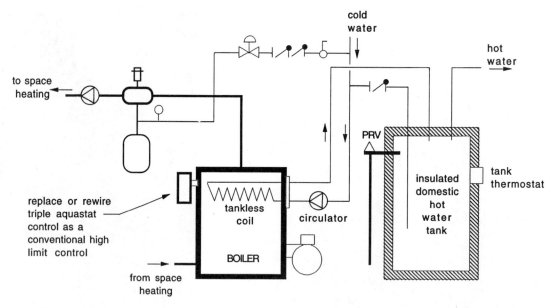

Figure 14–4 Adding a storage tank to a conventional tankless water heating system

temperature, the boiler and tank circulator are shut off. As domestic hot water is drawn from the storage tank, its temperature drops until the tank's setpoint control closes to begin the cycle again.

As an option, the tank circulator can be configured to run for several minutes after the boiler shuts off. This is accomplished using a delay on break time delay relay to operate the circulator. The temporarily sustained circulation purges residual heat from the boiler, moving it into the storage tank before it can escape up the chimney.

Since the tank circulator moves fresh (oxygenated) water, it must be suitable for use in an open system. Circulators constructed of stainless steel, bronze, or high temperature plastics are typical for such applications. Cast-iron circulators should never be used for this application.

The tank's setpoint control can be a conventional electro-mechanical control or an electronic control. The sensor bulb or thermistor should be mounted either in direct contact with the inner pressure vessel of an insulated storage tank or, preferably, into a sensor well inserted into the tank water.

14.4 INDIRECTLY-FIRED STORAGE WATER HEATERS

In recent years, the use of a well-insulated storage tank with its own internal heat exchanger has become an increasingly popular method of producing domestic hot water in combination with hydronic space heating. This has occurred largely as a result of the hydronics industry recognizing the problems and low efficiencies associated with tankless coil systems.

There are currently several manufacturers of so-called **indirectly fired storage water heaters**. These tanks contain an internal heat exchanger through which hot boiler water is circulated upon a call from the tank's setpoint control. The heat exchanger may be an internal coil as shown in Figure 14–5, or a tank-within-a-tank

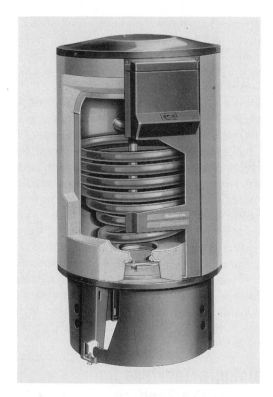

Figure 14–5 Example of an indirectly-fired storage water heater using an internal coil heat exchanger. Courtesy of Buderus Hydronic Systems, Inc.

Figure 14-6 Example of an indirectly-fired storage water heater using a tank-within-a-tank heat exchanger. Courtesy of Triangle Tube Co.

design as shown in Figure 14–6. In either case the heat exchanger can rapidly transfer heat from the boiler water to the domestic water.

To resist corrosion, the storage tank itself is often made of either stainless steel or ceramic-lined steel. It is surrounded by a well-insulated thermal jacket. The use of 2 inches or more of polyurethane foam insulation is typical. This minimizes standby heat loss and helps prevent short but frequent cycles of the boiler to make up for such losses.

The water in an indirectly fired storage water heater provides the thermal mass necessary to allow the boiler to operate for longer cycles and thus at higher efficiencies. When the tank setpoint control is satisfied, the boiler and tank circulator shut off. They can remain off, in some cases for several hours, while the hot water demands of the building are provided directly from the hot water stored in the tank. An example of how an indirectly-fired storage water heater tank is piped into a space heating system is shown in Figure 14–7.

Notice the zone supplying the domestic water heater is connected closest to the boiler. While this is not absolutely necessary, it does reduce heat loss from the outer portions of the common piping when only domestic water heating is required, such as during the summer.

A thermostatic mixing valve is also shown piped to the DHW tank. This valve allows the tank to be heated to a higher temperature than required at the fixtures. When this valve is used in combination with a tank thermostat having a wide differential, the length of the boiler firing cycle, and boiler cycle efficiency, can be increased. Since more energy is added to the tank during each firing cycle, the length of the off-cycle will also be extended.

Indirectly-fired storage water heaters can also be piped as a secondary circuit in a primary secondary system as shown in Figure 11–35.

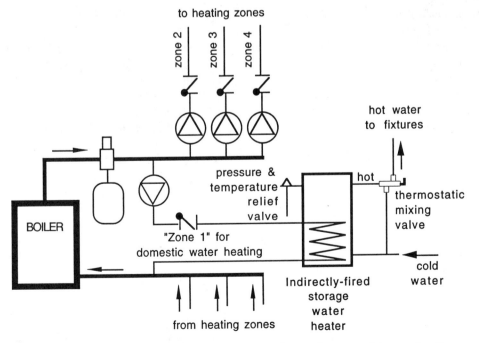

Figure 14–7 Piping an indirectly-fired storage water heater into a multi-zone heating system

	Specifications										
	Boiler Output (MBH)		Circulator Flow Rate (GPM)		Peak Flow (Gal/10 Min)		Continuous Flow (Gal/h)				
							First Hour		Next Hour		
Model No.	113 °F	140 °F	113 °F	140 °F	113 °F	140 °F	113 °F	140 °F	113 °F	140 °F
TR 13	79	66	5	4	46	29	155	110	148	87
TR 20	97	80	7	5	67	43	216	107	182	106
TR 30	119	97	8	7	83	61	251	145	225	128
TR 36	160	118	11	8	102	67	289	179	300	157
TR 45	190	141	13	9	126	79	395	232	357	186
TR 60	305	260	20	17	203	110	654	317	575	330
TR 75	327	324	22	22	226	131	700	401	618	428
TR 120	413	359	28	24	386	272	990	628	779	522

Conditions:
—200 °F boiler water supply
—50 °F water in inner tank
—113 and 140 °F thermostat setting
—Forced circulation and vertical installation only.

Figure 14–8 Example of performance data for an indirectly-fired storage water heater. Courtesy of Triangle Tube Co.

Performance of Indirectly-fired Storage Water Heaters

The thermal performance of indirectly-fired storage water heaters is usually stated in terms of the volume of hot water (of a specified temperature) the tank can produce in a given period of time. The hotter the boiler water supplied to the tank's heat exchanger, and the higher the flow rate through the heat exchanger, the greater the rate of hot water production.

Since the thermal performance of the tank depends on several factors, manufacturers often provide a "snapshot" of performance ratings at specified operating conditions as shown in Figure 14–8.

This particular data shows the maximum volume of hot water (at either 113 °F or 140 °F) the tank can produce during the first ten minutes of demand, as well as over the first hour of demand. These ratings assume the tank is fully heated to either 113 °F or 140 °F setpoint when the demand period begins. They are also based on boiler water entering the coil at 200 °F, and a cold water entering the tank at 50 °F. Changes to either temperature could increase or decrease the rate of hot water production. If the amount of water drawn from the tank exceeds these rated values, its outlet temperature will decrease. The last column of the table shows the *continuous* production rate of domestic hot water, assuming all input temperatures and flow rates remain unchanged. The boiler heating capacity required to yield these hot water production rates is listed in the second and third columns of the chart. Notice the required heating capacity often equals or exceeds the design heat loss of a typical house.

If the boiler has a lower heating capacity, the hot water production rate will be proportionally less.

Since a high production rate of DHW depends on a high and sustained rate of heat input, the hydronic system controls are often set up to treat domestic hot water production as a **priority load**. When the tank thermostat calls for heat, the circulators to other loads supplied by the boiler are temporarily shut off. This allows full boiler capacity to go toward domestic hot water production.

Sizing the Tank

A suggested equation for sizing indirectly-fired storage water heater tanks for typical *residential* applications is as follows:

(Equation 14.2)

$$V = 10(\# \text{ bathrooms with tub or shower}) + 10$$

where:

$V = minimum$ suggested tank volume (gals)

A house with two full bathrooms would thus require a minimum 30 gallon tank. If the house contains a large whirlpool tub, or other equipment likely to require large quantities of hot water in a short period of time, a larger tank may be required.

For non-residential water heating loads, the tank is usually sized based on a steady state **continuous flow load requirement**. Plumbing design manuals often contain data on how to estimate the equivalent continuous hot

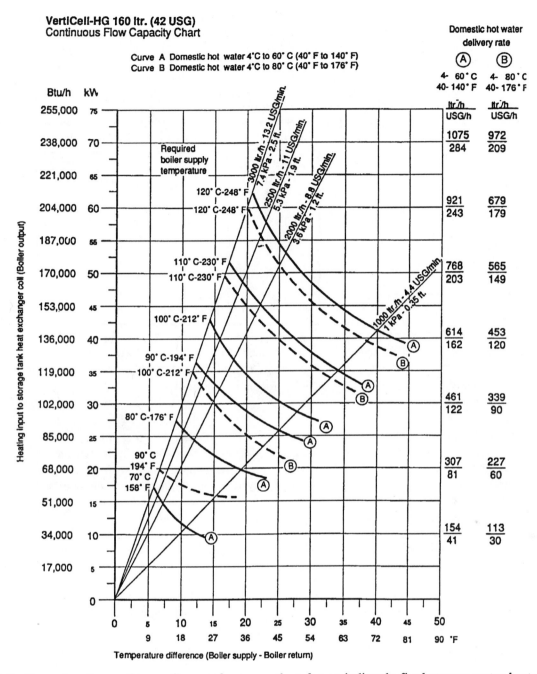

VertiCell-HG 160 ltr. (42 USG)
Continuous Flow Capacity Chart

Figure 14–9   Example of a continuous flow performance chart for an indirectly-fired storage water heater. Courtesy of Viessmann Manufacturing.

water flow requirement of commercial buildings based on their use. This approach tends to be conservative in that it ignores the heat stored in the tank water itself. This energy allows the tank to act as an energy reserve during periods of higher than normal demand.

Some tank manufacturers provide detailed sizing information in the form of graphs or software. Such information allows designers to evaluate the consequences of changing the coil inlet temperatures and flow rates. An example of a continuous flow performance chart is shown in Figure 14–9. An example of tank sizing software,

also based on continuous flow conditions, is shown in Figure 14–10.

14.5 GARAGE HEATING

The option of heating an attached garage is frequently requested by homeowners in cold and snowy northern climates. In some cases the desire is to keep the garage at a relatively constant temperature sufficient to melt ice

```
 F1-HELP                                                          F7-QUIT
┌─ - ISOCALC INPUTS - ═══════════════════════════════════════════════════┐
│ ENTER Temperature of incoming potable water,(44.6-100 deg.F).  50    deg F │
│ ENTER Temperature of outgoing potable water,(113-180 deg.F)..  140   deg F │
│ ENTER Flowrate of outgoing potable water, (1.1-30.8 gpm).....  2     gpm   │
│ ENTER Rate of heat loss from external piping,(0-34,130 BTUH).  5000  BTUH  │
└───────────────────────────────────────────────────────────────────────────┘
┌─ - ISOCALC RESULTS - ═══════════════════════════════════════════════════┐
│ THE SPECIFIED TANK LOAD IS.................................. 94110   BTUH  │
│                                                                           │
│                         INLET TEMP │FLOWRATE │OUTLET TEMP│PRESS.DROP      │
│                          (deg F)   │ ( gpm ) │ (deg F)   │(ft head)       │
│   »ST151/TT150  40 gal.«    176    │  25.1   │  168.4    │  7.44          │
│    ST201/TT200  53 gal.     185    │  13.2   │  170.6    │  2.25          │
│    ST301        79 gal.     194    │   8.9   │  172.6    │  1.08          │
│    ST401       106 gal.     203    │   6.6   │  174.0    │  0.62          │
│    ST551       146 gal.     212    │   5.1   │  174.5    │  0.38          │
│                                                                           │
└───────────────────────────────────────────────────────────────────────────┘
┌─ - ISOCALC MESSAGES - ═══════════════════════════════════════════════════┐
│ Use the UP ARROW or DOWN ARROW to highlight a tank. Coil requirements are │
│ shown at right. Press ESCape key to return to inputs. Press F1 for help.  │
└───────────────────────────────────────────────────────────────────────────┘
```

Figure 14-10 Example of selection and sizing software for indirectly-fired storage water heaters. Courtesy of Buderus Hydronic Systems, Inc.

from cars, or perhaps to allow the garage to be used as a workshop. In other cases, the owners want to heat the garage only when necessary, then shut the heat completely off to conserve fuel.

The intended usage of garage heating should be considered when selecting the type of heat emitter(s). For example, radiant slab heating is ideal in situations where a relatively constant temperature will be maintained. The warm floors will significantly improve comfort, as well as quickly melt snow and ice from vehicles and dry the floor slab. The large thermal mass of such systems, however, makes them very sluggish, and generally unsuitable for intermittent heating. An overhead unit heater would be a better choice for quickly warming the air in a garage.

Arguments can be made both for and against the use of antifreeze in garage heating applications. If the garage is to be constantly heated, the distribution system would ideally not be exposed to freezing conditions, and thus not require antifreeze. On the other hand, if the heat is to be totally shut off during cold weather, the distribution system will very likely experience subfreezing temperatures and therefore must contain antifreeze. The author's position is that antifreeze *must* be used in *all* hydronic garage heating subsystems installed in cold climates. Unanticipated events such as power outages or equipment breakdowns can lead to frozen pipes in a matter of hours. A hard freeze can burst piping, causing hundreds or even thousands of dollars in damage. The extra cost of using antifreeze in these applications is a decision the installer will not regret.

The simplest design approach is to use a separate zone circuit for the heat emitter(s) in the garage. This does, however, require antifreeze to be used in the entire hydronic system. An alternate approach is to separate the garage zone from the rest of the system using a heat exchanger as shown in Figure 14-11.

Notice the heat exchanger is piped into the system as a parallel branch circuit much like a zone circuit. The piping is arranged so boiler water and the antifreeze in the garage circuit flow through the heat exchanger in *opposite directions*. This **counterflow** arrangement will yield maximum heat transfer across the heat exchanger, and is always preferred over a one-directional flow arrangement.

The garage circuit is now an independent *closed* piping circuit with the heat exchanger as its heat source. As such it requires most of the usual piping components associated with a closed loop, namely an expansion tank, air separator, make-up water system, and most importantly a pressure relief valve. A flow check valve is not required on the antifreeze filled side of the garage circuit because heat will not be present in the heat exchanger when the garage zone is off. Inlet and outlet purging valves can also be added to this circuit if antifreeze is to be injected using the purge cart method described in Chapter 13.

Besides the cost of the heat exchanger, which may be several hundred dollars depending on capacity and operating conditions, there is a significant cost associated with these extra components. *For most residential applications, the total system volume is small enough that it is less expensive to add antifreeze to the entire system*

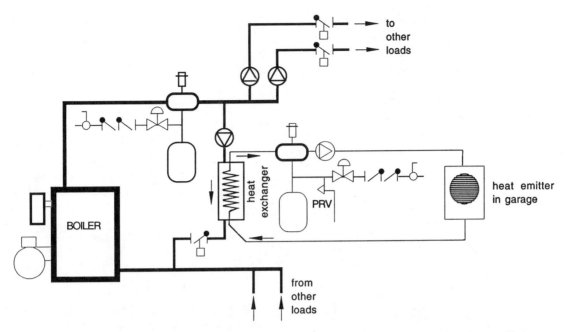

Figure 14–11   Use of a heat exchanger to separate the garage circuit that contains antifreeze

than to add an isolated zone using a heat exchanger. However, this may not be true of systems that contain larger volumes of fluid. The designer should compare these options on a case by case basis.

14.6 SPA AND HOT TUB HEATING

The same boiler that supplies space heating can sometimes be used to supply heat to a spa or hot tub. The common approach is to use a heat exchanger between the space heating system and the spa circuit. This prevents the treated water in the spa from mixing with the boiler water. A conceptual piping schematic is shown in Figure 14–12.

Notice again the heat exchanger is piped into the system as a parallel branch circuit much like a zone circuit. The piping is arranged so that boiler water and spa water flow through the heat exchanger in *opposite directions* and thus maximize the rate of heat transfer across the heat exchanger.

Only a portion of the spa water leaving the filter is routed through the heat exchanger. The amount routed

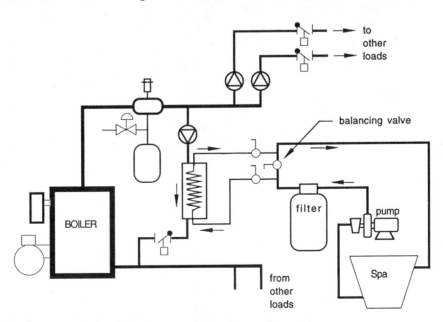

Figure 14–12   Conceptual piping schematic for adding spa or hot tub heating to a space heating system

through the heat exchanger is determined by the setting of the balancing valve between the tees that lead from the spa circuit to the heat exchanger. The more this valve is closed, the greater the flow rate of spa water through the heat exchanger. Ideally this flow rate should be high enough to produce good heat transfer in the heat exchanger, yet low enough to minimize the pressure loss caused by the partially closed balancing valve. The required flow rate can be estimated using the sensible heat rate equation discussed in Chapter 4. This equation is shown in rearranged form as Equation 14.3:

(Equation 14.3)

$$f = \frac{Q}{500(\Delta T)}$$

where:

f = flow rate of spa water through the heat exchanger (gpm)

Q = required rate of heat transfer across the heat exchanger (Btu/hr)

ΔT = temperature gain of the spa water through the heat exchanger (°F)

A design temperature gain of 5 °F to 10 °F across the heat exchanger is typical. Once the balancing valve is set to attain this temperature rise, it should not need further adjustment.

Heat input to the spa can be regulated using a setpoint control to sense the temperature of the water leaving the spa. When this temperature drops below the desired setpoint value, the zone circulator is operated sending boiler water through the heat exchanger. This same setpoint control must also fire the heat source when the spa needs heating.

Heating Loads of Spas

Because spas are usually kept covered and at a reduced temperature when not used, their standby heat loss is quite small, in some cases only a few hundred Btu/hr. This can easily be made up by even small heat sources. The majority of their load is determined by the temperature rise they require before use and how fast this rise must occur. This load can be estimated using a form of the sensible heat quantity equation from Chapter 4.

Example 14.2: Determine the average rate of heat flow necessary to bring a 500 gal spa from an initial temperature of 70 °F to a final temperature of 103 °F over a period of two hours.

Solution: The average rate of heat transfer can be determined by calculating the total energy required to raise the water temperature and dividing by the amount of time allowed for this to take place. The required energy can be calculated using the sensible heat quantity equation from Chapter 4.

$$q = (\# \text{ gallons})(8.33)(T_{\text{final}} - T_{\text{initial}})$$

$$= (500)(8.33)(103 - 70) = 137,400 \text{ Btu}$$

The average rate of heat transfer is found by dividing this energy requirement by the two-hour allowed time period:

$$\text{Average rate of heat transfer} = \frac{137,400 \text{ Btu}}{2 \text{ hr}}$$

$$= 68,700 \; Btu/hr$$

Discussion: This load is comparable to the design heating load of many houses. If the boiler is sized close to the design heat loss of the house, and if it is operating at or near design load conditions, it will not be able to simultaneously heat the spa in the time frame of two hours. There are, however, a number of ways the output of the heat source can be "shared" between the two loads.

One approach is to temporarily interrupt space heating for the time required to bring the spa up to temperature. This is quite feasible in buildings with high thermal mass, or those with high thermal mass heating systems such as radiant floor slabs. The heat stored in this mass will be released to partially compensate for the temporary lack of heat input from the heat source. The system's controls would be configured so spa heating is the priority load. Operation of all other zone circulators or zone valves would be halted whenever the spa setpoint control is calling for heat. This control strategy is routinely used for domestic water heating applications as previously discussed.

Another option would be to design the spa system for a longer warm-up period and thus a lower heat input rate. While this may not be as convenient for those wanting to use the spa on short notice, it significantly improves the feasibility of using extra boiler capacity to heat the spa with less need for priority control. A setpoint control could monitor the temperature of the water going to the space heating zones. If this temperature dropped below a preset value, spa heating could be temporarily suspended until the boiler water temperature recovers. This recovery might only take a few minutes, especially during mild weather.

It is not advisable to add the spa heating load to the design heat load of the building and select the heat source with this total capacity. This approach could result in a boiler with a capacity two or more times greater than the design heat load of the building. The resulting drop in seasonal efficiency will be substantial, as discussed in Chapter 3.

14.7 POOL HEATING

In some cases, hydronic space heating systems can also service swimming pools as auxiliary loads. As with spa heating, a heat exchanger is used to separate the pool water from that in the space heating system. The piping schematic shown in Figure 14–12 could also be used for pool heating.

The feasibility of heating a pool with the same heat source used for space heating depends on the magnitude of the pool heating load, and when it occurs relative to the space heating load. Outdoor pools that are only used during warm months can often use the same boiler that heats the house during cooler weather. Indoor pools that are heated year-round may also be candidates depending on their average heating load. Outdoor pools that are to be kept at comfortable conditions during cooler weather will likely have loads high enough to require a dedicated heat source.

Many factors affect the heating load of a swimming pool. These include:

- Radiation cooling to the sky (outdoor pools)
- Heat gain from solar radiation (outdoor pools)
- Convective heat loss to the air
- Conduction losses through the sides and bottom of the pool
- Evaporation losses from the water surface

The dominant heat loss from pools is caused by evaporation of water. Each pound of water that evaporates from the surface of the pool requires about 960 Btu to change it from a liquid to a vapor. A conservative assumption is that most of this energy is drawn from the water itself. Evaporation losses increase as wind speed increases and as the relative humidity of the air above the pool decreases. The magnitude of this loss is difficult to determine. Even if known at a given set of conditions, evaporation losses are likely to fluctuate as weather conditions change. Estimates range from 30 to well over 100 Btu/hr/ft^2 of pool surface area. *A pool cover will significantly reduce evaporation as well as convective losses, and is highly recommend for any heated pool application.*

The heat exchanger selected for a pool heating application should be capable of transferring the total heat output of the heat source to the pool, while operating within its normal range of temperatures and flow rates. This allows the maximum available heat input rate to the pool whenever other priority loads, such as space heating or domestic water heating, do not require heat input. The high thermal mass of even small swimming pools allows them to soak up heat, even in sporadic bursts, with very gradual temperature increases.

14.8 HEAT EXCHANGERS

Many of the auxiliary heating loads described in this chapter require a heat exchanger to separate boiler water from fresh or chemically treated water. The sizing and selection of heat exchangers require careful calculations. Both thermal and hydraulic characteristics of candidate heat exchangers must be evaluated. Chemical compatibility between the heat exchanger materials and the fluids should also be verified.

One type of heat exchanger often used in smaller hydronic heating applications is called a **coil-and-shell heat exchanger**. It contains a helical coil of finned copper tubing through which one fluid passes. The other fluid passes through a steel shell that surrounds this coil. This shell itself is surrounded by insulation that reduces heat loss to the surrounding air. An example of a coil-and-shell heat exchanger is shown in Figure 14–13.

Hot water from the heat source is pumped through the steel shell. Because the shell is made of steel, it *must* be part of a *closed-loop* hydronic system. Fresh water, treated water, or a mixture of water and antifreeze

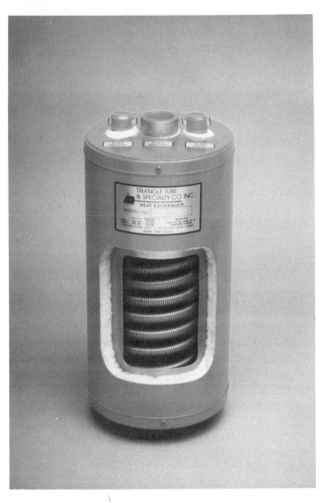

Figure 14–13   Example of a small coil-and-shell heat exchanger. Courtesy of Triangle Tube Co.

is pumped through the copper coil. This coil resists corrosion from air in the water, and other chemicals. It is always advisable, however, to verify the compatibility of the coil material and the intended fluid with the heat exchanger manufacturer.

Thermal Performance of Heat Exchangers

The rate of heat transfer across a heat exchanger depends on the inlet temperature and flow rate of both fluid streams, as well as the surface area separating them. The greater the temperature difference between the two entering fluids, and the greater their flow rates, the faster a given heat exchanger will move heat from one fluid to another.

Performance prediction methods vary from one manufacturer to another. Some provide tabular performance data. Others provide graphs or charts. The most common performance prediction method is based on the concept of heat exchanger **effectiveness**. This term is defined as the ratio of the actual rate of heat transfer across the heat exchanger divided by the theoretical maximum possible heat transfer that would be possible for a given set of operating conditions.

(Equation 14.4)

$$\text{Effectiveness} = E = \frac{\text{actual rate of heat transfer}}{\text{theoretical maximum possible rate of heat transfer}}$$

Effectiveness is usually determined from a graph that plots its value against something called the **capacitance rate ratio**. To determine the capacitance rate ratio, one must first determine the capacitance rate (CR) of each fluid entering the heat exchanger using Equations 14.5a and 14.5b.

(Equation 14.5a)

$$CR_{coil} = (f_{coil})(8.01)(c_{coil\ fluid})(D_{coil\ fluid})$$

(Equation 14.5b)

$$CR_{shell} = (f_{shell})(8.01)(c_{shell\ fluid})(D_{shell\ fluid})$$

where:

f_{coil} and f_{shell} = flow rates through the coil and shell respectively (gpm)

$C_{coil\ fluid}$ and $C_{shell\ fluid}$ = specific heats of the coil and shell fluids (Btu/lb/°F)

$D_{coil\ fluid}$ and $D_{shell\ fluid}$ = densities of the coil and shell fluids (lb/ft³)

The capacitance rate ratio is the ratio of the *smaller* capacitance rate divided by the *larger* capacitance rate:

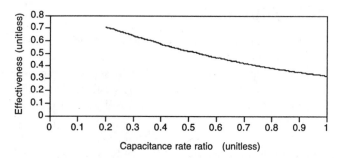

Figure 14–14   Example of a graph of effectiveness versus capacitance rate ratio for a coil-and-shell heat exchanger

(Equation 14.6a)

If CR_{coil} is smaller than CR_{shell} then: $CRR = \dfrac{CR_{coil}}{CR_{shell}}$

(Equation 14.6b)

If CR_{shell} is smaller than CR_{coil} then: $CRR = \dfrac{CR_{shell}}{CR_{coil}}$

Once the capacitance rate ratio is calculated, the effectiveness of the heat exchanger can be read from a graph supplied by the manufacturer. An example of such a graph is shown in Figure 14–14.

The actual rate of heat transfer across the heat exchanger can now be calculated using Equation 14.7.

(Equation 14.7)

$$Q = E(CR_{min})(T_{hot_{in}} - T_{cold_{in}})$$

where:

Q = rate of heat transfer across the heat exchanger (Btu/hr)

E = effectiveness from manufacturer's graph (unitless)

CR_{min} = *smaller* of either CR_{shell} or CR_{coil} (Btu/hr/°F)

$T_{hot\ in}$ = *inlet* temperature of the fluid from the heat source (°F)

$T_{cold\ in}$ = *inlet* temperature of the fluid being heated (°F)

Example 14.3: Boiler water enters the shell side of a heat exchanger at 4 gpm and 180 °F. A 50% mixture of ethylene glycol and water enters the coil side of the heat exchanger at 3 gpm and 120 °F. Assume the effectiveness of the heat exchanger is represented by the graph of Figure 14–14. Determine the actual rate of heat transfer across the heat exchanger.

Solution: The FLUIDS program in the Hydronics Design Toolkit is first used to determine the density and specific heat of both fluids at the indicated entering temperatures:

For 50% ethylene glycol:

$$C_{\text{coil fluid}} = 0.832 \text{ Btu/lb/}°\text{F}$$

$$D_{\text{coil fluid}} = 65.8 \text{ lb/ft}^3$$

For water:

$$C_{\text{shell fluid}} = 1.002 \text{ Btu/lb/}°\text{F}$$

$$D_{\text{shell fluid}} = 60.5 \text{ lb/ft}^3$$

The capacitance rate of each fluid stream is now calculated using Equations 14.5a and 14.5b:

$$CR_{\text{coil}} = (3)(8.01)(0.832)(65.8) = 1,316 \text{ Btu/hr/}°\text{F}$$

$$CR_{\text{shell}} = (4)(8.01)(1.002)(60.5) = 1,942 \text{ Btu/hr/}°\text{F}$$

Since the capacitance rate of the coil fluid is the smaller value, the capacitance rate ratio is:

$$CRR = \frac{CR_{\text{coil}}}{CR_{\text{shell}}} = \frac{1,316}{1,942} = 0.68$$

Entering the graph in Figure 14–14 at 0.68 on the horizontal axis, read up to the curve, then over to the right to get an effectiveness value of approximately 0.44.

The heat transfer across the heat exchanger is now determined using Equation 14.7:

$$Q = E(CR_{\text{min}})(T_{\text{hot}_{\text{in}}} - T_{\text{cold}_{\text{in}}}) = 0.44(1,316)(180 - 120)$$

$$= 34,740 \text{ Btu/hr}$$

SUMMARY

The ability to supply several different type of auxiliary heating loads from a single heat source is a unique feature of hydronic heating systems. It reduces installation costs compared to using separate heat sources for each load. It also gives the potential for increased boiler efficiency by creating greater run fractions.

Wherever possible the designer should try to maximize the use of the heat source by sharing its heat output as necessary. Priority control often provides a suitable method of accomplishing this.

By making use of the piping and control design information in previous chapters, a creative designer can often combine space heating with one or more auxiliary loads to yield an efficient and cost effective system.

KEY TERMS

Auxiliary heating loads
Capacitance rate ratio
Coil-and-shell heat exchanger
Continuous flow load requirement
Counterflow
DHWLOAD program
Domestic hot water
Domestic water heating
Effectiveness
Indirectly-fired storage water heater
Priority load
Tankless coil
Usage profile

CHAPTER 14 QUESTIONS AND EXERCISES

1. Estimate the cost of domestic water heating for a typical family of five assuming cold water enters the system at 45 °F and is supplied to the fixtures at 130 °F.
2. A family of four lives in a house with older water fixtures and tends to use liberal amounts of domestic hot water (20 gals/person/day). The cold water temperature averages 55 °F. The water is heated to 145 °F.
   a. Estimate the daily energy required for domestic water heating.
   b. What would be the daily DHW energy requirement if the family reduced water usage to an average of 15 gals/person/day by using flow restricting fixtures?
   c. What would be the daily DHW energy requirement if the family maintained the 15 gals/person/day usage, *and* reduced the setpoint temperature of the hot water tank to 120 °F?
   d. What would be the annual energy savings of case C compared to case A, assuming hot water is used 355 days per year?
3. Describe two disadvantages of tankless coil water heaters relative to systems with storage tanks.
4. Why do indirectly-fired storage water heaters have the potential to produce hot water much faster than conventional electric water heaters?
5. Describe priority heating control as it applies to domestic water heating as an auxiliary load to space heating.
6. Why is it a good idea to use antifreeze in garage heating circuits, even if the garage thermostat will be constantly kept above freezing?
7. Why is combining snow melting as an auxiliary load to space heating not a good idea?
8. Water enters the shell of a shell-and-coil heat exchanger at 175 °F and 10 gpm. The coil has water entering

at 135 °F and 8 gpm. The effectiveness of the heat exchanger is given in Figure 14–14. Determine the rate of heat transfer across the heat exchanger. Also determine the outlet temperature on the coil side of the heat exchanger.

9. Make a sketch of a piping schematic for a hydronic system having three space heating zones, an indirectly-fired storage water heater, and a spa heating sub-system. Describe how the system would be controlled.

10. Estimate the annual cost of heating hot water according to the usage and temperatures described in Exercise 14.1, with the following assumptions on fuel cost and system efficiency:

a. Using electricity at $0.10/kwhr and 100% efficiency

b. Using natural gas at $0.80/therm and 75% efficiency

c. Using #2 fuel oil at $1.00/gallon and 80% efficiency

Use the FUELCOST program in the Hydronics Design Toolkit to verify your calculations.

SCHEMATIC SYMBOLS A

SCHEMATIC SYMBOLS FOR PIPING COMPONENTS

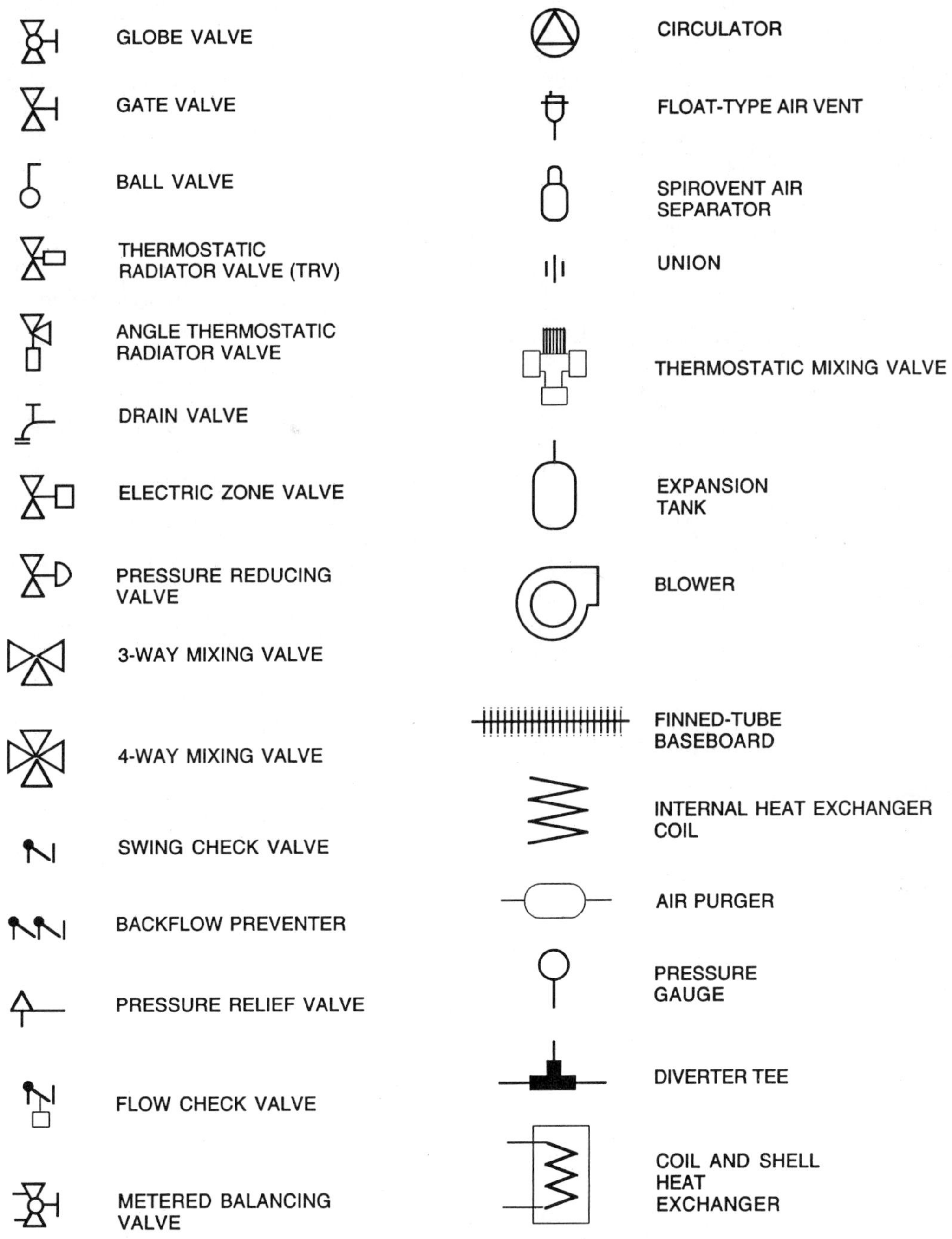

GLOBE VALVE

GATE VALVE

BALL VALVE

THERMOSTATIC RADIATOR VALVE (TRV)

ANGLE THERMOSTATIC RADIATOR VALVE

DRAIN VALVE

ELECTRIC ZONE VALVE

PRESSURE REDUCING VALVE

3-WAY MIXING VALVE

4-WAY MIXING VALVE

SWING CHECK VALVE

BACKFLOW PREVENTER

PRESSURE RELIEF VALVE

FLOW CHECK VALVE

METERED BALANCING VALVE

CIRCULATOR

FLOAT-TYPE AIR VENT

SPIROVENT AIR SEPARATOR

UNION

THERMOSTATIC MIXING VALVE

EXPANSION TANK

BLOWER

FINNED-TUBE BASEBOARD

INTERNAL HEAT EXCHANGER COIL

AIR PURGER

PRESSURE GAUGE

DIVERTER TEE

COIL AND SHELL HEAT EXCHANGER

SCHEMATIC SYMBOLS FOR ELECTRICAL COMPONENTS

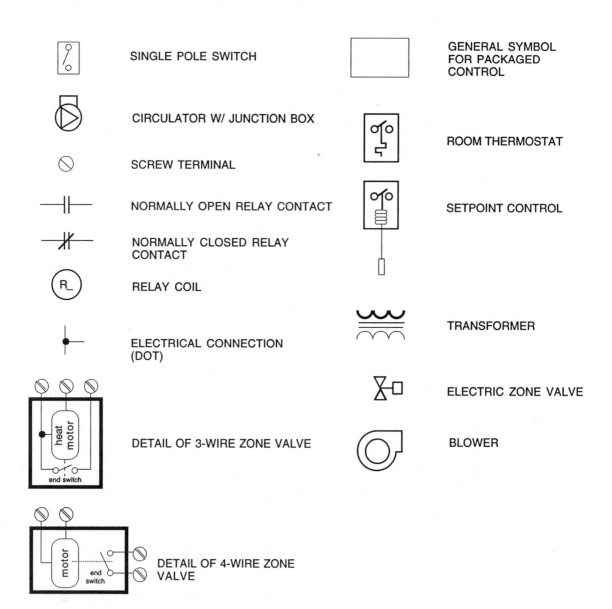

SINGLE POLE SWITCH

CIRCULATOR W/ JUNCTION BOX

SCREW TERMINAL

NORMALLY OPEN RELAY CONTACT

NORMALLY CLOSED RELAY CONTACT

RELAY COIL

ELECTRICAL CONNECTION (DOT)

DETAIL OF 3-WIRE ZONE VALVE

DETAIL OF 4-WIRE ZONE VALVE

GENERAL SYMBOL FOR PACKAGED CONTROL

ROOM THERMOSTAT

SETPOINT CONTROL

TRANSFORMER

ELECTRIC ZONE VALVE

BLOWER

R-VALUES OF COMMON BUILDING MATERIALS

B

MATERIAL	R-VALUE*	
INSULATIONS		
Fiberglass batts (standard density)	3.17	per inch
Fiberglass batts (high density)	3.5	per inch
Blown fiberglass	2.45	per inch
Blown cellulose fiber	3.1–3.7	per inch
Foam in place urethane	5.6–6.3	per inch
Expanded polystyrene panels (beadboard)	3.85	per inch
Extruded polystyrene panels	5.4	per inch
Polyicocyanurate panels (aged)	7.2	per inch
Phenolic foram panels (aged)	8.3	per inch
Vermiculite	2.1	per inch
MASONRY AND CONCRETE		
Concrete	0.10	per inch
8-in concrete block	1.11	for stated thickness
w/vermiculite in cores	2.1	for stated thickness
10-in concrete block	1.20	for stated thickness
w/vermiculite in cores	2.9	for stated thickness
12-in concrete block	1.28	for stated thickness
w/vermiculite in cores	3.7	for stated thickness
Common brick	0.2–0.4	per inch
WOOD AND WOOD PANELS		
Softwoods	0.9–1.1	per inch
Hardwoods	0.8–0.94	per inch
Plywood	1.24	per inch
Waferboard or oriented strand board	1.59	per inch
FLOORING		
Carpet (1/4-in nylon level loop)	1.36	for stated thickness
Carpet (1/2-in polyester plush)	1.92	for stated thickness polyurethane
Foam padding (8 lb density)	4.4	per inch
Vinyl tile or sheet flooring (nominal 1/8 in)	0.21	for stated thickness
Ceramic tile	0.6	per inch
MISCELLANEOUS		
Drywall	0.9	per inch
Vinyl clapboard siding	0.61	for all thicknesses
Fiberboard sheathing	2.18	per inch
Building felt (15 lb/100 ft^2)	0.06	for stated thickness
Polyolefin housewrap	~0	for all thicknesses
Poly vapor barriers (6-mil)	~0	for stated thickness

* The R-value for a specific thickness of a material may be obtained by multiplying the R-value per inch by the thickness in inches (or fractions of inches). The units on R-value are the standard U.S. units of °F × hr × ft^2/Btu.

The data in this table were taken from a number of sources including the *ASHRAE Handbook of Fundamentals* and literature from several material suppliers. It represents *typical* R-values for the various materials. In some cases a range of R-value is stated due to variability of the material. For more extensive data consult the *ASHRAE Handbook of Fundamentals* or contact the manufacturer of a specific product.

R-VALUES OF AIR FILMS

INSIDE AIR FILMS	R-VALUE*
Horizontal surface w/upward heat flow (ceiling)	0.61
Horizontal surface w/downward heat flow (floor)	0.92
Vertical surface w/horizontal heat flow (wall)	0.68
45-degree sloped surface w/upward heat flow	0.62

OUTSIDE AIR FILMS	
7.5 mph wind on any surface (summer condition)	0.25
15 mph wind on any surface (winter condition)	0.17

APPENDIX C
USEFUL CONVERSION FACTORS AND DATA

ENERGY:
1 Watt hour = 1 whr = 3.413 Btu
1 Kilowatt hour = 1 kwhr = 3,413 Btu
1 Therm = 100,000 Btu
1 MMBtu = 1,000,000 Btu

POWER AND HEAT FLOW:
1 kilowatt = 1 kw = 3,413 Btu/hr
1 Ton = 12,000 Btu/hr
1 Horsepower = 0.746 kilowatt = 2546 Btu/hr

LENGTH:
1 Foot = 12 inches = 0.3048 meters

AREA:
1 Square foot = 144 square inches = 0.092903 square meters

VOLUME:
1 Cubic foot = 7.49 gallons = 1728 cubic inches = 0.0028317 cubic meters

PRESSURE:
1 psi = 6894.76 Newtons/square meter = 6894.76 pascal
Absolute pressure (psia) = Gauge pressure (psig) + 14.7

TEMPERATURE:
$°C = (°F - 32)/1.8$ $°F = (°C \cdot 1.8) + 32$

CHEMICAL ENERGY CONTENT OF COMMON FUELS:
1 Cubic foot of natural gas = 1020 Btu (chemical energy content)
1 Gallon #2 fuel oil = 140,000 Btu (chemical energy content)
1 Gallon propane = 91,200 Btu (chemical energy content)

---

* The listed R-values are for nonreflective surfaces typical of most building surfaces. For reflective or partially reflective surfaces consult the *ASHRAE Handbook of Fundamentals*. The units on R-value are the standard English units of $°F \times hr \times ft^2/Btu$.

GLOSSARY

Above-deck dry system: A radiant floor heating system in which the tubing and heat transfer plates are installed above the plywood subflooring.

Absolute pressure: The pressure of a liquid or gas relative to a pure vacuum. It is often expressed in units of psia (pounds per square inch absolute).

Actuator: An electrically or pneumatically driven device that adjusts the position of a valve stem or damper based on the signal it receives from a controller.

Air binding: The inability of a circulator to dislodge a large air pocket at a high point in a piping system.

Air change method: A traditional method of expressing the rate of air leakage into a building. One air change per hour means the entire volume of air in the building is replaced with outside air each hour.

Air film resistance: The thermal resistance of the air film at the top of the finish floor.

Air handler: A generic name for a device consisting of a blower and a finned-tube coil. It is used to either heat or cool the air passing through it based on the temperature of the fluid circulating through the tubes of the coil.

Air purger: A device that separates air bubbles from the moving system fluid and ejects them from the system.

Air-side pressurization: The pressure on the air side of a diaphragm-type expansion tank before fluid enters the tank. This pressure is adjusted by adding or releasing air through the Schrader valve on the tank.

Air vent: A manual or automatic device that releases air bubbles from a piping system.

Ampacity: The maximum current that can be carried by a conductor in a given application as determined by the National Electrical Code.

Angle valve: A valve with its outlet port rotated 90 degrees from its inlet port.

Annual Fuel Utilization Efficiency (AFUE): An estimate of the seasonal efficiency of a heat source based on a federal government standard.

Aquastat: A device that measures the temperature of a liquid at some point in a system, and opens or closes electrical contacts based on that temperature and its setpoint temperature.

Atmospheric: A term often used to describe gas burners that are designed to operate directly exposed to (unpressurized) room air.

Automatic air vents: Air vents that can automatically eject air as it accumulates beneath them.

Auxiliary heating loads: Any other heating load served by a hydronic system that's main function is space heating.

Available floor space: The floor space in a room that can be used to release heat as part of a floor heating system.

Average flow velocity: The average speed of a fluid particle across a particular cross section in a piping component.

Backflow preventer: A specialized piping component that contains the functional equivalent of two check valves and a vent port. It prevents any fluid from a hydronic system from flowing backward and contaminating fresh water.

Ball valve: A valve containing a rotating ball with a machined hole through it. It can be used for component isolation or limited flow regulation.

Baseboard tee: A special fitting that resembles a 90 degree elbow with an additional threaded tapping. This fitting is frequently used to mount an air vent at the outlet end of finned-tube baseboard convectors.

Below-deck dry system: A radiant floor heating system in which the tubing and heat transfer plates are installed below the plywood subflooring.

Bend supports: Preformed elbows or clips that allow PEX tubing to be bent to a relatively small radius without kinking.

Bimetal element: An assembly composed of two dissimilar metal strips joined at their ends. It creates the movement of electrical contacts in a device such as a room thermostat.

Bin temperature data: A means of reporting the duration of ranges of outside temperature based on the number of hours in a year the temperature is within a given range.

Boiler drain: A valve that connects to a hose thread and can be placed at any point in a system requiring drainage.

Boiler feed water valve: Another name for a pressure reducing valve used to reduce water main pressure to the pressure required in a hydronic system. This valve admits water to the system as needed to make up for small water losses.

Boundary layer: A thin layer of fluid that moves slowly along the inside wall of a pipe. The thicker the boundary layer is, the poorer the heat transfer between the bulk of the fluid and the tube wall.

Brine: A generic term for a solution of water and an antifreeze such as propylene glycol, ethylene glycol, or calcium chloride.

British thermal unit: The amount of energy required to raise one pound of water by one degree Fahrenheit, also abbreviated Btu.

Buffer tank: An insulated, water-filled storage tank that forms the link between a heat source and a hydronic distribution system. This tank allows the rate of heat input from the heat source to be different than the rate of heat output to the distribution system.

Building heating load: The rate at which heat must be added to an entire building to maintain interior comfort.

Buoyancy: The upward force that causes a less dense material to rise above a more dense fluid.

Butt fusion: The process of joining polyethylene pipe by melting and then pushing together the two (butt) ends of piping.

Capacitance rate ratio: The ratio of the smaller fluid capacitance rate through a heat exchanger to the larger capacitance rate.

Cavitation: The formation of vapor pockets when the pressure on a liquid drops below its vapor pressure. Cavitation is very undesirable in circulators.

Central deareating device: A device mounted near the outlet of the heat source that attempts to collect air bubbles and eject them from the system.

Centrifugal pump: An electrically driven device that adds mechanical energy (e.g., head) to a fluid using a rotating impeller. In this text the word pump refers specifically to a centrifugal pump, and is used synonymously with the word circulator.

Chase: A passage through a building for accommodating mechanical equipment such as conduit, piping, or an exhaust flue.

Check valve: A valve that limits flow to one direction only.

Chemical energy content: The theoretical maximum energy content of a fuel based on its chemical composition.

Circulator: A term that is synonymous with the word pump. A device that creates fluid motion in a hydronic system by adding head energy to the fluid.

Close-coupling (of pumps): Bolting two or more inline circulators together end to end, in the same flow direction, to increase head.

Closed-loop system: A piping system that is sealed at *all points* from the atmosphere.

Coefficient of performance (COP): The ratio of the rate of heat output of a heat pump to the rate of electrical energy input, *in the same units*. High COPs are desirable.

Coil-and-shell heat exchanger: A type of heat exchanger in which an internal copper coil is surrounded by a steel shell.

Combustion efficiency: The efficiency of a heat source in converting the chemical energy content of its fuel into heat, based on measurements of exhaust gas temperature and carbon dioxide content.

Compressible fluids: Fluids, usually gases, that can be compressed into smaller volumes when pressure is applied to them.

Compressor: The device in which a refrigerant gas is compressed in volume, and at the same time, increased in temperature.

Condensate: The liquid formed as a gas, such as water vapor, condenses.

Condenser: The device in which a refrigerant gas releases heat, and in doing so changes from a vapor into a liquid.

Conduction: A natural process whereby heat is transferred by atomic vibrations through a solid or liquid material.

Confined space: Any space not meeting the definition of unconfined space in the National Fuel Gas Code.

Contactor: Another name for a relay that is rated for relatively high current/line voltage loads.

Continuous flow load requirement: The rate at which heat is required to supply a steady domestic hot water demand.

Control joint: A saw cut groove penetrating part way through a concrete slab for the purpose of forcing a crack to occur along a predetermined path.

Convection: A means by which heat is transferred between a solid surface and a fluid such as air or water.

Convector: A generic name for heat emitters that transfer the majority of their heat output by convection.

Copper water tube: Copper tubing specifically manufactured for use in conveying water for domestic use, and within hydronic heating systems.

Counterflow: A description for a heat exchanger where the two fluids flow in opposite directions.

Cross-linked polyethylene: Polyethylene in which long molecular chains have been bonded together to form a stronger and more temperature resistant material. The acronym *PEX* is often used to represent cross-linked polyethylene.

Cycle efficiency: The net efficiency of a heat source accounting for heat losses during off-cycles. Cycle efficiency decreases as run fraction decreases.

Deaerated: A term used to describe a liquid that has been processed to remove a portion of the air it contains. This air may be in the form of bubbles or dissolved in solution.

Deareating: The process of removing both bubbles and dissolved air from the fluid in a hydronic system.

Degree day: The difference between the outside average temperature and 65 °F over 24 hours. Daily degree days can be added together to obtain monthly and annual total degree days.

Demand charge: A charge billed by electric utilities based on the rate at which electrical energy is supplied to a customer, rather than just the quantity of energy used.

Density: In the context of this book, density refers to the weight of a substance divided by its volume. Some readers will recognize this as being the same as the *specific weight* of a substance. Common English units for density, as defined herein, are lb/ft³ (pounds per cubic foot). For the physicist, density is technically defined as the mass of a substance divided by its volume, in *slugs*/ft³.

Design dry bulb temperature (97.5%): The temperature the outside air is at, or above, 97.5% of the year. It is usually the outside temperature at which the design heating load is calculated.

Design heating load: The heating load of a building or room when the outside air temperature is at its design dry bulb value.

Dewpoint temperature: The temperature at which water vapor in a mixture of other gases begins to condense into liquid droplets. Other compounds in a mixture of gases also have an associated dewpoint temperature.

Diaphragm-type expansion tank: An expansion tank that contains a flexible diaphragm that separates the air and water in the tank.

Dielectric union: A union made of steel on one side and brass on the other. A dielectric (electrically insulating) material separates the two metals to prevent galvanic corrosion.

Differential: A *range* of temperature between the point where the electrical contacts of a control open and when they close. The differential of some controls is fixed, while on others it can be adjusted.

Differential pressure bypass valve: A valve that is designed to allow flow through it when the pressure difference across its ports reaches a specific value. It is commonly used in multi-zone parallel piping systems.

Diffuse: The migration of molecules of one substance through the molecular structure of another material. Oxygen, for example, can slowly diffuse through the molecular structure of certain polymers.

DIN rail: An industry standard mounting rail for relays and other modular controls.

Dissolved air: Air mixed with water at a molecular level. As such it cannot be seen.

Distribution system: An assembly of piping, fittings, and valves that conveys heated water from a heat source to heat emitters.

Diverter tee: A special fitting resembling a tee with an internal baffle. This fitting is used to force a portion of the flow entering the tee out through a branch circuit connected to its side port.

DOE heating capacity: The heat output rating of boilers up to 300,000 Btu/hr based on a federal government rating standard.

Domestic hot water: Heated water used for purposes of washing, cooking, or other uses. Domestic hot water is considered suitable for human consumption.

Domestic water: Water suitable for human consumption.

Domestic water heating: Producing hot water suitable for washing, cooking, or human consumption.

Double interpolation: A mathematical method of estimating a value that is known to lie between two other values of each of two independent parameters.

Draft proving switch: A device that measures the negative air pressure in an exhaust system and closes its electrical contacts whenever a safe negative pressure is maintained.

Draindown system: A type of active solar energy system in which freezing is avoided by having all fluid drain out of the collectors and exposed piping, *to waste*, at the end of each operating cycle.

Dry base boiler: A boiler design in which the boiler block rests on top of the combustion chamber.

Dry system: Any type of radiant floor heating system that does not use a poured material to embed the tubing.

Earth heat exchanger: A description of piping buried in the soil for the purpose of extracting or rejecting heat.

Effectiveness: The ratio of the actual heat transfer across a heat exchanger to the maximum possible rate of heat transfer. Effectiveness is a unitless number between 0 and 1. Higher values are more desirable.

Effective total R-value: The average thermal resistance of a wall, ceiling, floor, etc., after compensating for the presence of framing members.

Efficiency: The ratio of a desired output quantity or rate, divided by the necessary input quantity or rate. Both quantities must have the same mathematical units.

Electro-mechanical controls: Controls that use mechanical force generated by springs, bimetal elements, or pressure diaphragms to open and close electrical contacts.

Element: A name for the finned-tube component in a baseboard convector. The name also applies to the device that creates heat in an electric boiler or water heater.

End suction pump: A pump with an inlet connection perpendicular to the plane of its impeller.

End switch: A switch built into a zone valve that closes its electrical contacts when the valve reaches its fully open position.

Entrained air: Air bubbles that are carried along with the fluid flowing through a piping system.

EPDM: Ethylene Propylene Diene Monomer. A high quality rubber compound resistant to many chemicals.

Equivalent hydraulic resistance: The result of mathematically combining two or more hydraulic resistances into a single hydraulic resistance that represents their combined effect.

Equivalent length of a component: The concept of replacing a particular piping component by a straight length of pipe, of the same diameter, that would yield the same hydraulic resistance as the component being replaced.

Equivalent resistor: A symbol that represents the combined hydraulic resistance of two or more piping components.

Erosion corrosion: A process where metal is removed from the inside surfaces of piping, or piping components due to excessively high flow velocity.

Evaporator: The device in which a refrigerant absorbs heat, and in doing so changes from a liquid to a vapor.

Expansion compensator: A piping device that absorbs the expansion and contraction movement of piping as it changes temperature.

Expansion device: A device that causes entering liquid refrigerant to undergo a pressure drop and associated temperature drop.

Exposed surface: Any building surface exposed to outside air temperature.

EXPTANK program: A program in the Hydronics Design Toolkit that assists in sizing and determining the air-side pressurization of diaphragm-type expansion tanks.

Eye (of an impeller): The center opening where fluid flows into an impeller.

Fan-coil convector: A heat emitter that contains an internal blower or fan to force air through a finned-tube coil where it absorbs heat.

Feet of head: The common English units of expressing the head of a fluid. The units of feet represent the total mechanical energy content of each pound of fluid, and are derived from simplifying the units of Ft•lb/lb.

Finish floor resistance: The thermal resistance of any finish floor material that is placed on top of the heated slab.

Finned tube: A tube with added fins to enhance heat transfer.

Finned-tube baseboard convector: A type of convector that is usually composed of a copper tube with aluminum fins mounted in a steel enclosure located at the base of walls.

Fire-tube boiler: A boiler design where hot combustion products passing through tubes heat water on the outside of the tubes.

Flanges: The common means of connecting a circulator to piping. The pump's volute has integral cast flanges on its inlet and discharge ports. These bolt to matching flanges threaded onto the piping. An O-ring provides a pressure tight seal between the flanges.

Flat plate solar collectors: Devices that convert solar radiation to heat using a flat, fluid-cooled plate in an insulated and glazed housing.

Float-type air vents: A type of automatic air vent that uses an internal float to open and close the air venting valve.

Floor loading: The weight supported by each square foot of floor, expressed in units of pounds per square foot. Minimum floor loadings are specified in building codes.

Flow check valve: A valve with a weighted plug that prevents backward flow *and* forward thermosiphoning of hot water.

Flow coefficient (C_v): The flow rate, in gallons per minute, of 60 °F water needed to create a pressure drop of one psi across a piping component. C_v values for valves are often listed in manufacturer's specifications.

Flow rate: The volumetric rate of flow of a fluid. For liquids it is often expressed in units of gallons per minute (gpm). For gases it is often expressed in units of cubic feet per minute (cfm).

Flow velocity: The speed of an imaginary fluid particle at some point in a piping system. Common English units for flow velocity are feet per second.

Fluid properties factor (α): A number that combines the fluid properties of density and viscosity of a given fluid into a single index for use in determining hydraulic resistance.

Flux: A paste containing an acid that chemically cleans the surfaces of copper tubing and fittings being joined by soldering.

Forced convection: Heat transfer between a solid surface and a moving fluid where the fluid's motion is created by a fan, blower, or pump.

Forced-water purging: Using a forced-water stream to entrain air in a piping system and drive it out of a drain valve.

Free area: The *unobstructed* area for air to flow through a louvered panel, screen, etc.

Full port ball valve: A ball valve in which the hole through the ball is approximately the same diameter as the pipe size of the valve.

Full storage systems: A thermal storage system designed to store all the energy needed by the load, for a 24-hour period, during the preceding off-peak period.

Galvanic corrosion: Corrosion between dissimilar metals caused by the flow of electrons from a less noble metal (on the galvanic chart) to a more noble metal.

Gaseous cavitation: Cavitation caused by air bubbles entrained in the fluid rather than vapor pockets.

Gate valves: A type of valve designed for component isolation purposes.

Gauge pressure: The pressure of a liquid or gas measured relative to atmospheric pressure at sea level. Common English units for gauge pressure are psig (pounds per square inch gauge), or simply psi.

Globe valves: A type of valve designed for flow regulation purposes.

Gravity drainback system: A type of active solar energy system in which freezing is avoided by having all water drain out of the collectors and exposed piping *into a holding tank* at the end of each operating cycle.

Gravity purging: Filling a hydronic system from the bottom and allowing the air in the piping to rise up to, and out of, high point vents.

Gypsum-based underlayments: A pourable, self-leveling mixture containing gypsum-based cement and washed sand. It is specifically intended for use as a thin slab material.

Hard drawn tubing: Copper tubing that is sold in straight lengths.

Hard-wired control logic: Specific wiring connections between components that force a control system to operate in a predetermined manner.

Head: The total mechanical energy content of a fluid at some point in a piping system.

Heat anticipator: An adjustable resistor contained in a room thermostat that preheats the bimetal element to minimize room temperature overshoot.

Heat capacity: A material property indicating the amount of heat required to raise one cubic foot of the material by one degree Fahrenheit. Common English units for heat capacity are Btu/ft³/°F.

Heat emitter: A generic term for a device that releases heat from a circulating stream of heated water into the space to be heated. Examples include convectors, radiator, and fan-coils.

Heat flux: The rate of heat flow across a unit of area. The common English units for heat flux are Btu/hr/ft².

Heating capacity: The rate of heat output of a heat source, usually expressed in Btu/hr.

Heating curve: Another name for a reset line.

Heating effect factor: An allowance of 15% extra heat output, beyond the laboratory tested heat output, allowed by the IBR rating standard for finned-tube baseboard convectors.

Heat motor: A device that produces a small linear or rotary movement of a shaft when heated by an electrical current. Heat motors are commonly used to operate zone valves.

Heat pump: A device that uses the refrigeration cycle to move heat from an area of lower temperature to one of higher temperature.

Heat purging: Removing a portion of the heat stored in the thermal mass of a boiler after the burner has stopped firing.

Heat source: A device that supplies heat to a space heating distribution system.

Heat transfer plate: An aluminum plate that fits over floor heating tubing and provides lateral heat conduction away from the tubing.

High point vents: Devices located at the high point(s) of a piping system that can release air that may accumulate at these high points.

Horizontal panel radiator: A panel radiator that is wider than it is tall, and has tubes oriented in the horizontal direction.

Hydraulic equilibrium: The condition where the head added by the system's circulator exactly equals the head dissipated by viscous friction of the fluid flowing through the system.

Hydraulic resistance: A means of representing the ability of piping components to remove head energy from a flowing fluid. The greater the hydraulic resistance of a component, the more head it removes from the fluid.

Hydraulic resistance diagrams: A drawing composed of hydraulic resistors that represents an assembly of piping components.

Hydraulic resistor: A symbol that represents the hydraulic resistance of a piping component.

HYRES program: A program in the Hydronics Design Toolkit that estimates the hydraulic resistance of a piping component using flow rate and head loss data.

IBR heat output ratings: An industry standard rating system for the heat output of baseboard convectors at various water temperatures. This rating is determined by independent testing under the direction of the Hydronics Institute.

Impeller: A rotating disc with curved vanes contained within a centrifugal pump that adds head to a fluid as it is accelerated from its center toward its outer edge.

Implosion: A term describing the rapid and violent collapse of vapor pockets as the pressure around them rises above the vapor pressure of the fluid.

Incompressible: The ability of liquids to transmit large pressures without themselves being compressed into smaller volumes.

Indirectly-fired storage water heater: An insulated storage tank that contains an internal heat exchanger for heating the domestic water in the tank.

Infiltration: The unintentional leakage of outside air into heated space.

Infrared radiation: Another name for thermal radiation. It is technically electromagnetic radiation having wavelengths longer than can be seen by the eye.

Injection mixing control: A method of controlling the temperature in a hydronic distribution system by adding hot water in pulses.

Inline circulator: A circulator constructed such that its inlet and outlet ports lie on a common centerline.

Interpolation: A mathematical method of estimating a numerical value that is known to be between two other stated values. It is based on proportioning.

Inter-zone heat transfer: Heat movement across interior partitions, floors, and ceilings separating building areas that are kept at different temperatures.

Isolation flanges: Special flanges that contain a built-in shutoff valve.

Iteration: A design process that entails making successively refined estimates until a mathematically stable operating condition is determined.

Kick-space heater: A small horizontal fan-coil convector that mounts in the recessed space at the base of a kitchen or bathroom cabinet.

Ladder diagram: Electrical schematic drawings that are used to design and document a control system.

Laminar flow: A classification of fluid flow where streamlines remain parallel as the fluid moves along the pipe. Laminar flow is more typical at very low flow velocities.

Latent heat: Heat that is added or removed from a substance without any temperature change. This occurs as a material changes phase from either a solid to a liquid, or a liquid to a vapor.

Line voltage: A term that usually indicates a nominal 120 VAC relative to ground potential.

Load: A term that represents that heating requirement necessary to maintain a desired temperature in a building, room, or a volume of material such as a tank of water.

Lockshield/balancing valve: A valve that mounts on the outlet of a heat emitter and serves as a combination isolation valve, flow balancing valve, and drain valve.

Low voltage: For hydronic systems, most low voltage control circuits operate at 24 VAC.

Make-up water system: An assembly of components that automatically adds water to a hydronic system to make up for minor losses. It usually consists of a shutoff valve, feed water valve, and backflow preventer.

Manifold: A piping component that serves as a common beginning or ending point for two or more parallel piping circuits.

Manifold station: A combination of a supply manifold and return manifold that services two or more floor heating circuits.

Manometer: A device to measure the pressure difference between two points in a piping system.

Manual air vents: Air vents that must be opened and closed manually.

Metered balancing valves: Specialized valves that allow the flow rate in a branch piping circuit to be precisely adjusted to a desired value.

Microbubbles: Very small bubbles that give water a cloudy appearance.

Microbubble resorber: A type of deaerator that separates and collects microbubbles from the system's fluid. It lowers the dissolved air content of the water, enabling it to absorb any residual air in the system.

Microprocessor: A digital electronic device containing logic circuits that can be programmed to execute specific control operations.

Mixing valve: A valve that blends two fluid streams entering at different temperatures to achieve a desired outlet temperature.

MIX program: A program in the Hydronics Design Toolkit that predicts the blended temperature of two or more fluid streams entering a common point in a piping system at different temperatures and flow rates.

MMBtu: An abbreviation for one million Btus.

Modular boiler system: A group of two or more boilers that operate as needed to supply the current heating load of a building.

Modulating control: A control method that allows infinitely variable adjustments of a controlled device to track a variable heating load.

MONOFLO program: A program in the Hydronics Design Toolkit that assists in designing one-pipe distribution systems.

Multiple tube passes: When the tubing that makes up the coil of an air handler is arranged so there are two or more series-connected tube planes in the coil.

Multi-zone/multi-circulator system: A hydronic distribution system that uses a separate circulator for each zone circuit.

Multi-zone relay center: A specialized control consisting of multiple relays and a transformer. This control is used in multi-zone hydronic systems.

Multi-zone/zone valve system: A hydronic distribution system that uses a separate zone valve for each zone circuit. A single circulator supplies the entire system.

National Fuel Gas Code: A model code developed by the National Fire Protection Association that is widely recognized in the U.S. It specifically covers the fuel supply, ventilation, and exhaust requirements of combustion-type heat sources.

National pipe thread (NPT): The standard tapered threads used on piping, fittings, and valves in the U.S.

Natural convection: Heat transfer between a solid surface and a moving fluid when the fluid's motion is created only by natural processes such as the tendency of a warm fluid to rise.

Net Positive Suction Head Available (NPSHA): The total head available to push water into a circulator at a given point in a piping system. This indicator is totally determined by the piping system and fluid it contains, and does not depend on the circulator. It is expressed in feet of head. To avoid cavitation, the NPSHA must be equal to or greater than the NPSHR of the selected circulator.

Net Positive Suction Head Required (NPSHR): The minimum total head required at the inlet of the circulator to prevent cavitation. This value is specified by pump manufacturers.

Normally closed contacts: Electrical contacts in a relay that remain closed when the relay coil is deenergized.

Normally open contacts: Electrical contacts in a relay that remain open when the relay coil is deenergized.

Off-peak periods: Set periods of time, often at night, when electrical energy is sold at a reduced rate by the utility.

One-pipe systems: A hydronic distribution system that uses diverter tees to connect individual heat emitters to a main distribution piping circuit.

On/off control: Attempting to maintain a setpoint temperature by turning heat delivery on and off.

Open loop system: A piping system that either conveys fresh water or is exposed, *at any point*, to the atmosphere.

Operating mode: A specific condition or state of the controlled devices in a system that allows it to accomplish a specific objective. Hydronic heating systems can have several operating modes including off, on, priority domestic water heating, heat purging, etc.

Operating point: The point on a graph where the pump curve intersects the system resistance curve. This point indicates the flow rate at which the system will operate.

Outdoor reset control: A control method that increases the water temperature in a hydronic system as the outdoor temperature drops.

Oxygen diffusion barrier: One or more layers of special compounds that significantly reduce the rate at which oxygen can diffuse through a polymer material.

Packaged boilers: Boilers that are traditionally supplied with factory-mounted components such as high limit control, circulator, pressure relief valve, and oil or gas burners.

Panel radiator: A heat emitter that mounts on a wall and releases the majority of its heat as thermal radiation.

Parallel direct-return system: A hydronic distribution system consisting of two or more branch piping paths that begin from a common supply pipe and end at a common return pipe. The first branch connected to the supply pipe is also the first to connect to the return pipe.

Parallel piping: When two or more branch piping paths originate from a common starting point and end at a common ending point.

Parallel pumps: Two or more pumps that are mounted to a common inlet header and discharge into a common outlet header.

Parallel reverse-return system: A hydronic distribution system consisting of two or more branch piping paths that begin from a common supply pipe and end at a common return pipe. The first branch connected to the supply pipe is the *last* to connect to the return pipe.

Partial storage systems: A thermal storage system that stores only a portion of the energy needed by the load, for a 24-hour period, during the preceding off-peak period.

PEX tubing: An acronym for cross-linked polyethylene tubing.

Pickup allowance: A term used to describe extra boiler capacity allowed for quickly heating up a distribution system following an off cycle.

PIPELOSS program: A program in the Hydronics Design Toolkit that estimates the heat loss of copper tubing.

Pipe size: A term that, unless otherwise specified for pipes under 12 inches in diameter, refers to the approximate *inside* diameter of a pipe.

Piping offset: An assembly of pipe and fittings designed to absorb the expansion and contraction of piping as it changes temperature.

Point of no pressure change: The location where an expansion tank is attached to a hydronic system. The pressure at this point remains unchanged regardless of whether the circulator is on or off.

Poles (of a switch): The number of independent electrical circuits that can be simultaneously passed through a switch or relay.

Polybutylene: A polymer material suitable for fabrication into tubing, fittings, and valves.

Polymer: A hydrocarbon substance consisting of long molecular chains. Many polymers are commonly called plastics. Common examples include polyethylene, polybutylene, and polypropylene.

Power-venting: Using an electrically driven blower rather than a chimney to create the negative pressure required to safely vent combustion products from a heat source to outside air.

Pressure head: The mechanical energy possessed by a fluid due to its pressure.

Pressure reducing valve: A special valve that reduces the pressure of water as it flows from a water supply system into a hydronic system. This valve is also called a feed-water valve.

Pressure relief valve: A spring-load valve that opens to release fluid from the system whenever the pressure at its location exceeds its rated opening pressure.

Primary circuit: A main distribution circuit tied to the system's heat source, and supplying several secondary circuits.

Primary/secondary piping: A method of interfacing individual distribution circuits to a central main distribution circuit. The method involves the use of closely spaced tees in the primary circuit.

Primary side: The side of a transformer connected to line voltage.

Priority load: A control strategy in which some heating loads are temporarily suspended while the priority load is heated.

Pump (centrifugal): An electrically driven device that uses a rotating impeller to add mechanical energy (e.g., head) to a fluid. In this text the word pump refers specifically to a centrifugal pump, and is used synonymously with the word circulator.

PUMPCURV program: A program in the Hydronics Design Toolkit that generates a pump curve based on flow rate and head data.

Pump curve: A graph indicating the head energy added to the fluid by the pump versus the flow rate of the fluid through the pump. Such curves are provided by pump manufacturers.

Pump efficiency: The efficiency of transferring mechanical energy from the pump's shaft to the fluid.

Pump head: The head added to a fluid by the pump at some particular flow rate.

PUMP/SYS program: A program in the Hydronics Design Toolkit that determines the intersection of a specified pump curve and a specified system resistance curve.

Purge cart: A tool based on a high capacity pump that can purge air from a system, filter its water, and also add antifreeze to the system.

Purge valve: A specialized valve that allows easy forced-water purging of a hydronic system.

Push/pull arrangement: Pumps that are mounted in series with each other, with the same flow direction, but not in direct contact with each other.

RADFLOOR program: A program in the Hydronics Design Toolkit that predicts the outlet temperature and heat output of floor heating circuits.

Radiant baseboard: An extruded aluminum plate with integral tubes that mounts at the base of walls, and releases the majority of its heat as thermal radiation.

Radiator: A generic term for heat emitters that transfer the majority of their heat output by thermal radiation.

Radiator valve: A valve that regulates the flow of water through an individual heat emitter.

Refrigeration cycle: The process of moving heat from a region of lower temperature to one of higher temperature by repeatedly expanding and compressing a compound called the refrigerant.

Relay: An electrically operated switch.

Relay socket: A base a relay plugs into, and at which external wires leading to the relay are terminated.

Reset line: A line having a specific slope on a graph of supply water temperature versus outdoor air temperature.

Reset ratio: The ratio of the change in supply water temperature to the change in outdoor air temperature that a reset control will attempt to maintain. The reset ratio is also the mathematical slope of a reset line.

Reynold's number: A unitless number that can be used to predict whether flow will be laminar or turbulent.

Room air temperature profile: A graph showing how room air temperature varies from floor to ceiling.

Room heating loads: The rate at which heat must be added to an individual room to maintain interior comfort.

Room temperature unit (RTU): A device that detects when a room is sufficiently heated and prevents the heating system from adding further heat.

Room thermostat: An adjustable, temperature-operated switch, mounted in a room, that turns heat delivery on and off to regulate room temperature.

Run fraction: The ratio of the on time of a heat source to the total elapsed time. In general, short run fractions are undesirable.

Secondary circuit: An individual distribution piping circuit with its own circulator that receives heat from the building's primary circuit.

Secondary side: The side of a transformer connected to lower (e.g., control) voltage.

Sectional boiler: A boiler assembled from cast-iron sections.

Sections: The cast-iron assemblies that are joined together to form a boiler block.

Sensible heat: Heat, that when added or removed from a material, is evidenced by a change in the temperature of the material.

SERIESBB program: A program in the Hydronics Design Toolkit that determines the proper size of finned-tube baseboard convectors arranged in a series piping circuit.

Series circuit: An assembly of pipe and piping components that forms a complete loop.

Series piping path: An assembly of pipe and piping components connected end to end to form a path between two points.

Series pumps: Two or more pumps connected end to end and having the same flow direction.

Setback: Reducing the temperature setting of a setpoint control, such as a room thermostat, to conserve energy.

Setpoint control: A control that attempts to maintain a preset temperature at some location in a system by controlling the operation of other devices.

Setpoint temperature: A desired temperature that is to be maintained if possible, and is set on some type of control such as a thermostat.

Shaft seals: Metal or synthetic components of a pump that prevent fluid from leaking out where the impeller shaft penetrates the volute.

Single series circuit: A hydronic distribution system in which all heat emitters are connected inline with the distribution piping. The full system flow rate passes through each heat emitter.

Sink: In the context of a heat pump, the sink refers to the material the high temperature heat is released into.

Slab resistance: The thermal resistance offered by the combination of the tube wall and concrete surrounding the tube in a floor heating system.

Smooth pipe: A generic term for tubing with relatively smooth inner walls. Examples would include drawn copper tube, polybutylene tubing, and PEX tubing.

Socket fusion: The process of joining polybutylene piping by heating the inner surface of a fitting and the outer surface of the piping, then pushing the heated pieces together.

Soft temper tubing: Copper tubing that is sold in coils and can be bent with relative ease.

Source: In the context of a heat pump, the source refers to the material from which the low temperature heat is extracted.

Space heating: The process of adding heat to a building or room to maintain indoor comfort.

Specific heat: A material property indicating the amount of heat required to raise one pound of the material by one degree Fahrenheit, often expressed in Btu/lb/°F.

Spring-loaded check valve: A check valve that contains a spring to close the valve regardless of its mounting position.

Stack effect: The tendency of warm air to move upward in an enclosure similar to how smoke rises through a chimney.

Staged control: Using two or more stages or rates of heat input to maintain a setpoint temperature.

Standard expansion tank: An expansion tank that does not have a diaphragm.

Static pressure: The pressure at some point in a piping system measured while the fluid is at rest. Static pressure increases from a minimum at the top of a system to a maximum at the bottom. Static pressure is created by the weight of the fluid itself, plus any extra pressurization applied at the top of the fluid.

Stationary air pockets: Air that accumulates at the high points of piping systems and creates air binding.

Steady state efficiency: The efficiency of converting fuel into heat when a heat source operates continuously with nonvarying input and output conditions.

Stratification: The natural tendency for a warm fluid to rise above a cooler fluid due to density differences. This term is commonly used to describe warm air rising toward the ceiling of a room while cooler air descends toward the floor.

Swing check: A check valve with an internal disc that swings on a pivot.

System heat output curve: A graph that shows the total heat output of a hydronic distribution system based on the average water temperature supplied to it.

System resistance curve: A curve plotted on a graph that indicates the head loss of a piping system versus the flow rate through it.

Tankless coil: A finned copper coil that is inserted into a boiler to provide domestic hot water.

Telestat: A specialized type of electrically operated heat motor designed to screw onto a supply manifold.

Template block: A wooden block fabricated to allow heating tubing to neatly penetrate the surface of a floor slab.

Therm: A quantity of energy equal to 100,000 Btus. The therm is the traditional unit in which natural gas is sold.

Thermal break: The placement of insulation between two materials to prevent rapid heat transfer between them.

Thermal conductivity: A physical property of a material that indicates how fast heat can move through the material by conduction. The higher the thermal conductivity, the faster heat can move through the material.

Thermal energy: Energy in the form of heat. The term thermal energy is used synonymously with the word heat.

Thermal envelope: The shell formed by all building surfaces that separates heated space from unheated space. Exterior walls, windows, and exposed ceilings are all parts of the thermal envelope of a building.

Thermal equilibrium: The condition of a system when the rate of energy loss from the system is exactly the same as the rate of energy input to the system.

Thermal expansion: The natural tendency of materials to expand as their temperature increases.

Thermal mass: The tendency of a material to store heat due to its mass and specific heat.

Thermal radiation: Heat that is transferred in the infrared portion of the electromagnetic radiation. This radiation cannot be seen by the eye, but, like visible light, travels in straight lines, and can be absorbed, reflected, and transmitted by various surfaces.

Thermal resistance: The tendency of a material to resist heat flow. Insulation materials, for example, tend to have high thermal resistance. Other materials such as copper and glass have low thermal resistance.

Thermistor: A solid state temperature sensor that changes resistance as its temperature changes.

Thermosiphon: A term describing the natural tendency of heated water to rise in a hydronic system, even when the circulator is off.

Thermostatic mixing valve: A mixing valve with its own internal thermostat.

Thermostatic operator: A nonelectric device that mounts to a radiator valve and opens and closes the valve as necessary to maintain a set room temperature.

Thermostatic radiator valve: The assembly of a thermostatic operator and a radiator valve.

Thin slab: A slab of gypsum-based underlayment or lightweight concrete poured over floor heating tubing. Typical thin slab thicknesses range from 1.25 inches to 1.5 inches.

Three-piece circulator: A circulator in which the wetted parts are separated from the motor by a coupling assembly.

Throws (of a switch): The number of settings of a switch or relay for which a circuit can be completed through the switch or relay.

Time delay relay: A relay that's contacts do not open or close at the same time coil voltage is applied or interrupted.

Time-of-use rate: A rate structure offered by an electric utility that bases the price of electricity on the time of day it is supplied.

Ton (of capacity): A term representing a rate of energy flow of 12,000 Btu/hr. It is frequently used to describe the capacity of heat pumps and air conditioners.

Total equivalent length: The sum of the equivalent lengths of all piping, fittings, valves, and other components in a piping circuit.

Total head: The total mechanical energy content of a fluid at some point in a piping system.

Total R-value of an assembly: The overall total thermal resistance of an assembly of materials that are fastened together.

Towel warmer: A specialty radiator that is designed to have towels draped over it for warming and drying.

Transformer: An electrical component for reducing or increasing voltage in an AC circuit. Its common application in hydronic systems is to reduce line voltage (120 VAC) to control voltage (24 VAC).

Triple action control: A preassembled control that provides high limit, low limit, and circulator control in hydronic systems using tankless water heaters.

Turbulent flow: A classification of flow where streamlines repeatedly cross over each other as the fluid flows along the pipe. Turbulent flow is the dominant type of flow in hydronic heating systems.

Unconfined space: A space in a building that has a minimum of 50 cubic feet of volume per 1,000 Btu/hr of gas input rating of all equipment in that space.

Unit heater: A generic name for a fan-coil that is usually mounted near the ceiling of a room, with downward directed air flow.

Unsaturated state of air solubility: When the amount of dissolved air in the system water is low enough to enable the water to resorb any residual air trapped in other parts of the system.

Usage profile: A graph that shows the quantity of domestic hot water required at any hour of the day.

Vapor pressure: The minimum (absolute) pressure that must be maintained on a liquid to prevent it from changing to a vapor. The vapor pressure of most liquids increases as their temperature increases.

VA rating: Volt-amp rating. A common output rating for control transformers.

Variable speed injection mixing: A method of controlling the temperature in a hydronic heating system by using a variable speed pump to add hot water to a continuously circulating distribution system.

Velocity head: The mechanical energy possessed by a fluid due to its motion.

Velocity profile: A sketch that shows how the velocity of the fluid varies from the centerline of a pipe out to the pipe wall.

Vertical panel radiator: A panel radiator that is taller than it is wide, and has tubes oriented in a vertical direction.

Viscosity: A means of expressing the natural tendency of a fluid to resist flow.

Volute: The chamber that surrounds the impeller in a centrifugal pump.

Warm weather shut down: A control mode in which the system's circulator is automatically turned off during warm weather.

Water-logged: A condition in which the captive air in an expansion tank is replaced by water.

Wet base boiler: A term used to describe boiler sections that totally surround the combustion chamber.

Wet rotor circulator: A circulator in which the motor's armature and its impeller are an integral unit that is surrounded and cooled by the fluid passing through the circulator.

Working pressure: The allowable upper pressure limit at which a pipe or piping component can be used.

Zone header: A preassembled combination of two or more zone valves and associated heat motors that is designed to be mounted as a single unit.

Zone valve: Valves used to allow or prevent the flow of hot water from a heat source through individual zone distribution circuits.

Zoning: Dividing a hydronic distribution system into two or more independently controlled distribution circuits.

INDEX

Hydronics Design Toolkit User's Manual

MARIO RESTIVE

JOHN SIEGENTHALER, P.E.

Hydronics Design Toolkit User's Manual

MARIO RESTIVE
JOHN SIEGENTHALER, P.E.

HYDRONICS DESIGN TOOLKIT

User's Manual

Mario Restive and John Siegenthaler
Mohawk Valley Community College
Utica, New York

CONTENTS

TERMS AND CONDITIONS

The following terms and conditions apply to the use of the Hydronics Design Toolkit. Any variation of these terms or conditions must be approved by Delmar Publishers in writing.

ALLOWED USES OF THE HYDRONICS DESIGN TOOLKIT

1. You may use the product on only one computer at a time.

2. You may make a single copy of the software for backup or archival purposes, provided all copyright notices are reproduced on the copy.

3. You may NOT distribute, loan, rent, lease, resell for profit, modify, adapt, or translate the product, nor create derivative works based on the product, or any part thereof.

4. You may NOT transfer by any means, printed or electronic, copies of the product to another person, or remove any copyright notices or other identification from the product.

5. You may NOT decompile, reverse engineer, disassemble, or otherwise reduce the software to a human perceivable form.

LIMITED WARRANTY

Delmar Publishers warrants to the original purchaser that the disk on which the Hydronics Design Toolkit program is supplied will be free from defects in materials and workmanship under normal use, for a period of 90 days from the date of purchase. At its option, Delmar Publishers will either replace the disk or refund the purchase price.

The foregoing is in lieu of all other warranties, expressed or implied, including the warranties of merchantability and fitness for a particular purpose. Your sole remedy is replacement of a defective disk as explained previously.

Some states do not allow the exclusion of implied warranties, so the above exclusion may not apply to you. This warranty gives you specific rights; other rights vary from state to state.

LIMITATION OF LIABILITY

This product is sold *as is*. Delmar Publishers or the software authors shall not be held liable for any direct, indirect, incidental, or consequential damages such as, but not limited to, loss of profits or benefits resulting from the use of the product, even if Delmar Publishers or the authors have been informed of the possibility of such damages. The user assumes all responsibilities for any decisions made or actions taken based on information obtained using the product.

Some states do not allow the limitation or exclusion of liability for incidental or consequential damages so the previous limitation may not apply to you.

INTRODUCTION AND STARTUP

1.1 INTRODUCTION

The Hydronics Design Toolkit is a compilation of 19 individual programs that assist in the design of hydronic heating systems for residential and light commercial buildings. These programs are based on the methods and data in Modern Hydronic Heating for Residential and Light Commercial Buildings, by John Siegenthaler (Delmar Publishers).

The toolkit is designed as a teaching tool for use in a course on hydronic heating, as well as a design assistant for those working in the trade. It greatly expands the ability of a system designer to explore design options using a "what if" approach. Design decisions that effect both the cost, performance, and life span of a hydronic system can be investigated.

The software should be thought of as the design equivalent of an installation toolbox. For example, during the installation of a hydronic system, an installer uses tools such as a wrench, screwdriver, measuring tape, and so forth, for specific small tasks. An experienced installer quickly determines which tools are needed and the order in which to use them. The installation process requires frequent swapping of tools. The collection of tools in the toolbox, and the sequence in which they are used, allows the installer to complete the job.

Likewise, the Hydronics Design Toolkit contains individual *design tools* that perform a specific, but limited, task. However, with a small investment of learning time, the designer will quickly see how to move in and out of various design tools so even complex hydronic systems can be designed. This approach affords considerably more flexibility than would be possible with a single program that completely designs a certain type of system from beginning to end but cannot be modified as required for special circumstances.

1.2 COMPUTER REQUIREMENTS

The Hydronics Design Toolkit runs on MS-DOS-compatible personal computers running DOS 3.0 or higher, with a minimum of 512K RAM. The program supports the following video display formats: Hercules Monochrome Graphics Card (HGC), CGA, EGA, VGA, and SVGA. The program does not require a coprocessor but will use one if present. The program does not support input from a mouse. The colors used for the display screens were selected so that the program can also be used on a laptop PC with a monochrome LCD display.

1.3 MAKING A BACKUP COPY OF THE PROGRAM

Before using the Hydronics Design Toolkit, be sure to make a backup copy of the program disk. For systems with have two floppy disk drives (A: drive and B: drive), use the following procedure:

1. Insert the original 3.5" program disk into the A: drive, and close the lever.

2. Insert a blank, formatted disk into the B: drive, and close the lever.

3. Type copy A:*.* B: being sure to leave a blank space before the drive designations *A* and *B*. Press return.

Note: If the drives have other designations, substitute these designations into this procedure.

If your computer has only one floppy disk drive, use the following procedure:

1. At the A: > prompt type *diskcopy A: A:* being sure to leave a space before the second *A*.

2. The computer will prompt you to insert the source disk, which is the original program disk. After reading some data from the source disk, the computer will prompt you to insert the target disk, which is the disk to which you are copying. Remove the source disk from the drive and insert a blank disk. Press return, and

continue to follow the screen prompts, exchanging the disks as necessary.

This procedure will format the blank disk, and make an exact duplicate of the original program disk. When the process is finished, remove the target disk and attach a label. By law, the copyright notice must be written on this label. Store the original program disk in a safe location, and use the duplicate disk for routine work with the program.

1.4 INSTALLING THE PROGRAM ON A HARD DISK

The most convenient way to run the toolkit is from a hard disk. This method allows slightly faster program operation than other methods.

For the following installation procedure, it is assumed that the hard disk drive to which the Hydronics Design Toolkit will be written is designated as the C: drive. It is also assumed that the system has a 3.5" floppy disk drive designated as the A: drive. If the disk drives are designated otherwise, substitute the appropriate drive names into the installation procedure.

1. Start up your computer and go to the C: prompt.
2. Type *MD\HDT* and press return.
3. Type *CD\HDT* and press return.
4. Insert the original program disk into the A: drive, and close the lever.

5. Type *copy A:*.* C:* and press return. Be sure to leave a blank space before the drive designations A and C.

This procedure creates a new directory on your hard drive named HDT (short for Hydronics Design Toolkit). It then installs all files on the original program disk into this directory.

Note: If you choose to install the files in another directory on the hard disk, be sure all files on the original disk are copied into the same directory as the program (executable) file.

1.5 STARTING THE PROGRAM FROM A FLOPPY DISK

After the system has been booted, insert the program disk into the 3.5" floppy disk drive and close the lever (if there is one). Make sure this drive is logged as the currently active drive. For example, if you inserted the disk into the A: drive, be sure the A: > prompt is being displayed next to the blinking screen cursor. If it is not, type *A:* and press return to make it the active disk drive.

Type *TOOLKIT* and press return. In a short time, the startup screen should appear on your display as shown in Figure 1-1. Pressing any key at this point will bring up the main menu screen shown in Figure 1-2. The program is now ready to go to work.

```
          THE HYDRONICS DESIGN TOOLKIT 1.0




   By   Mario J. Restive
        John Siegenthaler
        Mohawk Valley Community College




   No part of this software or its documentation may be copied
   without the written permission of Delmar Publishers.
   Copyright 1995, Delmar Publishers, All rights reserved.

           PRESS ANY KEY TO CONTINUE.
```

Figure 1-1 The startup screen.

```
MAIN MENU  - HYDRONICS DESIGN TOOLKIT -    F1-HELP    F10-EXIT
┌─────────────────────────────────────────────────────────────────┐
│  1. ROOMLOAD    Calculates design heating load of a room          │
│  2. BSMTLOAD    Calculates design heating load of a basement       │
│  3. PIPEPATH    Determines characteristics of user-defined piping path│
│  4. PUMPCURV    Generates a pump curve from manufacturer's data    │
│  5. PUMP/SYS    Determines intersection of pump and system curves  │
│  6. SERIESBB    Analyzes a series string of finned-tube baseboards │
│  7. PIPELOSS    Calculates heat loss from a pipe                   │
│  8. MONOFLO     Calculates flow division through Monoflo fittings  │
│  9. MIX         Calculates blended temperature of merging fluid streams│
│ 10. HYRES       Calculates the hydraulic resistance of a device   │
│ 11. EXPTANK     Calculates size and pressurization of an expansion tank│
│ 12. PARALLEL    Calculates equivalent resistance of parallel piping paths│
│ 13. RADFLOOR    Finds heat output of a specified floor heating circuit│
│ 14. FLUIDS      Calculates the properties of several fluids        │
│ 15. PIPESIZE    Selects pipe size based on flow rate and flow velocity│
│ 16. BINEFF      Calculates true seasonal efficiency of a boiler    │
│ 17. DHWLOAD     Calculates domestic hot water energy usage         │
│ 18. FUEL COST   Calculates cost of delivered energy                │
│ 19. ECONOMICS   Calculates total owning and operating cost of a system│
└─────────────────────────────────────────────────────────────────┘
┌─MESSAGE─────────────────────────────────────────────────────────┐
│ Use the UP or DOWN ARROW keys to select a procedure and press <Enter>.│
└─────────────────────────────────────────────────────────────────┘
```

Figure 1-2 The main menu screen.

If you have difficulty starting the Hydronics Design Toolkit using this procedure, check your DOS user's manual for possible differences in procedure.

Note: When the program is run from the floppy disk it is important that the disk remain in the drive while the program is being used. The program will need to read the disk whenever the help function is used or other data files are needed.

1.6 STARTING THE PROGRAM FROM A HARD DISK

After the system is booted, go to the directory in which the toolkit was installed. If you followed the installation procedure described in Section 1.4, all program files should be installed in a directory named HDT. You can go to this directory by typing *CD\ HDT* at the C: > prompt and pressing return. If you did not follow this procedure, consult your DOS manual for the appropriate path routing procedure.

Once in the appropriate directory, type *TOOLKIT* and press return. In a short time, the startup screen shown in Figure 1-1 should appear on your display. Pressing any key at this point will bring up the main menu screen shown in Figure 1-2. The program is now ready to go to work.

PROGRAM STRUCTURE 2

2.1 THE MAIN MENU SCREEN

The main menu screen shown in Figure 1-2 is the "hub" from which the user may select any of the 19 programs in the toolkit. A short description of each program follows its name on the main menu. For a more detailed description of any one program, use the up or down arrow key to place the white highlighting bar over the program's name, and then press the F1 key. This opens a help screen with a more complete description of the program. After reading this help screen, press any key to return to the main menu screen.

2.2 LAUNCHING A SPECIFIC PROGRAM FROM THE MAIN MENU

To begin working with a given program, use the up or down arrow keys to place the white highlighting bar over the name of the program and press return. The display screen of the selected program will momentarily appear.

2.3 RETURNING TO THE MAIN MENU FROM A SPECIFIC PROGRAM

To return to the main menu from any of the programs, press the F10 key while the main screen of any of the programs is displayed.

The ability to quickly move back and forth between the main menu and any of the individual programs will prove to be a key feature in the usefulness of the toolkit.

2.4 SCREEN LAYOUT

The screens displayed by the programs in the toolkit are divided into four distinct areas, each serving a specific function. These areas are (from top to bottom):

- the F-KEY line
- the INPUT area
- the RESULTS area
- the MESSAGE area

These areas are illustrated in Figure 2-1.

The F-KEY line at the very top of a program's screen lists the F-keys that are active within that program. To a large extent, the function of the F-keys is consistent from one program to the next with some minor exceptions. For example, pressing the F1 key will always display a help screen applicable to the line on which the white highlighting bar is located. The F2 key is always used to bring up a file access window for loading or saving files to disk.

The INPUT area of the screen is where information the program needs is entered or selected.

In the RESULTS area, all calculated results are displayed.

In the MESSAGE area, the user reads a prompt describing what actions are appropriate at that point in the program.

2.5 EDIT TABLES

Two of the 19 programs, PIPEPATH and SERIES-BB, have a special type of input screen in addition to their main screen. This special input screen is called an edit table. It automatically appears whenever the white highlighting bar is moved over the bottom line of the input area designated by the prompt line beginning with the words EDIT TABLE.

2.6 HELP SCREENS

A major feature of the Hydronics Design Toolkit is an extensive system of context-sensitive help screens. These screens should be thought of as a mini-user manual on disk. They contain text information related to the particular line on which the white highlighting bar is located. The appropriate help screen can be viewed at any time by pressing the F1 key. After reading the

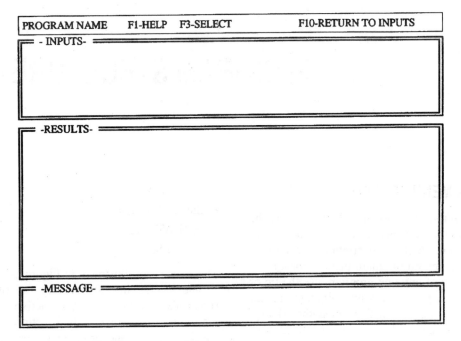

Figure 2-1 The four areas of the program screens.

help screen, press any key to return to the same point from which the help screen was opened.

2.7 PRINTING

Almost all screens within the program are text-based screens that can be sent to a printer by pressing the print screen key while the screen is displayed. Make sure the printer is turned on and is online before pressing the print screen key.

Note: All of the screens in the Hydronics Design Toolkit use special characters to create borders and other effects. For these screens to print properly, the printer must be set to print the upper ASCII character set. Nearly all printers can create these characters provided they are properly configured by the user. In some cases this requires the setting of DIP switches within the printer. In other cases, it may require the printer to emulate another printer. Consult the owner's manual for your printer for instructions on how to set it to print these characters. If, after printing a screen, you notice the borders or other characters on the screen do not appear as they do on the display, the printer is not properly set for the upper ASCII characters.

After pressing the print screen key, it may be necessary, on some printers, to perform a form feed to advance the paper through the printer. Consult the owner's manual for your printer for specific instructions on performing a form feed.

Note: The graphs created by the PIPEPATH, PUMPCURV, and PUMP/SYS programs are not totally text-based screens and, therefore, cannot be printed using the print screen key.

2.8 PROGRAM STRUCTURE CHART

The chart shown in Figure 2-2 depicts the overall structure of the Hydronics Design Toolkit. All 19 programs are accessible from the main menu. All programs have their own context-sensitive help files. All programs also have built-in error detection and recovery systems and can generate printed output using the print screen key (with the exception of graphs).

The PIPEPATH and PUMPCURV programs can create disk files. The PUMP/SYS program can read these disk files.

The solid black arrows show possible paths by which results from one program may be used as input to another program as the analysis of a system progresses. This exchange of data will be discussed more thoroughly in Section 5.

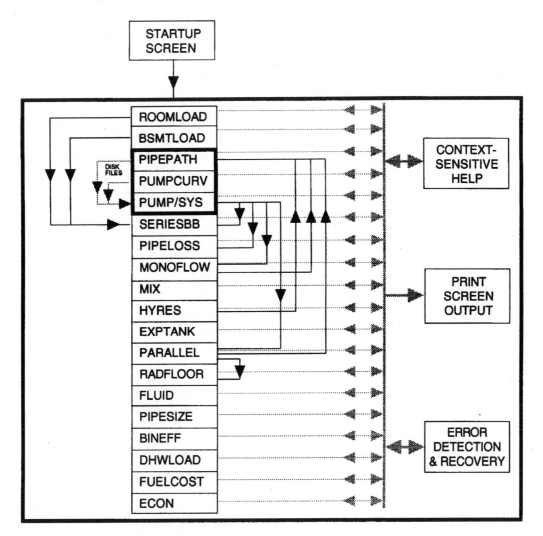

Figure 2-2 Program structure chart.

ENTERING INFORMATION INTO A PROGRAM 3

3.1 INPUT TYPES

The input area of the various programs contains three possible types of inputs. Each input type is identified by the first word in the prompt message for the input line. The three key words are:

- ENTER
- SELECT
- QUESTION

These words are used to maintain consistency throughout the input screens, as well as to minimize the chance of erroneous data being entered. Each type requires a different form of response from the user.

3.2 ENTER-TYPE INPUTS

Any input line beginning with the word *ENTER* allows the user to type in a number within the range specified in parentheses at the end of the input text line. Only values within this range will be accepted by the program. Should the user attempt to enter a value outside the designated range, the program will beep and the display will return to the original value.

Because of space limitations, some programs do not show the allowable range of each input. However, the range protection is still active; the maximum and minimum values of the range are listed in the help screen associated with the particular input line.

The value on an ENTER-type input line when the screen first appears is called a default value. Such default values exist for all ENTER-type inputs in the program. They have been carefully selected to help minimize input time for routine systems. They also ensure the program will not stop at any point due to lack of data. If the default value appearing on a particular line when the screen appears is appropriate for the system currently being analyzed, there is no need to do anything at that input line.

To enter a different value on any lines beginning with ENTER, use the arrow keys to move the white highlighting bar over the current value, and then type in the new value. There is no need to delete the current value first, since the program will automatically overtype it with the new value. If desired, the delete and backspace keys can be used to make changes to the value. Once the desired number has been typed in, the program will test its value for acceptance when any of the following take place:

- the user presses return
- the user attempts to move the highlighting bar using the arrow keys
- the user attempts to move to a different screen

If the program determines the input is either invalid or out of range, it will beep and the original value will reappear. The user should then enter another value.

3.3 SELECT-TYPE INPUTS

Any input line beginning with the word SELECT allows the user to choose a selection from a pop-up selection window. Such selections may be for pipe sizes, types of fluid, fittings, or even geographic locations. Only selections appearing within the associated pop-up windows may be used by the program.

A default selection will always appear on each SELECT-type input line when the program is started. If this selection is appropriate for the system being analyzed, there is no need to alter that particular line. If the default selection needs to be changed, the user should position the white highlighting bar over the line, and then press the F3 key to display the associated selection window. The up or down arrow keys are then used to place the white highlighting bar over the desired selection. Pressing the return key accepts the selection and returns the user to the input screen with the new selection being displayed on the input line.

Most selection windows are completely displayed on a single screen. However, if all selections cannot fit vertically within the screen, the word *MORE* will appear on the lower border of the selection window. In such cases, use the page up or page down keys to view the remaining selections.

If the user has opened one of the pop-up selection windows and then decides not to make any change, pressing the ESC key will return the program to the input line from which the pop-up selection window was opened, with the original selection still in place.

3.4 QUESTION-TYPE INPUTS

Any input line beginning with the word *QUESTION* allows the user to type in only one of two possible inputs: a *Y* for yes, or an *N* for no. Only these two characters will be accepted as valid answers to the question.

3.5 THE FUNCTION KEYS

This section briefly describes the operation of the function keys. In most cases, the action initiated by a function key remains the same from one program to the next. However, some exceptions follow.

F1-HELP

By pressing the F1 key, the user can access the on-line help windows built into the program. Each help window gives a brief summary of the particular input or output line from which it was accessed, and directions on how to proceed at that point in the program. After reading the help window, press any key to return to the program.

F2-FILE

Pressing F2 from any of the program screens on which it appears opens a file access window over that screen. Within this window are prompt lines containing the words *LOAD*, *SAVE*, and *EXIT*.

If you press F2 and then decide not to save or load a file, use the arrow keys to place the white highlighting bar over the word *EXIT* and press return. This returns you back to the program screen.

To save all data in the current program as a disk file, use the up or down arrow keys to place the white highlighting bar over the SAVE line and press return. The program will open another window asking you to name the file. Type in a file name up to eight characters long, containing only letters or numerals, and then press return. The program automatically tags the file name with an extension and saves it to the currently logged disk. A small message will momentarily appear near the bottom of the screen indicating the file has been saved. The new file name will appear in the upper right corner of the screen. If saving to a floppy disk, be sure a formatted disk is in the drive BEFORE attempting to save a file.

To load a previously created file into a program, press the F2 key to open the file access window. Use the arrow keys to move the white highlighting bar over the prompt line containing the word *LOAD* and press return. Another window containing the names of all files the program can open (on the currently active disk) will appear. Again, use the up or down arrow keys to move the white highlighting bar over the desired file name and press return. All data from the file will now be loaded into the program, and the file access window will disappear.

F3-SELECT

Pressing F3 whenever the white highlighting bar is positioned over a SELECT-type input opens the associated selection window.

F4-GRAPH

Three of the programs—PIPEPATH, PUMPCURV, and PUMP/SYS—can display graphs. Within these programs, the graphs may be viewed by pressing the F4 key.

F4-FIND RESULTS

Within the program SERIESBB, pressing the F4 key will run the program with the currently specified inputs. This is the only program in the toolkit for which the F4 key is used for this function.

F5-ZERO ALL

This key is used in the PIPEPATH program only. Its function is to zero all entries in a large array of numbers, much like the clear key zeroes a calculator.

F10-EXIT TO MAIN MENU

The F10 key is used to return the user from any program to the main menu. It is also used in some programs to return from a special screen to the main program screen. The prompt in the message area will always indicate when the key is used for this function. This key is also used to return to the DOS prompt when finished using the toolkit. The user will be asked to confirm that they want to quit the program before this final action takes place.

PROGRAM DESCRIPTIONS 4

This section presents a brief description of each program along with a picture of the program's screens. The programs are presented in the same order they appear on the main menu. After reading the description of each program, the reader is encourage to run the program and experiment with its inputs.

4.1 ROOMLOAD

The ROOMLOAD program is a simple tool for obtaining the design heating load of a room using the methods presented in Chapter 2 of the book. It is intended to be used on a repeated basis, room-by-room, until the load of each room of the building is determined. See Figure 4-1. These results can later be used to help design a hydronic system.

The program works somewhat like a spreadsheet in that results are updated instantly whenever any of the inputs are changed. The design heating load of the room currently specified by the inputs, always is displayed in the results area. A bar graph is displayed on the right side of the screen. The bars are aligned with the corresponding input lines on the left side of the screen and indicate the percent of the room's design heat loss that each component (windows, doors, walls, infiltration, etc.) represents. The length of the bars also changes whenever an input is changed. This gives the user instant feedback on how the heating load of the currently specified room is proportioned.

After determining the design heating load of a given room, it is a good idea to print out the screen before proceeding to the next room. Begin the next room by entering a new room name at the top of the input screen. This name is very helpful for later identification of multiple, printed outputs. In many cases, some or all of the R-value data from one room also will apply to other rooms and thus does not need to be changed.

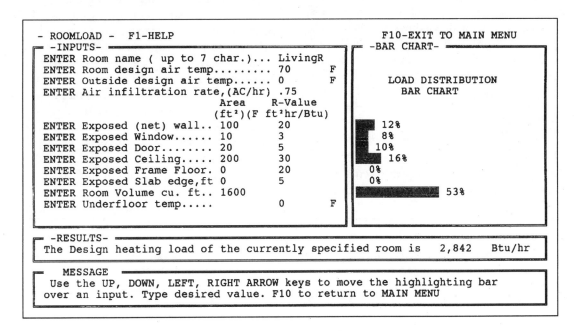

Figure 4-1 Example of a screen from the ROOMLOAD program.

```
- BSMTLOAD -  F1-HELP                          F10-EXIT TO MAIN MENU
  -INPUTS-
  ENTER Basement design air temp,(40 to 85)..............  65    deg. F
  ENTER Outside design air temp,(-30 to 40)..............   0    deg. F
  ENTER Air infiltration rate, (0.1 to 2.0)..............  .2    AC/HR.
  ENTER Basement wall height,(6 to 12)...................   8    ft
  ENTER Height of wall exposed above grade,(.5 to 6)......   1    ft
  ENTER Perimeter of exposed basement wall, (10 to 400).... 140   ft.
  ENTER R-value of additional wall insulation,(0 to 25)....  10   F ft²hr/Btu
  ENTER Floor area of basement, (100 to 4000)............. 1200   ft²

                                              Area          R-Value
                                              (ft²)       (F ft²hr/Btu)
  ENTER Exposed framed wall..............     0              20
  ENTER Exposed window..................    10               3
  ENTER Exposed door....................    20               5
  ENTER Exposed slab edge... (linear ft.)    0               5

  -RESULTS-
  The Design heating load of the currently specified basement   9,099 Btu/hr

    MESSAGE
  Use the UP, DOWN, LEFT, RIGHT ARROW keys to move the highlighting bar
  over an input. Type desired value. F10 to return to MAIN MENU
```

Figure 4-2 Example of a screen from the BSMTLOAD program.

4.2 BSMTLOAD

The BSMTLOAD program is a companion to the ROOMLOAD program. See Figure 4-2. It is used to estimate the design heating load of a basement using the methods presented in Chapter 2 of the book. It allows design heating loads to be calculated for basements having both standard concrete or masonry walls, as well as framed walls with windows and doors. The latter type of walls are often present in a "walk-out" basement design. The design heat load of the basement, displayed in the results area, is instantly updated whenever any of the inputs are changed.

4.3 PIPEPATH

The PIPEPATH program allows the user to completely specify a hydronic piping path or circuit containing a wide variety of smooth pipe, fittings, valves, and even custom components. Using methods presented in Chapter 6 of the text, the program determines the head loss and associated pressure drop through the specified piping path at the specified flow rate. It also determines the hydraulic resistance of the specified piping path. This resistance can be used in other toolkit programs such as PARALLEL and MONOFLO.

Other results include the Reynolds number in the largest and smallest specified piping. In addition to water, the program can work with various concentrations of both ethylene glycol and propylene glycol. A screen shot of the main screen of the PIPEPATH program is shown in Figure 4-3a.

Placing the white highlighting bar on the bottom input line beginning with the words EDIT TABLE, automatically opens an extensive table of pipe, fittings, valves, and other components for copper pipe sizes from 1/2" through 2". The arrow keys move the white highlighting bar around this table. When the highlighting bar is positioned in the appropriate row for a given fitting or valve and under the desired pipe size column, the user simply types in the number of such components in the piping path being described. Recall that the F5 key can be used to quickly zero the count of all such components. This feature is helpful when beginning to specify a new piping path, or when a predetermined hydraulic resistance will be the only resistance describing the piping path. A screen shot of the edit table is shown in Figure 4-3b.

Each input is range protected to prevent invalid or unrealistic numbers from being entered. Each line also has an associated help screen that can be accessed by pressing the F1 key.

```
- PIPEPATH -  F1-HELP  F2-FILE  F4-GRAPH    F10-EXIT TO MAIN MENU File: Samplpip
┌─ -INPUTS- ══════════════════════════════════════════════════════════════════
│ ENTER Average fluid temperature,(80-250)....................  180     deg.F.
│ ENTER Flow rate (if known) otherwise ignore input,(0.25-90).  1       gpm
│ ENTER Additional Hydraulic Resistance,(0-5.0)...............  0
│ SELECT Fluid type (press F3)................................  water
│ EDIT TABLE OF PIPE, FITTINGS, AND VALVES....................  ▓Go To▓
│
┌─ -RESULTS- ═══════════════════════════════════════════
│ Head loss of piping path     =     0.28    ft. head
│ Pressure drop of piping path =     0.12    psi
│ Hydraulic Resistance         =     0.275
│ Reynolds No. largest pipe    =    10,806
│ Reynolds No. Smallest pipe   =    10,806
│
│
│
│
┌─ MESSAGE ═════════════════════════════════════════════════════════════════
│ Use the UP or DOWN keys to move the highlighting bar over an input. For
│ ENTER inputs type desired value. SELECT inputs press F3.
```

Figure 4-3a Example of the main screen in the PIPEPATH program.

```
- PIPEPATH -  F1-HELP      F5-ZERO ALL       F10-RETURN TO INPUTS  File: Samplpip
┌─ -EDIT TABLE- ═══════════════════════════════════════════════════════════════
│                        1/2"   3/4"   1"   1 1/4"  1 1/2"   2"   5/8"PEX
│ Straight pipe(lin. ft.) 0     100    0     0       0       0      0
│ 90 degree ells          0      0     0     0       0       0
│ 45 degree ells          0      0     0     0       0       0
│ Tees (straight through) 0      0     0     0       0       0
│ Tees (side port)        0      0     0     0       0       0
│ Monoflo tees                   0     0     0       0       0
│ Gate valves             0      0     0     0       0       0
│ Globe valves            0      0     0     0       0       0
│ Angle valves            0      0     0     0       0       0
│ Ball valves             0      0     0     0       0       0
│ Swing check valves      0      0     0     0       0       0
│ Flow check valves              0     0     0       0       0
│ Butterfly valves        0      0     0     0       0       0
│ Deareator                      0     0             0       0
│ Reducer Coupling        0      0     0     0       0       0
│ Boiler (C.I. sectional)                            0
│
┌─ MESSAGE ═════════════════════════════════════════════════════════════════
│ Use the UP, DOWN, LEFT, RIGHT ARROW keys to move the highlighting bar
│ over an input. Type desired value. F10 to return to input screen.
```

Figure 4-3b Example of the edit table in the PIPEPATH program.

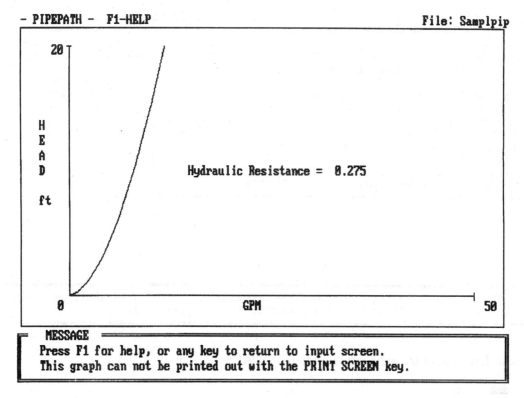

Figure 4-3c Example of a system resistance curve produced by the PIPEPATH program.

By pressing the F4 key, the user can view a graph of the system resistance curve for the currently specified piping system. This curve is discussed thoroughly in Chapter 6 of the book. It is a standard graphical description of the relationship between head loss and flow rate for any closed-loop piping system. An example of this graph is shown in Figure 4-3c. Press F10 to exit the graph and return to the main program screen.

The PIPEPATH program can create a disk file of the piping path currently described by the inputs. The user should follow the instructions in Section 3 of this manual to do so. This piping file can be called up for later modifications. It can also be read by the PUMP/SYS program to determine how the piping circuit will operate with a wide variety of circulators.

4.4 PUMPCURV

The PUMPCURV program generates a mathematical pump curve for any circulator for which flow rate versus head data is available. The concept and use of pump curves is covered in Chapter 7 of the book.

The user types in three data points for the flow rate

and associated head produced by a given circulator. These can usually be taken directly from the manufacturer's literature. For the most accurate results, these data points should be widely separated along the pump curve. For example, one data point can be at the point of zero flow rate, another at a minimum value of head, and the third roughly half way between the first two.

A second-order polynomial equation is fit to this data. This equation has the following form:

$$\text{Head} = b_0 + (b_1) \times (\text{gpm}) + (b_2) \times (\text{gpm})^2$$

where:
Head = the head produce by the pump (in feet of head)
gpm = the flow rate through the pump (in gpm)
b_0, b_1, and b_2 = the coefficients of the polynomial
    equation determined by the program.

The results area displays the coefficients of this equation, as well as the complete equation itself. The results are instantly updated whenever one or more of the inputs are changed. An example of the program's screen is shown in Figure 4-4a.

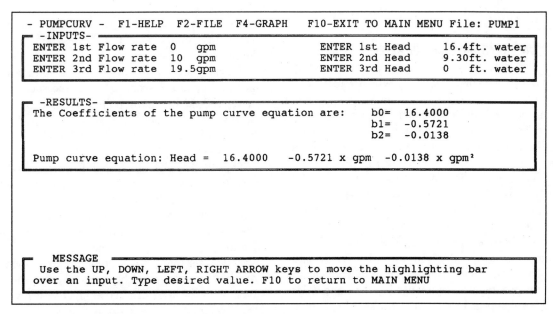

Figure 4-4a Example of a screen from the PUMPCURV program.

The resulting pump curve can be saved as a disk file for later use by the PUMP/SYS program. The F2 key invokes the disk access routine to save the file. This is the most convenient way to reuse the results of this program.

Pressing the F4 key displays a graph of the pump curve and the three input points that were entered to generate it. This graph corresponds to the graph that usually appears in manufacturer's literature. An example of this graph is shown in Figure 4-4b.

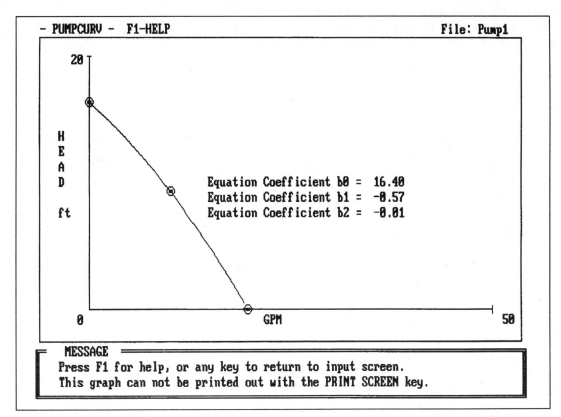

Figure 4-4b Example of a pump curve graph produced by the PUMPCURV program.

4.5   PUMP/SYS

The PUMP/SYS program allows the user to determine the flow rate and head at which a specific piping circuit will operate when combined with a specific circulator.

When the PUMP/SYS screen first appears, the hydraulic resistance and file name listed for the piping circuit file are the same as the hydraulic resistance and file name currently specified in the PIPEPATH program. Likewise, the data and file name appearing for the pump curve file are the same as the data and file name currently in the PUMPCURV program. This data linking between programs allows the PUMP/SYS program to find the intersection of the system curve and pump curve specified in the PIPEPATH and PUMPCURV programs instantly. This is convenient because the programs are often used in the same order in which they appear in the main menu: for example, PIPEPATH, then PUMPCURV, and finally PUMP/SYS.

Another way to use the PUMP/SYS program is to load previously created disk files for both the piping circuit and pump curve. Both types of files are loaded into PUMP/SYS by pressing the F2 key and selecting the appropriate file category, such as pump or piping, and then selecting the appropriate file names. The program instantly determines the intersection between the system curve of the specified piping circuit and specified pump curve. This intersection, called the operating point, is described in Chapter 7 of the book. A screen shot of the PUMP/SYS program is shown in Figure 4-5a.

Pressing the F4 key displays a graph showing the specified system resistance curve, the specified pump curve, and the operating point. Numerical values for the flow rate and head at the operating point are also displayed. An example of this graph is shown in Figure 4-5b.

The PIPEPATH, PUMPCURV, and PUMP/SYS programs work together to analyze and display the key concepts in the hydraulic design of a closed-loop hydronic heating system. The number of individual pipe circuits and pump files that can be created is limited to 99 of each type. As the user accumulates files, especially for frequently used pumps, the time required to use these programs becomes shorter and shorter. Eventually the user can accomplish in a few minutes what could take hours to do by hand.

4.6   SERIESBB

The SERIESBB program sizes and analyzes a series-connected finned-tube baseboard heating system. The user specifies data for each of up to 15 baseboards. This data includes the design heating load assigned to each baseboard, room air temperature, flow rate, and thermal output rating of the baseboard. To speed input on routine systems, default values for room air temperature, flow rate, and thermal rating of the

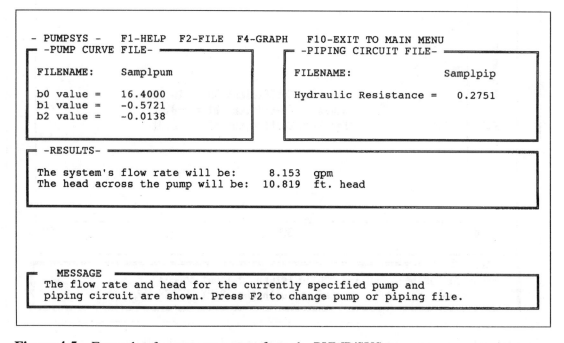

Figure 4-5a Example of a program screen from the PUMP/SYS program.

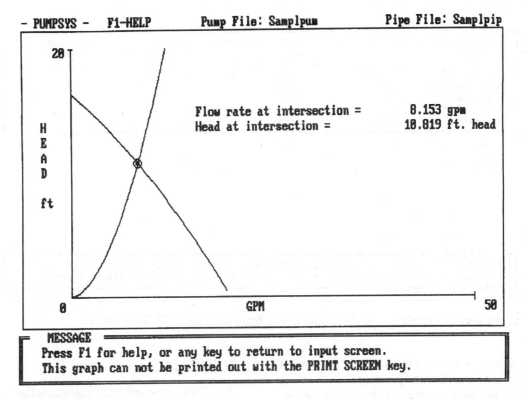

Figure 4-5b Example of a graph showing a pump curve, system resistance curve, and their intersection as produced by the PUMP/SYS program.

baseboard are set in the input area. These default values automatically carry forward into an edit table where they can be modified as required for special circumstances. This feature eliminates the need to make repetitive inputs when all the baseboards are the same.

The system can also be designed around several fluids, including water and various concentrations of antifreezes. A display of the program screen and edit table are shown in Figures 4-6a and 4-6b, respectively.

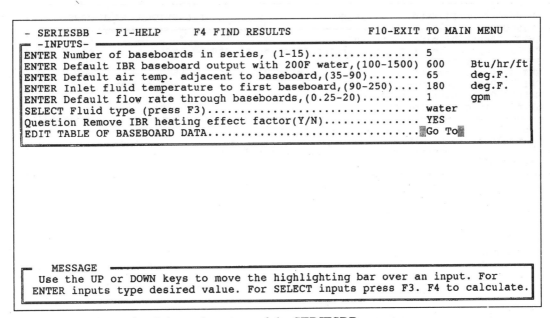

Figure 4-6a Example of the main screen of the SERIESBB program.

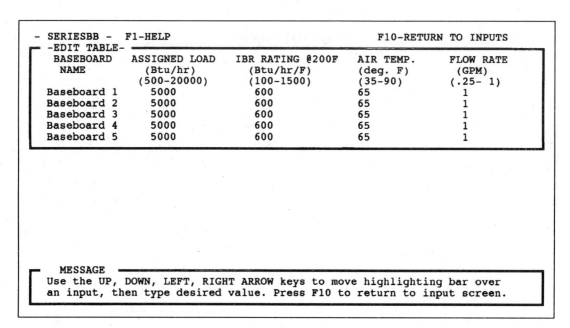

```
- SERIESBB -  F1-HELP                           F10-RETURN TO INPUTS
 -EDIT TABLE-
   BASEBOARD     ASSIGNED LOAD    IBR RATING @200F    AIR TEMP.    FLOW RATE
     NAME          (Btu/hr)         (Btu/hr/F)        (deg. F)      (GPM)
                 (500-20000)       (100-1500)         (35-90)      (.25- 1)
   Baseboard 1      5000              600                65           1
   Baseboard 2      5000              600                65           1
   Baseboard 3      5000              600                65           1
   Baseboard 4      5000              600                65           1
   Baseboard 5      5000              600                65           1




   MESSAGE
   Use the UP, DOWN, LEFT, RIGHT ARROW keys to move highlighting bar over
   an input, then type desired value. Press F10 to return to input screen.
```

Figure 4-6b Example of the edit table screen from the SERIESBB program.

This program is a strong stand-alone design tool. It accounts for details, such as temperature drop from one baseboard to the next and the effect of flow rate on thermal performance, that are not factored into other commercially available programs.

Keep in mind that the ROOMLOAD program can be used to generate the heating loads needed as inputs to this program.

4.7 PIPELOSS

The PIPELOSS program calculates the rate of heat loss from copper tubing in size ranges from 1/2" to 2". It can be used to demonstrate the effect of several factors such as air temperature, fluid type, flow rate, insulation, and length on pipe heat loss. This program is especially valuable when trying to correct the design of a hydronic distribution system for long lengths of pipe passing through cool spaces.

```
- PIPELOSS -  F1-HELP     F3-SELECT              F10-EXIT TO MAIN MENU
 -INPUTS-
  ENTER Length of pipe,(1 to 500)...................... 10         ft.
  ENTER Temperature of fluid entering pipe,(80 to 250).. 150       deg. F
  ENTER Flow rate in pipe,(.5 to 98.5).................. 1         gpm
  ENTER Air Temperature adjacent to pipe, (-20 to 80)... 55        deg. F
  SELECT Pipe size,(press F3 )......................... 1/2" copper
  SELECT Pipe insulation,(press F3 ).................... none
  SELECT Fluid type,(press F3 )........................ water

 -RESULTS-
  Pipe inlet temperature.............        150.0 deg. F
  Pipe outlet temperature............        149.3 deg. F
  Heat loss from pipe................          322 BTU/hr




   MESSAGE
   Use the UP or DOWN ARROW keys to move the highlighting bar over an input,
   then type in a value. Press F10 to exit to MAIN MENU. Press F3 to SELECT.
```

Figure 4-7 Example of a screen from the PIPELOSS program.

4.8 MONOFLO

The MONOFLO program determines the equivalent hydraulic resistance of a heat emitter piped into a 1-pipe distribution system using Bell and Gossett Monoflo® tees as shown in Figure 4-8a. This configuration of a branch circuit connected to a main circuit using one or two Monoflo tees will be referred to as a "monoflo block." The resulting single hydraulic resistance of the monoflo block can be entered back into the PIPEPATH program for further analysis.

The program first requires the user to define the branch piping path in terms of a hydraulic resistance using the PIPEPATH program. This approach allows the branch path to consist of almost any combination of piping, fittings, and heat emitters. The user also specifies the length of pipe between the tees connecting the branch path to the main circuit, as well as the specific combination of Monoflo-brand fittings being used. The program uses methods discussed in Chapter 6 of the book for reducing parallel hydraulic resistances into a single equivalent resistance. The program also determines the portion of the system flow rate that is diverted through the branch piping circuit. See Figure 4-8b.

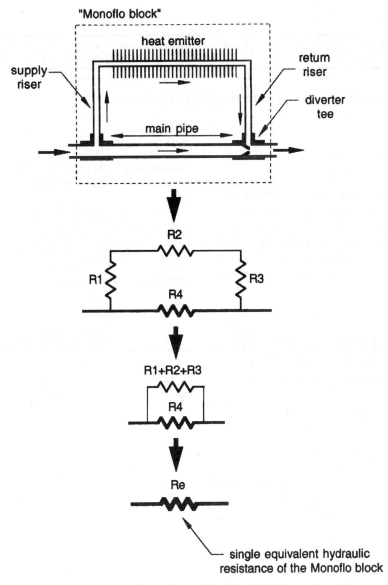

Figure 4-8a Concept of a "monoflo block" being reduced to a single equivalent resistance.

```
- MONOFLO -    F1-HELP    F3-SELECT                F10-EXIT TO MAIN MENU
  -INPUTS-
   ENTER Hydraulic resistance of branch circuit,(.01 to 10)..  .1
   ENTER Temperature of system fluid,(80 to 250)............  150     deg. F
   ENTER Length of pipe between tees,(1 to 100)...............  10      ft.
   ENTER System flow rate into tee,(.5 to 100)...............  10      gpm
   SELECT Monoflo tee configuration,(press F3 )..............  single 3/4"
   SELECT System fluid,(press F3 ).........................  water

  -RESULTS-
   Equivalent hydraulic resistance of Monoflo configuration..  0.044
   Flow rate in branch circuit.............................  5.73 gpm
   Flow rate between tees..................................  4.27 gpm




     MESSAGE
   Use the UP or DOWN ARROW keys to move the highlighting bar over an input,
   then type in a value. Press F10 to exit to MAIN MENU. Press F3 to SELECT.
```

Figure 4-8b Example of a screen from the MONOFLO program.

4.9 MIX

The MIX program determines the temperature and flow rate of a fluid stream formed by merging two to six incoming fluid streams, each having their own temperature and flow rate. Examples of where this occurs include return manifolds on hydronic radiant heating circuits, outlet tees in 1-pipe (diverter tee) systems, return manifolds for multi-zone systems, or any place a fitting is used to combine two or more fluid streams into one.

The program also allows a variety of systems fluids to be simulated. It fully accounts for the variation of fluid density and specific heat with temperature, and thus can handle incoming fluids over a wide temperature range. See Figure 4-9.

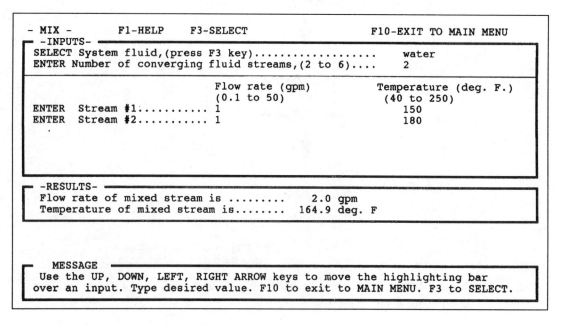

```
- MIX -        F1-HELP    F3-SELECT                F10-EXIT TO MAIN MENU
  -INPUTS-
   SELECT System fluid,(press F3 key)..................    water
   ENTER Number of converging fluid streams,(2 to 6)....    2

                              Flow rate (gpm)        Temperature (deg. F.)
                              (0.1 to 50)            (40 to 250)
   ENTER   Stream #1.......... 1                       150
   ENTER   Stream #2.......... 1                       180



  -RESULTS-
   Flow rate of mixed stream is ........    2.0 gpm
   Temperature of mixed stream is........  164.9 deg. F


     MESSAGE
   Use the UP, DOWN, LEFT, RIGHT ARROW keys to move the highlighting bar
   over an input. Type desired value. F10 to exit to MAIN MENU. F3 to SELECT.
```

Figure 4-9 Example of a screen from the MIX program.

4.10 HYRES

The HYRES program allows the user to determine the hydraulic resistance of various devices used in a hydronic piping circuit. Examples of such devices include panel radiators, fan-coil convectors, heat exchangers, or other devices not found in the edit table of the PIPEPATH program.

The user must supply data for the head loss of the device at three different flow rates of 60°F water. The following equation is fit to this data:

$$H_L = (r)f^{1.75}$$

where:
H_L = the head loss of the component (in feet of head)
r = The hydraulic resistance of the components
f = the flow rate through the component (in gpm)

The program calculates a hydraulic resistance value (r) for the device. This resistance can then be entered into the PIPEPATH program to include it within the description of a piping circuit.

If the Cv flow coefficient of a device is known, it can be converted into an approximate hydraulic resistance by the HYRES program. In such a case, data for head loss versus flow rate is not needed. See Figure 4-10.

The underlying theory behind the hydraulic resistance of a device is explained in Chapter 6 of the book.

4.11 EXPTANK

The EXPTANK program determines the proper size diaphragm-type expansion tank for a specified system. See Figure 4-11. It also determines the required pressurization of the air chamber in the tank. Features of the program include the ability to adjust expansion tank size for fluids such as glycol-based antifreezes that have higher rates of expansion compared to water. Tank size is determined so that at maximum system temperature, the pressure at the pressure relief valve is 5 psi below its rated opening pressure. The program also estimates system volume and the amount of antifreeze required, when applicable. The program is based on methods and equations presented in Chapter 12 of the book.

4.12 PARALLEL

The PARALLEL program reduces a group of up to six parallel-connected hydraulic resistances into a single equivalent hydraulic resistance. See Figure 4-12. An example of such a situation would be several radiant floor heating circuits connected to a common manifold assembly. Each parallel circuit is specified by its hydraulic resistance, which must be previously determined using the PIPEPATH program. This program calculates the equivalent hydraulic resistance of the parallel piping paths using methods from Chapter 6 of the book.

```
 - HYRES -      F1-HELP                        F10-EXIT TO MAIN MENU
   -INPUTS-
   ENTER Maximum flow rate for the device,(2 to 100)........ 10     gpm
   ENTER Average fluid temperature at the device,(60-250).... 120    deg. F.
   SELECT System fluid,(press F3 )........................... water
   QUESTION Do you know the Cv value of the device?.......... NO

                                   Flow rate @60 F.     Head Loss @60 F.
                                   (.01 to 100 gpm)   (.01 to 100 ft. head)
   ENTER First data point............... 2                   1.68
   ENTER Second data point.............. 4                   5.66
   ENTER Third data point............... 8                   19.03


   -RESULTS-
   The fitting value of "a" =                       0.499
   The fitting value of "b" =                       1.751
   Hydraulic resistance of device at specified temp. =  0.420



     MESSAGE
    Use the UP, DOWN, LEFT, RIGHT ARROW keys to move the highlighting bar
   over an input, then type in a value. F10 to exit to MAIN MENU
```

Figure 4-10 Example of a screen from the HYRES program.

```
┌─────────────────────────────────────────────────────────────────────────┐
│ - EXPTANK -    F1-HELP    F3-SELECT              F10-EXIT TO MAIN MENU     │
│ ┌─ -INPUTS- ──────────────────────────────────────────────────────────┐  │
│ │ ENTER Height of uppermost piping above boiler,(.5 to 100).  25    ft. │  │
│ │ ENTER Height of expansion tank above boiler,(.5 to 10)....  7     ft. │  │
│ │ ENTER Highest operating temperature of system,(125 to 250) 200   deg. F│
│ │ ENTER Boiler volume + Misc. volumes,(0 to 1000)..........  12    gal  │  │
│ │ ENTER Pressure relief valve setting,(15 to 60)....        30    psi  │  │
│ │ SELECT System fluid,(press F3 to select).................  water      │  │
│ │ ENTER Length of 1/2 in. copper pipe in system,(0 to 1000). 0     ft. │  │
│ │ ENTER Length of 3/4 in. copper pipe in system,(0 to 1000). 200   ft. │  │
│ │ ENTER Length of  1  in. copper pipe in system,(0 to 1000). 0     ft. │  │
│ │ ENTER Length of  1¼ in. copper pipe in system,(0 to 1000). 0     ft. │  │
│ │ ENTER Length of  1½ in. copper pipe in system,(0 to 1000). 0     ft. │  │
│ │ ENTER Length of  2  in. copper pipe in system,(0 to 1000). 0     ft. │  │
│ └──────────────────────────────────────────────────────────────────────┘  │
│ ┌─ -RESULTS- ─────────────────────────────────────────────────────────┐  │
│ │ Estimated system volume.............................  17.4 gal       │  │
│ │ Required (MINIMUM) expansion tank volume.............   2.3 gal       │  │
│ │ Required volume of specified antifreeze..............   0.0 gal       │  │
│ │ Required pressurization of tank diaphagm.............  12.8 psi       │  │
│ └──────────────────────────────────────────────────────────────────────┘  │
│ ┌─ MESSAGE ───────────────────────────────────────────────────────────┐  │
│ │ Use the UP or DOWN ARROW keys to move the highlighting bar over an input,│
│ │ then type in a value. Press F10 to exit to MAIN MENU. Press F3 to SELECT.│
│ └──────────────────────────────────────────────────────────────────────┘  │
└─────────────────────────────────────────────────────────────────────────┘
```

Figure 4-11 Example of a screen from the EXPTANK program.

```
┌─────────────────────────────────────────────────────────────────────────┐
│ - PARALLEL -  F1-HELP                           F10-EXIT TO MAIN MENU     │
│ ┌─ -INPUTS- ──────────────────────────────────────────────────────────┐  │
│ │ ENTER Number of parallel piping circuits,(2 to 6)..............  6     │  │
│ │ ENTER Flow rate in common piping,(.5 to 100)...................  12  gpm│
│ │ ENTER Hydraulic resistance of circuit #1 (.001 to 10)..........  .1    │  │
│ │ ENTER Hydraulic resistance of circuit #2 (.001 to 10)..........  .1    │  │
│ │ ENTER Hydraulic resistance of circuit #3 (.001 to 10)..........  .1    │  │
│ │ ENTER Hydraulic resistance of circuit #4 (.001 to 10)..........  .1    │  │
│ │ ENTER Hydraulic resistance of circuit #5 (.001 to 10)..........  .1    │  │
│ │ ENTER Hydraulic resistance of circuit #6 (.001 to 10)..........  .1    │  │
│ └──────────────────────────────────────────────────────────────────────┘  │
│ ┌─ -RESULTS- ─────────────────────────────────────────────────────────┐  │
│ │ Equivalent hydraulic resistance of all parallel circuits,..0.0043     │  │
│ │ Flow rate in circuit # 1 ...............................  2.00  gpm    │  │
│ │ Flow rate in circuit # 2 ...............................  2.00  gpm    │  │
│ │ Flow rate in circuit # 3 ...............................  2.00  gpm    │  │
│ │ Flow rate in circuit # 4 ...............................  2.00  gpm    │  │
│ │ Flow rate in circuit # 5 ...............................  2.00  gpm    │  │
│ │ Flow rate in circuit # 6 ...............................  2.00  gpm    │  │
│ └──────────────────────────────────────────────────────────────────────┘  │
│ ┌─ MESSAGE ───────────────────────────────────────────────────────────┐  │
│ │ Use the UP or DOWN ARROW keys to move the highlighting bar over an input,│
│ │ then type in a value. Press F10 to exit to MAIN MENU.                 │  │
│ └──────────────────────────────────────────────────────────────────────┘  │
└─────────────────────────────────────────────────────────────────────────┘
```

Figure 4-12 Example of a screen from the PARALLEL program.

If the flow rate entering the common point of the parallel circuits is known, the program can also determine how the flow will divide up among the parallel paths.

4.13 RADFLOOR

The RADFLOOR program estimates the temperature drop and heat output of a one-way, serpentine-shaped radiant floor heating circuit as shown in Chapter 10 of the book. The circuit is modeled as a linear element with an exponential function describing the temperature drop as a function of circuit length, floor construction, flow rate, and room temperature. See Figure 4-13.

The user selects a "core" floor construction that fixes the tube size, spacing, and embedment material. The user also selects a finish floor system that adds thermal resistance to the upward heat flow.

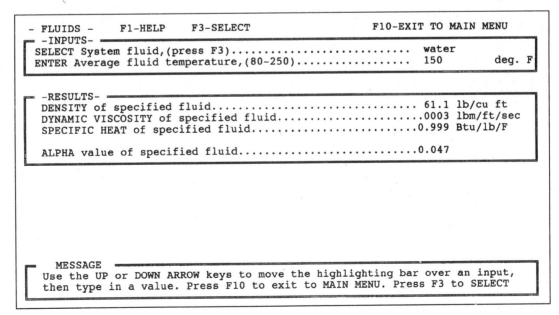

```
- RADFLOOR -  F1-HELP     F3-SELECT              F10-EXIT TO MAIN MENU
  -INPUTS-
  ENTER Fluid temperature entering floor circuit,(85to150) 100      deg. F.
  ENTER Flow rate through the floor circuit,(.25 to 10)... 1        gpm
  ENTER Room air temperature,(50 to 85).................... 70      deg. F.
  ENTER Active floor area served by floor circuit(50to1000) 500    ft²
  ENTER Required upward heat output of floor,(500to25000).. 10000   Btu/hr
  ENTER Estimated downward+edge floor heat loss(100to10000) 500     Btu/hr
  ENTER Length of tubing in floor circuit,(25 to 500)...... 250     ft.
  ENTER R-value of air film at top of floor(.5 to .65)..... .61     Fhrft²/Btu
  SELECT Floor core type,( press F3)....................... CORE #2
  SELECT Finish flooring material(s),( press F3)........... FINISH #1
  SELECT Fluid type,( press F3)........................... water

  -RESULTS-
  Floor circuit outlet temperature................. 90.4 deg. F.
  Temperature drop through floor circuit........... 9.6 deg. F.
  Actual UPWARD heat output from floor area........ 4,512 Btu/hr
  DOWNWARD + EDGEWISE heat output from floor area.. 226 Btu/hr
  UPWARD heat flux output from floor area.......... 9.02 Btu/hr/sq ft.

  MESSAGE
  Use the UP or DOWN ARROW keys to move the highlighting bar over an input.
  ENTER lines, type in a value, SELECT lines, press F3. MAIN MENU Press F10.
```

Figure 4-13 Example of a screen from the RADFLOOR program.

Before using this program, the user should use the PARALLEL program to determine the flow rates in the various parallel floor circuits. These floor circuits are then analyzed, one at a time, using RADFLOOR program.

The program determines both the upward and downward heat output of the floor served by the piping circuit. It can be used to demonstrate the excessive downward heat flow of a radiant heating system with poor underside insulation. The effect of high thermal resistance floor coverings can be shown also.

4.14 FLUIDS

The FLUIDS program determines the density, dynamic viscosity, specific heat, and alpha value of water- and glycol-based antifreeze solutions. It incorporates the same fluid models used by the other programs in the toolkit. It is useful for demonstrating the effect of temperature on fluid properties and for comparing these properties among several fluids. See Figure 4-14.

```
- FLUIDS -   F1-HELP     F3-SELECT              F10-EXIT TO MAIN MENU
  -INPUTS-
  SELECT System fluid,(press F3)......................... water
  ENTER Average fluid temperature,(80-250)............... 150      deg. F

  -RESULTS-
  DENSITY of specified fluid............................. 61.1 lb/cu ft
  DYNAMIC VISCOSITY of specified fluid...................0003 lbm/ft/sec
  SPECIFIC HEAT of specified fluid.......................0.999 Btu/lb/F

  ALPHA value of specified fluid.........................0.047

  MESSAGE
  Use the UP or DOWN ARROW keys to move the highlighting bar over an input,
  then type in a value. Press F10 to exit to MAIN MENU. Press F3 to SELECT
```

Figure 4-14 Example of a screen from the FLUIDS program.

The alpha value of the fluid is derived from a combination of density and dynamic viscosity based on the definition of Reynolds number and the Darcy equation. It is used along with other data to calculate the head loss of fluid flowing through a specified pipe. It is defined in Chapter 6 of the book.

4.15 PIPESIZE

The PIPESIZE program determines the minimum size of copper tubing necessary to convey water at a specified flow rate, so that the flow velocity does not exceed a specified value. It also displays the flow velocity associated with the specified flow rate in copper tube sizes from 1/2" to 2". It is very helpful for preliminary pipe sizing during system design. The program's help screens also contain data on suggested pipe sizes. See Figure 4-15.

4.16 BINEFF

The BINEFF program estimates the seasonal efficiency of a boiler based on data defining its cycle efficiency as a function of run time. Equations are fit to

this data and then combined with bin temperature data for one of seventeen selected geographic locations. The program combines the efficiency relationship with the bin temperature data to estimate the true seasonal efficiency of a boiler in a given system at a specified location. It also estimates the total amount of fuel consumed over a heating season. See Figure 4-16.

The program is very useful for demonstrating the detrimental effect of boiler oversizing on seasonal efficiency. It also can be used to compare low mass and high mass boilers in the same application, provided cycle efficiency data is available. This program shows that true seasonal efficiency is not as simple as a single AFUE value might suggest.

4.17 DHWLOAD

The DHWLOAD program computes daily and seasonal energy usage for domestic water heating. It also includes an estimate for standby losses from the tank based on the R-value of its insulation. For simplicity, the program assumes a cylindrical domestic hot water storage tank with a height to diameter ratio of 2.5. See Figure 4-17.

```
- PIPESIZE -   F1-HELP                           F10-EXIT TO MAIN MENU
  -INPUTS-
  ENTER flow rate,(.25 to 100)................................... 10    gpm
  ENTER maximum flow velocity,(2 to 15)....................... 5      ft/sec

  -RESULTS-
  Velocity for 1/2" pipe size..  12.62 ft/sec
  Velocity for 3/4" pipe size..   6.21 ft/sec
  Velocity for  1 " pipe size..   3.67 ft/sec   This is the recommended size.
  Velocity for  1¼" pipe size..   2.45 ft/sec
  Velocity for  1½" pipe size..   1.75 ft/sec
  Velocity for  2 " pipe size..   1.01 ft/sec




     MESSAGE
     Use the UP or DOWN ARROW keys to move the highlighting bar over an input,
     then type in a value. Press F10 to exit to MAIN MENU.
```

Figure 4-15 Example of a screen from the PIPESIZE program.

```
- BINEFF -      F1-HELP     F3-SELECT              F10-EXIT TO MAIN MENU
  -INPUTS-
   ENTER Outside design air temperature,(-30 to 40).......  0            deg. F
   ENTER Desired inside air temperature(50 to 85).........  70           deg. F
   ENTER Design heating load of building,(10K to 300K)....  50000        Btu/hr
   ENTER Heating balance point temp of building,(40 to 65)  65           deg. F
   ENTER D.O.E. heating capacity of boiler,(25K to 300K)..  70000        Btu/hr
   ENTER Steaty state boiler efficiency,(65 to 95)........  84           %
   ENTER Cycle efficiency of boiler at 75% run fraction,..  81           %
   ENTER Cycle efficiency of boiler at 50% run fraction,..  77           %
   ENTER Cycle efficiency of boiler at 25% run fraction,..  60           %
   ENTER Cycle efficiency of boiler at 10% run fraction,..  35           %
   SELECT Fuel type,(press F3).........................    #2 Fuel Oil
   SELECT Weather data location,(press F3)...............  Syracuse   NY

  -RESULTS-
   Total seasonal heat delivered by boiler........  135  MMBtu/season
   Total seasonal energy content of fuel consumed.  218  MMBtu/season
   Seasonal efficiency of boiler..................  62.0 %
   Total units of fuel consumed...................  1556 gal

    MESSAGE
   Use the UP or DOWN ARROW keys to move the highlighting bar over an input,
   then type in a value. Press F10 to exit to MAIN MENU. Press F3 to SELECT.
```

Figure 4-16 Example of a screen from the BINEFF program.

```
- DHWLOAD -    F1-HELP                              F10-EXIT TO MAIN MENU
  -INPUTS-
   ENTER Average cold water inlet temperature,(35 to 110)... 50     deg. F
   ENTER Desired hot water supply temperature,(110 to 200).. 120    deg. F
   ENTER # gallons per day hot water usage,(0 to 500)....... 50     gal/day
   ENTER Tank volume,(20 to 250)............................ 42     gal
   ENTER R-value of tank insulation,(2 to 20).............. 11     F hr ft²/Btu
   ENTER Air temperature adjacent to DHW tank,(35 to 100)... 60     deg. F

  -RESULTS-
   Estimated DAILY water heating energy requirements........  29,155  Btu/day
   Estimated DAILY tank standby heat loss...................   2,121  Btu/day
   Estimated DAILY Total water heating energy usage.........  31,276  Btu/day

   Estimated ANNUAL water heating energy requirements.......   10.6   MMBtu/yr
   Estimated ANNUAL tank standby heat loss..................    0.8   MMBtu/yr
   Estimated ANNUAL Total water heating energy usage........   11.4   MMBtu/yr

    MESSAGE
   Use the UP or DOWN ARROW keys to move the highlighting bar over an input,
   then type in a value. Press F10 to exit to MAIN MENU.
```

Figure 4-17 Example of a screen from the DHWLOAD program.

4.18 FUELCOST

The FUELCOST program computes the delivered cost of energy from several common fuels. The user inputs the fuel cost in actual purchase units. They also input the conversion efficiency associated with the heating system, such as seasonal average efficiency. The program then determines the delivered cost in dollars per million Btus. This program is useful for comparing the cost of various fuels on a unit price basis. It can also demonstrate the cost savings associated with higher efficiency equipment. See Figure 4-18.

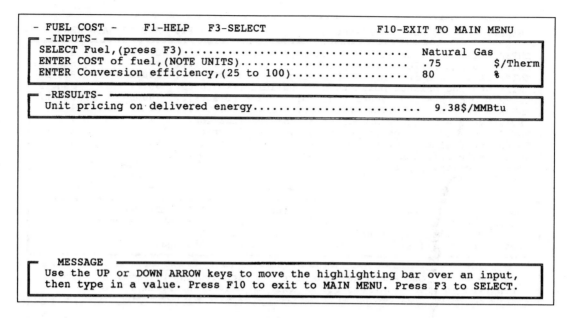

Figure 4-18 Example of a screen from the FUELCOST program.

4.19 ECONOMICS

The ECONOMICS program performs a simplified life-cycle cost analysis on the specified system. It computes the total operating cost of the system over the life of the loan by inflating the annual fuel cost at a constant specified rate. It computes the total owning cost by calculating the amortization of the loan used to finance the system over the specified term. The total owning and operating cost is reflective of the trade-off between less expensive systems with higher operating costs, such as electric resistance heating, versus more expensive systems with lower operating costs. The program also includes the option of determining the seasonal energy use of the system based on standard degree-day calculations. See Figure 4-19.

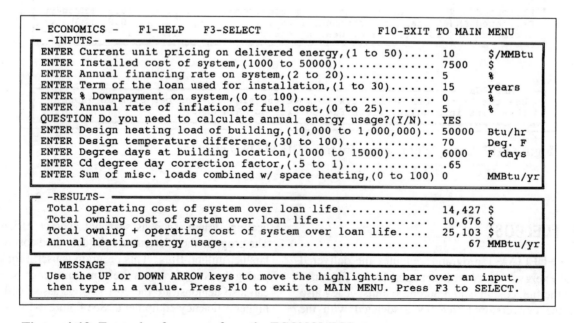

Figure 4-19 Example of a screen from the ECONOMICS program.

USING THE PROGRAMS TO DESIGN A SYSTEM 5

The Hydronics Design Toolkit was designed as a group of individual programs that can be used individually or jointly depending on the needs of the designer. This "modular" nature allows the toolkit to systematically reduce even complex piping systems to manageable calculations.

The example that follows shows how the various programs in the toolkit can be used to analyze and size a small series-loop hydronic heating system.

First, assume that the ROOMLOAD program was used in conjunction with the plans of the home and a tabulation of its construction materials. Repeated use of the program produced the following design heating loads:

- Living room: 8,500 Btu/hr
- Dining room: 7,000 Btu/hr
- Kitchen: 5,000 Btu/hr
- Bedroom 1: 4,500 Btu/hr
- Bathroom: 3,000 Btu/hr
- Bedroom 2: 5,500 Btu/hr
- Master bathroom: 3,500 Btu/hr
- Master bedroom: 4,200 Btu/hr

The BSMTLOAD program was used to estimate the design heating load of the basement at 9,000 Btu/hr. Thus the total design heating load of the home, including the basement, is 50,200 Btu/hr.

At this stage, the designer contemplates using a single-series loop of finned-tube baseboard. After examining the floor plan, the approximate distance around the perimeter of the house, where the piping circuit will be routed, is 185 ft. The designer contemplates using 3/4 in. copper tubing for the distribution circuit. Considering the riser lengths to the individual baseboards, the total piping circuit length will be about 210 ft. Estimated quantities of fittings, valves, and other components are determined also.

The designer now uses the PIPEPATH program to determine the approximate hydraulic resistance of this circuit. The data input to the edit table of the PIPEPATH program is as follows:

- 210 ft. of 3/4 in. copper tubing
- 48, 3/4 in. x 90 degree copper elbows
- 10, 3/4 in. copper tees (straight through flow)
- 4, 3/4 in. gate valves for isolating circulator and boiler, if necessary
- 1, 1 in. deaerator
- 1, 1 in. flow check valve
- 1 cast iron sectional boiler

The designer plans to supply water to the distribution circuit at 180°F and expects the average water temperature in the circuit to be about 170°F.

A display of the main screen and edit table from the PIPEPATH program are shown in Figures 5-1a and 5-1b.

Note: Since the flow rate through the circuit is not known at this point, the default value of 1 gpm is simply left in place, but ignored.

The PIPEPATH program immediately determines the hydraulic resistance of the proposed circuit to be 0.925. The circuit is given the name "loop1" and stored to disk.

The designer also proposes to use a circulator with the following head versus flow rate data (taken from the manufacturer's literature):

Flow rate	Head
0 gpm	16.4 ft.
10 gpm	9.3 ft.
20 gpm	0 ft.

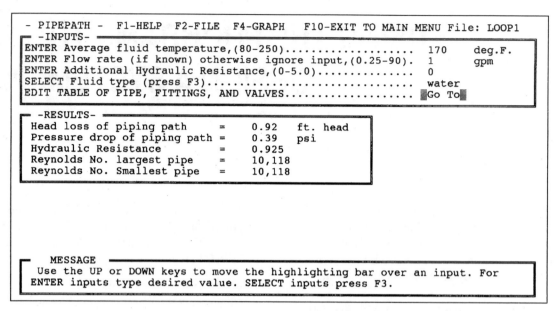

```
- PIPEPATH -  F1-HELP  F2-FILE  F4-GRAPH    F10-EXIT TO MAIN MENU File: LOOP1
  ┌─ -INPUTS-
  │ ENTER Average fluid temperature,(80-250)...................  170    deg.F.
  │ ENTER Flow rate (if known) otherwise ignore input,(0.25-90). 1      gpm
  │ ENTER Additional Hydraulic Resistance,(0-5.0)...............  0
  │ SELECT Fluid type (press F3)...............................  water
  │ EDIT TABLE OF PIPE, FITTINGS, AND VALVES...................  ▌Go To▌

  ┌─ -RESULTS-
  │ Head loss of piping path    =     0.92    ft. head
  │ Pressure drop of piping path =    0.39    psi
  │ Hydraulic Resistance        =     0.925
  │ Reynolds No. largest pipe   =    10,118
  │ Reynolds No. Smallest pipe  =    10,118


  ┌─ MESSAGE ═══════════════════════════════════════════════════════════
  │  Use the UP or DOWN keys to move the highlighting bar over an input. For
  │ ENTER inputs type desired value. SELECT inputs press F3.
```

Figure 5-1a The main screen from the PIPEPATH program with data related to the example system.

```
- PIPEPATH -  F1-HELP    F5-ZERO ALL        F10-RETURN TO INPUTS  File: LOOP1
  ┌─ -EDIT TABLE-
  │                         1/2"   3/4"   1"   1 1/4"  1 1/2"  2"   5/8"PEX
  │ Straight pipe(lin. ft.) 0      210    0    0       0       0    0
  │ 90 degree ells          0      48     0    0       0       0
  │ 45 degree ells          0      0      0    0       0       0
  │ Tees (straight through) 0      10     0    0       0       0
  │ Tees (side port)        0      0      0    0       0       0
  │ Monoflo tees                   0      0    0       0       0
  │ Gate valves             0      4      0    0       0       0
  │ Globe valves            0      0      0    0       0       0
  │ Angle valves            0      0      0    0       0       0
  │ Ball valves             0      0      0    0       0       0
  │ Swing check valves      0      0      0    0       0       0
  │ Flow check valves              0      1    0       0       0
  │ Butterfly valves        0      0      0    0       0       0
  │ Deareator                      0      1            0       0
  │ Reducer Coupling        0      0      0    0       0       0
  │ Boiler (C.I. sectional)                            1

  ┌─ MESSAGE ═══════════════════════════════════════════════════════════
  │  Use the UP, DOWN, LEFT, RIGHT ARROW keys to move the highlighting bar
  │ over an input. Type desired value. F10 to return to input screen.
```

Figure 5-1b The edit table from the PIPEPATH program with data related to the example system.

To establish a pump curve, this data is input to the PUMPCURV program. The resulting pump curve is:

$$\text{Head} = 16.4 + (-0.6) \times (gpm) + (-0.011) \times (gpm)^2$$

This pump curve is named "pump1," and saved as a disk file. A display of the PUMPCURV program screen with this data and result is shown in Figure 5-2.

To determine the performance of the proposed piping circuit with this circulator, the designer now uses the PUMP/SYS program. Since the data for the piping circuit is still present in the PIPEPATH program and the data for the pump is still present in the PUMPCURV program, the flow rate and head at the

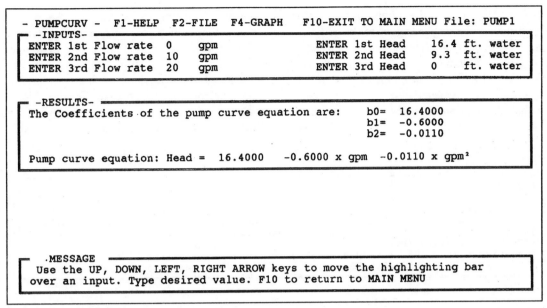

Figure 5-2 A screen shot of the PUMPCURV program with the data and results of the example system.

operating point of the system instantly appears when the PUMP/SYS screen appears.

flow rate = 4.608 gpm

head across pump = 13.401 ft.

Notice that it was not necessary to load either of these files before the operating point could be determined by PUMP/SYS. See Figure 5-3. However, since the data was saved as disk files, these files can be loaded at any future time to get the same result.

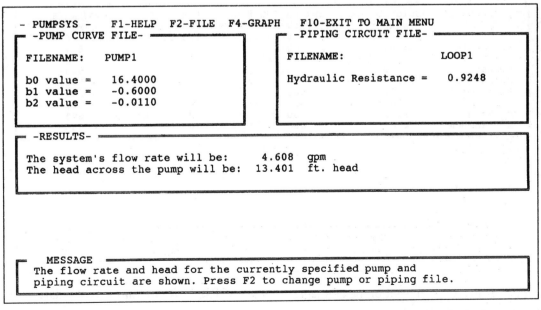

Figure 5-3 A screen shot of the PUMP/SYS program with the results of the example

Now that the flow rate in the proposed distribution system is estimated, the designer goes to the SERIES-BB program to size the individual finned-tube baseboards. A display of the main screen as well as the edit table from the SERIESBB program are shown in Figures 5-4a and 5-4b, respectively.

Notice that all room names and heating loads have been loaded into the edit table. *The order in which the heating loads are listed must be the same order in which they are piped into the distribution circuit.* Also, notice that the default room air temperature of 70°F, the baseboard output rating of 600 Btu/hr/ft. at 200°F, and the flow rate of 4.608 gpm have all carried forward from the input area to the edit table. Thus, the only data that needed to be entered into the edit table was the room names and design heating loads.

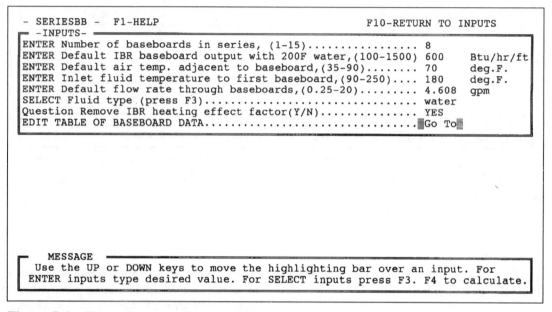

```
 - SERIESBB -   F1-HELP                            F10-RETURN TO INPUTS
 -INPUTS-
ENTER Number of baseboards in series, (1-15)................. 8
ENTER Default IBR baseboard output with 200F water,(100-1500) 600    Btu/hr/ft
ENTER Default air temp. adjacent to baseboard,(35-90)........ 70     deg.F.
ENTER Inlet fluid temperature to first baseboard,(90-250).... 180    deg.F.
ENTER Default flow rate through baseboards,(0.25-20)......... 4.608  gpm
SELECT Fluid type (press F3)................................. water
Question Remove IBR heating effect factor(Y/N).............. YES
EDIT TABLE OF BASEBOARD DATA.................................║Go To║


   MESSAGE
   Use the UP or DOWN keys to move the highlighting bar over an input. For
 ENTER inputs type desired value. For SELECT inputs press F3. F4 to calculate.
```

Figure 5-4a The main screen from the SERIESBB program with the data for the example system.

```
 - SERIESBB -   F1-HELP                            F10-RETURN TO INPUTS
 -EDIT TABLE-
  BASEBOARD    ASSIGNED LOAD   IBR RATING @200F   AIR TEMP.    FLOW RATE
    NAME         (Btu/hr)        (Btu/hr/F)       (deg. F)      (GPM)
               (500-20000)      (100-1500)        (35-90)     (.25- 4.608)
  Living rm.      8500             600               70          4.608
  Dining rm.      7000             600               70          4.608
  Kitchen         5000             600               70          4.608
  Bedroom 1       4500             600               70          4.608
  Bathroom        3000             600               70          4.608
  Bedroom 2       5500             600               70          4.608
  M. Bath         3500             600               70          4.608
  M. Bedroom      4200             600               70          4.608




   MESSAGE
   Use the UP, DOWN, LEFT, RIGHT ARROW keys to move highlighting bar over
 an input, then type desired value. Press F10 to return to input screen.
```

Figure 5-4b A screen shot of the edit table from the SERIESBB program with the data form the example system.

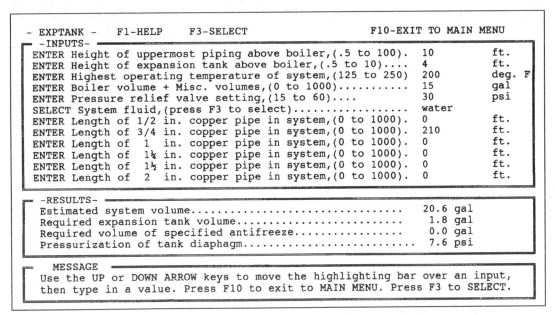

```
- SERIESBB -  F1-HELP                                  F10-RETURN TO INPUTS
 -RESULTS-
 BASEBOARD    ASSIGNED LOAD   LENGTH    INLET TEMP.   OUTLET TEMP.   HEAT OUTPUT
   NAME         (Btu/hr)       (ft.)     (deg. F.)     (deg. F.)      (Btu/hr)
 Living Rm.      8,500          16          180           176           8,635
 Dining Rm.      7,000          14          176           173           7,215
 Kitchen         5,000          11          173           171           5,453
 Bedroom 1       4,500          10          171           169           4,803
 Bathroom        3,000           7          169           167           3,270
 Bedroom 2       5,500          12          167           165           5,474
 M. Bath         3,500           8          165           163           3,535
 M. Bedroom      4,200          10          163           162           4,314




Total heat = 42,699 Btu/hr    Total length =    88 ft.  Total temp. drop = 18 F.

  MESSAGE
  Press PRINT SCREEN to print. Press F10 to return to INPUT screen.
  Press F1 for HELP.
```

Figure 5-5 A screen shot of the results of the SERIESBB program for the example

After returning to the main screen by pressing F10, the user presses F4 to calculate the required baseboard lengths for each room. A display of the results screen is shown in Figure 5-5. Notice that a total of 88 ft. of baseboard is required. The temperature drop around the distribution circuit will be 18°F. This is fairly typical and confirms that the estimated average water temperature in the circuit (170°F) was a good estimate.

At this point, the designer could use the proposed design or experiment with other types of baseboard, circulators, and pipe sizes to find how they will effect the performance and sizing of the system. For example, by simply routing the water through the circuit in the opposite direction, thus reversing the positions of the room names and loads in the SERIESBB edit table, the total length of baseboard increases to 89 ft. The number of combinations the designer can experiment with is limited only by their imagination.

Assuming this design will be used, the designer goes on to size an expansion tank for the system using the EXPTANK program. Specific data related to the system has been input on the program screen shown in Figure 5-6.

```
- EXPTANK -   F1-HELP    F3-SELECT               F10-EXIT TO MAIN MENU
 -INPUTS-
 ENTER Height of uppermost piping above boiler,(.5 to 100).  10      ft.
 ENTER Height of expansion tank above boiler,(.5 to 10)....   4      ft.
 ENTER Highest operating temperature of system,(125 to 250)  200     deg. F
 ENTER Boiler volume + Misc. volumes,(0 to 1000)..........   15      gal
 ENTER Pressure relief valve setting,(15 to 60)....         30      psi
 SELECT System fluid,(press F3 to select)..................  water
 ENTER Length of 1/2 in. copper pipe in system,(0 to 1000).   0      ft.
 ENTER Length of 3/4 in. copper pipe in system,(0 to 1000).  210     ft.
 ENTER Length of  1  in. copper pipe in system,(0 to 1000).   0      ft.
 ENTER Length of  1¼ in. copper pipe in system,(0 to 1000).   0      ft.
 ENTER Length of  1½ in. copper pipe in system,(0 to 1000).   0      ft.
 ENTER Length of  2  in. copper pipe in system,(0 to 1000).   0      ft.

 -RESULTS-
 Estimated system volume.................................  20.6 gal
 Required expansion tank volume..........................   1.8 gal
 Required volume of specified antifreeze.................   0.0 gal
 Pressurization of tank diaphagm.........................   7.6 psi

  MESSAGE
  Use the UP or DOWN ARROW keys to move the highlighting bar over an input,
  then type in a value. Press F10 to exit to MAIN MENU. Press F3 to SELECT.
```

Figure 5-6 A screen shot of the EXPTANK program showing data for the example house.

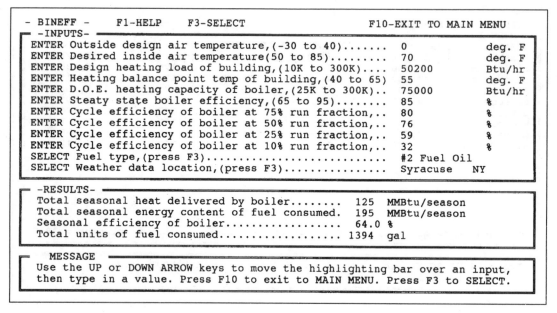

```
- BINEFF -     F1-HELP    F3-SELECT                 F10-EXIT TO MAIN MENU
  -INPUTS-
  ENTER Outside design air temperature,(-30 to 40).......  0          deg. F
  ENTER Desired inside air temperature(50 to 85).........  70         deg. F
  ENTER Design heating load of building,(10K to 300K)....  50200      Btu/hr
  ENTER Heating balance point temp of building,(40 to 65)  55         deg. F
  ENTER D.O.E. heating capacity of boiler,(25K to 300K)..  75000      Btu/hr
  ENTER Steaty state boiler efficiency,(65 to 95)........  85         %
  ENTER Cycle efficiency of boiler at 75% run fraction,..  80         %
  ENTER Cycle efficiency of boiler at 50% run fraction,..  76         %
  ENTER Cycle efficiency of boiler at 25% run fraction,..  59         %
  ENTER Cycle efficiency of boiler at 10% run fraction,..  32         %
  SELECT Fuel type,(press F3)...........................  #2 Fuel Oil
  SELECT Weather data location,(press F3)...............  Syracuse   NY

  -RESULTS-
  Total seasonal heat delivered by boiler........  125   MMBtu/season
  Total seasonal energy content of fuel consumed.  195   MMBtu/season
  Seasonal efficiency of boiler..................  64.0  %
  Total units of fuel consumed...................  1394  gal

   MESSAGE
  Use the UP or DOWN ARROW keys to move the highlighting bar over an input,
  then type in a value. Press F10 to exit to MAIN MENU. Press F3 to SELECT.
```

Figure 5-7 A screen shot of the BINEFF program with data for the selected boiler and installation.

The results indicate an expansion tank with a minimum volume of 1.8 gallons is required. The air side of the tank should be pressurized to 7.6 psi, and the feed water valve adjusted to produce this pressure at the location of the tank.

The designer selects a boiler with a DOE heating capacity of 75,000 Btu / hr for this system. The boiler manufacturer has supplied part load efficiency data for the boiler as follows:

Percent run time	Efficiency
100	85
75	80
50	76
25	59
10	32

By using the BINEFF program, the designer can estimate the seasonal efficiency of this boiler, in this particular building, in a particular location. Assume the building is in Syracuse, New York. Also, assume it has a heating balance point of 55°F, and that the design heating load is based on an outdoor design temperature of 0°F. A display of the BINEFF program screen with the appropriate data is shown in Figure 5-7.

Notice that the seasonal efficiency of the boiler is estimated to be 64 percent, and that 1,394 gallons of #2 fuel oil is the estimated fuel consumption. Out of 195 MMBtu of fuel consumed, 125 MMBtu of heat was actually delivered to the system.

The designer can now estimate the cost of this energy using the FUELCOST program. After selecting #2 fuel oil, which will be assumed to cost $0.90 per gallon at this location, and the estimated seasonal efficiency of 64 percent, the program estimates the cost of delivered energy at $10.04 per MMBtu. A quick multiplication of $10.04 per MMBtu of delivered energy times 125 MMBtu delivered to the house, yields a seasonal heating cost of:

$$\left(10.04 \ \frac{\$}{\text{MMBtu}}\right)\left(125 \ \frac{\text{MMBtu}}{\text{season}}\right) = 1,255 \ \$ \ / \ \text{season}$$

To evaluate this design against other options, the designer could go on to use the ECONOMICS program.

SUMMARY

This brief example shows how information can be moved, manually or through disk files, from one program to another. It shows how the toolkit can be used to explore detailed design options, followed by estimates of the operating cost of these options. The text shows several other examples of how the programs can be used for specific design answers.

It is our hope that you will enjoy the Hydronics Design Toolkit and use it for routine as well as customized design assistance.

Hydronics Design Toolkit User's Manual

MARIO RESTIVE
JOHN SIEGENTHALER, P.E.

Written by John Siegenthaler, Professional Engineer and Assistant Professor at Mohawk Valley Community College, *Modern Hydronic Heating* is your complete source for designing, installing and maintaining hydronic heating systems!

Combined, *The Hydronics Design Toolkit* and *Modern Hydronic Heating* is the ULTIMATE package!

The *Hydronics Design Toolkit Software*, written and developed cooperatively by John Siegenthaler, P.E. and Mario Restive, was created specifically for hydronic system design calculation. This revolutionary, ground breaking software:

- Calculates design heating load of a room and/or basement
- Determines characteristics of user-defined piping path
- Generates a pump curve from manufacturers data
- Calculates heat loss from a pipe
- Calculates the properties of several fluids
- Calculates true seasonal efficiency of a boiler AND MORE!

Modern Hydronic Heating may be purchased without the software. Order # 0-8273-6595-0.

Don't miss these other exciting titles from Delmar Publishers:

Refrigeration and Air Conditioning Technology, 3E
by Bill Whitman and Bill Johnson,
Order # 0-8273-5646-3

Practical Heating Technology
by Bill Johnson,
Order # 0-8273-4881-9

Heat Pumps: Theory and Service
by Lee Miles,
Order # 0-8273-4956-4

TO ORDER OTHER DELMAR TITLES, CALL 1-800-347-7707.

"*Modern Hydronic Heating is very accurate and contains the specific information anyone interested in hydronics heating needs to know. It is especially informative and useful with the associated computer disk and its program calculating capabilities.*"

- Steve Lapp
Grundfos PumpS
Corporation

"*I know of no other text that covers modern hydronic system design so completely and in such depth ...when you add the Hydronics Design Toolkit Software, the combination is unbeatable!*"

- Roy Collver
Mechanical Systems
Marketing Ltd.

"*Destined to be a bible of the industry - the software is indispensable.*"

- George Oulton
Euro-Tech Inc.

Delmar Publishers

ISBN 0-8273-6816-X

90000

9 780827 368163

Hydronics Design Toolkit User's Manual

MARIO RESTIVE
JOHN SIEGENTHALER, P.E.

Written by John Siegenthaler, Professional Engineer and Assistant Professor at Mohawk Valley Community College, *Modern Hydronic Heating* is your complete source for designing, installing and maintaining hydronic heating systems!

Combined, *The Hydronics Design Toolkit* and *Modern Hydronic Heating* is the ULTIMATE package!

The *Hydronics Design Toolkit Software*, written and developed cooperatively by John Siegenthaler, P.E. and Mario Restive, was created specifically for hydronic system design calculation. This revolutionary, ground breaking software:

- Calculates design heating load of a room and/or basement
- Determines characteristics of user-defined piping path
- Generates a pump curve from manufacturers data
- Calculates heat loss from a pipe
- Calculates the properties of several fluids
- Calculates true seasonal efficiency of a boiler AND MORE!

Modern Hydronic Heating may be purchased without the software. Order # 0-8273-6595-0.

Don't miss these other exciting titles from Delmar Publishers:

Refrigeration and Air Conditioning Technology, 3E
by Bill Whitman and Bill Johnson,
Order # 0-8273-5646-3

Practical Heating Technology
by Bill Johnson,
Order # 0-8273-4881-9

Heat Pumps: Theory and Service
by Lee Miles,
Order # 0-8273-4956-4

TO ORDER OTHER DELMAR TITLES, CALL 1-800-347-7707.

"Modern Hydronic Heating is very accurate and contains the specific information anyone interested in hydronics heating needs to know. It is especially informative and useful with the associated computer disk and its program calculating capabilities."

- Steve Lapp
Grundfos PumpS
Corporation

"I know of no other text that covers modern hydronic system design so completely and in such depth ...when you add the Hydronics Design Toolkit Software, the combination is unbeatable!"

- Roy Collver
Mechanical Systems
Marketing Ltd.

"Destined to be a bible of the industry - the software is indispensable."

- George Oulton
Euro-Tech Inc.

Delmar Publishers

ISBN 0-8273-6816-X
90000

9 780827 368163